Lecture Notes in Computer Science 13233

More information about this series at https://link.springer.com/bookseries/558

Stan Sclaroff · Cosimo Distante · Marco Leo ·
Giovanni M. Farinella · Federico Tombari (Eds.)

Image Analysis and Processing – ICIAP 2022

21st International Conference
Lecce, Italy, May 23–27, 2022
Proceedings, Part III

Springer

Editors
Stan Sclaroff
Boston University
Boston, MA, USA

Cosimo Distante
National Research Council
Lecce, Italy

Marco Leo
National Research Council
Lecce, Italy

Giovanni M. Farinella
University of Catania
Catania, Italy

Federico Tombari
Technische Universität München
Garching, Germany

ISSN 0302-9743 ISSN 1611-3349 (electronic)
Lecture Notes in Computer Science
ISBN 978-3-031-06432-6 ISBN 978-3-031-06433-3 (eBook)
https://doi.org/10.1007/978-3-031-06433-3

This Springer imprint is published by the registered company Springer Nature Switzerland AG
The registered company address is: Gewerbestrasse 11, 6330 Cham, Switzerland

Preface

The International Conference on Image Analysis and Processing (ICIAP) is an established biennial scientific meeting promoted by the Italian Association for Computer Vision, Pattern Recognition and Machine Learning (CVPL - formerly GIRPR), which is the Italian IAPR Member Society, and covers topics related to theoretical and experimental areas of Computer Vision, Image Processing, Pattern Recognition, and Machine Learning with emphasis on theoretical aspects and applications. The 21st International Conference on Image Analysis and Processing (ICIAP 2022) was held in Lecce, Italy, during May 23–27, 2022 (postponed from 2021 due to the COVID-19 pandemic), in the magnificent venue of the Teatro Apollo, (http://www.iciap2021.org). It was organized by the Lecce Institute of Applied Sciences and Intelligent Systems, an Institute of CNR, the National Research Council of Italy. ICIAP 2022 was sponsored by NVIDIA, E4 Computer Engineering s.p.a., and ImageS s.p.a., and was endorsed by the Apulia Region and the Province of Lecce.

The conference covered both established and recent scientific trends, with particular emphasis on Video Analysis and Understanding, Pattern Recognition and Machine Learning, Deep Learning, Multiview Geometry and 3D Computer Vision, Image Analysis, Detection and Recognition, Multimedia, Biomedical and Assistive Technology, Digital Forensics and Biometrics, Image Processing for Cultural Heritage, and Robot Vision.

The ICIAP 2022 main conference received 297 paper submissions from all over the world, including Austria, Azerbaijan, Bangladesh, Belgium, Brazil, Canada, China, the Czech Republic, the UK, Finland, France, Germany, Greece, Hungary, India, Ireland, Japan, Latvia, Lebanon, Mongolia, New Zealand, the Netherlands, Italy, Pakistan, Peru, Poland, Portugal, Russia, Syria, Spain, South Korea, Sri Lanka, Luxembourg, South Africa, Sweden, Switzerland, Turkey, the United Arab Emirates, and the USA. Two rounds of submissions were introduced, independent from each other, in order to improve the quality of papers. Paper selection was carried out by 25 expert researchers who acted as area chairs, together with the International Program Committee and an expert team of reviewers. The rigorous peer-review selection process, carried out by three distinct reviewers for each submission, ultimately led to the selection of 162 high-quality manuscripts, with an overall acceptance rate of 54%.

The main conference program included 45 oral presentations, 145 posters, and five invited talks. The invited talks were presented by leading experts in computer vision and pattern recognition: Larry S. Davis, University of Maryland and Amazon (USA), Roberto Cipolla, University of Cambridge (UK), Dima Damen, University of Bristol (UK), and Laura Leal-Taixe, Technische Universität München (Germany).

ICIAP 2022 also included 11 tutorials and hosted 16 workshops, seven competitions, and a special session on topics of great relevance with respect to the state of the art. The tutorial and workshop organizers came from both industry and academia.

Several awards were presented during the ICIAP 2022 conference. The Best Student paper award was supported by the MDPI Journal of Imaging, and several other prizes were conferred under the Platinum level sponsorship provided by NVIDIA.

The success of ICIAP 2022 is credited to the contribution of many people. Special thanks should be given to the area chairs, who did a truly outstanding job. We wish to thank the reviewers for the immense amount of hard work and professionalism that went into making ICIAP 2022 a successful meeting. Our thanks also go to the Organizing Committee for their unstinting dedication, advice, and support.

We hope that ICIAP 2022 has helped to build a piece of the future, a future where technologies can allow people to live comfortably, healthily, and in peace.

May 2022

Cosimo Distante
Stan Sclaroff
Giovanni Maria Farinella
Marco Leo
Federico Tombari

Organization

General Chairs

Cosimo Distante	National Research Council, Italy
Stan Sclaroff	Boston University, USA

Technical Program Chairs

Giovanni Maria Farinella	University of Catania, Italy
Marco Leo	National Research Council, Italy
Federico Tombari	Google and TUM, Germany

Area Chairs

Lamberto Ballan	University of Padua, Italy
Francois Bremond	Inria, France
Simone Calderara	University of Modena and Reggio Emilia, Italy
Modesto Castrillon Santana	University of Las Palmas de Gran Canaria, Spain
Marco Cristani	University of Verona, Italy
Luigi Di Stefano	University of Bologna, Italy
Sergio Escalera	University of Barcelona, Spain
Luiz Marcos Garcia Goncalves	UFRN, Brazil
Javier Ortega Garcia	Universidad Autonoma de Madrid, Spain
Costantino Grana	University of Modena and Reggio Emilia, Italy
Tal Hassner	Facebook AML and Open University of Israel, Israel
Gian Luca Marcialis	University of Cagliari, Italy
Christian Micheloni	University of Udine, Italy
Fausto Milletarì	NVIDIA, USA
Vittorio Murino	Italian Institute of Technology, Italy
Vishal Patel	Johns Hopkins University, USA
Marcello Pelillo	Università Ca' Foscari Venice, Italy
Federico Pernici	University of Florence, Italy
Andrea Prati	University of Parma, Italy
Justus Piater	University of Innsbruck, Austria
Elisa Ricci	University of Trento, Italy
Alessia Saggese	University of Salerno, Italy
Roberto Scopigno	National Research Council, Italy

Filippo Stanco	University of Catania, Italy
Mario Vento	University of Salerno, Italy

Workshop Chairs

Emanuele Frontoni	Università Politecnica delle Marche, Italy
Pier Luigi Mazzeo	National Research Council, Italy

Publication Chair

Pierluigi Carcagni	National Research Council, Italy

Publicity Chairs

Marco Del Coco	National Research Council, Italy
Antonino Furnari	University of Catania, Italy

Finance and Registration Chairs

Maria Grazia Distante	National Research Council, Italy
Paolo Spagnolo	National Research Council, Italy

Web Chair

Arturo Argentieri	National Research Council, Italy

Tutorial Chairs

Alessio Del Bue	Italian Institute of Technology, Italy
Lorenzo Seidenari	University of Florence, Italy

Special Session Chairs

Marco La Cascia	University of Palermo, Italy
Nichi Martinel	University of Udine, Italy

Industrial Chairs

Ettore Stella	National Research Council, Italy
Giuseppe Celeste	National Research Council, Italy
Fabio Galasso	Sapienza University of Rome, Italy

North Africa Liaison Chair

Dorra Sellami	University of Sfax, Tunisia

Oceania Liaison Chair

Wei Qi Yan	Auckland University of Technology, New Zealand

North America Liaison Chair

Larry S. Davis	University of Maryland, USA

Asia Liaison Chair

Wei Shi Zheng	Sun Yat-sen University, China

Latin America Liaison Chair

Luiz Marcos Garcia Goncalves	UFRN, Brazil

Invited Speakers

Larry S. Davis	University of Maryland and Amazon, USA
Roberto Cipolla	University of Cambridge,UK
Dima Aldamen	University of Bristol, UK
Laura Leal-Taixe	Technische Universität München, Germany

Steering Committee

Virginio Cantoni	University of Pavia, Italy
Luigi Pietro Cordella	University of Napoli Federico II, Italy
Rita Cucchiara	University of Modena and Reggio Emilia, Italy
Alberto Del Bimbo	University of Firenze, Italy
Marco Ferretti	University of Pavia, Italy
Fabio Roli	University of Cagliari, Italy
Gabriella Sanniti di Baja	National Research Council, Italy

Endorsing Institutions

International Association for Pattern Recognition (IAPR)
Italian Association for Computer Vision, Pattern Recognition and Machine Learning (CVPL)
Springer

Institutional Patronage

Institute of Applied Sciences and Intelligent Systems (ISASI)
National Research Council of Italy (CNR)
Provincia di Lecce
Regione Puglia

Contents – Part III

Pattern Recognition and Machine Learning

Hangul Fonts Dataset: A Hierarchical and Compositional Dataset for Investigating Learned Representations

Jesse A. Livezey[1,2(✉)], Ahyeon Hwang[3], Jacob Yeung[1], and Kristofer E. Bouchard[1,2,4,5]

[1] Biological Sciences and Engineering Division, Lawrence Berkeley National Laboratory, Berkeley, CA, USA
{jlivezey,kebouchard}@lbl.gov, jacobyeung@berkeley.edu

[2] Redwood Center for Theoretical Neuroscience, University of California, Berkeley, CA, USA

[3] Mathematical, Computational and Systems Biology, University of California, Irvine, CA, USA
ahyeon.hwang@uci.edu

[4] Helen Wills Neuroscience Institute, University of California, Berkeley, CA, USA

[5] Computational Research Division, Lawrence Berkeley National Laboratory, Berkeley, CA, USA

Abstract. Hierarchy and compositionality are common latent properties in many natural and scientific image datasets. Determining when a deep network's hidden activations represent hierarchy and compositionality is important both for understanding deep representation learning and for applying deep networks in domains where interpretability is crucial. However, current benchmark machine learning datasets either have little hierarchical or compositional structure, or the structure is not known. This gap impedes precise analysis of a network's representations and thus hinders development of new methods that can learn such properties. To address this gap, we developed a new benchmark dataset with known hierarchical and compositional structure. The Hangul Fonts Dataset (HFD) is comprised of 35 fonts from the Korean writing system (Hangul), each with 11,172 blocks (syllables) composed from the product of initial, medial, and final glyphs. All blocks can be grouped into a few geometric types which induces a hierarchy across blocks. In addition, each block is composed of individual glyphs with rotations, translations, scalings, and naturalistic style variation across fonts. We find that both shallow and deep unsupervised methods show only modest evidence of hierarchy and compositionality in their representations of the HFD compared to supervised deep networks. Thus, HFD enables the identification of shortcomings in existing methods, a critical first step toward developing new machine learning algorithms to extract hierarchical and compositional structure in the context of naturalistic variability.

Supplementary Information The online version contains supplementary material available at https://doi.org/10.1007/978-3-031-06433-3_1.

S. Sclaroff et al. (Eds.): ICIAP 2022, LNCS 13233, pp. 3–14, 2022.
https://doi.org/10.1007/978-3-031-06433-3_1

Keywords: Representation learning · Hierarchy · Compositionality

1 Introduction

Advances in machine learning, and representation learning in particular, have long been accompanied by the creation and detailed curation of benchmark image datasets [5,16]. Often, such datasets are created with particular structure believed to be representative of the types of structures encountered in the world. For example, many image datasets have varying degrees of hierarchy and compositionality, as exemplified by parts-based decompositions, learning compositional programs, and multi-scale representations [6,15,17]. In contrast, synthetic image datasets often have known, (at least partial) factorial latent structure [22]. Having a detailed understanding of how the structure of a dataset relates to representations that are learned by any machine learning algorithm, whether linear (e.g., independent components analysis) or non-linear (e.g., deep networks) is crucial. Indeed, one of the desired uses of machine learning in scientific applications is to learn latent structure from complex datasets that provide insight into the data generation process [29].

Benchmark image datasets such as MNIST (Fig. 1A) and CIFAR10 [14,16] enabled research into early convolutional architectures. Large image datasets like ImageNet (Fig. 1B) [5] have fueled the development of networks that can solve complex tasks like pixel-level segmentation and image captioning. Although these datasets occasionally have known semantic hierarchy (ImageNet classes are derived from the WordNet hierarchy) or labeled attributes which may be part of a compositional structure (attributes like "glasses" or "mustache" in the CelebA dataset [18]), the complexity of these images prevents a quantitative understanding of how the hierarchy or compositionality is reflected in the data. On the other hand, synthetic benchmark datasets such as dsprites (Fig. 1C), and many similar variations [2,22], have known factorial latent structure. However, these datasets typically do not have (known) hierarchy or compositionality. Thus, benchmark datasets, which have known hierarchical and compositional structure with naturalistic variability are lacking.

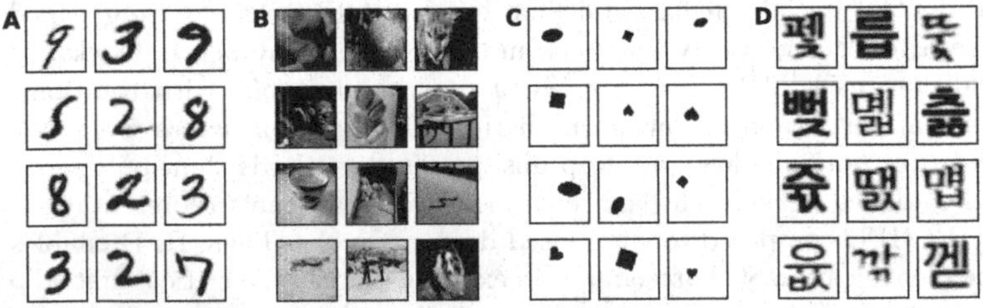

Fig. 1. Ground-truth hierarchy and compositionality are lacking in benchmark machine learning datasets. A Samples from the MNIST dataset. **B** Samples from the ImageNet dataset. **C** Samples from the dsprites dataset. **D** Samples from the Hangul Fonts Dataset.

Machine learning and deep learning methods have been applied to a variety of handwritten and synthetic Hangul datasets with a focus on glyph recognition applications, font generation, and mobile applications [10,11,25–27,31]. HanDB is an early handwritten Hangul dataset [30] and contains approximately 100 samples of each of the 2350 most commonly used blocks. The similarly named Hangul Font Dataset packages a number of open fonts for potential machine learning applications with a focus on the vectorized contour information for the blocks rather than understanding the latent structure of the blocks [13]. As far as we are aware, the Hangul Fonts Dataset presented here is the only Hangul dataset that includes compositional and hierarchical annotations which are required for representation learning research.

A number of methods have been proposed to uncover "disentangled" latent structure from images [1,4,8,17]. While factorial representations are useful for certain tasks like sampling [28], they do not generally capture hierarchical or compositional structures. Deep networks can learn meaningful feature hierarchies, wherein features from higher levels of the hierarchy are formed by the composition of lower level features [9,19,24,32]. Although hierarchy and compositionality are common structures in scientific datasets, benchmark datasets for assessing whether representation learning methods can recover them are lacking.

In this work, we present the new Hangul Fonts Dataset (HFD) (Fig. 1D) designed for investigating hierarchy and compositionality in representation learning methods. The HFD contains a large number of data samples (391,020 samples across 35 fonts), annotated hierarchical and compositional structure, and naturalistic variation. Together these properties address a gap in benchmark datasets for deep learning, and representation learning research more broadly. To give examples of the potential use of the HFD, we explore whether typical deep learning methods can be used to uncover the underlying generative model of the HFD. We find that deep unsupervised networks do not recover the hierarchical or compositional latent structure, and supervised deep networks are able to partially recover the hierarchical latent structure. Thus, the Hangul Fonts Dataset exposes limitations in existing methods for learning hierarchical and compositional structure and will be useful for future investigations of representation learning methods.

2 The Hangul Fonts Dataset

The Korean writing system (Hangul) was created in the year 1444 to promote literacy. Since the Hangul writing system was partially motivated by simplicity and regularity, the rules for creating "blocks" are regular and well specified. The Hangul alphabet consists of "glyphs" broken into 19 initial glyphs, 21 medial glyphs, and 27+1 final glyphs (including no final glyph) which generate $19 \times 21 \times 28 = 11,172$ possible combinations of glyphs which are grouped into initial-medial-final (IMF) blocks. The Hangul Fonts Dataset (HFD) uses this prescribed structure as annotations for the image of each block. The dataset consists of images of all blocks drawn in 35 different open-source fonts for a total of 391,020 annotated images. Appendix 1 contains detailed definitions of blocks, glyphs, and atoms and their linguistic meaning. Each Hangul block can be annotated as having initial, medial, and final (IMF) independent generative variables which can be represented as IMF class labels associated with each block. In addition, there are variables corresponding to a geometric hierarchy and variables corresponding to compositions of glyphs. Together, these different descriptions of the data facilitate investigation into what aspects of this known structure representation learning methods will learn when trained on the HFD.

The geometric rules for creating a block from glyphs induce a hierarchy over the blocks. The initial glyph is located on the left or top of the block as either single or double glyphs (ㄱ or ㄲ in Fig. 2A). There are 5 possible medial glyph geometries: below, right-single, right-double, below-right-single, or below-right-double (ㅗ, ㅏ, ㅖ, ㅘ, or ㅞ in Fig. 2A). The final glyph is at the bottom of the block as single, double, or absent glyphs (ㄱ or ㄳ in Fig. 2A). Grouping the blocks by the 30 geometric possibilities together induce a 2-level hierarchy based on their IMF class labels. The geometric variables describe the coarse layout (high level) of a block which is shared by many IMF combinations (low level) (Fig. 2B and C, bottom and middle levels). Additionally, the 30 geometric categories can be split into their initial, medial or final geometries (Fig. 2B and C, bottom and middle levels). The geometric context of a glyph can change the style of the glyph within a block for a specific font, which is relevant for the representation analysis in Sect. 4. The medial glyph geometry can have a large impact on how an initial glyph is translated and scaled in the block. Similarly, the final geometry can impact the scaling of the initial and medial glyphs. These contextual dependencies can be searched for in learned representations of the data. For example, a supervised deep network trained to predict the initial glyph class may use information from the medial geometry early in the network but then eventually discard that information when predicting the initial glyph class.

Since each block is composed of initial, medial, and final glyphs, the blocks can also be annotated with compositional variables. There is a base set of atomic glyphs (atoms) from which all IMF glyphs are created (Fig. 2D, Atom row). Then, one initial, one medial, and one final glyph are composed into a block (Fig. 2D, IMF and Block rows). In this view, each block is built from a composition of a base set of atomic glyphs, potentially composed with a rotation, translation, or scaling, which are then laid out according to the geometric rules. The underlines in the Atom and IMF rows of Fig. 2D correspond to inclusion in the final colored blocks in the bottom row. In this paper, for comparisons with learned representations, the composition variables are encoded in 2 ways (although the full structure is available in the dataset). The first is a "bag-of-atoms mod rotations" variable where each block is given a vector of binary features which contains a 1 if the block contains at least one atom from the top row of Fig. 2D in any position with any rotation and a 0 otherwise (16 features). The second is a similar "bag-of-atoms" variable where the same atomic glyph with different rotations is given different feature elements (24 features). These

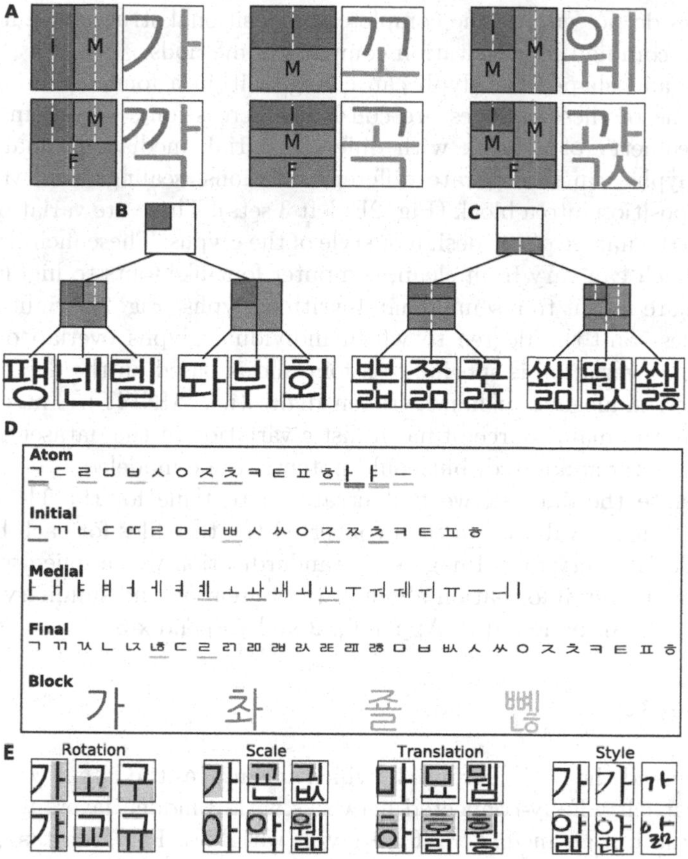

Fig. 2. Hierarchy and composition in the Hangul Fonts Dataset. A, Hierarchy: Each block can be grouped by the initial, medial, and/or final geometry. Six block geometries and example blocks are shown (left and right in each pair). Blue indicates the possible locations of initial glyphs, orange indicates the possible locations of medial glyphs, and green indicates the possible locations of final glyphs. A white dashed line indicates that either a single or double glyph can appear. **B, C, Example hierarchies:** The bottom row of the hierarchy are individual blocks. Each triplet of blocks fall under one of the geometric categories from **A** (middle row) which defined the 2-level hierarchy. Then, a third level can be defined for initial, medial, or final geometric categories (top row). **D, Composition:** Each block is composed of a set of atomic glyphs. The Atom row shows the atomic set of glyphs when scale, translations, and rotations are modded out. The Initial, Medial, and Final (IMF) rows show all IMF glyphs. The Block row shows four example blocks with different types of structure. The color of the block is used to underline the IMF glyphs that compose the block and Atoms that compose the IMFs. **E, Variability:** Two example glyphs (rows) across three different IMF contexts (columns) are shown for each type of variation. **Rotation:** Left-most block is rotated once counterclockwise in the next block, then twice counterclockwise in the final block. **Scale:** Size of initial glyph decreases from left to right as highlighted in red. **Translation:** Highlighted glyph takes on various shapes as it is translated to different regions of the block. **Style:** Less to more stylized from left to right. (Color figure online)

two variables do not encode the complete compositional structure, but they are amenable to common representation comparison methods.

The size and shape of a glyph can change within a font depending on the context. Some of these changes are consistent across fonts and stem from the changing geometry of a block with different initial, medial, or final contexts (Fig. 2). Glyphs can incorporate different rotations, scalings, and translation during composition into a block (Fig. 2E, left 3 sets). There are variations across fonts due to the nature of the design or style of the glyphs. These include the style of glyphs which can vary from clean, computer font-like fonts to highly stylized fonts which are meant to resemble hand-written glyphs (Fig. 2E, rightmost set). Line thickness and the degree to which individual glyphs overlap or connect also vary. This variation is specific to a font and is based on the decision of the font designer, analogous to hand-written digits (i.e., MNIST). These types of variation are the main source of naturalistic variation in the dataset since they cannot be exactly annotated, but could potentially be modeled.

To generate the dataset, we first created a text file for the 11,172 blocks using their Unicode values. We then converted the text files for each block into an image file for every font. Image size standardization was applied within and across fonts. Further information about dataset creation and summary statistics for the dataset can be found in Appendix 2 and Appendix 3.

3 Methods

We implemented the β-VAE from [3], which encourages the latent codes to have a specific capacity. Fully-connected networks with 3 hidden layers were trained on one of the initial, medial, or final glyph variables. For each task, 100 sets of hyperparameters were used for training. Additional details along with the hyperparameters and their ranges are listed in Appendix 4. The model with the best validation accuracy was chosen and the downstream analysis was done on the test set representations (test accuracies reported in Appendix 3).

The 35 fonts were used in a 7-fold cross validation loop for the machine learning methods. The fonts were randomly permuted and then 5 fonts were used for each of the non-overlapping validation and test sets. The analysis of representations was done on the test set representations. For the supervised deep networks, the Kmeans clustering analysis and sparse logistic regression analysis were applied to the activations of every layer both before and after the ReLU nonlinearities. For the unsupervised VAEs, they were applied to samples from the latent layer.

To compare accuracies (and chance accuracies) across models with differing numbers of classes (between 2 and 30), we 0-1 normalize the accuracies across models to make comparisons more clear. Specifically, for a model with accuracy $= a$ and chance $= c$, we report Norm. acc. $= \frac{a-c}{1-c}$ which is 0 when $a = c$ and is 1 when $a = 1$, independent of the number or distribution of classes.

4 Results

Both shallow and deep network models learn representations of the input data. Here, we compare the learned representations in deep variational autoencoders and deep feedforward classifiers (for unsupervised shallow methods, see Appendix 4). We consider whether the learned representations are organized around any of the categorical labels and hierarchy variables with an unsupervised KMeans analysis. Then, we investigate whether the hierarchy or compositionality variables can be decoded with high accuracy from a small set of features in the representations.

It is currently not known whether deep network representations are typically organized around the generative variables of a dataset. In order to understand this, we test whether the latent hierarchical structure of the Hangul blocks is a major component of the learned representations using unsupervised clustering of the representations. We compare the hierarchy geometry classes from Fig. 2A to KMeans clusterings of the test set representations (where k is set to the number of class in consideration, for more details, see Appendix 4). For the deep unsupervised methods (Fig. 3A), we find that the medial label and geometry, final label, and all_geometry variables are all marginally present ($0 <$ normalized accuracy ≤ 0.25) in the representations. The other variables are not recovered by the unsupervised methods (normalized accuracy ≈ 0). This shows that while VAE variants may be able to disentangle factorial structure in data, they are not well suited to extracting geometric hierarchy from the HFD with high fidelity.

Supervised deep networks cleanly extract and recover the label they are trained on (Fig. 3B-D, first 3 columns) with increasing accuracy across layers (Norm. acc. > 0.25). When trained on the initial label, the initial, medial, and all_geometry variables can all be marginally recovered, highlighting the contextual dependence of the initial glyph on the medial geometry. The medial_geometry variable can be decoded with accuracy significantly above chance across all layers ($p < 0.01$, 1-sample t-test). However, the normalized accuracy drops from about 0.22 in the first layer to less than .01 by the last layer. This indicates that although the network may be using the medial geometry context in the early layers, it is compressed out of the representation by the final layers. The initial geometry is not present in the first 2 layers, but becomes marginally present in the final layers. When trained on the medial labels, the medial geometry is present with high accuracy and the all_geometries labels are marginally present. When trained on the final labels, the final geometry becomes present by the last 2 layers. There is a small amount of interaction with the medial geometry, but it is not as large as the initial-medial interaction. Several normalized accuracies are less than zero, however this is likely due to statistical fluctuations around 0 (p-values > 0.01, 1 sample t-test). These results indicate that supervised deep networks do learn representations that mirror aspects of the hierarchical structure of the dataset that are most relevant for the task, and generally do not extract non-relevant hierarchy information.

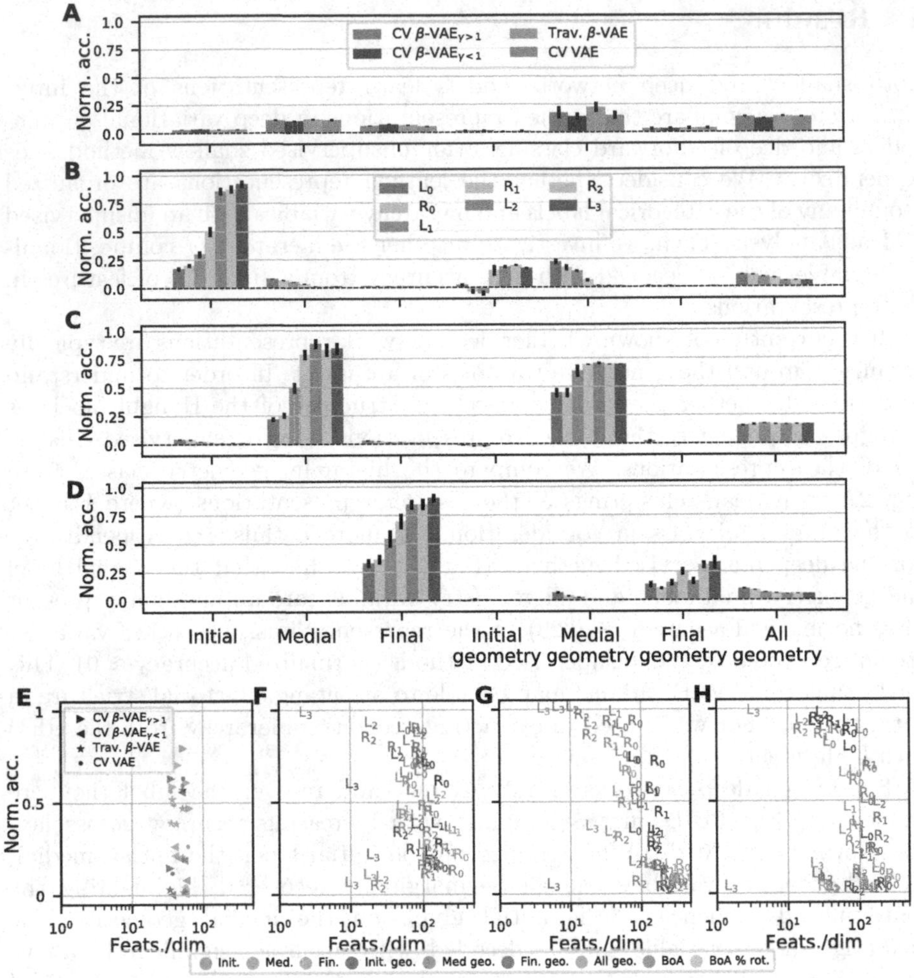

Fig. 3. Hierarchy and compositionality are not typically represented in deep networks. A–D: Normalized clustering accuracy ± s.e.m. is shown across training targets, latent generative variables, layers (L is the linear part, R is after the ReLU), and model types. **A** Normalized clustering accuracies for representations learned with various deep VAE models. **B–D** Normalized clustering accuracies for deep representations trained to predict the initial, medial, and final label, respectively. **E–H:** Held-out logistic regression normalized accuracy versus the ratio of the number of features selected to the variable dimensionality is shown. Color indicates latent variable type. **E**: Results from the VAE model variants. Shape is model type. **F–H:** Results from supervised deep networks trained on the initial, medial, and final tasks, respectively. Letters correspond to the layers from A–D.

Understanding whether deep network representations tend to be more distributed or local is an open area of research [23]. We investigated whether deep networks learn a local representation by training sparse logistic regression models to predict the latent hierarchy and compositionality variables from the representations. If the representation of a hierarchy or compositionality variable is present and simple (linear), we would expect the normalized accuracy to be high (near 1 on the y-axis of the plots in Fig. 3E–H). If a representation of a variable is "local", we would expect the variable to be decoded using approximately the same number of features as it has dimensions (near 10^1 on the x-axis of Fig. 3E–H) and "distributed" representation to have a much higher ratio. To test this, we compare these two measures across models and target variables and also across layers for the supervised deep networks.

We find that unsupervised (β-)VAEs (Fig. 3E) learn consistently distributed representations of the latent variables (typically 30–60$\times$ more features than the variable dimension are selected). In terms of the prediction accuracy, the β-VAEs selected by cross-validation tend to have higher accuracy across variables than the VAE and the β-VAE selected for traversals, although there is a fair amount of heterogeneity. For supervised deep networks (Fig. 3F–H), the supervision variable (initial, medial, final, respectively), has high accuracy across layers, and moves from a more distributed to a more local representation at deep layers. For the initial and medial labels, the medial geometry can also be read out with high accuracy and an increase in localization across layers. The initial geometry is not read out with high accuracy in the initial and medial label networks, and the final geometry variable can only be predicted well for the final label network. The all_geometry variable can be predicted at marginal accuracy for all networks. The compositional Bag-of-Atoms (BoA) features cannot be predicted well (often at or below chance) for any network and the BoA mod rotations can only be read out with marginal accuracy for the initial label network. These results suggest that standard, fully-connected deep networks do not typically learn local representations for variables except for those they are trained on (and correlated variables).

5 Discussion

The Hangul Fonts Dataset (HFD) presented here has hierarchical and compositional latent structure that allows each image to have ground-truth annotations, making the HFD well suited for deep representation research. Using a set of unsupervised and supervised methods, we are able to extract a subset of the variables from the representations of deep networks. Several VAE variants have relatively poor variable recovery from their latent layers, while supervised deep networks have clear representation of the variables they are trained on (and closely related variables). In many scientific domains like cosmology, neuroscience, and climate science, deep learning is being used to make high accuracy predictions given growing dataset sizes [12,20,21]. However, deep learning is not commonly used to directly test hypotheses about dataset structure. This is partially because the nonlinear, compositional structure of deep networks, which is conducive to high accuracy prediction from complex data, is not ideal for interrogating hypotheses about data. Understanding how to better recover such structure from deep network representations will broaden the application of deep learning in science.

In this work, relatively small fully-connected and convolutional networks were considered. However, these techniques can be applied to larger feedforward networks, recurrent networks, or networks with residual layers to understand the impact on learned representations. Compared to disentangling [7], relatively little work addresses how to define and evaluate hierarchy and compositionality in learned representations. Furthermore, unsupervised or semi-supervised cross-validation metrics that can be used for model selection across a range of structure recovery tasks (e.g., disentangling, hierarchy recovery, compositionality recovery) are lacking. We hope the HFD can be used to develop unsupervised and semi-supervised cross-validation metrics.

The HFD is based on a set of fonts which provide some naturalistic variation. However, the amount of variation is likely much smaller than what would be found in a handwritten dataset of Hangul blocks. One benefit to using fonts is that the dataset can be easily extended as new fonts are created. To this end, we release the entire dataset creation pipeline to aid in future expansion of the HFD or the creation of similar font-based datasets. Another potential limitation and area of future work is determining how to encode variables like hierarchy and compositionality. In this dataset, there is a natural class-based encoding for the shallow geometry hierarchy. The Bag-of-Atoms composition encoding ignores structure that is potentially relevant for recovering compositionality (much like Bag-of-Words features discard potentially useful structure in natural language processing). The specific compositional and hierarchical structure in the HFD, the particular encodings used, and the corresponding analysis may not be applicable across all types of compositionality or hierarchy, for instance some hierarchy may be fuzzy, rather than discrete and tree-like. For example, the KMeans clustering analysis was applied to all variables with mutually-exclusive class structure, but could not be applied to the bag-of-atoms feature vectors. However, we hope that the HFD inspires more research into tools for extracting these features from learned representations.

Acknowledgements. JAL, AH, and KEB were supported by the Deep Learning for Science LBNL LDRD. We are grateful for the feedback on the project from the Neural Systems and Data Science Lab.

References

1. Bell, A.J., Sejnowski, T.J.: The "independent components" of natural scenes are edge filters. Vision Res. **37**(23), 3327–3338 (1997). https://doi.org/10.1016/S0042-6989(97)00121-1
2. Burgess, C., Kim, H.: 3D shapes dataset (2018). https://github.com/deepmind/3dshapes-dataset/

3. Burgess, C.P., et al.: Understanding disentangling in β-VAE. arXiv preprint arXiv:1804.03599 (2018)
4. Cheung, B., Livezey, J.A., Bansal, A.K., Olshausen, B.A.: Discovering hidden factors of variation in deep networks. arXiv preprint arXiv:1412.6583 (2014)
5. Deng, J., Dong, W., Socher, R., Li, L.J., Li, K., Fei-Fei, L.: Imagenet: a large-scale hierarchical image database. In: 2009 IEEE Conference on Computer Vision and Pattern Recognition, pp. 248–255. IEEE (2009). https://doi.org/10.1109/CVPR.2009.5206848
6. Denton, E., Chintala, S., Szlam, A., Fergus, R.: Deep generative image models using a laplacian pyramid of adversarial networks. arXiv preprint arXiv:1506.05751 (2015)
7. Higgins, I., et al.: Towards a definition of disentangled representations. arXiv preprint arXiv:1812.02230 (2018)
8. Higgins, I., et al.: beta-VAE: learning basic visual concepts with a constrained variational framework. In: International Conference on Learning Representations, vol. 3 (2017)
9. Kell, A.J., Yamins, D.L., Shook, E.N., Norman-Haignere, S.V., McDermott, J.H.: A task-optimized neural network replicates human auditory behavior, predicts brain responses, and reveals a cortical processing hierarchy. Neuron **98**(3), 630–644 (2018). https://doi.org/10.1016/j.neuron.2018.03.044
10. Kim, I.-J., Choi, C., Lee, S.-H.: Improving discrimination ability of convolutional neural networks by hybrid learning. Int. J. Doc. Anal. Recognit. (IJDAR) **19**(1), 1–9 (2015). https://doi.org/10.1007/s10032-015-0256-9
11. Kim, I.J., Xie, X.: Handwritten hangul recognition using deep convolutional neural networks. Int. J. Doc. Anal. Recognit. (IJDAR) **18**(1), 1–13 (2015). https://doi.org/10.1007/s10032-014-0229-4
12. Kim, S., et al.: Deep-hurricane-tracker: tracking and forecasting extreme climate events. In: 2019 IEEE Winter Conference on Applications of Computer Vision (WACV), pp. 1761–1769. IEEE (2019). https://doi.org/10.1109/WACV.2019.00192
13. Ko, D.H., Lee, H., Suk, J., Hassan, A.U., Choi, J.: Hangul font dataset for Korean font research based on deep learning. KIPS Trans. Softw. Data Eng. **10**(2), 73–78 (2021)
14. Krizhevsky, A., Nair, V., Hinton, G.: The CIFAR-10 dataset, vol. 55 (2014). http://www.cs.toronto.edu/kriz/cifar.html
15. Lake, B.M., Salakhutdinov, R., Tenenbaum, J.B.: Human-level concept learning through probabilistic program induction. Science **350**(6266), 1332–1338 (2015). https://doi.org/10.1126/science.aab3050
16. LeCun, Y., Bottou, L., Bengio, Y., Haffner, P.: Gradient-based learning applied to document recognition. Proc. IEEE **86**(11), 2278–2324 (1998)
17. Lee, D.D., Seung, H.S.: Learning the parts of objects by non-negative matrix factorization. Nature **401**(6755), 788–791 (1999). https://doi.org/10.1038/44565
18. Liu, Z., Luo, P., Wang, X., Tang, X.: Deep learning face attributes in the wild. In: Proceedings of International Conference on Computer Vision (ICCV), December 2015. https://doi.org/10.1109/ICCV.2015.425
19. Livezey, J.A., Bouchard, K.E., Chang, E.F.: Deep learning as a tool for neural data analysis: speech classification and cross-frequency coupling in human sensorimotor cortex. PLoS Comput. Biol. **15**(9), e1007091 (2019). https://doi.org/10.1371/journal.pcbi.1007091

20. Livezey, J.A., Glaser, J.I.: Deep learning approaches for neural decoding across architectures and recording modalities. Brief. Bioinform. **22**(2), 1577–1591 (2021). https://doi.org/10.1093/bib/bbaa355
21. Mathuriya, A., et al.: Cosmoflow: using deep learning to learn the universe at scale. In: SC 2018: International Conference for High Performance Computing, Networking, Storage and Analysis, pp. 819–829. IEEE (2018). https://doi.org/10.1109/SC.2018.00068
22. Matthey, L., Higgins, I., Hassabis, D., Lerchner, A.: dSprites: disentanglement testing sprites dataset (2017). https://github.com/deepmind/dsprites-dataset/
23. Nguyen, A., Yosinski, J., Clune, J.: Multifaceted feature visualization: uncovering the different types of features learned by each neuron in deep neural networks. arXiv preprint arXiv:1602.03616 (2016)
24. Nickel, M., Kiela, D.: Poincaré embeddings for learning hierarchical representations. In: Advances in Neural Information Processing Systems 30 (2017)
25. Oktaviani, S., Sari, C.A., Rachmawanto, E.H., et al.: Optical character recognition for hangul character using artificial neural network. In: 2020 International Seminar on Application for Technology of Information and Communication (iSemantic), pp. 34–39. IEEE (2020). https://doi.org/10.1109/iSemantic50169.2020.9234215
26. Park, G.R., Kim, I.J., Liu, C.L.: An evaluation of statistical methods in handwritten hangul recognition. Int. J. Doc. Anal. Recognit. (IJDAR) **16**(3), 273–283 (2013). https://doi.org/10.1007/s10032-012-0191-y
27. Purnamawati, S., Rachmawati, D., Lumanauw, G., Rahmat, R., Taqyuddin, R.: Korean letter handwritten recognition using deep convolutional neural network on android platform. In: Journal of Physics: Conference Series, vol. 978, p. 012112. IOP Publishing (2018). https://doi.org/10.1088/1742-6596/978/1/012112
28. Schmidhuber, J.: Learning factorial codes by predictability minimization. Neural Comput. **4**(6), 863–879 (1992). https://doi.org/10.1162/neco.1992.4.6.863
29. Stevens, R., Taylor, V., Nichols, J., Maccabe, A.B., Yelick, K., Brown, D.: AI for science (2020)
30. Korea University: HanDB: PE92 and SERI95 (2017). https://github.com/callee2006/HangulDB
31. Van Eck, P.: Handwritten Korean character recognition with tensorflow and android (2017). https://developer.ibm.com/patterns/create-mobile-handwritten-hangul-translation-app/
32. Yamins, D.L., Hong, H., Cadieu, C.F., Solomon, E.A., Seibert, D., DiCarlo, J.J.: Performance-optimized hierarchical models predict neural responses in higher visual cortex. Proc. Natl. Acad. Sci. **111**(23), 8619–8624 (2014). https://doi.org/10.1073/pnas.1403112111

Out-of-Distribution Detection Using Outlier Detection Methods

Jan Diers(✉) and Christian Pigorsch

Friedrich-Schiller-University Jena, Fürstengraben 1, 07743 Jena, Germany
{jan.diers,christian.pigorsch}@uni-jena.de
https://www.wss.uni-jena.de/

Abstract. Out-of-distribution detection (OOD) deals with anomalous input to neural networks. In the past, specialized methods have been proposed to identify anomalous input. Similarly, it was shown that feature extraction models in combination with outlier detection algorithms are well suited to detect anomalous input. We use outlier detection algorithms to detect anomalous input as reliable as specialized methods from the field of OOD. No neural network adaptation is required; detection is based on the model's softmax score. Our approach works unsupervised using an Isolation Forest and can be further improved by using a supervised learning method such as Gradient Boosting.

Keywords: Out-of-distribution detection · Outlier detection · Isolation Forest

1 Introduction

Machine learning methods and in particular neural networks are the backbone of many modern applications in research and industry. To apply these methods, at first, training data is collected, then models are trained and evaluated on validation or test data. If the error is low and the model makes reliable predictions, the model is released to be used in software. In the software, for example, it controls autonomous systems, detects production errors, or analyzes text.

But what happens if the data from the real environment does not match the training data well? What happens if the real data does not match the training data, e.g. if targets have changed? The model will not refuse to predict, but will fail the task undetected. It cannot signal the user that the new input is an unknown class or from a different distribution of the class.

Most of the modern models are unable to detect this change of environment (called "domain adaptation" or "concept drift"). Depending on how the distribution of the test data changes, it is possible to reweight the training data in the model's objective function to reflect the domain change [11]. However, this

Supplementary Information The online version contains supplementary material available at https://doi.org/10.1007/978-3-031-06433-3_2.

S. Sclaroff et al. (Eds.): ICIAP 2022, LNCS 13233, pp. 15–26, 2022.
https://doi.org/10.1007/978-3-031-06433-3_2

does not work if the data and labels change arbitrarily, i.e., no knowledge about the distribution of the test data exists in advance. In practice, this is actually the case because it is impossible to know in advance what data points the model will be queried on. It is an open challenge for AI safety to teach models not to make a decision if they are unsure about the decision [1]. Current models lack this capability.

Recently, the research field of out-of-distribution detection (OOD) has emerged, which deals with processing data points that do not match the distribution of the training data. The goal is to detect anomalous input and deny an invalid prediction. The raw softmax score is not suitable in these cases for neural networks [4] as well as for other machine learning methods [15] - although it is supposed to represent the confidence of the decision.

For image classification, the problem is formulated as follows. A model is trained to distinguish between M different classes $K_1, K_2, ..., K_M$. At inference time, data appears which originates from a different domain, i.e. not from the M classes already known before, but from a new class. The model can therefore only misclassify, because the true label K_{M+1} is unknown to the model. So how do we need to change models to let them handle anomalous input? How can we find out if an input is from an unknown class?

Hendrycks and Gimpel [8] provide a baseline against which misclassified and out-of-domain input can be detected. They empirically find that the predicted confidence of a neural network is lower when the input contains a foreign class K_{M+1}.

It has been shown that extracted features of a neural network, trained on ImageNet, provide an effective way to detect anomalous images in the input [2]. In addition, there is evidence that anomalous input in image data can also be reliably detected with supervised learning methods [9,17]. The label for the supervised methods is generated automatically based on arbitrary images that represent the distributions different to the normal distribution.

We follow up this work and show that outlier detection algorithms also reliably detect anomalous input when the neural network has been specifically trained for a particular task. The detection of out-of-distribution data becomes even more reliable when a supervised learning method is used instead of the unsupervised outlier-detection algorithm.

Our proposed methodology does not require re-training of models and can be applied to any existing models. We build on the work of [8] and detect OOD input based solely on predicted class probabilities. To do this, we fit an Isolation Forest that separates expected class probabilities from abnormal class probabilities. This enables the detection of the OOD input. Validation data, which should be available for every model anyway, is sufficient for fitting the Isolation Forest.

Our contributions to the research area are:

1. We propose an unsupervised method to distinguish in-distribution from out-of-distribution input. The results indicate that the assumptions and methods of outlier and deep anomaly detection are also relevant to the field of out-of-distribution detection.

2. The method works on the basis of an Isolation Forest. It can be applied to existing models, requires no special training and no special architecture of the model. The employed loss function also does not need to be adapted. Classification accuracy does not suffer since no change is made to the model.
3. We show empirically, using common benchmark datasets, that our approach leads to a better detection of out-of-domain input than with current OOD detection techniques. This is especially the case when the classification accuracy of the network is already low.
4. We present that the results can be further improved if a supervised learning method is used instead of the Isolation Forest to detect the anomalies. The supervised learning method generalizes well to previously unknown OOD data.

2 Related Literature

The related literature for this work is primarily guided by two research areas. The first is the field of out-of-distribution detection. The second research field covers the area of outlier detection and has received little attention in the field of out-of-distribution detection. We will give a brief overview of the two research fields in the following section.

In the following, we assume that a model f has been trained to predict the probability $f(x_{in}) = p(y|x_{in})$. The estimation of the likelihood will succeed if $x \in X_{in}$, i.e., the input x corresponds to the distribution of the training data X_{in}.

2.1 Out-of-Distribution Detection

Out-of-distribution detection considers the case where $p(y|x_{out})$ is estimated, where x_{out} is taken from a distribution that has no correspondence with X_{in}. In particular, the class probabilities $p(y|x_{out})$ must be false, since none of the y in X_{out} exists.

Consider the example, that f was trained to distinguish cats from dogs. X_{in} then includes images of cats and dogs. It follows that X_{out} represents all other images, that neither contain cats nor dogs. The target $y \in \{0, 1\}$ indicates whether the image contains a cat or a dog. The predicted probability $p(y|x_{out})$ is then false in all cases, since X_{out} does not include images of cats or dogs.

An approach to determine whether $p(y|\cdot)$ was estimated based on x_{in} or x_{out} is provided by [8]. The paper finds that the overall estimated confidence for X_{out} is lower than the estimated confidence for X_{in}. This allows to define a threshold that detects out-of-domain input: If $p(y|\cdot)$ is too low, then assume x_{out}, otherwise it is x_{in}. Liang et al. [12] follow up on this work and propose their method called ODIN. ODIN uses temperature scaling to obtain a better calibrated model that estimates $p(y|\cdot)$ more reliably. This further increases the difference between predictions on X_{in} and X_{out}. In addition, the approach also makes changes to the input to obtain a more robust estimate. To achieve this, the gradient w.r.t. input is calculated and the input is changed so that $p(y|x)$ increases:

$$\tilde{x} = x - \epsilon \cdot \text{sign}(-\nabla_x \log(f(x)_y)). \tag{1}$$

These studies rely on the difference of the confidences to be sufficiently large for detecting OOD input. In addition, there are other approaches that do not rely on the predicted confidences, such as the one proposed by DeVries and Taylor [5]. They add an output to the model to represent the confidence of the decision. This gives the model the ability to output a confidence for which it expects the estimated probabilities to be correct.

Ren et al. [16] work with two different models. For this purpose, they decompose $p(x)$ into a semantic part $p(x_S)$, which contains the relevant information for the class membership y. The irrelevant part (noise), which is not necessary for estimating $p(y|x)$, is subsumed under $p(x_B)$. The joint occurrence is then modeled as $p(x) = p(x_S) \cdot p(x_B)$. The decomposition is used to estimate two different models. The models allow separating between semantics and noise, which enables the detection of OOD input.

Another line of research has emerged under the name Outlier Exposure. This refers to methods that have already been trained on data from X_{out}. It turns out that neural networks usually generalize to other, previously unknown data from X_{out} and thus provide reliable detection.

Hendrycks et al. [9] use Outlier Exposure to adjust the loss function L of the model so that the entropy $p(y|x_{out})$ is high. To achieve this, they introduce an additional term into the loss function for the classification task:

$$\min \mathbb{E}_{(x,y)\sim X_{in}}[L(f(x), y)] + \lambda \mathbb{E}_{x\sim X_{out}}[H(f(x), U)], \tag{2}$$

where H corresponds to the cross entropy and U corresponds to the uniform distribution over M classes. The change of the loss function encourages the model to predict a uniform distribution when there is input from x_{out}. The authors also present ways to apply the method when the problem is not a classification task.

Previous work uses datasets X_{out}, which are semantically different from X_{in}, to evaluate their methods. Chen et al. [3] note that this evaluation is incomplete. They show that most methods inadequately recognize when OOD is input generated based on adversarial attacks. The authors therefore extend the approach of [9] to include data that was generated using adversarial attacks. This produces more robust detection of X_{out} when X_{out} is minimally modified data from X_{in}. They use the following adjusted loss function:

$$\min \mathbb{E}_{(x,y)\sim X_{in}}[\max_{\delta \in B(x,\epsilon)}[-\log(f(x+\delta)_y)]] + \lambda \mathbb{E}_{x\sim X_{out}} \max_{\delta \in B(x,\epsilon)}[H(f(x), U)]]. \tag{3}$$

The set of changes to the input that deviate at most by ϵ from the original input is denoted by $B(x, \epsilon) = \{\delta \in \mathbb{R}^n : |\delta||_\infty \leq \epsilon \wedge x + \delta \text{ is valid}\}$. $x + \delta$ is valid if the minimum and maximum pixel values of the image are maintained. The generation is thus similar to $\tilde{x}$ from (1), with the change in maximum distance from the original ϵ.

Sun et al. [18] highlight our observation that OOD input can be detected in the extracted features of trained networks. The authors find that activations of neurons in the penultimate layer show a changed distribution when OOD input is present. The work proposes to threshold activations of neurons to a maximum value. This way, the authors succeed in distinguishing OOD input more clearly from in-distribution data.

2.2 Outlier Detection

Our work uses outlier detection methods to detect the OOD input of neural networks. A fixed definition of outliers does not exist, however, the community has widely agreed on the definition of Hawkins [7]: *"An outlier is an observation which deviates so much from the other observations as to arouse suspicions that it was generated by a different mechanism"*. The distinction between X_{in} and X_{out} thus fulfills all requirements to be an application of outlier detection.

Outlier detection methods search for points that are located in areas of low density. If the data generating distribution is known, it is easy to calculate $p(x)$. For empirical data, methods must be found to estimate $p(x)$.

Common methods for outlier detection are based on kernel density estimates, k-nearest neighbor methods (kNN), dimension reduction or support vector machines, among others. Methods from the domain of neural networks rely on, among others, generative adversarial networks or autoencoders for this purpose. See [23] for a comprehensive overview of the recent progress in the field.

In our work, we use the Isolation Forest [13] as a method for outlier detection. As a tree-based method, it has similarity to the implicit density estimation of kNN-based methods, with the difference that it is a model-driven approach. The Isolation Forest scales linearly with the size of the data set, which is a major advantage over instance-based kNN methods.

The Isolation Forest is composed of multiple Isolation Trees that perform random splits to features. The objective is to isolate individual data points in the nodes. The fewer splits are required before a point can be isolated into a terminal node, the larger is the outlier factor of that point. The outlier factor is defined as the expected path length of a point in all trees of the forest.

Combining outlier detection methods with neural network features has already been used in the community. For example, [2] use the features of a network trained on ImageNet to detect anomalies using kNN. Our approach is similar. However, unlike [2], we do not use features based on ImageNet but train each of the networks for the dataset. Furthermore, we do not use intermediate features, but rather the output of the network. Both of these are motivated by the fact that we are interested in detecting OOD input when the network has been trained for a specific task.

The traditional methods of outlier detection work unsupervised. The reason is that outliers from the past are not necessarily representative for outliers in the future. This prohibits the reliable use of supervised learning methods. However, it is not true for every anomaly detection task that the distribution of outliers may change over time. There are some anomalies, e.g. in medicine, whose origin is well understood, and therefore it can be assumed with a high degree of certainty that anomalies will not change in the future.

If this is the case, then anomaly detection can be transformed into a supervised problem. The task then corresponds to an ordinary classification problem. Usually, the data is highly unbalanced, since the outliers represent, by definition, a minority of the data points.

For OOD detection, there is evidence that anomalies also do not change over time. For example, Ruff et al. [17] and Hendrycks et al. [9] report that neural networks also detect previously unknown OOD data well when trained with arbitrary, other OOD data. This also justifies the use of outlier exposure models for the task and suggests that supervised learning methods may be well-suited for OOD detection. In our experiments, we use Gradient Boosting [6] for this purpose and confirm that reliable OOD detection is possible with supervised learning methods.

3 Method

Our method uses the softmax values ($\hat{y}$) of a neural network f. The values in $\hat{y}$ follow a distribution, which we denote as D_{in}, so that $f(X_{in}) \sim D_{in}$. We further assume that different input leads to different output. OOD-Input X_{out} therefore follows a different distribution, i.e. $f(X_{out}) \sim D_{out}$. We will come back to this topic in the discussion of our empirical results in Sect. 6.

We use different approaches to distinguish elements of D_{in} from elements of D_{out}. Our first approach is based on an unsupervised Isolation Forest. For this, we use the Isolation Forest to estimate the normal distribution D_{in}, which we can use to estimate $p(f(x))$. If the Isolation Forest signals that $p(f(x))$ is low, we assume that x does not come from X_{in} and instead represents OOD input.

We can determine the distribution D_{in} by splitting a validation set from the training data. On the validation set, we can apply f to obtain data that follow the distribution D_{in}. We then fit the Isolation Forest on this data. See also Algorithm 1. We note that Isolation Forest was chosen because this method scales linearly with the size of the data set. This is an advantage over other methods of outlier detection. However, in experiments not shown here, we were able to successfully detect OOD input using other outlier detection methods, but the runtime was increased.

Algorithm 1: OOD-Detection using an Isolation Forest

Result: Isolation Forest for OOD-Detection
Input: Trained neural network, validation set, Isolation Forest
begin
  1. make predictions on validation set;
  2. fit Isolation Forest on predictions;
  3. apply Isolation Forest to new predictions;
end

As a second approach, we propose to use supervised learning to distinguish D_{in} and D_{out}. To do this, we also use the validation data from the training set and take completely different data to simulate D_{out}. In our case, the Food101 dataset [14] is the basis for determining D_{out}. The Food101 dataset contains images from different food categories and has no similarity to the other datasets we use.[1] This procedure is identical to the outlier exposure models. We then fit a Gradient Boosting classifier that separates between points $f(X_{in})$ and $f(X_{out})$. The choice of Gradient Boosting is not of particular importance; many classification models are suitable for this purpose. The use of supervised anomaly detection seems feasible because known OOD input generalizes well to unknown OOD input [9,17]. That is, although we use Food101 as an exemplary OOD dataset to train the model, we can also detect OOD input from other datasets. The details are shown by Algorithm 2.

Algorithm 2: OOD-Detection with a Gradient Boosting Classifier

Result: Gradient Boosting Classifier for OOD-Detection
Input: Trained neural network,
validation set,
any OOD-dataset (e.g. Food101),
Gradient Boosting Classifier (GBM)
begin
1. $\hat{y}_{in} \leftarrow$ make predictions on validation set;
2. $\hat{y}_{out} \leftarrow$ make predictions on OOD set;
3. fit the GBM to learn difference between $\hat{y}_{in}$ and $\hat{y}_{out}$;
4. apply GBM to new predictions;

end

4 Experiment Setup

For our experiments, we use 8 different datasets as in- and OOD data. The images are all cropped to a size of 224 × 224 pixels and enriched with smaller augmentation steps (flip and contrast). The details of the datasets can be found in Table 1. We use EfficientNets [20] pre-trained on ImageNet data to perform the transfer learning on the respective dataset. We set the label smoothing parameter [19] to 0.2. We first train the front layers for three epochs with a learning rate of 0.01, before further fine-tuning the model starting from the second block with a learning rate of 0.001. Adam [10] is used as optimizer. We apply early stopping and reduce the learning rate by a factor of 0.6 if the validation loss could not be reduced 3 epochs in a row. The classification error of the models on the different datasets is 0.01% for CatsVsDogs, 0.03% for Cifar10, 0.15% for Cifar100, 0.34%

[1] Previously published work in the area of OOD often uses the Tiny Images dataset [22] to calibrate the models. This dataset has been withdrawn by the authors [21]. Instead, we use the Food101 [14] dataset to train outlier exposure models and the Gradient Boosting classifier.

for Cars196 and 0.14% for Cassava. Regardless of the OOD method used, the classifier results are the same.

Table 1. Details for the datasets used in this study. Textures and SVHNCropped are only used as out-of-distribution datasets and therefore do not have a training or validation split. The Food101 dataset is only used as training and validation set to train Outlier Exposure and Gradient Boosting models. It is not used for out of distribution detection on test images.

Name	Number of			
	Classes	Train samples	Val samples	Test samples
Cars196	196	6.515	1.629	8.041
Cassava	5	5.656	1.889	1.885
CatsVsDogs	2	13.957	4.653	4.652
Cifar10	10	40.000	10.000	10.000
Cifar100	100	40.000	10.000	10.000
Food101	101	60.600	15.150	–
Textures	47	–	–	5.640
SVHNCropped	10	–	–	10.000

In the benchmark we compare 5 different methods: the Baseline approach [8], ODIN [12], Outlier Exposure [9] and our proposed methods based on Isolation Forest and Gradient Boosting. Our code is publicly available on GitHub.[2]

5 Results

For the metrics, we stay consistent with other work in the field. We consider the out-of-distribution error (OOD error), the area-under-curve value (AUC), and the false precision rate when the true precision rate is 95% (FPR at 95% TPR). We want to point out that OOD detection is a topic with a high level of practical relevance. The system must make a *binary* decision whether the input is anomalous or not - the score of a point is not sufficient for this task. Therefore, we emphasize the importance of the OOD error, which is the number of misclassified OOD images. In practice, we would like to minimize this error. The AUC is a metric that does not require a threshold, but works based on scores.

Table 2 shows the results averaged over 5 in-distribution datasets, each tested against 6 other OOD datasets. The results with an Isolation Forest provide comparable results to the methods designed specifically for out-of-distribution detection. The advantage of Isolation Forests is that no adjustment to the model is necessary. If an existing model is already in use, only a validation set is needed

[2] https://github.com/jandiers/ood-detection.

to apply our method. ODIN can also be applied to existing models, but this requires two runs: In the first run, the gradient to the input data must be computed, and then in the second run, the confidence on the modified images must be obtained. This significantly increases the runtime of the inference and thus complicates the practical use of the method. The Baseline approach can also be used without modification, but falls behind the results of Isolation Forest and Gradient Boosting. Outlier exposure approaches cannot be applied to existing models because special loss functions must be used during training.

Table 2. Results averaged over all out-of-distribution datasets and all in-distribution datasets. The supervised OOD-method based on Gradient Boosting classification outperforms all other methods in all metrics. Especially in terms of OOD-error, which is the most important metric when OOD is applied in practice, the supervised method is clearly the best. Also the OOD detection based on Isolation Forests works well. Note that Isolation Forests work completely unsupervised and therefore solve a much more difficult task to detect OOD-Input.

Average over 5 in-distribution datasets	Average over 6 OOD datasets		
	OOD error	OOD AUC	FPR at 95% TPR
Baseline	0.30	0.85	0.51
ODIN	0.25	0.82	0.44
Outlier exposure	0.30	0.86	0.50
Isolation forest	0.26	0.85	0.56
Gradient boosting classifier	0.14	0.92	0.25

The supervised method based on Gradient Boosting works particularly reliably. In all metrics this approach performs best, especially with respect to the OOD error, the superiority of the approach is clear. On average, 25% misclassified OOD input from ODIN can be reduced by Gradient Boosting to an error rate of 14%, which is an improvement of 44%.

The AUC tests different thresholds for classification. Most models yield similar AUC values with different OOD error values. This suggests that the threshold for binary classification (OOD vs. in-distribution) is difficult to choose for many methods. Gradient Boosting is again the superior methodology, but by a smaller margin compared to the other metrics.

The third metric of FPR at 95% TPR is also borrowed from common literature in this research field. It measures the False Positive Rate when the True Positive Rate is 95%. A lower value is better here. In this metric, other methods perform slightly better than Isolation Forest. Again, the best model by far is the supervised OOD detection via Gradient Boosting Classifier.

Table 3 summarizes the results in more detail. Shown are the metrics, each as an average over 6 out-of-distribution datasets. For example, the Isolation Forest has an OOD error of 18% on the Cifar10 dataset. This means on average, 18% of the images were misidentified as Cifar10 images, even though they were from a different dataset. The baseline approach also has an error rate of 18% on Cifar10. The error rate for ODIN is 21%, outlier exposure has 15%, Gradient Boosting has 20%. We provide detailed results for each dataset as supplementary material.

Table 3. Results for out of distribution detection based on 6 different datasets. Values represent averages. While ODIN and Outlier Exposure require explicit optimization for out of distribution detection, our approach only requires a validation set to learn the outlier distribution. It works with any existing classifier without modification.

Dataset	In-distribution method	Average over 6 OOD datasets		
		OOD error	OOD AUC	FPR at 95% TPR
CatsVsDogs	Baseline	0.04	0.98	0.02
	ODIN	0.05	0.98	0.04
	Outlier Exposure	0.03	0.99	0.02
	Isolation Forest	0.09	0.98	0.03
	Gradient Boosting Classifier	0.04	0.98	0.07
Cifar10	Baseline	0.18	0.89	0.37
	ODIN	0.21	0.81	0.42
	Outlier Exposure	0.15	0.93	0.29
	Isolation Forest	0.18	0.92	0.30
	Gradient Boosting Classifier	0.20	0.91	0.29
Cifar100	Baseline	0.25	0.86	0.50
	ODIN	0.21	0.86	0.40
	Outlier Exposure	0.23	0.88	0.47
	Isolation Forest	0.38	0.71	0.94
	Gradient Boosting Classifier	0.27	0.87	0.43
Cars196	Baseline	0.45	0.66	0.96
	ODIN	0.43	0.61	0.93
	Outlier Exposure	0.45	0.68	0.95
	Isolation Forest	0.44	0.78	0.75
	Gradient Boosting Classifier	0.05	0.99	0.04
Cassava	Baseline	0.57	0.85	0.69
	ODIN	0.35	0.86	0.43
	Outlier Exposure	0.62	0.80	0.76
	Isolation Forest	0.22	0.86	0.77
	Gradient Boosting Classifier	0.12	0.88	0.40

6 Discussion and Conclusion

An open question is why the anomalous input can be detected in the softmax values of the neural network. Although it is consistent with previous research that the softmax values allow good generalization to other OOD input [9,17], there is a lack of arguments to support these observations. In the introduction of our method, we stated that $f(X_{in}) \sim D_{in}$ and consequently $f(X_{out}) \sim D_{out}$ holds, and that the corresponding softmax values allow for a distinction between the elements of these distributions. To support this assumption and to visualize the distributions, we compute the t-SNE visualization of the softmax activations of in-distribution data (D_{in}) and plot the softmax activations of out-of-distribution data (D_{out}) against it. The Figure is included in the supplementary material of this work.

It can be observed that even previously unknown OOD data is projected where already known OOD data is located. While t-SNE does not reflect a global topology, it is a clear indication that the clusters of in-distribution and OOD data are distinctly separated. This explains the absence of the distribution shift in the neural network output and enables supervised detection of OOD input. The performance of the Isolation Forest is explained by the well-defined clusters. The cluster structure represents a fundamental assumption of outlier detection that is exploited by the Isolation Forest. In existing clusters (in-distribution clusters), it is difficult for the Isolation Forest to isolate individual points. In widely spread clusters, this is easier (out-of-distribution data).

When looking at the results, it is also noticeable that Gradient Boosting is superior to the other methods. One reason might be, that datasets that are difficult to classify (Cassava, Cars196) do not form a good precondition for existing OOD methods. If general classification accuracy already suffers, then OOD detection is particularly difficult. It also turns out that using a model on the entire softmax distribution is beneficial. The input to the Baseline approach is exactly the same as for Gradient Boosting and Isolation Forest, however these methods are superior to the simple Baseline.

With our work, we show that outlier detection methods are well suited to detect OOD input in neural networks. It is also possible to apply our method to existing methods like ODIN or Outlier Exposure. It is expected that this will further improve performance. The extensive experiments on various datasets encourage us to further explore methods that combine traditional methods of outlier detection with the more modern application of out-of-distribution detection.

References

1. Amodei, D., Olah, C., Steinhardt, J., Christiano, P., Schulman, J., Mané, D.: Concrete Problems in AI Safety (2016)
2. Bergman, L., Cohen, N., Hoshen, Y.: Deep Nearest Neighbor Anomaly Detection (2020)
3. Chen, J., Li, Y., Wu, X., Liang, Y., Jha, S.: Robust out-of-distribution detection for neural networks. https://arxiv.org/pdf/2003.09711
4. Guo, C., Pleiss, G., Sun, Y., Weinberger, K.Q.: On calibration of modern neural networks. In: International Conference on Machine Learning, pp. 1321–1330 (2017). http://proceedings.mlr.press/v70/guo17a.html
5. DeVries, T., Taylor, G.W.: Learning Confidence for Out-of-Distribution Detection in Neural Networks (2018)
6. Friedman, J.H.: Stochastic gradient boosting. Comput. Stat. Data Anal. **38**(4), 367–378 (2002). https://doi.org/10.1016/S0167-9473(01)00065-2. https://www.sciencedirect.com/science/article/pii/S0167947301000652

7. Hawkins, D.M.: Identification of Outliers, vol. 11. Chapman and Hall, London (1980). https://link.springer.com/content/pdf/10.1007/978-94-015-3994-4.pdf
8. Hendrycks, D., Gimpel, K.: A Baseline for Detecting Misclassified and Out-of-Distribution Examples in Neural Networks (2016)
9. Hendrycks, D., Mazeika, M., Dietterich, T.: Deep Anomaly Detection with Outlier Exposure (2018)
10. Kingma, D.P., Ba, J.: Adam: a method for stochastic optimization. https://arxiv.org/pdf/1412.6980
11. Zhang, K., Schölkopf, B., Muandet, K., Wang, Z.: Domain adaptation under target and conditional shift. In: International Conference on Machine Learning, pp. 819–827 (2013). http://proceedings.mlr.press/v28/zhang13d.html
12. Liang, S., Li, Y., Srikant, R.: Enhancing The Reliability of Out-of-distribution Image Detection in Neural Networks (2017)
13. Liu, F.T., Ting, K.M., Zhou, Z.H.: Isolation forest. In: Giannotti, F. (ed.) Eighth IEEE International Conference on Data Mining, 2008. IEEE, Piscataway (2008). https://doi.org/10.1109/icdm.2008.17
14. Bossard, L., Guillaumin, M., Van Gool, L.: Food-101 – mining discriminative components with random forests. In: Fleet, D., Pajdla, T., Schiele, B., Tuytelaars, T. (eds.) ECCV 2014. LNCS, vol. 8694, pp. 446–461. Springer, Cham (2014). https://doi.org/10.1007/978-3-319-10599-4_29
15. Niculescu-Mizil, A., Caruana, R.: Predicting good probabilities with supervised learning. In: Dzeroski, S. (ed.) Proceedings of the 22nd International Conference on Machine Learning. ACM, New York (2005). https://doi.org/10.1145/1102351.1102430
16. Ren, J., et al.: Likelihood ratios for out-of-distribution detection. https://arxiv.org/pdf/1906.02845
17. Ruff, L., Vandermeulen, R.A., Franks, B.J., Müller, K.R., Kloft, M.: Rethinking assumptions in deep anomaly detection. https://arxiv.org/pdf/2006.00339
18. Sun, Y., Guo, C., Li, Y.: React: out-of-distribution detection with rectified activations. arXiv abs/2111.12797 (2021)
19. Szegedy, C., Vanhoucke, V., Ioffe, S., Shlens, J., Wojna, Z.: Rethinking the inception architecture for computer vision. In: 29th IEEE Conference on Computer Vision and Pattern Recognition. IEEE, Piscataway (2016). https://doi.org/10.1109/cvpr.2016.308
20. Tan, M., Le, V.Q.: Efficientnet: rethinking model scaling for convolutional neural networks. In: International Conference on Machine Learning (2019). https://arxiv.org/pdf/1905.11946
21. Torralba, A., Fergus, R., Freeman, W.T.: 80 million tiny images (01072020). https://groups.csail.mit.edu/vision/TinyImages/
22. Torralba, A., Fergus, R., Freeman, W.T.: 80 million tiny images: a large data set for nonparametric object and scene recognition. IEEE Trans. Pattern Anal. Mach. Intell. **30**(11), 1958–1970 (2008). https://doi.org/10.1109/tpami.2008.128
23. Wang, H., Bah, M.J., Hammad, M.: Progress in outlier detection techniques: a survey. IEEE Access **7**, 107964–108000 (2019)

Relaxation Labeling Meets GANs: Solving Jigsaw Puzzles with Missing Borders

Marina Khoroshiltseva[1,2], Arianna Traviglia[1,2], Marcello Pelillo[1,2], and Sebastiano Vascon[1,2](✉)

[1] Università Ca' Foscari, Dorsoduro 3246, 30123 Venice, Italy
{m.khoroshiltseva,sebastiano.vascon}@unive.it
[2] Istituto Italiano di Tecnologia, CCHT, Via Torino 155, 30100 Mestre, Venice, Italy

Abstract. This paper proposes JiGAN, a GAN-based method for solving Jigsaw puzzles with eroded or missing borders. Missing borders is a common real-world situation, for example, when dealing with the reconstruction of broken artifacts or ruined frescoes. In this particular condition, the puzzle's pieces do not align perfectly due to the borders' gaps; in this situation, the patches' direct match is unfeasible due to the lack of color and line continuations. JiGAN, is a two-steps procedure that tackles this issue: first, we repair the eroded borders with a GAN-based image extension model and measure the alignment affinity between pieces; then, we solve the puzzle with the relaxation labeling algorithm to enforce consistency in pieces positioning, hence, reconstructing the puzzle. We test the method on a large dataset of small puzzles and on three commonly used benchmark datasets to demonstrate the feasibility of the proposed approach.

Keywords: Jigsaw puzzles · Image extension · Relaxation labeling

1 Introduction

The jigsaw puzzle is a well-known game where small (and often irregular) pieces must be fitted together to reconstruct the complete image or shape. Despite its entertaining and educational origins, solving a puzzle has numerous applications in different fields, such as image editing, reconstruction of broken artifacts [9], shredded documents [7], genome biology [27]. In its simplest version, which is known as the *square jigsaw puzzle*, the square pieces should be reordered on a 2D grid to form a coherent image. Formally, one should look for a permutation matrix that encodes such reordering and represents the correct solution of the puzzle. Although demonstrated to be NP-complete [8], the automatic puzzle-solving problem puzzles the minds of researchers in computer science, mathematics, and engineering for years. Numerous approaches tackled the problem, involving functional optimization [1,5,13], greedy algorithm [10,11,20,22,24], and machine learning [4,15,18].

S. Sclaroff et al. (Eds.): ICIAP 2022, LNCS 13233, pp. 27–38, 2022.
https://doi.org/10.1007/978-3-031-06433-3_3

A more complex task concerns finding a solution when pieces are missing or eroded. Many real-world problems, such as recovering of ancient documents and broken artifacts [9], can be seen as jigsaw puzzles with missing information (boundaries or entire pieces). This task has been only partially explored in the last year due to its complexity [4,18].

In this paper, we propose to extend [13] for the case where the borders of the patches are ruined. To simulate the erosion in the puzzle, we create gaps between pieces removing pixels lying on the borders. The gaps interrupt the color and the line continuation between patches, making compatibility functions unusable or highly inaccurate.

To alleviate this problem, we adopt an image extension technique; the idea is to extend the patches borders to cover the eroded parts in the picture with synthetically generated pixels. Image inpainting and extension are broadly studied in computer vision, and various techniques were proposed [2,6,16,25,26]. We consider that the image extension model is more suitable for our task, as we want to extend the images outside the original border rather than filling missing parts inside of each patch. The GAN-based model for image extension proposed in [25] shows impressive results, hence we adopt their model for our procedure: first, we recover the eroded borders of each patch by extending it in all directions and we compute the pairwise compatibility on repaired patches; then we apply the solver [13] to reconstruct the image.

The paper is organized as follow: in Sect. 2 we discuss the state-of-the-art of puzzle-solving methods; Sect. 3 details our model, Sects. 3.1, 3.2 discus the image extension model and the compatibility computation, respectively; in Sect. 3.3 we recipe our puzzle solver, and finally we discuss the experiments and present our results in Sect. 4.

2 Related Works

In recent years, the image jigsaw problem has been tackled with different computational approaches proposing a variety of solutions. Cho et al. [5] presented a graphical model based on the patch transform and proposed an algorithm that minimizes a probability function via loopy belief propagation. Pomeranz et al. [20] introduced the first fully automatic puzzle solver proposing a greedy placer and a novel prediction-based dissimilarity. Their approach relies on finding pairs of pieces with a very high probability of being together. Sholomon et al. [22] proposed a solver based on a genetic algorithm that can solve large puzzles. Paikin et al. [17] extended the work in [20] by solving puzzles with unknown orientations and with missing pieces, introducing new affinity measures. Son et al. [24] considerably improved solving puzzles with unknown orientation by using loop constraints. Andalo et al. [1] presented a global formulation for jigsaw problems, optimizing the affinity between adjacent pieces by numerically solving a constrained quadratic program. Gallagher et al. [10] represented a puzzle as a graph, their algorithm considers edges connecting all pieces in all possible geometric combinations and then trims edges by finding a Minimum Spanning Tree.

Brandao et al. [3] extended the work introduced in [10] by modeling the jigsaw problem as an edge selection problem in a graph, where the nodes represented the various tile orientations.

In [13] the puzzle-solving problem is tackled as a problem of finding a consistent labeling that satisfies certain compatibility relations. The problem is solved using the classical relaxation labeling algorithm coupled with the Sinkhorn-Knop matrix normalization procedure [23], while adopting the Mahalanobis gradient compatibility function [10] to calculate the affinity of the parts.

Only a few papers addressed solving jigsaw puzzles when borders are missing. Paumard et al. [18] tackled the 3×3 puzzle problem with a probabilistic model; to emulate the erosion, they randomly cropped a fragment inside each piece; then, given a central fragment, they used a neural network to predict the relative positions of the remaining fragments and computed the shortest path in the graph to reassemble the puzzle. Bridger et al. [4] proposed a method to solve the puzzle with ruined regions; first, they recovered the missing parts using a GAN-based model and then reconstructed the image using greedy solver form [17]. Although the method works nicely, it is computationally intensive since it considers all the possible combinations of patches pairs and their relations. Ru Li et al. [15] introduce JigsawGAN, a self-supervised GAN-based approach, that combines global semantic information and edge information of each piece, to solve 3×3 puzzle. The output of the model is then a permutation matrix of all the pieces.

Similarly to Bridger et al. [4], this paper tackles the puzzle problem with ruined regions; however, their work differs from ours in two crucial points: *i)* [4] fills in the gaps in the image by applying inpainting algorithm to each pair of patches for all possible transformations; instead, we recover the damaged borders of each single patch using image extension algorithm. That is more convenient from a computational point of view. *ii)* [4] uses a solver based on naive greedy placer; instead, we cast the problem as a consistent labeling problem [12], and solve the puzzle using the relaxation labeling algorithm that enjoys excellent theoretical properties [19].

To summarize, the contributions of this paper are three-fold:

1. This is the first paper proposing a model that exploits generative adversarial networks and relaxation labeling processes together
2. We extended a previous model to handle a more complex task, such as jigsaw puzzles with eroded borders
3. We show the feasibility of our model on a variety of different datasets.

3 Model

In this section, we introduce JiGAN, our GAN-based approach to solving jigsaw puzzles. Suppose we are given N images, that represent the patches of the puzzle; the borders of the patches are eroded implying the gaps between parts in the puzzle. The goal is to reassemble the original image or, saying differently,

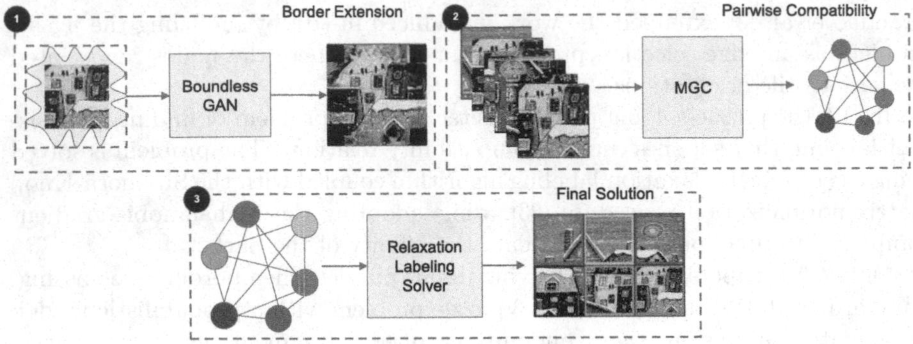

Fig. 1. Pipeline of the algorithm. ① Given a patch, we extend its borders using Boundless GAN [25]. ② We exploited the generated borders and compute pairwise compatibility between all the patches using Mahalanobis Gradient Compatibility (MGC) [10]. ③ Relaxation Labeling is then used to find a consistent labeling (positioning) of each piece.

assign a position in a 2-dimensional grid assemble plane to each patch of the puzzle. As in previous works, we assume that the patches are of the same size, the orientation is known, and the gaps created by eroded borders are of the same regular size. Our model is illustrated in Fig. 1 and is based on three following key ideas: 1) extending the eroded patches border using a GAN model; 2) computing dissimilarity score for each pair of patches and transforming dissimilarity scores in the matrix of compatibility coefficients; 3) given the compatibility map, running the relaxation labeling puzzle solver and reconstructing the image.

3.1 Border Extension

The various methods for compatibility computation, discussed in previous works [5,17,20,24], are normally based on the color gradient and the continuation of the edge, and perform well for puzzles without erosion. However, the gaps created by erosion, will make any of these functions inaccurate and unreliable. For this reason we first repair the eroded edges by generating the band of new pixels all around the given patch. To do this we use an image extension technique called Boundless [25]. The idea is to extrapolate the image of the patch in all directions, to cover the void created by the erosion. The Boundless is a GAN-based model tailored to extend the image content along any direction, i.e. to fill the image content outside the original boundaries. The extended regions are expected to match the original area on a structural, textual, and semantic level. For our task, we use the pre-trained model on Places [28] provided by Google[1]. The limitation of the model is that it is trained to extend the image in one direction (right). In order to extend the images of the puzzle pieces all around, we pass each piece through the generator four times by rotating it 90°.

[1] Pretrained Boundless model from TensowrflowHub.

Formally, given the $\tilde{i}$-th piece of a puzzle, its extended version is denoted by

$$i = \Phi(\tilde{i}, \beta, \theta) \tag{1}$$

where β is the percentage of image extension, and $\Phi(...)$ is the Boundless model parametrized by θ. Once the damaged borders get repaired, we can use the reconstructed patches to calculate the patch compatibility.

3.2 Pairwise Compatibility

The compatibility measure quantifies the affinity between pieces and predicts the likelihood of two patches to be neighbors. We measure the piece affinity by computing the dissimilarity between the abutting boundary pixels of two adjacent pieces; to this end, we adopt the *Mahalanobis Gradient Compatibility* (MGC) developed by Gallagher [10] and further improved by Son et al. [24]. MGC considers both the color differences across pieces borders and the directional derivative differences along the borders. Assuming that the two candidate pieces are positioned such that piece i is placed to the left of piece j, the dissimilarity measure $\Gamma_R(i,j)$ is defined as:

$$\Gamma_R(i,j) = D_R(i,j) + D_L(j,i) + D'_R(i,j) + D'_L(j,i). \tag{2}$$

The first two terms, D_R and D_L, penalize the changes in the pixel values across the boundary in the following way:

$$D_R(i,j) = \sum_{s=1}^{S} (\Lambda_R^{(ij)}(s) - E_R^{(ij)}(s)) V_{iR}^{-1} (\Lambda_R^{(ij)}(s) - E_R^{(ij)}(s))^\top \tag{3}$$

where $E_R^{(ij)}(s)$ is the expected change across the boundary, $\Lambda_R^{(ij)}(s)$ is the pixel intensity change across the boundary and V_{iR} is a sample covariance calculated from samples of the border pixels. D'_R and D'_L are calculated by replacing $i(u,v)$ with the directional derivatives $\delta(u,v) = i(u,v) - i(u-1,v)$.

Once the pairwise dissimilarity scores are calculated for each pair of pieces in all possible neighboring relationships (right, up, left, down), we convert them to normalized compatibility values, as follows:

$$C_{\mathcal{R}}(i,j) = \max\left(1 - \frac{\Gamma_R(i,j)}{K_{min_{\mathcal{R}}}(i)}, 0\right) \tag{4}$$

where $K_{min_{\mathcal{R}}}(i)$ is the K-min value of the dissimilarity between all other pieces in relation $\mathcal{R}$ to piece i. The smaller the value of K, the more sparse $C_{\mathcal{R}}(i,j)$ becomes, leading to a more efficient relaxation labeling process.

3.3 Relaxation Labeling Puzzle Solver

In our formulation, the puzzle pieces are considered as a set of objects and their possible positions as a set of labels, the puzzle problem is viewed as the problem of finding consistent labeling that satisfies certain compatibility relations, with an additional requirement for one-to-one correspondences between the puzzle's tiles and their positions. We solve the puzzle using classical relaxation labeling algorithm [19] that, starting from the uniform probability (barycentre point) distribution, progressively updates the assignment matrix till it converges to the consistent labeling, which in our case corresponds to a permutation matrix.

Consistent Labeling Problem. In this section we recap some basic concepts of relaxation labeling. Suppose we are given a set of objects $B = \{b_1, \ldots, b_n\}$ and a set of labels $\Lambda = \{\lambda_1, \ldots, \lambda_m\}$, the task is to assign a label to each object in B. To this end two sources of information are available: (1) local measurements, which capture the characteristic features of each object, (2) contextual information, quantitatively expressed a matrix of *compatibility coefficients* $R = [r_{ij\lambda\mu}]$. The coefficient $r_{ij\lambda\mu}$ measures the strength of compatibility between the hypotheses "b_i has label λ" and "b_j has label μ".

The label assignments for object b_i is represented by a probability distribution $\mathbf{p}_i$ over all possible labels. Formally, $\mathbf{p}_i \in \Delta^m$, where where

$$\Delta^m = \left\{ \mathbf{x} \in \mathbb{R}^m \mid x_\lambda \geq 0 \;\; \wedge \;\; \sum_{\lambda=1}^{m} x_\lambda = 1 \right\} \tag{5}$$

The compatibility model $\mathbf{R}$ is considered "contextual" because it naturally leads to measures of *contextual support* (i.e., how much the context supports the assignment of a particular label λ to object b_i) and defined [12] as

$$q_{i\lambda} = \sum_{j,\mu} r_{ij\lambda\mu} p_{j\mu}. \tag{6}$$

A process that relaxes a given inconsistent assignment $\mathbf{p}$ towards a more consistent one, will increase $p_{i\lambda}$ when $q_{i\lambda}$ is high and decrease it when $q_{i\lambda}$ is low. The best-known update rule, that guarantees the converge to a consistent labeling [19] under non-negativity and symmetry conditions on $\mathbf{R}$, is defined by the following iterative procedure [19,21]:

$$p_{i\lambda}(t+1) = \frac{p_{i\lambda}(t) q_{i\lambda}(t)}{\sum_\mu p_{i\mu}(t) q_{i\mu}(t)} \quad \forall i, \lambda \tag{7}$$

The initial labeling is a starting point of the process and corresponds to a set of assignments for the entire set of objects. It can be initialized in different ways depending on whether some prior knowledge exists or not. If prior knowledge is not available, the object is assigned the same probability for all labels.

The relaxation algorithm takes as input an initial (imperfect) labeling assignment and progressively updates it according to the compatibility model $\mathbf{R}$. The process continuous until the fixed point is reached, that correspond to a consistent labeling (when every object chooses his best label).

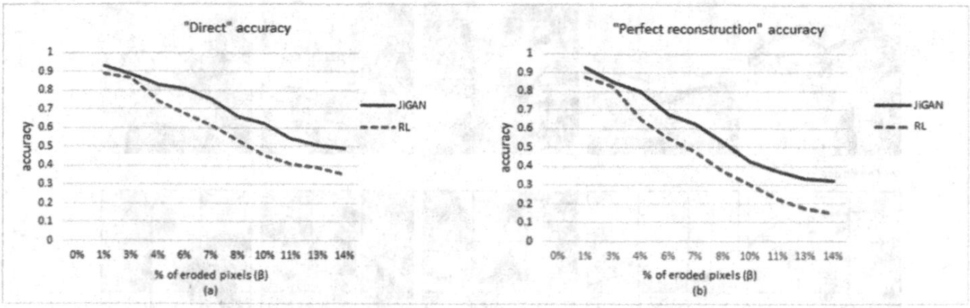

Fig. 2. JiGAN (blue) vs RL (red) models: average Direct (a) and Perfect (b) accuracy then increasing the erosion gaps β. (Color figure online)

Relaxation Labeling Algorithm for Puzzle Solving. We cast jigsaw puzzle solving as a consistent labeling problem. The set of objects B represents the puzzle pieces, the labels Λ are the positions in the reconstruction plane (hence $m = n$), and the task is to assign a different position from Λ to each puzzle piece from B. The $\mathbf{P} \in \Delta^{n\times m}$ is a soft assignment matrix (where each row represents a probability distribution of the positions for a piece and each column represents a probability distribution of the pieces for a position), $\Delta^{n\times m}$ is the multi-simplex with $\Delta^m = \{\mathbf{p}_i \mid p_{i\lambda} \geq 0 \wedge \sum_\lambda p_{i\lambda} = 1\}$ and $\Delta^n = \{\mathbf{p}_\lambda \mid p_{i\lambda} \geq 0 \wedge \sum_i p_{i\lambda} = 1\}$, where $p_{i\lambda}$ is the probability of piece i to choose position λ. Thus $\mathbf{P} = p_{i\lambda}$ is doubly stochastic matrix such that $\sum_\lambda p_{i\lambda} = \sum_i p_{i\lambda} = 1$.

The relaxation labeling update rule guarantees that $\mathbf{P}$ is a stochastic matrix (i.e., rows sum to 1) but does not enforce the same constraint for its columns. Therefore, the optimization process can converge to a labeling that does not represent a permutation (producing a solution with multiple pieces assigned the same position and vice versa). To enforce one-to-one correspondence constraints, we endow the relaxation process with matrix balancing algorithm, adopting Sinkhorn-Knopp (SK) normalization [23]. SK algorithm transforms a given non-negative square matrix to its related doubly stochastic version, by alternately normalizing the rows and columns. SK is incorporated in our algorithm as an additional balancing step in each iteration.

4 Experiments and Results

Datasets. We assessed the performance considering two benchmarks. First, we test our method on a large dataset of small (synthetic) images. Following JigsawGAN [15] we create our collection of 1600 images randomly picked up from PACS dataset [14]. Our collection is divided into 4 object categories (elephant, guitar, person, house), each of which covers 4 image styles (paintings, photos, cartoons, and sketches). Each of 1600 images is cut into 72×72 pixels size pieces generating a 9-pieces puzzle (3×3). For the second test, we apply our method to three datasets [5,20], widely used as performance benchmarks; each contains 20 images of increasing size. We cut the images into equal size pieces, generating puzzles of 70, 88, and 150 pieces(for the 1st, 2nd, and 3rd data sets respectively).

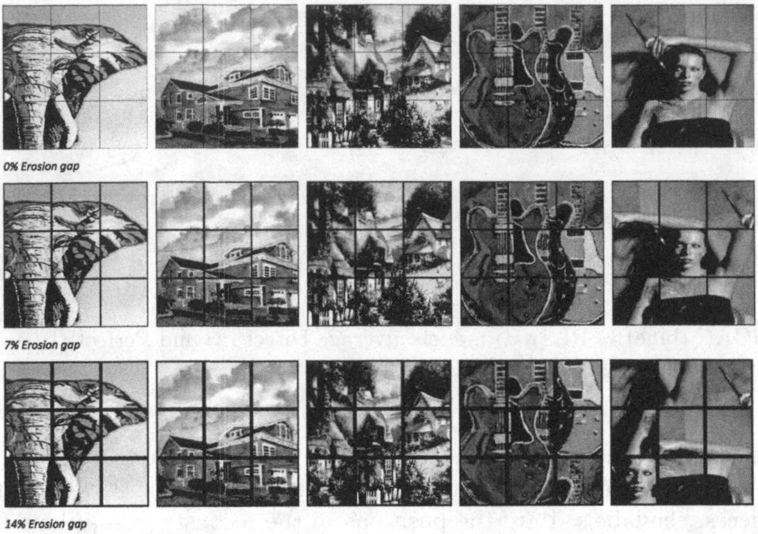

Fig. 3. Qualitative results for small puzzles from Pacs dataset (0%, 7%, 14% erosion of piece size)

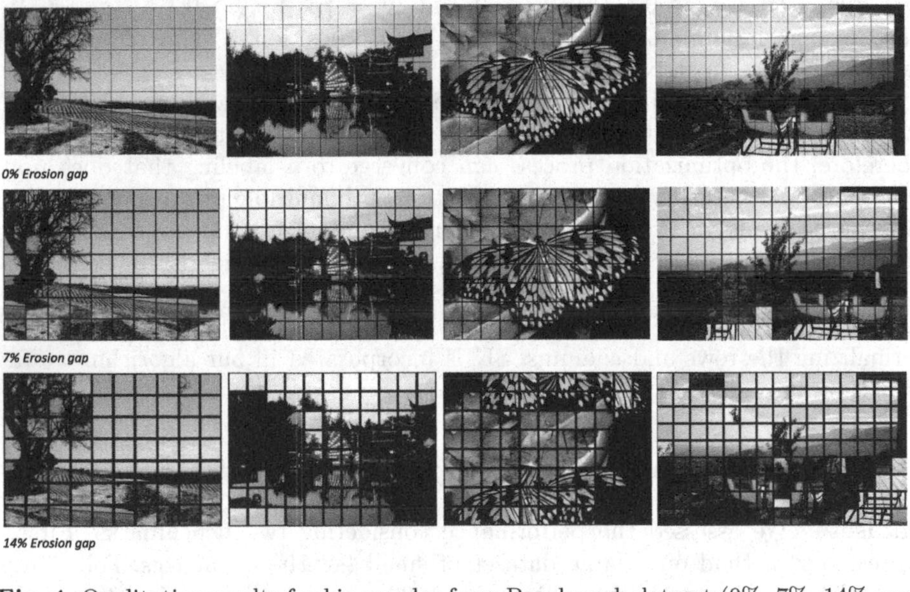

Fig. 4. Qualitative results for big puzzles from Benchmark dataset (0%, 7%, 14% erosion of piece size)

Accuracy Metrics. To evaluate the performance of the algorithm we adopt three accuracy measures, widely used in literature: *Direct Comparison* metric, which measures the ratio of pieces placed in the correct position; the *Neighbor Comparison* metric that measures the ratio of correctly assigned neighbors in the solution, and the *Perfect Reconstruction* metric that is a binary indicator of whether all pieces in the puzzle are in the correct position; applied to a dataset, the *Perfect Reconstruction* is a ratio of perfectly solved puzzles.

Experiments. We performed experiments on the two aforementioned benchmarks considering the three different metrics and an increasing level of border erosion, $\beta \in \{0\%, 7\%, 14\%\}$. Without erosion ($\beta = 0\%$) the performance of JiGAN and RL [13] are the same.

We compare our result to [13] that is our direct competitor, as our model is an extension of it. Concerning [4], although the idea is similar to ours, their model involves much more information (all possible pairing and rotation of puzzle's pieces), thus a direct comparison would not be fair.

Experiments with PACS Dataset (Small Puzzles): using the PACs dataset, we conduct two types of experiments: first, we generate 3×3 puzzles without any gap between pieces and run the relaxation labeling (RL) solver [13]; second, to simulate the erosion of the boards, we generate the puzzles with gaps between pieces with two different levels of erosion 7% and 14% gaps. We compare two methods: the RL algorithm without the image extension step, and our JiGAN procedure that involves the completion of the eroded border.

Table 1. RL [13] vs. JiGAN (our model). PACS datasets

	Direct accuracy					Perfect reconstruction				
	No gap	7% gap		14% gap		No gap	7% gap		14% gap	
	RL	RL	JiGAN	RL	JiGAN	RL	RL	JiGAN	RL	JiGAN
House	0.92	0.57	**0.74**	0.41	**0.60**	0.90	0.46	**0.64**	0.26	**0.42**
Elephant	0.88	0.51	**0.74**	0.30	**0.54**	0.86	0.41	**0.64**	0.16	**0.36**
Guitar	0.83	0.42	**0.65**	0.26	**0.48**	0.77	0.33	**0.49**	0.13	**0.27**
Person	0.90	0.56	**0.72**	0.40	**0.58**	0.89	0.51	**0.65**	0.28	**0.43**
Mean	**0.88**	0.50	**0.70**	0.32	**0.53**	**0.85**	0.41	**0.60**	0.19	**0.35**

Table 1 shows the results of puzzle reconstruction in terms of direct comparison accuracy measure and perfect reconstruction ratio. It can be seen that, for the case without gaps, our solver performs well in all categories. While in the cases with erosion, the performance of the solver algorithm decreases as the level of erosion increases. However, the image extension step is beneficial to puzzle reconstruction concerning the algorithm without extension.

Nevertheless, the performance of the model degrades with a larger gap and negatively influences the accuracy of the solver. To further investigate this degradation effect, we perform the experiments by gradually increasing the erosion gaps and observing the accuracy of the algorithm with and without extension steps. The plots in Fig. 2 illustrate the performances of the solver applied to 400 randomly selected puzzles with different levels of erosion. As expected, the larger the erosion, the less accurate the results. A qualitative analysis is reported in Fig. 3.

Experiments with Benchmark Datasets: for further evaluation, we apply our method to the large puzzles generated from the three benchmark datasets. As before, we conduct two experiments applying erosion of 7% and 14% of piece size. Tables 2 shows the results of the RL solver run without reconstruction of the eroded border and the results of the puzzle solver after the GAN image extension algorithm is applied. As in the case with small puzzles, the larger erosion gaps, the lower the accuracy of the puzzle solution. The performance of the GAN model gradually degrades with the larger area of generated pixels. However, applying the inpainting algorithm significantly increases the accuracy of puzzle reconstruction concerning the results of the solver without image extension.

Table 2. RL [13] vs. JiGAN (our model). Benchmark datasets

	Direct accuracy					Neighbour accuracy				
	No gap	7% gap		14% gap		No gap	7% gap		14% gap	
	RL	RL	JiGAN	RL	JiGAN	RL	RL	JiGAN	RL	JiGAN
70 pieces	0.97	0.22	**0.51**	0.11	**0.32**	0.97	0.46	**0.66**	0.35	**0.45**
88 pieces	0.99	0.23	**0.59**	0.07	**0.31**	1.00	0.46	**0.65**	0.30	**0.40**
150 pieces	0.99	0.12	**0.38**	0.06	**0.15**	0.99	0.41	**0.54**	0.28	**0.33**
Mean	**0.98**	0.19	**0.49**	0.08	**0.26**	**0.98**	0.45	**0.62**	0.31	**0.39**

Figure 4 illustrates some qualitative results of reconstruction results for puzzles with different levels of erosion. It can be seen that without erosion we obtain the perfect reconstruction in most of the cases; for images with 7% of erosion gap, the overall result is good, however, the images have some errors most of which are minor and negligible to human eyes. As it can be expected, the results of reconstruction of images with 14% of erosion are less accurate than those with 7% of erosion. Though in some examples the misplaced patches make it difficult the perception the image; in other cases, the reconstruction results are acceptable for the human eye.

5 Conclusion

In this paper, we extend the method proposed in [13] to handle the challenging task of solving a puzzle with ruined borders. The previous methods, based on the compatibility calculated on the color gradient across the edges, effectively solve the puzzles without gaps, but the performance immediately drops in the presence of erosion gaps.

We introduce the idea of repairing damaged patches by involving the GAN model for image extension. We apply the extension procedure on each patch separately, thus avoiding expensive inpainting for all combinations in pairs. The main idea is to regenerate the missing pixels around each patch. Then we calculate the compatibility between the repaired patch and apply the puzzle-solving algorithm.

We show that combining of solving algorithm and deep learning model can be a viable solution to the problem of a puzzle with ruined regions. Our two-step procedure produces better results compared to the previous method. However, the quality of the final reconstruction depends on the level of degradation; the larger the erosion gap, the worse the final result. However, the overall results with a moderate level of erosion are generally acceptable to human eyes.

Acknowledgements. This work has received funding from the European Union's Horizon 2020 research and innovation programme under grant agreement No 964854.

References

1. Andaló, F.A., Taubin, G., Goldenstein, S.: PSQP: puzzle solving by quadratic programming. IEEE TPAMI **39**(2), 385–396 (2017)
2. Barnes, C., Shechtman, E., Finkelstein, A., Goldman, D.B.: PatchMatch: a randomized correspondence algorithm for structural image editing. ACM Trans. Graph. (Proc. SIGGRAPH) **28**(3) (2009)
3. Brandão, S., Marques, M.: Hot tiles: a heat diffusion based descriptor for automatic tile panel assembly. In: Hua, G., Jégou, H. (eds.) ECCV 2016. LNCS, vol. 9913, pp. 768–782. Springer, Cham (2016). https://doi.org/10.1007/978-3-319-46604-0_53
4. Bridger, D., Danon, D., Tal, A.: Solving jigsaw puzzles with eroded boundaries (2019)
5. Cho, T.S., Avidan, S., Freeman, W.T.: A probabilistic image jigsaw puzzle solver. In: Proceedings of CVPR, pp. 183–190 (2010)
6. Clevert, D.A., Unterthiner, T., Hochreiter, S.: Fast and accurate deep network learning by exponential linear units (ELUs) (2016)
7. Deever, A., Gallagher, A.: Semi-automatic assembly of real cross-cut shredded documents. In: Proceedings of ICIP, pp. 233–236 (2012)
8. Demaine, E.D., Demaine, M.L.: Jigsaw puzzles, edge matching, and polyomino packing: Connections and complexity. Graphs Comb. **23**(Suppl. 1), 195–208 (2007)
9. Derech, N., Tal, A., Shimshoni, I.: Solving archaeological puzzles. CoRR abs/1812.10553 (2018)
10. Gallagher, A.C.: Jigsaw puzzles with pieces of unknown orientation. In: Proceedings of CVPR, pp. 382–389 (2012)
11. Gur, S., Ben-Shahar, O.: From square pieces to brick walls: the next challenge in solving jigsaw puzzles. In: ICCV, pp. 4029–4037 (2017)
12. Hummel, R.A., Zucker, S.W.: On the foundations of relaxation labeling processes. IEEE TPAMI **5**(3), 267–287 (1983)
13. Khoroshiltseva, M., Vardi, B., Torcinovich, A., Traviglia, A., Ben-Shahar, O., Pelillo, M.: Jigsaw puzzle solving as a consistent labeling problem. In: Tsapatsoulis, N., Panayides, A., Theocharides, T., Lanitis, A., Pattichis, C., Vento, M. (eds.) CAIP 2021. LNCS, vol. 13053, pp. 392–402. Springer, Cham (2021). https://doi.org/10.1007/978-3-030-89131-2_36

14. Li, D., Yang, Y., Song, Y.Z., Hospedales, T.M.: Deeper, broader and artier domain generalization. In: Proceedings of the IEEE International Conference on Computer Vision (ICCV), October 2017
15. Li, R., Liu, S., Wang, G., Liu, G., Zeng, B.: Jigsawgan: self-supervised learning for solving jigsaw puzzles with generative adversarial networks. CoRR abs/2101.07555 (2021)
16. van den Oord, A., Kalchbrenner, N., Kavukcuoglu, K.: Pixel recurrent neural networks (2016)
17. Paikin, G., Tal, A.: Solving multiple square jigsaw puzzles with missing pieces. In: Proceedings of CVPR, pp. 4832–4839 (2015)
18. Paumard, M., Picard, D., Tabia, H.: Deepzzle: solving visual jigsaw puzzles with deep learning and shortest path optimization. CoRR abs/2005.12548 (2020)
19. Pelillo, M.: The dynamics of nonlinear relaxation labeling processes. J. Math. Imag. Vis. **7**(4), 309–323 (1997)
20. Pomeranz, D., Shemesh, M., Ben-Shahar, O.: A fully automated greedy square jigsaw puzzle solver. In: Proceedings of CVPR, pp. 9–16 (2011)
21. Rosenfeld, A., Hummel, R.A., Zucker, S.W.: Scene labeling by relaxation operations. IEEE Trans. Syst. Man Cybern. **6**, 420–433 (1976)
22. Sholomon, D., David, O.E., Netanyahu, N.S.: A generalized genetic algorithm-based solver for very large jigsaw puzzles of complex types. In: Proceedings of AAAI, pp. 2839–2845 (2014)
23. Sinkhorn, R., Knopp, P.: Concerning nonnegative matrices and doubly stochastic matrices. Pacific J. Math. **21**(2), 343–348 (1967)
24. Son, K., Hays, J., Cooper, D.B.: Solving square jigsaw puzzle by hierarchical loop constraints. IEEE TPAMI **41**(9), 2222–2235 (2018)
25. Teterwak, P., et al.: Boundless: generative adversarial networks for image extension (2019)
26. Yu, J., Lin, Z., Yang, J., Shen, X., Lu, X., Huang, T.: Free-form image inpainting with gated convolution (2019)
27. Zhao, F., et al.: A jigsaw puzzle inspired algorithm for solving large-scale no-wait flow shop scheduling problems. Appl. Intell. **50**(1), 87–100 (2019). https://doi.org/10.1007/s10489-019-01497-2
28. Zhou, B., Lapedriza, A., Khosla, A., Oliva, A., Torralba, A.: Places: a 10 million image database for scene recognition. IEEE Trans. Pattern Anal. Mach. Intell. **40**(6), 1452–1464 (2018). https://doi.org/10.1109/tpami.2017.2723009

Computationally Efficient Rehearsal for Online Continual Learning

Charalampos Davalas(✉), Dimitrios Michail, Christos Diou, Iraklis Varlamis, and Konstantinos Tserpes

Department of Informatics and Telematics, Harokopio University of Athens, 17778 Athens, Greece
cdavalas@hua.gr

Abstract. Continual learning is a crucial ability for learning systems that have to adapt to changing data distributions, without reducing their performance in what they have already learned. Rehearsal methods offer a simple countermeasure to help avoid this catastrophic forgetting which frequently occurs in dynamic situations and is a major limitation of machine learning models. These methods continuously train neural networks using a mix of data both from the stream and from a rehearsal buffer, which maintains past training samples. Although the rehearsal approach is reasonable and simple to implement, its effectiveness and efficiency is significantly affected by several hyperparameters such as the number of training iterations performed at each step, the choice of learning rate, and the choice on whether to retrain the agent at each step. These options are especially important in resource-constrained environments commonly found in online continual learning for image analysis. This work evaluates several rehearsal training strategies for continual online learning and proposes the combined use of a drift detector that decides on (a) when to train using data from the buffer and the online stream, and (b) how to train, based on a combination of heuristics. Experiments on the MNIST and CIFAR-10 image classification datasets demonstrate the effectiveness of the proposed approach over baseline training strategies at a fraction of the computational cost.

Keywords: Catastrophic forgetting · Continual learning · Online learning

1 Introduction

Continual learning aims at developing methods for adapting to new data distributions without dropping their performance on previously learned tasks. Decrease of performance on previous tasks, also known as *catastrophic forgetting*, occurs due to the fact that data is presented to the model incrementally and drawn from different distributions, essentially violating the i.i.d. assumption [1]. Several methods have been proposed to address the catastrophic forgetting

S. Sclaroff et al. (Eds.): ICIAP 2022, LNCS 13233, pp. 39–49, 2022.
https://doi.org/10.1007/978-3-031-06433-3_4

problem, including (a) regularization methods [9] that pose constraints on the parameter update mechanism during training, (b) rehearsal/replay methods [14], that keep a set of representative past samples to be used along with new data in training, and (c) parameter isolation methods that train only a subset of model parameters with the new data [10]. Despite their relative simplicity, rehearsal methods seem to work surprisingly well, as shown in a recent survey [3]. Most methods presented in the bibliography capitalize on "offline" (batch) learning, in which case all the data for all tasks are available for training in each iteration. Online continual learning introduces several additional challenges, including the continuously changing effectiveness in the current task, the forgetting of previous tasks, and the computational complexity of the training procedure. Edge and/or real-time application environments, can impose significant restrictions in the available computation resources for model training [4,12,13]. In this paper, we consider online continual learning for image classification under the prism of computational efficiency. For this purpose, we explore training strategies that can be applied at each training step, and balance the trade-off between model effectiveness and computational complexity. A rehearsal-based online continual learning setup is used to evaluate several training strategies, which involve decisions on *when* and *how* to train.

2 Related Work

Rehearsal-based methods in continual learning [3] mix selected samples from previous tasks with samples from new tasks during training. In task incremental continual learning, we assume clearly divided task boundaries, with all data of each task provided incrementally, i.e. all training data from the first task, then all data from the second task etc. [3,14]. A way out of this relaxation to the more general problem of online continual learning may be offered in the approach of [11], where a Bayesian approach to infer the task context is suggested. Another approach by [17] presents an algorithm that uses the Shannon entropy as a measure to select task samples that are representative of previously seen classes without being affected by the fact that task boundaries are unknown. One prominent method for the Incremental Learning in Online Scenario has recently been introduced in [5]. This method combines various techniques to avoid catastrophic forgetting, including an adaptation of iCaRL [14] to the online learning case. A common attribute of these incremental learning methods is that they use a "static" training strategy. The training procedure takes place at predefined intervals (e.g., after a fixed number of samples is observed), while the number of training iterations/epochs at each training step is also fixed. In many occasions, this can lead to unnecessary training iterations, which add to the time complexity and can lead to overfitting. Our work is targeting the general online continual learning setting, where the task boundaries in the online stream of input images are not known in advance. The timeliness and memory objectives are considered together with complexity, plasticity, scalability and accuracy. Those characteristics allow our work to depart from previous research and deliver the following contributions:

- We propose a *decision mechanism* for determining *when* to train. This mechanism offers a significant advantage according to the time complexity. Another advantage is that this mechanism is suitable for detecting any changes in concept, therefore dealing with the issue of task agnosticism.
- We propose a *dynamic rehearsal strategy* as a solution of *how* to train, in an unpredictable, online setup. We propose strategies for dynamically determining the number of training iterations and learning rate based on the error of the model, as well as based on the convergence of the model parameters.

3 Online Continual Learning

3.1 Scenario

In a real-world setting, boundaries between different image classification problems, are not known in advance. We consider the following motivating scenario, where a constant stream of annotated data is used for model training in an online fashion as shown in Fig. 1. The stream of data is non-i.i.d., since it is sampled from different tasks in each time period. The task boundaries and identities are unknown. Each task is a sequence of annotated samples from a set of n classes $C_1, \ldots, C_n$. For simplicity, we assume that data comes in fixed-size batches (e.g. 32 samples per batch), so the stream is considered as a sequence of batches, which comprises of a set of samples $B_t = (X_t, Y_t)$, with $t \geq 1$ representing the batch index. The goal is to train a sequence of models, where each model h_t for $t \geq 1$ is trained with B_t in an online manner.

3.2 The Proposed Online Rehearsal Method Framework

This section describes the general rehearsal framework, which is more formally defined in Algorithm 1. During training, we maintain two buffers P and R, called the *Postponed* and *Rehearsal* buffers respectively. Let's assume that at time-step t a batch of new samples $B_t = (X_t, Y_t)$ is acquired from the stream, h_{t-1} is the model from the previous time-step, and P_t and R_t are the current state of the two buffers. The Postponed buffer is initially empty and the Rehearsal buffer initially contains a user-selected fixed number $q = |R|/n$, where n is the number of classes, of exemplars per class sampled uniformly at random. Algorithm 2 covers a wide range of training strategies for the online continual learning scenario.

In the one extreme the model is trained in each iteration by stacking the new batch B_t with a single (sampled uniformly at random) batch from the Rehearsal buffer [2]. In the other extreme, we may assume that all task data is available at a single moment in time, and an oracle correctly predicts the task boundaries and retrains at the end of each task using all the data. Between these two extreme cases, without any knowledge about the task boundaries, we are forced to deal with several issues such as: (a) when to train, (b) how to mix samples from the Postponed and Rehearsal buffers, (c) how many iterations to train at the current time-step, and (d) what learning rate schedule to use. At each time-step the algorithm first decides whether to train the model with a new batch

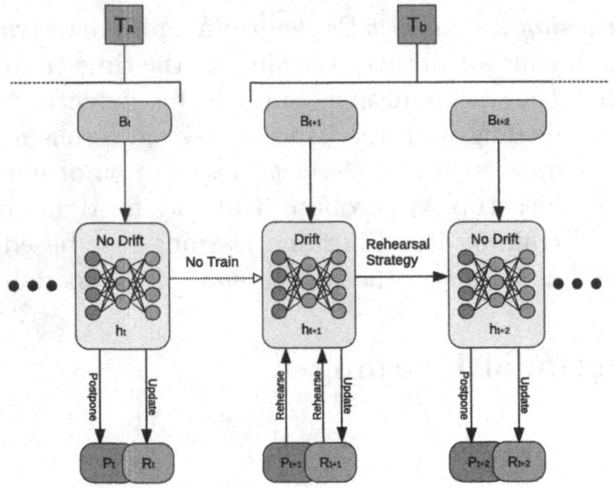

Fig. 1. Proposed strategy outline. Each batch $B_t, B_{t+1}, B_{t+2}, \ldots$ belongs to a specific classification task $T_a, T_b, \ldots$ and $h_t, h_{t+1} \ldots$ is the sequence of produced models after each step. A new batch B_t is added to a Postponed buffer P_t if the model doesn't need any updates. When training is needed in the next batch (i.e. B_{t+1}), both the new Rehearsal buffer R_{t+1} and a new Postponed buffer P_{t+1} are mixed together with B_{t+1} for rehearsal training. In every step the Rehearsal buffer is being updated, regardless of training schedule.

or not. If not, the new batch is appended to the Postponed buffer P_t and the algorithm returns. If the algorithm decides to train the model, it employs a *rehearsal strategy* (see Sect. 4 for the supported options). Part of the strategy is to create mini-batches by combining new samples (i.e. from $P_t \cup B_t$) with samples from the Rehearsal buffer R_t. If training took place in the previous time-step, then $P_t = \emptyset$. After each training step, the Rehearsal buffer R_t is updated (as explained in the following) and the Postponed buffer P_t is cleared.

Algorithm 1: Streaming rehearsal framework single timestep t.

Input: Model h_t, State S_t (global variables from previous timesteps), Postponed buffer P_t, Rehearsal buffer R_t, New batch $B_t = (X_t, Y_t)$ from stream S

Output: updated h'_t, S'_t, P'_t, and R'_t

$train \leftarrow decideToTrain(\{h_t, S_t, P_t \cup B_t, R_t\})$;
if $train = False$ **then**
    update $P'_t \leftarrow P_t \cup B_t$ and state S'_t;
    return $\{h_t, S'_t, P'_t, R_t\}$;
$h'_t, S'_t \leftarrow RehearsalTraining(\{h_t, S_t, P_t \cup B_t, R_t\})$;
$R'_t \leftarrow updateRehearsalBuffer(R_t, P_t \cup B_t)$;
$P'_t \leftarrow \emptyset$;
return $\{h'_t, S'_t, P'_t, R'_t\}$;

The updating of the Rehearsal buffer is performed using the state-of-the-art algorithm proposed by [5]. It is a simplified variant of the *prioritized example selection* algorithm [14] that is based on herding. Its main difference, which is in-line with our computational complexity requirements, is that it maintains a running average estimator for each class, in all occasions the buffer is constant and always contain $q = |R|/n$ exemplars per class.

Algorithm 2: RehearsalTraining

Input: model h, state S (global variables), Postponed buffer P, Rehearsal buffer R

Output: updated model h' and state S'

$iter \leftarrow decideIterations(S, P, R)$;

$lrs \leftarrow decideLearningRateSchedule(S, P, R)$;

initialize optimizer with lrs;

forall $i \in 1, \ldots, iter$ **do**

  $D \leftarrow createMixedMiniBatches(i, S, P, R)$;

  foreach *mini-batch* $b \in D$ **do**

    perform one optimizer (e.g. SGD) step with mini-batch b;

return *updated model* h' *and state* S';

4 Alternative Rehearsal Strategies

This section presents a number of alternative ways to perform the rehearsal during training.

4.1 Continuous Rehearsal

The baseline strategy, referred as *continuous rehearsal* in the following, assumes that training takes place in each time-step. In each iteration, the latest batch B_t is combined with a different batch $r_j \in R_t$ in order to produce two new batches containing samples from B_t and r_j in a 50-50 ratio. A different batch of samples, r_j, is selected from R_t at each time step t in a round-robin fashion.

A simple experience replay method [2] can be achieved by setting the number of iterations to one. As shown in Sect. 5, this proves quite effective w.r.t. forgetting, but does not learn new tasks fast enough. This can be solved by increasing the number of iterations at each time-step at the cost of more training times. In what follows, we use `CONR-n` to denote this baseline method with n iterations (`CONR-1` for a single iteration).

4.2 Drift Activated Rehearsal

In the task-agnostic scenario, where the boundaries of tasks are unknown, we rely on a *concept drift detector* [18] in order to decide *when* to train. The ECDD detector [16] that we employ uses exponentially weighted moving average charts (EWMA) as an indicator of divergence between samples. This single pass method, with $\mathcal{O}(1)$ update in each time-step, is suitable for performance-critical streaming applications.

$\hat{\mu}_t$ and variance of the misclassification error U_t. It also maintains a more slowly updated running average Z_t, of the error. The following rules are used to decide whether to train:

(i) $Z_t > \hat{\mu}_t + L_t\hat{\sigma}_{Z_t}$, i.e., the running average error estimate must not exceed L_t times the standard deviations above the mean, where L_t is a dynamically updated control limit computed by ECDD.
(ii) $U_t > \hat{\mu}_t + 2\hat{\sigma}_{U_t}$, i.e., the current batch error U_t must not be too high (2 standard deviations above the estimated mean).
(iii) $Z_t > E$, the running average of the error must not exceed a threshold E.
(iv) no training during the last ρ time-steps.

Reasonable choices are $E = 0.2$ for the error threshold, and $\rho = 20$ for the no-training time-steps threshold. Assuming a constant number of iterations for training, the main difference with the `CONR-n` method is that when triggered, the Postponed buffer P will likely contain multiple batches. In each training iteration we iterate over all batches of P. For each batch $p_i \in P$ we read a batch $r_j \in R$ and create two batches which contain 50% from each. Again the buffer R is used in a round-robin fashion using a position pointer that is updated at each time-step. We use `DRIFTA-n` to denote the method which performs a constant number of n iterations when the drift detector is triggered. A variation of this method is to use two detectors, one for the stream B_t as above, and another for the Rehearsal buffer R, by sampling from the Rehearsal buffer uniformly at random. Training takes place when any of the two detectors is triggered, based on the above mentioned heuristics. Let `2DRIFTA-n` denote this strategy with a constant number of n iterations for training.

4.3 Dynamic Number of Training Iterations

So far we have assumed a fixed number of n training iterations per time-step. A more adaptive strategy is to dynamically compute the number of training iterations based on the rate of wrong predictions in the stream. We denote as `DRIFTA-DYN-n` the strategy which relates the estimator Z_t values to the number of iterations. We have experimented with a simple rule which computes the number of iterations as $\lceil 2 * n * log_2(1 + Z_t) \rceil$ where n is a user chosen constant. Similarly, `2DRIFTA-DYN-n` does the same using both drift-detectors by using the maximum of Z_t and $\hat{Z}_t$.

4.4 Iterate Until Convergence

A different approach in deciding the number of training iterations is to rely on the convergence of the model. In this approach we monitor the loss function $\mathcal{L}$ during the iterations and keep two exponential moving averages, one short and one long. They are updated as

$$\mathcal{A}_s = (1 - \alpha_s)\mathcal{A}_s + \alpha_s\mathcal{L} \tag{1}$$

$$\mathcal{A}_l = (1 - \alpha_l)\mathcal{A}_l + \alpha_l\mathcal{L} \tag{2}$$

with $\alpha_s = 0.5$ and $\alpha_l = 0.05$ respectively. We stop training when the two values converge, i.e. $|\mathcal{A}_l - \mathcal{A}_s| < \epsilon$ for some hyper-parameter value ϵ. This approach can be combined with both continuous and drift-activated approaches resulting in `CONR-CONV`, `DRIFTA-CONV` and `2DRIFTA-CONV`.

4.5 Adjust Learning Rate

A last action that can affect the efficiency of the rehearsal strategy is to adjust the learning rate across iterations.

The simplest approach is to keep the learning rate η constant to a pre-defined value. Another approach is to use a decay mechanism, e.g. an inverse time decay, that modifies the learning rate through the training iterations. Finally, the drift-activated methods can also use Z_t and a predefined η value to dynamically adjust the initial learning rate. We concluded that a good choice of learning rate can be achieve by using the following function:

$$\mathcal{L}\mathcal{R}_{new} = \mathcal{L}\mathcal{R}_0 * min(100, 5 * e^{3Z_t}) \tag{3}$$

where LR_{new} is the new initial learning rate and LR_0 is the global, pre-defined learning rate.

5 Experiments

5.1 Experimental Setup

Dataset. To evaluate the online continual learning strategies of Section we use both MNIST digits [8] and the CIFAR-10 image classification dataset [7].

Following the online continual learning scenario described in Sect. 3.2, we split the training data into five tasks, where each task contains images from two classes, i.e., $T_i = \{(\mathbf{x}, y) | y \in \{2i, 2i + 1\}\}$, $i = 0, 1, 2, 3, 4$. An online annotated image stream, S, is generated by first sampling multiple images from the first task, then the second task and so on

$$S = (T_0^{(0)}, T_1^{(0)}, T_2^{(0)}, T_3^{(0)}, T_4^{(0)}, T_0^{(1)}, T_1^{(1)}, \cdots) \tag{4}$$

For our experiments, each set $T_i^{(j)}$ has a fixed size of 3200 images, grouped into 100 batches. The model does not know when this change occurs, relying only to the drift detector for feedback. Each batch consists of 32 images sampled randomly from the classes of the current task, T_i. The stream for each experiment has a total length of 1500 batches, with each task appearing three times ($j = 0, 1, 2$). To keep results comparable across experiments, we use the same random seed for sampling, as well as for the initialization of model parameters.

Model and Pre-training. Following [14] and [5], we use the ResNet32 model for the experiments, and specifically its adaptations to the MNIST and CIFAR-10 dataset, as described by [6]. The MNIST model is trained offline for 15 epochs with 500 images while the CIFAR-10 variant is trained for 100 epochs with a random subset of 15000 images. In both occasions there is an equal number of training images for each of the ten classes. This leads to a model that has gone through a "warm-up" training stage but has not yet been fully trained.

Metrics. To evaluate the effectiveness and efficiency of each training strategy, we use the following metrics:

- *Accuracy* (A_t): Accuracy of the model evaluated in the held-out test set, averaged across all tasks.
- *Current task accuracy* (C_t): Accuracy of the model evaluated in the held-out test set, but only for the images belonging to classes of the current task, T_i.
- *Online accuracy* (O_t): Accuracy in each batch of S, evaluated right before it is used for training. This approach was also used in the work of [5].
- *Training iterations* (N_t): Cumulative number of iterations performed during stochastic gradient descent optimization. Given that the model and the batch size is the same across all experiments, this metric that can be used to compare the computational complexity of different training strategies.

Each metric is computed at every step, t, while we report averages across t, e.g., $\bar{A}_t$.

5.2 Experiment 1: The Need for Continuous Rehearsal

This experiment demonstrates the need for an adaptive training strategy in online settings. In Fig. 2, we compare three training strategies which include (i) no rehearsal (ii) continuous rehearsal with 1 training iteration/experience replay [15] (`CONR-1`), and (iii) 50 training iterations (`CONR-50`). In all cases the learning rate was fixed to $\eta = 0.01$. For the MNIST dataset, it is obvious that `CONR-1` and `CONR-50` are very close in terms of average accuracy, albeit the computational cost of `CONR-50` is 50 times higher. Experiments on the current task accuracy C_t on the CIFAR-10 dataset (Fig. 3) pinpoint more accurately the core of the problem for continuous rehearsal strategies. We can see some improvement by `CONR-50`, yet the computational cost is 50 times higher than `CONR-1` which can be prohibitive in resource constrained environments.

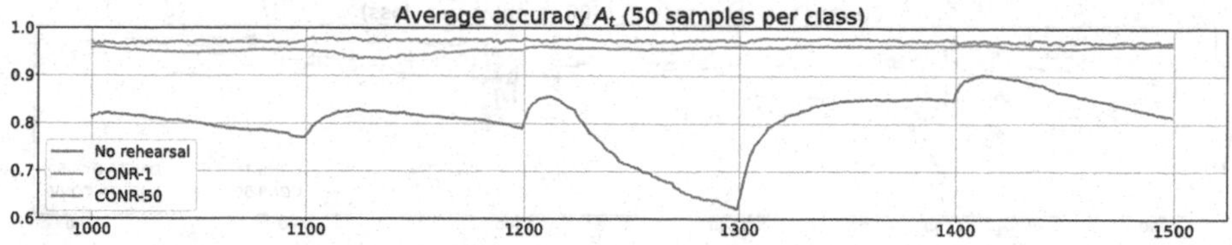

Fig. 2. Results of average accuracy (A_t) for the MNIST dataset. The buffer includes 50 samples per class in the Rehearsal buffer. Continuous rehearsal with one rehearsal `CONR-1` is sufficient. `CONR-50` is slightly better but at the cost of a more resource intensive implementation

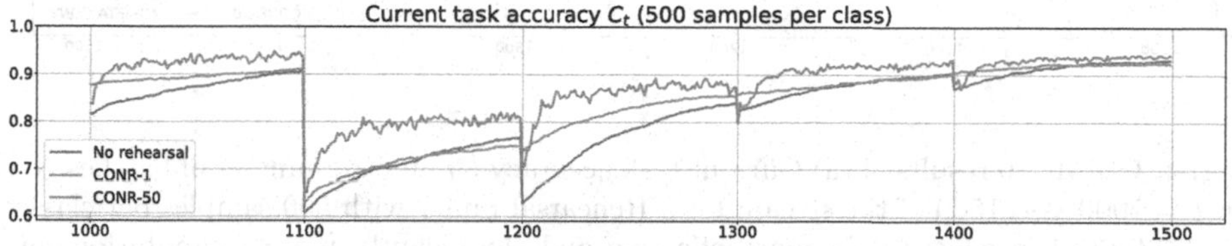

Fig. 3. Results of current task accuracy (C_t) for the CIFAR-10 dataset. The buffer includes 500 samples per class. `CONR-1` performs similar to having no rehearsal at all. `CONR-50` iterations has a clear benefit, but requires 50 times higher computational cost

Table 1. Comparison of the different online training strategies for the MNIST dataset in terms of average values of the metrics across the entire stream. Results are provided for a Rehearsal buffer with 50 samples per class.The highest effectiveness per metric is shown in bold. Numbers in red and blue show the sum of training iterations of the best continuous rehearsal methods and the best proposed methods respectively

Strategy	N_t	$\bar{A}_t$	$\bar{C}_t$	$\bar{O}_t$
`NO REHEARSAL`	1500	0.788	0.93	0.924
`CONR-1`	3000	0.943	0.964	0.963
`CONR-25`	75000	**0.963**	0.984	0.982
`CONR-50`	150000	**0.963**	0.985	0.984
`DRIFTA-DYN-50`	9562	0.961	**0.986**	0.984
`2DRIFTA-DYN-50`	10366	0.962	**0.986**	**0.985**
`DRIFTA-CONV`	3808	0.92	0.983	0.981
`2DRIFTA-CONV`	4982	0.949	0.984	0.982

5.3 Experiment 2: Choosing the Best Rehearsal Strategy

This experiment evaluates the proposed training strategies. Table 1 contains MNIST results and compares all methods in terms of the training batches (N_t) as well as in terms of the average accuracies $\bar{A}_t$, $\bar{C}_t$ and $\bar{O}_t$. Note that we have added a decay factor for the learning rate (d = 0.05) in all methods for a more

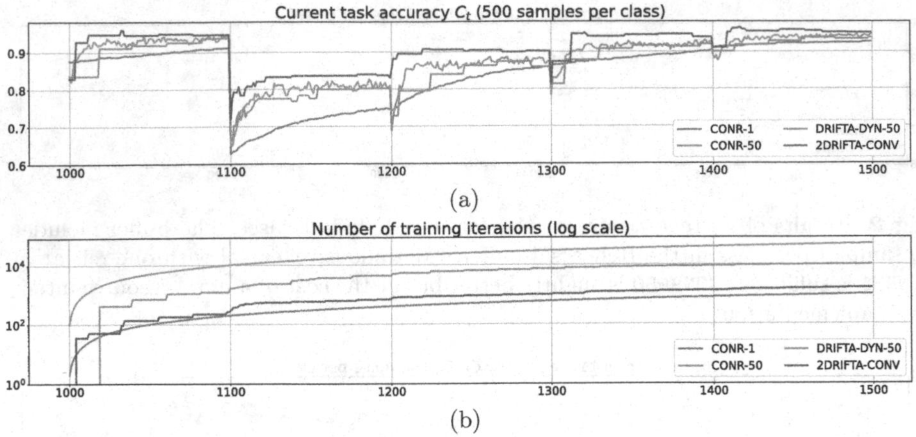

Fig. 4. CIFAR-10 results of (a) Current task accuracy C_t and (b) number of batches N_t for $t = 1000, \cdots, 1500$ of the stream for a Rehearsal buffer with 500 samples per class. `2DRIFTA-CONV` is consistently more effective and significantly more computationally efficient in this experimental setup.

fair comparison. According to the results for the MNIST dataset, the proposed drift activated dynamic rehearsal methods are most of the times almost equal whilst decreasing training iterations significantly. In CIFAR-10 tests, illustrated in Fig. 4, convergence-based methods are even better in terms of computational efficiency with very little compromise in the average metrics. The results clearly demonstrate the benefit of using the proposed drift activated dynamic and convergence rehearsal (especially `2DRIFTA-DYN-n` and `2DRIFTA-CONV`) for the CIFAR-10 dataset. The results show both the limitations of training with continuous rehearsal methods in an online setting and the benefits of a dynamic approach.

6 Conclusions

This work examined the problem of online continual learning from non i.i.d image streams, with unknown task boundaries, and introduced a generic rehearsal strategy that decides when, as well as how to train. The proposed strategy combines drift detection for the early detection of the task change with methods for determining training parameters (number of training iterations and learning rate) at each training step. The combination of these techniques achieves almost identical current and online task accuracy compared to the static rehearsal strategy baselines, while being more efficient in the use of rehearsal samples, leading to significantly reduced computational cost.

Acknowledgment. This work is supported by the "TEACHING" project that has received funding from the European Union's Horizon 2020 research and innovation programme under the grant agreement No 871385. The work reflects only the author's view and the EU Agency is not responsible for any use that may be made of the information it contains.

References

1. Bottou, L., Bousquet, O.: The tradeoffs of large-scale learning. In: Optimization for Machine Learning, p. 351 (2011)
2. Chaudhry, A., et al.: Continual Learning with Tiny Episodic Memories. CoRR, vol. abs/1902.10486 (2019)
3. Delange, M., et al.: A continual learning survey: defying forgetting in classification tasks. IEEE Trans. Pattern Anal. Mach. Intell., 1 (2021). https://doi.org/10.1109/TPAMI.2021.3057446
4. Demosthenous, G., Vassiliades, V.: Continual Learning on the Edge with TensorFlow Lite. arXiv preprint arXiv:2105.01946 (2021)
5. He, J., Mao, R., Shao, Z., Zhu, F.: Incremental learning in online scenario. In: Proceedings of the IEEE/CVF Conference on Computer Vision and Pattern Recognition, pp. 13926–13935 (2020)
6. He, K., Zhang, X., Ren, S., Sun, J.: Deep Residual Learning for Image Recognition. arXiv [cs.CV] (2015)
7. Krizhevsky, A.: Learning multiple layers of features from tiny images (2009)
8. Lecun, Y., Bottou, L., Bengio, Y., Haffner, P.: Gradient-based learning applied to document recognition. Proc. IEEE **86**(11), 2278–2324 (1998)
9. Li, Z., Hoiem, D.: Learning without Forgetting. CoRR, vol. abs/1606.09282 (2016)
10. Mallya, A., Lazebnik, S.: Packnet: adding multiple tasks to a single network by iterative pruning. In: Proceedings of the IEEE Conference on Computer Vision and Pattern Recognition, pp. 7765–7773 (2018)
11. Milan, K., Veness, J., Kirkpatrick, J., Hassabis, D., Koop, A., Bowling, M.: The forget-me-not process. In: Proceedings of the 30th International Conference on Neural Information Processing Systems, pp. 3709–3717 (2016)
12. Pellegrini, L., Graffieti, G., Lomonaco, V., Maltoni, D.: Latent Replay for Real-Time Continual Learning. CoRR, vol. abs/1912.01100 (2019)
13. Pellegrini, L., Lomonaco, V., Graffieti, G., Maltoni, D.: Continual Learning at the Edge: Real-Time Training on Smartphone Devices. arXiv preprint arXiv:2105.13127 (2021)
14. Rebuffi, S.-A., Kolesnikov, A., Sperl, G., Lampert, C.H.: iCaRL: incremental classifier and representation learning. In: Proceedings of the IEEE Conference on Computer Vision and Pattern Recognition, pp. 2001–2010 (2017)
15. Rolnick, D., Ahuja, A., Schwarz, J., Lillicrap, T.P., Wayne, G.: Experience Replay for Continual Learning. CoRR, vol. abs/1811.11682 (2018)
16. Ross, G.J., Adams, N.M., Tasoulis, D.K., Hand, D.J.: Exponentially weighted moving average charts for detecting concept drift. Pattern Recogn. Lett. **33**(2), 191–198 (2012)
17. Wiewel, F., Yang, B.: Entropy-based sample selection for online continual learning. In: 28th European Signal Proceedings Conference (EUSIPCO), pp. 1477–1481 (2021)
18. Widmer, G., Kubat, M.: Learning in the presence of concept drift and hidden contexts. Mach. Learn. **23**(1), 69–101 (1996)

Recurrent Vision Transformer for Solving Visual Reasoning Problems

Nicola Messina(✉), Giuseppe Amato, Fabio Carrara, Claudio Gennaro, and Fabrizio Falchi

Institute of Information Science and Technologies (ISTI), Italian National Research Council (CNR), Via G. Moruzzi 1, 56124 Pisa, Italy
{nicola.messina,giuseppe.amato,fabio.carrara,claudio.gennaro, fabrizio.falchi}@isti.cnr.it

Abstract. Although convolutional neural networks (CNNs) showed remarkable results in many vision tasks, they are still strained by simple yet challenging visual reasoning problems. Inspired by the recent success of the Transformer network in computer vision, in this paper, we introduce the Recurrent Vision Transformer (RViT) model. Thanks to the impact of recurrent connections and spatial attention in reasoning tasks, this network achieves competitive results on the *same-different* visual reasoning problems from the SVRT dataset. The weight-sharing both in spatial and depth dimensions regularizes the model, allowing it to learn using far fewer free parameters, using only 28k training samples. A comprehensive ablation study confirms the importance of a hybrid CNN + Transformer architecture and the role of the feedback connections, which iteratively refine the internal representation until a stable prediction is obtained. In the end, this study can lay the basis for a deeper understanding of the role of attention and recurrent connections for solving visual abstract reasoning tasks. The code for reproducing our results is publicly available here: https://tinyurl.com/recvit.

Keywords: Visual reasoning · Transformer networks · Deep learning

1 Introduction

Deep learning methods largely reshaped classical computer vision, solving many tasks impossible to face without learning representations from data. Convolutional neural networks (CNNs) obtained state-of-the-art results in many computer vision tasks, such as image classification [13,35], or object detection [6,26,27]. Recently, a novel promising architecture took hold in the field of image processing: the Transformer. Initially developed for solving natural language processing tasks, it found its way into the computer vision world, capturing the interest of the whole community. These Transformer-based architectures already proved their effectiveness in many image and video processing tasks [2,7,11,23,24]. The Transformer's success is mainly due to the power of the self-attention mechanism, which can relate every visual token with all the others, creating a powerful relational understanding

S. Sclaroff et al. (Eds.): ICIAP 2022, LNCS 13233, pp. 50–61, 2022.
https://doi.org/10.1007/978-3-031-06433-3_5

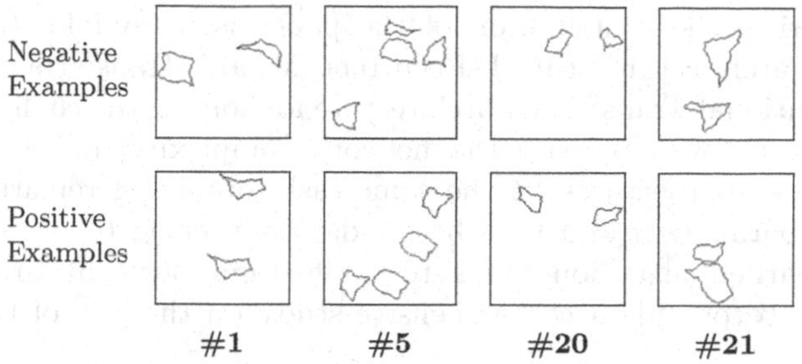

Fig. 1. Positive and negative examples from the considered SVRT problems: P.1: same shapes; P.5: two twisted pairs of same shapes; P.20: same shapes reflected along an unknown symmetry axis; P.21: same shapes but rotated and scaled.

pipeline. In this paper, we aim at studying the relational understanding capabilities of Vision Transformers in the context of an apparently simple yet non-trivial task, called *same-different* task. In short, the same-different task consists in understanding if two shapes in an image satisfy a certain rule. In the simpler case, the rule is merely that *the two shapes must be equal*; however, the rule is not known a priori and must be internally understood from the provided positive and negative examples. An example is given in Fig. 1. Humans perceive the world as a complex set of patterns composite together to form higher-level structures, such as the repeating chorus in a song. Through the same-different task, we can better understand the abstract abilities of current deep neural network models, even outside the computer vision world. The long-term results from these studies can be applied in a wide range of disciplines, from robotics and intelligent video surveillance to cultural heritage preservation.

The same-different task can be framed as a binary classification problem, and it has been partially solved with state-of-the-art convolutional architectures, particularly with ResNets [3,14,22,25]. From these studies, it has been observed that (a) deep CNNs are needed, with lots of free parameters, to relate distant zones of the image in search of matching patterns, and (b) usually, a lot of data is needed to learn the underlying rule, while humans can spot it with only a few samples. Furthermore, some works [18] emphasized the role of recurrent connections, which can iteratively refine the visual input until an optimal and stable conclusion is drawn. In the light of these observations, in this paper, we introduce a novel architecture, called Recurrent Vision Transformer (RViT), for solving the same-different problems. It is inspired by both the recent Vision Transformer (ViT) model [11] and by a recurrent version of the Transformer architecture called Universal Transformer [8]. The introduced architecture can understand and relate distant parts in the image using the powerful Transformer's attentive mechanism and iteratively refine the final prediction using feedback connections. Notably, we find that the base ViT model cannot learn any of the same-different tasks, suggesting that both a hybrid architecture (upstream CNN + downstream Transformer) and feedback connections can be the keys for solving the task.

To summarize, the contribution of the paper is many-fold: (a) we introduce a novel architecture, called Recurrent Vision Transformer (RViT), a hybrid Convolutional-Transformer architecture for solving the challenging same-different tasks; (b) we compare the network complexity and accuracy with respect to other architectures on the same task, obtaining remarkable results with less free parameters and thus better data efficiency; (c) we qualitatively inspect the learned attention maps to understand how the architecture is behaving, and we provide a comprehensive study on the role of the recurrent connections.

2 Related Work

Vision Transformers. The massive engagement of the Transformer architecture [33] in the Natural Language Processing community grew at the point that it trespassed the boundaries of language processing, finding wide applications in computer vision. In fact, it is possible to subdivide images into *patches* which can be fed as input to a Transformer encoder for further processing. Some of the Transformer-based architectures for vision, like Cross Transformers [10] or DETR [4], use the regular grid of features from the last feature map of a CNN as visual tokens. More recently, fully-transformer architectures, first among which ViT [11], have taken root. For the first time, no convolutions are used to process the input image. In particular, the ViT architecture divides the image in patches using the grid approach; the RGB pixel values from every patch are concatenated, and they are linearly projected to a lower-dimensional space to be used as visual tokens. The BERT-like [CLS] token [9] is then used as the classification head. Similarly, the TimeSformer [2] redefined attention both in space and time to understand long-range space-time dependencies in videos.

Same-Different Task. Many tasks have been proposed in computer vision to tackle abstract visual reasoning abilities of machine learning models, like CLEVR and Sort-of-CLEVR [17], Raven's Progressive Matrices (RPM), or Procedurally Generated Matrices (PGMs) [28]. In [12], the authors introduced the *Synthetic Visual Reasoning Test* (SVRT) dataset, composed of simple images containing closed shapes. It was developed to test the relational and comparison abilities of artificial vision systems. The work in [30] first showed, using the SVRT dataset, that the tasks involving comparisons between shapes were difficult to solve for convolutional architectures like LeNet and GoogLeNet [31]. The authors in [20] drawn a similar conclusion, introducing a variation of the SVRT dataset – the Parametric SVRT (PSVRT) for solving some shortcomings of the SVRT dataset – and concluding that the Relation Network [29] is also strained on the same-different judgments. Similarly, [25] developed a more controlled visual dataset to evaluate the reasoning abilities of deep neural networks on shapes having different distributions. The authors in [3,14] found that deep CNNs, like ResNet-50, can solve the SVRT problems even with a relatively small amount of samples (28k images). The authors in [21,22] demonstrated that also many other state-of-the-art deep learning architectures for classifying images (ResNet, DenseNets,

CorNet) models can learn this task, generalizing to some extent. Recently, [32] discussed the important role of attention in the same-different problems.

Recurrent Models. Recurrent models – LSTMs [16] and GRUs [5], to name a few – have been widely used for dealing with variable-length sequences, especially in the field of natural language processing. However, recently, many neuroscience and deep-learning works claimed the importance of recurrent connections outside the straightforward text processing, as they could have an essential role in recognition and abstract reasoning. The work in [18] claimed that the visual cortex could be comprised of recurrent connections, and the visual information is refined in successive steps. Differently, many works in deep learning tried to achieve Turing-completeness by creating recurrent architectures with dynamic halting mechanisms [1,8,15]. Although our work does not include dynamic halting mechanisms, it partially embraces these ideas, experimenting with recurrent connections for iteratively refining the final prediction.

3 The Recurrent Vision Transformer Model

The proposed model is based on the recent Vision Transformer – in particular, the ViT model [11]. The drawback of CNNs in solving the same-different problems is that sufficiently deep networks are needed to correlate distant zones in the image. The Transformer-like attention mechanism in ViT helps in creating short paths between image patches through the self-attention mechanism. Furthermore, inspired by the role of recurrent connections in the human's visual cortex [18], we modify the ViT Transformer encoder module by sharing the encoder weights among all the T layers (i.e., along the depth dimension), effectively creating a recurrent Transformer encoder model, similar to [8]. This has the effect of sharing weights not only in the sequence dimension as in standard Transformers, but also in the depth dimension, further constraining the model complexity. As a feature extractor, we use a small upstream CNN that outputs $N \times N$ $D-$dimensional features used as visual tokens in input to the Transformer encoder. The overall architecture is shown in Fig. 2.

By leveraging the recurrent nature of the architecture, we avoid explicitly tuning the depth of the network (i.e., the total number of recurrent iterations) by forcing the architecture to perform a prediction at each time step, using the CLS token. The most likely outcome among the predictions from all the time steps is then taken as the final prediction. More in detail, the model comprises T binary classification heads, one for each time step. During training, the binary cross-entropy loss at each time step is computed as $\mathcal{L}_t = \text{BCE}(y_t, \hat{y})$, where y_t is the network output from the t-th time step, and $\hat{y}$ is the ground-truth value. The various losses are then aggregated to obtain the final loss $\mathcal{L}_{\text{total}}$. We noticed that a simple average $\frac{1}{T}\sum_{t=1}^{T}\mathcal{L}_t$ already led to good results. However, we obtained the best results by using the automatic loss-weighting scheme proposed in [19]:

$$\mathcal{L}_{\text{total}} = \frac{1}{2}\sum_{t=1}^{T}\left(\frac{1}{e^{s_t}}\mathcal{L}_t + s_t\right), \tag{1}$$

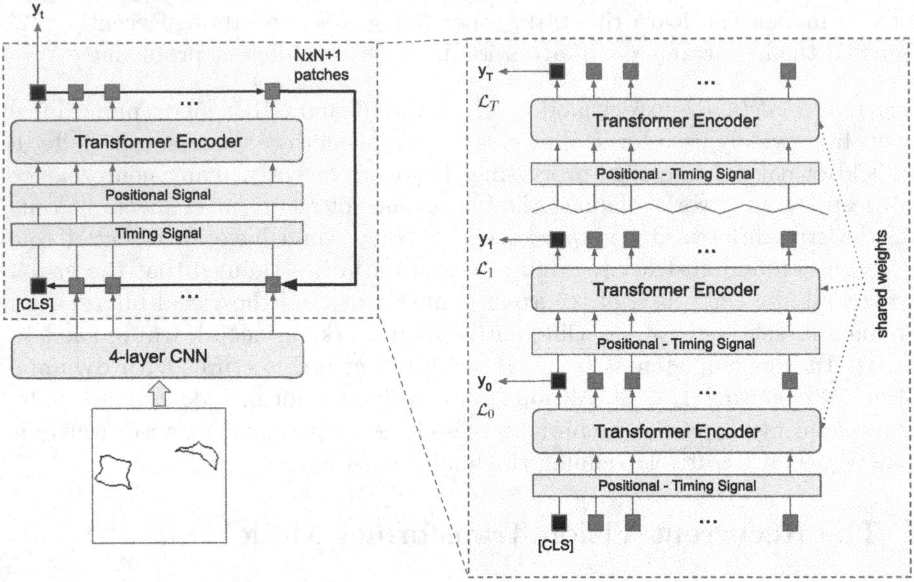

Fig. 2. The RViT architecture. The image is processed by a 4-layer CNN, outputting a 8×8 grid of visual features. The CLS token is added to this set, and the tokens are processed multiple times by the recurrent module. At each time step, the binary cross-entropy loss is computed against the ground-truth labels.

where s_t is a free scalar parameter that encodes the predicted uncertainty of the classification at the t-th time step, and the model automatically learns it during the training phase. We refer readers to [19] for more detailed derivation and discussion.

During inference, the maximum-likelihood prediction is taken as the final network output. In particular, the time step $\bar{t}$ at which the network reaches the maximum confidence is the one where the output probability is farthest from the pure chance in a binary classification setup ($p = 0.5$):

$$\bar{t} = \arg\max_t |y_t - 0.5|. \tag{2}$$

At this point, the final output is simply $y = y_{\bar{t}}$.

4 Experiments

In this section, we briefly introduce the SVRT dataset used in the experiments, and we present and discuss the performance of the Recurrent Vision Transformer on these problems.

4.1 Dataset

In this work, we use the *Synthetic Visual Reasoning Test* (SVRT) benchmark to test our proposed architecture. SVRT comprises 23 different sub-problems; each sub-problem comprises a set of positive and negative samples generated using a problem-specific rule. The objective of any classifier trained on a problem is to distinguish the positive and negative samples, and the only way to succeed is to discover the underlying rule.

From previous works [3,20] it is clear that relational problems – the ones involving shape comparisons under different geometric transformations – are the most difficult to solve for Deep Neural Networks. Thus, as in [21,22], we focus the attention on four of these problems: **Problem 1 (P.1)** - detecting the very same shapes, randomly placed in the image, having the same orientation and scale; **Problem 5 (P.5)** - detecting two pairs of identical shapes, randomly placed in the image. **Problem 20 (P.20)** - detecting the same shape, translated and flipped along a randomly chosen axis; **Problem 21 (P.21)** - detecting the same shape, randomly translated, orientated, and scaled. Positive and negative samples from each of these visual problems are shown in Fig. 1.

4.2 Setup

For the upstream CNN processing the pixel-level information, we used a 4-layer *Steerable CNN* [34]. A Steerable CNN describes E(2)-equivariant (i.e., rotation- and reflection-equivariant) convolutions on the image plane $\mathbb{R}^2$; in contrast to conventional CNNs, E(2)-equivariant models are guaranteed to generalize over such transformations other than simple translation and are therefore more data-efficient. In the ablation study in Sect. 4.4, we will give more insights on the role of Steerable CNNs over standard CNNs in solving the same-different task.

We forged two different versions of the RViT, a *small* and a *large* version, having the same structure but a different number of hidden neurons in the core layers: the small RViT produces 256-dimensional keys, queries, and values and outputs 256-dimensional visual features from the CNN, while the large RViT has these two parameters set to 512. We used the Adam optimizer; after a minor hyper-parameter tuning, we set the learning rate for all the experiments to 1e–4, and the number of attention heads to 4; we let the models train for 200 epochs, decreasing the learning rate to 1e–5 after 170 epochs. We tested the models using the snapshot with the best accuracy measured on the validation set.

In order to better compare with the ResNet-50 experiments in [3], we also tried to use as up-stream CNN the first two or three layers of a ResNet-50 pre-trained on ImageNet. For the image resolution, we mainly used $N = 16$, outputting 16×16 visual tokens from the CNN. During the pre-training experiments, instead, we used $N = 8$ for accommodating the output feature map resolution of the pre-trained model and also for performance reasons. During training, we set the maximum time steps $T = 9$.

We collected results using both 28k training images, following [3], and 400k training images, for comparing our proposed architectures with convolutional

Table 1. Accuracy (%) of our method, trained from-scratch, with respect to the baselines. #pars indicate the number of free parameters of the model.

Model	400k training samples				28k training samples				#pars ↓
	P.1 ↑	P.5 ↑	P.20 ↑	P.21 ↑	P.1 ↑	P.5 ↑	P.20 ↑	P.21 ↑	
RN [29]	50.0	50.0	50.0	50.0	50.0	50.0	50.0	50.0	0.4M
ViT [11]	50.0	50.0	50.0	50.0	50.0	50.0	50.0	50.0	26M
ResNet-18 [22]	99.2	**99.9**	95.5	96.2	99.2	98.4	93.7	50.0	11M
ResNet-50 [3]	–	–	–	–	95.4	89.9	92.9	72.6	23M
DenseNet-121 [22]	99.6	98.2	94.2	95.1	73.9	54.7	94.4	**85.8**	6.9M
CorNet-S [22]	96.9	96.8	95.0	**96.9**	98.8	97.1	92.3	82.5	52M
RViT-small	**99.9**	99.4	**98.9**	95.7	**99.6**	98.0	93.9	78.6	**0.9M**
RViT-large	**99.9**	99.0	98.8	96.4	**99.6**	**99.3**	**95.3**	77.8	3.1M

Table 2. Accuracy (%) of RViT-small, with the first layers of a ResNet-50 pre-trained on ImageNet, with respect to the full ResNet-50 baseline. In ResNet-50/11 we kept the first 11 layers, while in ResNet-50/23 the first 23.

Model	P.1 ↑	P.5 ↑	P.20 ↑	P.21 ↑	#pars ↓
ResNet-50 [3]	99.5	98.7	98.9	**92.5**	23M
RViT ResNet-50/11	99.6	98.6	94.5	91.6	**2.3M**
RViT ResNet-50/23	**99.7**	**99.7**	**99.4**	85.2	9.5M

networks trained in [21,22]. We used 18k images both for validation and testing. The images were generated with the SVRT original code, available online[1].

4.3 Results

We compared our model with other key architectures: the Relation Network (RN) [29] which by design should be able to correlate distant zones of the image; the Vision Transformer (ViT) [11] which recently achieved remarkable performance on classification tasks, although it is very data-hungry, and some state-of-the-art convolutional models—ResNet18, ResNet50, CorNet-S and DenseNet121—trained on the same task in [3,21,22]. Notably, CorNet-S also implements feedback connections, although it is much more complex, in terms of number of parameters, than our RViT architecture.

Looking at Table 1, we can see how neither the Relation Network nor the ViT converges on the four visual problems, for both 400k and 28k data regimes. The ViT probably needs more architectural inductive biases to understand the rules, while the relational mechanism of Relation Network is probably too simple for understanding the objects in the image and their relationships. Instead, our RViT model can obtain very competitive results on all tasks and on both data regimes,

[1] https://fleuret.org/git-tgz/svrt.

often outperforming the baselines. Noticeably, the RViT-small can learn all the four problems using only 0.9M free parameters, about 8 times fewer parameters than the smallest convolutional network able to solve the task (DenseNet121). This suggests that the model has the correct structure for understanding the visual problems, without having the possibility to memorize the patterns.

In Table 2, we instead report the accuracy of the small RViT model, where the upstream path is pre-trained on the classification task on ImageNet, following the work in [3]. Even in this case, the RViT achieves competitive results, but with much fewer free parameters and using only a slice – the first 11 and 23 layers – of the pre-trained ResNet-50 architecture.

4.4 Ablation Study

Following, we report some in-depth analysis of the RViTs performed with 28k training images.

The Role of Recurrent Connections and Steerable Convolution. In Table 3, we experimented with some variations of the RViT to understand the roles of recurrent connections and the employed 4-layers steerable CNN. The basic configuration is Conv. ViT, which is the same as the standard ViT from [11] but with an upstream CNN as the visual feature extractor. In contrast to the original ViT formulation, the Conv. ViT can improve significantly on P.1, P.20, and P.21, moving away from the chance accuracy. However, the most significant jump in accuracy happens when recurrent connections are introduced (Conv. RViT). In this case, the same model can learn all the visual problems, with an improvement of 67% on P.1 and 7% on P.21. Another improvement is obtained when using the Steerable CNNs [34]. This kind of CNN produces features equivariant to rotations and reflections. For this reason, it has a wider impact on P.20 and P.21, where shapes are reflected and rotated, respectively.

Recurrent connections seem to have critical importance. They highly regularize the model, making it more data-efficient and performing a dynamic iterative computation that procedurally refines both the previous internal representations and the previous predictions. To better appreciate this aspect, in Fig. 3 we show the mean time step $\bar{t}$, for each problem, where the model reaches the maximum confidence. Interestingly, P.1 and P.5 reach the best confidence in few iterations, while the more challenging P.20 and P.21 need much more pondering before stabilizing. More in detail, it can be noticed that although there is not too much difference considering the size of the models (Fig. 3a), the network seems majorly strained when the shapes are the *same* (Fig. 3b). This is reasonable: it is heavier to be sure that shapes coincide in every point, while it takes little to find even a single non-matching pattern to output the answer *different*.

Table 3. Ablation study on Convolutional ViT (Conv. ViT), on Convolutional Recurrent ViT (Conv. RViT), and Equivariant Convolutional Recurrent ViT (Eq. Conv. RViT). The last one is the model effectively employed in Tables 1 and 2. Accuracy (%) is in this case measured on the validation set.

Model	P.1 ↑	P.5 ↑	P.20 ↑	P.21 ↑
Conv. ViT	59.5	50.0	88.5	62.5
Conv. RViT	99.9	99.0	93.9	66.8
Eq. Conv. RViT	99.8	99.4	95.6	77.3

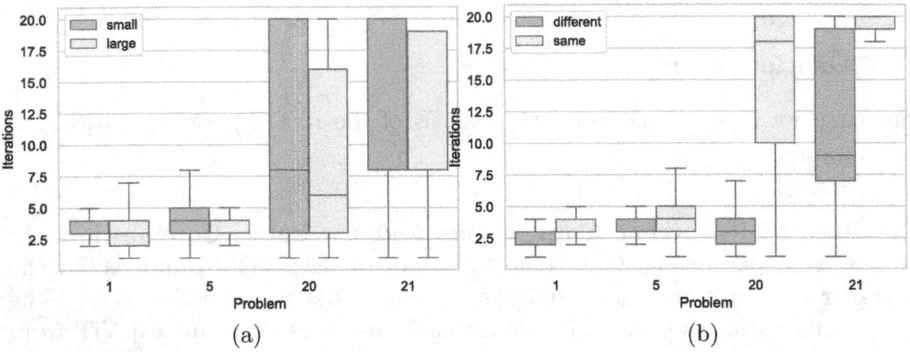

Fig. 3. The distribution of the best time step $\bar{t}$ grouped by (a) the two different RViT sizes (small, large), and (b) by the same-different label.

Visualizing the Attention. In Fig. 4, we reported a visualization of the self-attention maps learned by the trained models, computed in specific points (marked with red dots) in the image, and by averaging the four attention heads. The 16 × 16 grid allows us to appreciate fine details; in particular, we can see what parts of the shapes the model is attending to for producing the final answer. In most cases, the model correctly attends the other shape in search of the corresponding edges. In some instances, the attention map is not so neat (e.g., in (d) and (f)), emphasizing the intrinsic complexity of the tasks. Furthermore, in Fig. 5 we report the evolving attention maps at different time steps. The map is initially very noisy, but it is slowly refined as the number of iterations increases to create a stable representation.

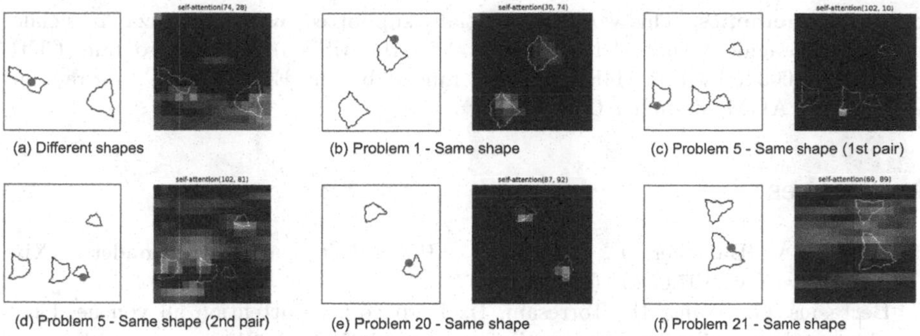

Fig. 4. Attention visualization on the different visual problems. The red dot shows the point in space with respect to which the self-attention is computed. (Color figure online)

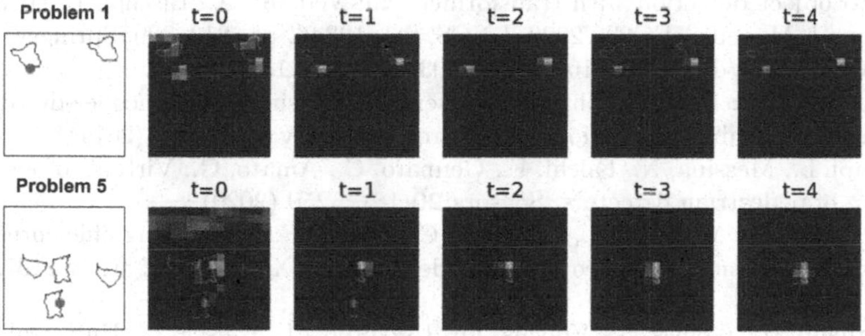

Fig. 5. Evolving attention maps at different time steps.

5 Conclusions

In this work, we leveraged the power of Vision Transformer and recurrent connections to create the Recurrent Vision Transformer Model (RViT) capable of solving some of the most challenging same-different tasks from the SVRT dataset. We showed that this architecture can defeat current methods on the same dataset, while being simpler, more data-efficient, and explainable to some extent. The experiments confirm the hypothesis that recurrent connections provide helps for understanding these visual problems, and the Transformer-like spatial attention enabled us to visualize what parts of the image the model is attending during the inference. The model outperforms the basic ViT model on this task, as well as other relation-aware architectures such as Relation Networks. In the future, we plan to transfer the seeds of this research to real use cases, where multiple possibly distant inputs need to be related and analyzed to draw a conclusion. For example, in surveillance applications, it may be useful to recognize the *same* person across multiple cameras or, in audio processing, recognize repeating patterns in a song for clustering or retrieval.

Acknowledgements. This work was partially supported by "Intelligenza Artificiale per il Monitoraggio Visuale dei Siti Culturali" (AI4CHSites) CNR4C program, CUP B15J19001040004, by the AI4EU project, funded by the EC (H2020 - Contract n. 825619), and AI4Media under GA 951911.

References

1. Banino, A., Balaguer, J., Blundell, C.: PonderNet: learning to ponder. arXiv preprint arXiv:2107.05407 (2021)
2. Bertasius, G., Wang, H., Torresani, L.: Is space-time attention all you need for video understanding? arXiv preprint arXiv:2102.05095 (2021)
3. Borowski, J., Funke, C.M., Stosio, K., Brendel, W., Wallis, T., Bethge, M.: The notorious difficulty of comparing human and machine perception. In: 2019 Conference on Cognitive Computational Neuroscience, pp. 2019–1295 (2019)
4. Carion, N., Massa, F., Synnaeve, G., Usunier, N., Kirillov, A., Zagoruyko, S.: End-to-end object detection with transformers. In: Vedaldi, A., Bischof, H., Brox, T., Frahm, J.-M. (eds.) ECCV 2020. LNCS, vol. 12346, pp. 213–229. Springer, Cham (2020). https://doi.org/10.1007/978-3-030-58452-8_13
5. Cho, K., et al.: Learning phrase representations using RNN encoder-decoder for statistical machine translation. arXiv preprint arXiv:1406.1078 (2014)
6. Ciampi, L., Messina, N., Falchi, F., Gennaro, C., Amato, G.: Virtual to real adaptation of pedestrian detectors. Sensors **20**(18), 5250 (2020)
7. Coccomini, D., Messina, N., Gennaro, C., Falchi, F.: Combining efficientnet and vision transformers for video deepfake detection. arXiv preprint arXiv:2107.02612 (2021)
8. Dehghani, M., Gouws, S., Vinyals, O., Uszkoreit, J., Kaiser, L.: Universal transformers. arXiv preprint arXiv:1807.03819 (2018)
9. Devlin, J., Chang, M., Lee, K., Toutanova, K.: BERT: pre-training of deep bidirectional transformers for language understanding. In: NAACL-HLT 2019, pp. 4171–4186. Association for Computational Linguistics (2019)
10. Doersch, C., Gupta, A., Zisserman, A.: Crosstransformers: spatially-aware few-shot transfer. arXiv preprint arXiv:2007.11498 (2020)
11. Dosovitskiy, A., et al.: An image is worth 16x16 words: transformers for image recognition at scale. arXiv preprint arXiv:2010.11929 (2020)
12. Fleuret, F., Li, T., Dubout, C., Wampler, E.K., Yantis, S., Geman, D.: Comparing machines and humans on a visual categorization test. Proc. Natl. Acad. Sci. **108**(43), 17621–17625 (2011)
13. Foret, P., Kleiner, A., Mobahi, H., Neyshabur, B.: Sharpness-aware minimization for efficiently improving generalization. arXiv preprint arXiv:2010.01412 (2020)
14. Funke, C.M., Borowski, J., Stosio, K., Brendel, W., Wallis, T.S., Bethge, M.: Five points to check when comparing visual perception in humans and machines. J. Vis. **21**(3), 16–16 (2021)
15. Graves, A., Wayne, G., Danihelka, I.: Neural turing machines. arXiv preprint arXiv:1410.5401 (2014)
16. Hochreiter, S., Schmidhuber, J.: Long short-term memory. Neural Comput. **9**(8), 1735–1780 (1997)
17. Johnson, J., Hariharan, B., van der Maaten, L., Fei-Fei, L., Lawrence Zitnick, C., Girshick, R.: CLEVR: a diagnostic dataset for compositional language and elementary visual reasoning. In: Proceedings of IEEE CVPR, pp. 2901–2910 (2017)

18. Kar, K., Kubilius, J., Schmidt, K., Issa, E.B., DiCarlo, J.J.: Evidence that recurrent circuits are critical to the ventral stream's execution of core object recognition behavior. Nat. Neurosci. **22**(6), 974–983 (2019)
19. Kendall, A., Gal, Y., Cipolla, R.: Multi-task learning using uncertainty to weigh losses for scene geometry and semantics. In: Proceedings of the IEEE Conference on Computer Vision and Pattern Recognition, pp. 7482–7491 (2018)
20. Kim, J., Ricci, M., Serre, T.: Not-so-CLEVR: learning same-different relations strains feedforward neural networks. Interface Focus **8**(4), 20180011 (2018)
21. Messina, N., Amato, G., Carrara, F., Falchi, F., Gennaro, C.: Testing deep neural networks on the same-different task. In: 2019 International Conference on Content-Based Multimedia Indexing (CBMI), pp. 1–6. IEEE (2019)
22. Messina, N., Amato, G., Carrara, F., Gennaro, C., Falchi, F.: Solving the same-different task with convolutional neural networks. Pattern Recogn. Lett. **143**, 75–80 (2021)
23. Messina, N., Amato, G., Esuli, A., Falchi, F., Gennaro, C., Marchand-Maillet, S.: Fine-grained visual textual alignment for cross-modal retrieval using transformer encoders. arXiv preprint arXiv:2008.05231 (2020)
24. Messina, N., Falchi, F., Esuli, A., Amato, G.: Transformer reasoning network for image-text matching and retrieval. In: 2020 25th International Conference on Pattern Recognition (ICPR), pp. 5222–5229. IEEE (2021)
25. Puebla, G., Bowers, J.S.: Can deep convolutional neural networks learn same-different relations? bioRxiv (2021)
26. Redmon, J., Farhadi, A.: YOLOv3: an incremental improvement. arXiv preprint arXiv:1804.02767 (2018)
27. Ren, S., He, K., Girshick, R., Sun, J.: Faster R-CNN: towards real-time object detection with region proposal networks. Adv. Neural. Inf. Process. Syst. **28**, 91–99 (2015)
28. Santoro, A., Hill, F., Barrett, D., Morcos, A., Lillicrap, T.: Measuring abstract reasoning in neural networks. In: International Conference on Machine Learning, pp. 4477–4486 (2018)
29. Santoro, A., et al.: A simple neural network module for relational reasoning. In: Advances in Neural Information Processing Systems, pp. 4967–4976 (2017)
30. Stabinger, S., Rodríguez-Sánchez, A., Piater, J.: 25 years of CNNs: can we compare to human abstraction capabilities? In: Villa, A.E.P., Masulli, P., Pons Rivero, A.J. (eds.) ICANN 2016. LNCS, vol. 9887, pp. 380–387. Springer, Cham (2016). https://doi.org/10.1007/978-3-319-44781-0_45
31. Szegedy, C., et al.: Going deeper with convolutions. In: Proceedings of IEEE CVPR, pp. 1–9 (2015)
32. Vaishnav, M., Cadene, R., Alamia, A., Linsley, D., Vanrullen, R., Serre, T.: Understanding the computational demands underlying visual reasoning. arXiv preprint arXiv:2108.03603 (2021)
33. Vaswani, A., et al.: Attention is all you need. In: Advances in Neural Information Processing Systems, pp. 5998–6008 (2017)
34. Weiler, M., Cesa, G.: General E(2)-equivariant steerable CNNs. In: Conference on Neural Information Processing Systems (NeurIPS) (2019)
35. Xie, S., Girshick, R., Dollár, P., Tu, Z., He, K.: Aggregated residual transformations for deep neural networks. In: Proceedings of the IEEE Conference on Computer Vision and Pattern Recognition, pp. 1492–1500 (2017)

Metric Learning-Based Unsupervised Domain Adaptation for 3D Skeleton Hand Activities Categorization

Yasser Boutaleb[1,2(✉)], Catherine Soladié[2], Nam-duong Duong[1], Amine Kacete[1], Jérôme Royan[1], and Renaud Seguier[2]

[1] IRT b-com, 1219 Avenue des Champs Blancs, 35510 Cesson-Sevigne, France
{yasser.boutaleb,nam-duong.duong,amine.kacete,jerome.royan}@b-com.com
[2] IETR/CentraleSupelec, Avenue de la Boulaie, 35510 Cesson-Sevigne, France
{catherine.soladie,Renaud.seguier}@centralesupelec.fr

Abstract. First-person hand activity recognition plays a significant role in the computer vision field with various applications. Thanks to recent advances in depth sensors, several 3D skeleton-based hand activity recognition methods using supervised Deep Learning (DL) have been proposed, proven effective when a large amount of labeled data is available. However, the annotation of such data remains difficult and costly, which motivates the use of unsupervised methods. We propose in this paper a new approach based on unsupervised domain adaptation (UDA) for 3D skeleton hand activity clustering. It aims at exploiting the knowledge-driven from labeled samples of the source domain to categorize the unlabeled ones of the target domain. To this end, we introduce a novel metric learning-based loss function to learn a highly discriminative representation while preserving a good activity recognition accuracy on the source domain. The learned representation is used as a low-level manifold to cluster unlabeled samples. In addition, to ensure the best clustering results, we proposed a statistical and consensus-clustering-based strategy. The proposed approach is experimented on the real-world FPHA data set.

Keywords: Unsupervised domain adaptation · Metric learning · Hand activities clustering

1 Introduction

Understanding first-person hand activity is a trending topic in computer vision with exhaustive research and practical applications, such as human-computer interaction [1], humanoid robotics [2], virtual/augmented reality [3], and multimedia for automated video analysis [4].

3D skeletal data is one of the most potent modalities for human activity recognition. Indeed, it provides a robust high-level description regarding common RGB imaging problems, such as background subtraction and light variation. To this end, several 3D skeleton-based activity recognition approaches have

S. Sclaroff et al. (Eds.): ICIAP 2022, LNCS 13233, pp. 62–74, 2022.
https://doi.org/10.1007/978-3-031-06433-3_6

been proposed. Most of them are based on supervised Neuronal Network (NN) architectures [5], which has been proven to be effective when a large amount of data is available. However, in real-world activity recognition applications, labeled 3D skeleton samples are usually difficult or expensive to obtain as they require the efforts of human expert annotators. In contrast, unlabeled activity samples are easy to obtain from routine experiments because they do not require human annotation. This has motivated researchers to make use of these unlabeled samples by applying fully unsupervised learning to categorize them [6]. However, clustering 3D hand activities without prior knowledge is still a challenging problem, especially when not enough data is available. This is due to the high intra-class variation and inter-class similarity that hand activities exhibit.

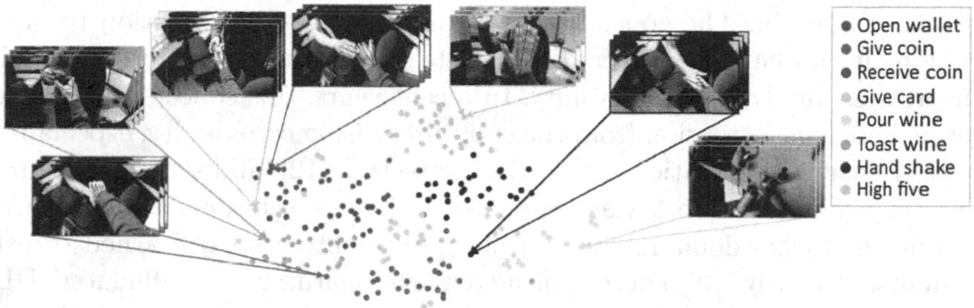

Fig. 1. 2D features embedding of unlabeled hand activities using our proposed method on the "social" scenario of the FPHA data set [7].

Therefore, in this paper we propose an approach for an unlabeled 3D skeleton-based hand activities categorization (clustering), which follows the UDA paradigm. First, we pre-train a supervised NN on labeled samples from the source domain. Next, we try to solve the UDA by using the pre-trained NN model feature space as a mapping manifold to categorize unlabeled target domain samples based on classical clustering methods. Better usage of the prior knowledge from the source domain is crucial for UDA. Therefore, the pre-trained model, which relies on the source domain, must map hand activities to a highly discriminative features space (Fig. 1). This requirement is highly discussed in the field of open-set face recognition [8–10]. Thus, we concluded that the desired feature space must satisfy two main objectives of metric learning: (1) maximizing inter-class distances and (2) minimizing intra-class distances for the mapped hand activities. This is one of the main issues we have addressed in this paper. On the other hand, we tackled a problem of the high sensitivity of clustering algorithms regarding the feature space provided by the pre-trained model, which we noticed through our experiments.

We highlight the contributions as follows: (1) to our humble knowledge, we are the first to propose a comprehensive and generic UDA approach applied to the categorization of unlabeled 3D skeleton-based hand activities. The proposed approach is experimented with and evaluated on a real-world data set. (2) We introduced a new automated penalty-based loss function for metric learning-based NN that we denote by APML. It aims at learning a discriminative features space while keeping a good recognition accuracy. The proposed loss is compared with four state-of-the-art losses. (3) To solve the sensitivity problem of clustering algorithms, which to our knowledge is not addressed in the state-of-the-art, we proposed a statistical strategy based on repetition and consensus clustering that we denote by SCC.

2 Related Work

2.1 Unsupervised Domain Adaptation (UDA)

UDA methods are based on the assumption that there are no available labels in the target domain. The concept is that only the label information from the source domain can be exploited to disseminate shared knowledge across domains and improve model transfer capability. In recent years, this concept of UDA has attracted increasing attention from the computer vision community, especially in the field of image classification [11], object detection [12], and semantic segmentation [13]. A reference work was proposed by [14], which aims at aligning the source and the target domains by projecting data onto a set of learned transfer components. Recently, the success of adversarial learning has influenced UDA proposed methods in the way that they mainly aim at simultaneously minimizing the domain discrepancy between the source and the target domain while learning a domain discriminator by maximizing its loss with respect to the discriminant features learner [15].

Generally, most of the introduced UDA methods tries to find the best mapping manifold that discriminantly represents unlabeled target domain samples. To this end, several metric learning-based UDA methods have been proposed [16]. Similar to our work, [17] proposed a metric-based adversarial discriminative domain adaptation for image tasks. First, they used a metric learning approach to train a supervised model on the source data set by optimizing the triplet loss function [18], which results in a pre-trained model that provides a discriminative embedding space. Next, they used the adversarial approach (same as in [15]) to make the extracted features from the source and target data sets indistinguishable. We do not need to align the target with the source domain in our UDA application. We assume that the learning process is continuous, and we may discover new class categories in the target domain, which do not match those in the source domain. Moreover, we have adopted our proposed metric learning loss to balance between classification accuracy on the source domain and UDA performance on the target domain.

Despite the progress in UDA for third-person activity recognition [19,20], there have been few works for the first-person hand activity recognition, which are based on RGB images modality [21]. To our knowledge, we are the first to address the UDA for hand activity recognition based on the 3D hand skeleton modality. Most of the introduced methods share the same goal of learning discriminative features space by focusing on low-level spatio-temporal features learning. 3D skeleton-based hand activity recognition is a well-studied problem. Various efficient low-level Spatio-temporal features learning techniques have been proposed, e.g., [22]. Knowingly, in this work, we adopted a state-of-the-art solution for the low-level spatio-temporal features, but we focused more on the high-level embedding features.

2.2 Metric Learning

Metric learning aims to learn a similarity function. It is combined with NN to enhance the discriminative power of the feature space by simultaneously enforcing intra-class compactness and inter-class discrepancy in the feature space. A triplet loss [18], contrastive loss [23], and center loss [24] are classically used to increase the Euclidean margin for better feature embedding, which is the same purpose as our proposed loss function. However, center loss only explicitly encourages intra-class compactness. Unlike our loss function that constrains each sample, contrastive/triplet losses require a carefully designed pair/triplet mining procedure, which is both time-consuming and performance-sensitive. More recently and closer to our proposed APML loss, [8–10] applied a penalty margin to penalize a modified standard softmax logit in the angular/cosine space, which showed a good performance in the field of open-set face recognition. In Sect. 3.4, we will give a more detailed comparison with these three methods.

3 Proposed Method

For a given set of unlabeled hand activitie samples $\{x^u\}$ of a target domain, we exploit the knowledge from labeled samples $\{x^l, y^l\}$ of the source domain to find

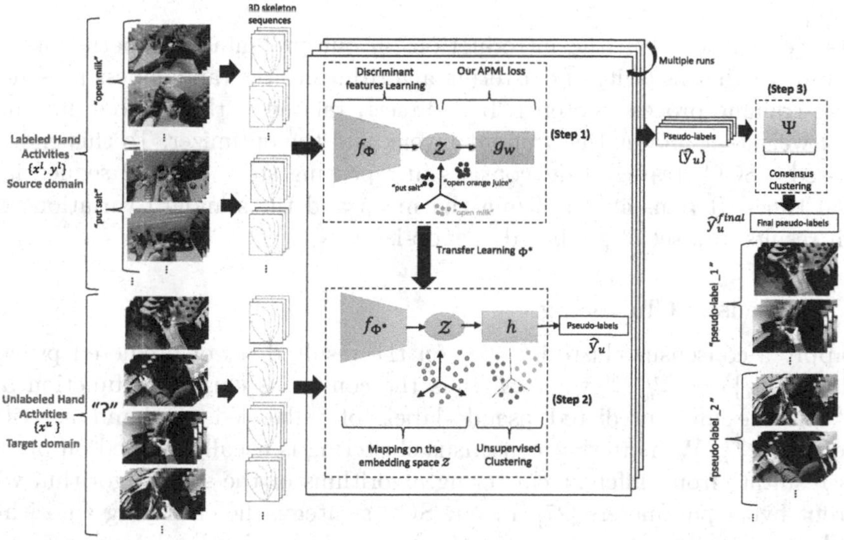

Fig. 2. The proposed method. We exploit the knowledge from labeled samples of the source domain $\{x^l, y^l\}$ to categorize the unlabeled ones of the target domain $\{x^u\}$.

the correspondent unknown set of labels $\mathcal{Y}_u = \{y^u\}$. Where x^u is a particular unlabeled sample and y^u is its correspondent label that we seek to find. As illustrated in Fig. 2, the method proceeds in three main steps that we describe in Sects. 3.1, 3.2 and 3.3 respectively. In Sect. 3.4, we introduce our proposed APML loss function.

3.1 Supervised Learning for Discriminative Features Space Creation

First, we pre-train the 3D skeleton-based hand activity recognition NN $g_w(f_\phi(.))$ where $g_w(.)$ is our APML loss function and $f_\phi(.)$ is the backbone spatio-temporal feature learner, with w and ϕ learnable parameters. The training is preformed on labeled 3D hand activities of the source domain $\{x^l, y^l\}$ where y^l is the label of the particular sample x^l. This is done in a classical supervised manner as shown in Fig. 2 (step 1). We have adopted the NN architecture proposed by [22] as our backbone $f_\phi(.)$. We chose this NN for its outstanding recognition performance and its reproducibility convenience.

Thanks to our APML loss function, this first step result in a pre-trained model $f_{\phi^*}(.)$ with optimized parameters ϕ^*, which maps activities on a discriminative features space $\mathcal{Z} \subset \mathbb{R}^d$ where d is the embedding dimension.

3.2 Unlabeled Activities Clustering

In this second step, we map the set of unlabeled 3D hand activities sequences $\{x^u\}$ into the $\mathcal{Z}$ space. This is done by passing these activities trough the pre-trained NN $f_{\phi^*}(x^u) = z^u \in \mathcal{Z}$, which results in a set of embedded activities $\{z^u\}$. Next, we apply a classical clustering that we formulate as $h(\{z^u\}) = \hat{\mathcal{Y}}_u$ where $h(.)$ is the clustering function and $\hat{\mathcal{Y}}_u = \{\hat{y}^u\}$ are the predicted pseudo-labels for the unlabeled set of activities $\{x^u\}$. Figure 2 (step 2) gives an illustration for this step.

As we mentioned in the introduction, in our preliminary experiments, we observed a high sensitivity of clustering algorithms to the randomness present in the pre-training process of step 1. It is directly related to the random initialization of weights ϕ and w, the dropout layers, and the optimizer. To this end, we proposed a SCC strategy that consists in repeating steps 1 and 2 sequentially several times. It aims at stabilizing the mean and the standard variation [25], which results in a set of predicted pseudo-labels $\{\hat{\mathcal{Y}}_u\}$.

3.3 Consensus Clustering

We apply a consensus clustering [26] on the resulted set of predicted pseudo-labels $\Psi(\{\hat{\mathcal{Y}}_u\}) = \hat{\mathcal{Y}}_u^{final}$ where $\Psi(.)$ is the consensus clustering function and $\hat{\mathcal{Y}}_u^{final}$ is the final predicted pseudo-labels of unlabeled 3D hand activities sequences $\{x^u\}$. We note that consensus clustering is usually applied on predictions resulting from different clustering algorithms or the same algorithm with different hyper-parameters [27]. In our SCC strategy, the clustering algorithms and their hyper-parameters remain the same while only the data distribution changes.

3.4 Automated Penalty-Based Metric Learning (APML) Loss Function

First, let formally describe the well-known classification loss function. Namely Softmax that we denote by $\mathcal{L}_{Softmax}$ and we formulate in combination with the cross-entropy loss as follows:

$$\mathcal{L}_{Softmax} = -\frac{1}{N}\sum_{i=1}^{N} log\frac{e^{w_{y_i} z_i + b_{y_i}}}{\sum_{j=1}^{k} e^{w_j z_i + b_j}} \quad (1)$$

where N is the training batch size, k is the number of labeled classes. $z_i \in \mathcal{Z}$ is the embedding vector of the i-th sample x_i where $f_\phi(x_i) = z_i$. w_j is the j-th wight of the classifier layer $g_w(z_i)$ and and $b_j \in R^k$ is the bias. In other words, the dot product-based logit, which is formulated as $w_{y_i} z_i + b_{y_i}$, is passed through a softmax function that normalizes and results in a soft probabilistic affinity score. The Softmax loss separates embedding vectors from different classes by maximizing the posterior probability of the positive ground-truth class. However, it does not explicitly imply any constraints on the learning process to satisfy the metric learning objectives.

In response to softmax limitations, [8] implied heavy constraints by proposing a manually penalized softmax logit. First, they fixed $b_i = 0$ and normalized embedding vectors $\|z_i\| = 1$ and the weights $\|w_j\| = 1$. Then, the logit is transformed as $s.cos(m_1\theta_{y_i})$ where $\theta_{y_i} = arcos(\frac{w_{y_i} z_i}{\|w_{y_i}\|\|z_i\|})$ is the angle between the embedding and the weight vectors. The m_1 is the manually given multiplicative angular margin penalty parameter that seeks to insure maximum inter-class separation, and finally, s is the re-scaling hyper-parameter. Consequently, the posterior softmax probability will merely relies on cosine of $m_1\theta$, which penalizes the logit according to ground-truth labels. Similarly [10], proposed an additive angular margin m_2 by transforming the softmax logit as $s.cos(m_2 + \theta_j)$, which shown to be more effective. On the other hand, [9] defined the margin m_3 in the cosine space rather than the angle space to penalize the logit by transforming it as $s(cos(\theta_j) - m_3)$.

The various manually assigned margin penalties, whether added on the angle or cosine space, all enhance intra-class compactness and inter-class diversity by penalizing the target logit [28]. However, it mainly depends on the manually assigned values of the margin and the re-scaling hyper-parameters. The choice of these values relies on data distribution and the dimension of embedding space of the used NN backbone. We also note that [8–10] attributed the margin penalty only to the positive ground-truth class, while the target logit for false classes is set to $cos(\theta_i)$. This aim improves intra-class compactness, but on the other hand, it favorites the over-fitting, which limits the classification performance.

Based on these observations, we propose a fully automated penalty-based loss function that requires no additional hyper-parameters. It aims at satisfying the metric learning goals while preserving the classification performance. Unlike previous methods that focused on the angle between the embedding vectors z_i and the weights w_j, we focused on the Euclidean distance between these vectors. We denote it by $e_{i,j}(.,.)$ and we formulate it as follow:

$$e_{i,j}(z_i, w_j) = \|z_i - w_j\|_2 \tag{2}$$

Without normalizing w_j and z_i, the Euclidean norm (L_2-norm) is adaptively learned for minimizing the overall loss. So the features are learned in the Euclidean space, a widely recognized state-of-the-art choice.

To imply penalization that enhances the Euclidean margin, we interpret the distance e_j by a Student t-distribution. We fix the degree of freedom to 1, which is the same as a Cauchy heavy-tailed distribution in the low-dimensional map manifold $\mathcal{Z}$. We denote this distance probability by $p_{i,j}(.)$ and we formulate it as follow:

$$p_{i,j}(e_{i,j}) = (1 + e_{i,j})^{-1} \tag{3}$$

Following [29], we use the Student t-distribution with a single degree of freedom because it has the particularly nice property of approximating an inverse square law for large pairwise distances $e_j(w_j, z_i)$ in the $\mathcal{Z}$ space. This makes the map's representation of joint probabilities almost invariant to changes in the map's scale for embedded samples z_i that are far a par (e.g., noisy samples).

In contrast to [8–10] who use the softmax function to produce the probabilistic affinity score, we directly normalize the computed probabilities $p_{i,j}$. This is justified by the fact that our target logit is based on a probability interpretation. It is also justified because of the incompatibility of losses based on the Euclidean margin with the Softmax loss as reported in [8,24]. We formulate the probability normalization as follow:

$$\hat{y}^l_{i,j} = \frac{p_{i,j}}{\sum_{j=1}^{k} p_{i,j}} \tag{4}$$

This results in a probability prediction score that we denote by $\hat{y}^l_i = \{\hat{y}^l_{i,1}, ..., \hat{y}^l_{i,k}\}$. Finally, our loss function that we denote by $\mathcal{L}_{APLM}$, is formulated in combination with the cross-entropy loss as follow:

$$\mathcal{L}_{APLM} = -\frac{1}{N}\sum_{i=1}^{N} y^l_i log(\hat{y}^l_i) \tag{5}$$

where y^l_i is the one-hot encoding ground-truth label of the i-th sample x_i. In Sect. 4.4 we show the efficiency of the proposed loss function quantitatively.

4 Experiments

4.1 Data Set

To validate our proposed method on a real-world application, we used **FPHA data set** [7]. It is a publicly available data set for 3D skeleton-based first-person hand activity recognition, which provides 21 3D hand joint sequences. It contains 45 different activity categories distributed on three different scenarios: "kitchen" (24), "office" (13), and "social" (8), which can be partitioned as source/target domains. Table 1 shows the two data configurations that we have adopted for the experiments and which we denote by *Config 1* and *Config 2* respectively.

Table 1. The adopted data configurations based on the FPHA data set [7].

	Config 1			*Config 2*		
	Scenarios	Samples	Classes	Scenarios	Samples	Classes
Source	Kitchen+office	Train: 496 Test: 468 Train+Test: 964	37	Kitchen+social	Train: 431 Test: 412 Train+Test: 843	32
Target	Social	211	8	Office	332	13

4.2 Evaluation Protocols

We adopted two protocols (*P1* and *P2*) to evaluate the clustering of the unlabeled activities of the two target domains "social" and "office" that belong to the two configurations *Config 1* and *Config 2* respectively. We describe the two protocols as follows: In *P1*, we assume that the number of possible categories in the target domains is unknown, and that we seek to find. Consciously, we performed clustering without giving the ground-truth number of clusters k by using the Affinity Propagation clustering [30], which we evaluated using V-measure (Vm) [31] and Adjusted Mutual Information (AMI) [32] metrics. The Vm metric represents the harmonic mean of the homogeneity and the completeness metrics. The homogeneity metric (Vm-h) provides the rate of meaningful detected clusters. In *P2*, we gave the ground-truth number of clusters $k = 8$ and $k = 13$ for the "social" and the "office" target domains respectively, and we used the Agglomerative clustering [33], which we evaluated using the Unsupervised Clustering Accuracy (UCA) [34].

We also evaluated the activity recognition accuracy on the two source domains that belong to the two configurations *Config 1* and *Config 2*.

4.3 Implementation Details

Supervised Learning (Step 1). As we mentioned in Sect. 3.1, we pre-trained the NN architecture proposed by [22]. We kept the same implementation and data normalization settings for all the experiments, except for the following changes: the softmax layer is replaced by our layer that relies on our APML loss function. We used two stacked LSTMs layers of 100 units instead of one layer and reduced the number of training epochs to 500 to gain computation time. At the end of the training, a pre-trained model is obtained that maps the 3D skeleton-based hand activities into a discriminative feature space of dimension 128. Tensorflow framework is used for the implementations.

Clustering of Unlabeled Hand Activities (Step 2). We recover the pre-trained model, discard its last layer, and we use it to project unlabeled samples into the 128D feature space. Next, we perform clustering on these mapped samples adopting the protocols *P1* and *P2*. Following our SCC strategy, steps 1 and 2 are repeated 100 times. Therefore, in the end, we have 100 predicted sets of labels. Scikit-learn framework is used for the clustering implementations.

Consensus Clustering (Step 3). We adopted the approach proposed by [26]. It combines three efficient heuristics to solve the consensus clustering problem. The consensus clustering function takes the 100 predicted sets of labels as an input and outputs the final prediction set of labels that belong to the unlabeled activity samples. We adopted the python implementation proposed by [26].

4.4 State-of-the-Art Comparison

Tables 2 and 3 show the clustering results comparison. We compared the impact of our APML loss function with four state-of-the-art losses. As expected, the softmax loss performs very poorly against metric learning-based losses. We can see that our proposed loss is equivalent to the state-of-the-art. On *Config 1*, we achieve the best Vm score of 71.49% by adopting the protocol *P1* and UCA of 75.82% while adopting the protocol *P2*. On *Config 2*, the best UCA is dropped to 50.60% achieved by using [7] loss function, which is slightly better than our result. This is due to the extreme challenge that *Config 2* represents to the UDA compared to *Config 1*. Indeed, the number of labeled samples is decreased, while the number of unlabeled samples and their classes are increased.

Table 2. Clustering results adopting the protocol *P1*. The clustering is performed on unlabeled activities of the two target domains "social" and "office" that belong to *Config 1* and *Config 2* respectively. The $\hat{k}$ is the number of detected clusters. The AMI, Vm and Vm-h metrics are computed on the final prediction of our SCC strategy.

	Config 1				*Config 2*			
	AMI score	Vm score	Vm-h score	$\hat{k}$	AMI score	Vm score	Vm-h score	$\hat{k}$
Softmax	62.36	61.55	64.41	11	35.16	45.97	43.77	17
Liu et al. [8]	66.14	69.54	77.00	13	38.62	46.39	48.33	17
Wang et al. [9]	66.33	**69.98**	**78.98**	14	**40.07**	**47.99**	**50.46**	18
Deng et al. [10]	**66.37**	69.91	77.83	14	39.63	**47.87**	**50.59**	19
Our	**68.29**	**71.49**	**79.56**	13	**40.16**	46.09	49.97	17

Table 3. UCA results adopting the protocol *P2*. The min, max and mean values are computed on a population of 100 tries. The CC refers to the consensus clustering results based on the 100 tries following our SCC strategy. The clustering is performed on unlabeled activities of the two target domains "social" and "office" that belong to *Config 1* and *Config 2* respectively.

	Config 1				*Config 2*			
	Min UCA (%)	Max UCA (%)	Mean UCA (%)	CC UCA (%)	Min UCA (%)	Max UCA (%)	Mean UCA (%)	CC UCA (%)
Softmax	48.34	79.62	64.22	65.87	29.51	49.69	39.37	45.18
Liu et al. [8]	55.45	76.77	67.26	73.45	34.63	49.09	41.38	**50.60**
Wang et al. [9]	52.60	79.62	67.37	**73.93**	32.53	46.08	39.59	47.59
Deng et al. [10]	54.02	79.62	67.12	71.09	32.83	46.98	39.32	47.28
Our	54.97	88.62	67.83	**75.82**	34.03	47.59	40.37	**49.39**

Table 3 confirms the advantage of our SCC strategy. Indeed, for 100 tries, the final result of the consensus clustering is far from the minimum, above the average, and not so far from the maximum UCA. The SCC strategy allows selecting a good clustering prediction among multiple tries with respect to the unsupervised learning constraint.

Table 4 shows that our proposed loss overcomes all the metric learning-based losses by more than 6% of accuracy, while it is still equivalent to the softmax loss function. These results validate our APML loss function's main contribution, which aims at balancing between learning a clustering-friendly manifold that facilitates the clustering of the embedded unlabeled samples of the target domain and good recognition accuracy for the labeled ones of the source domain.

4.5 Unlabeled Activities Clustering Analysis

Confusion matrices in Fig. 3 show the highly meaningful quality of clustering provided by our method. Indeed, most of the confused activity classes are meaningfully too close for e.g., "give coin" and "give card" classes where the performed action "give" remain the same, while only the objects "card" and "coin" change. We also observe a high confusion between "toast wine" and "pour wine" classes where the action changes, but the social activity context is still meaningfully close. Even for the very challenging *Config 2*, the confusions are still significant, e.g., "read letter" and "take letter".

Table 4. The average recognition accuracy results over 100 tries. The recognition train/test is performed on the data of the two source domains that belong to *Config 1* and *Config 2*.

	Softmax	Liu et al. [8]	Wang et al. [9]	Deng et al. [10]	Our
Config 1	**94.80**	86.67	84.72	82.53	**95.46**
Config 2	**96.24**	90.30	88.88	86.62	**96.51**

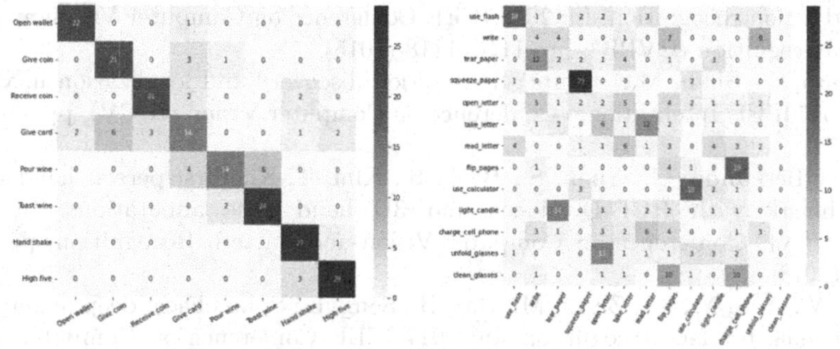

Fig. 3. The left and right matrices refer to the confusions of unlabeled activities clustering of the two target domains "social" and "office". The confusions are computed based on the protocol *P2*.

When adopting the protocol *P1* (k not given), we still get a high homogeneity score Vm-h of 79.56% (Table 2). This means that regrouped samples in each detected clusters are mostly similar, confirming the meaningful clustering quality.

5 Conclusion

We presented an application of UDA for unlabeled 3D skeleton-based hand activity clustering through this work. The experiments, based on a real-world data set, show that the feature space learned using our APML loss function allows meaningful clustering. We also confirmed that, in contrast to the state-of-art metric learning-based losses, our APML loss preserve a good recognition accuracy. We solved the clustering sensitivity problem using our SCC strategy, allowing the best clustering result selection.

As future improvements, we plan to design a NN architecture that provides a clustering-friendly feature space while stabilizing the clustering sensitivity since the SCC strategy can be computationally expensive for large NN architectures and large amounts of data.

References

1. Sridhar, S., Feit, A.M., Theobalt, C., Oulasvirta, A.: Investigating the dexterity of multi-finger input for mid-air text entry. In: CHI 2015 (2015)
2. Ramirez-Amaro, K., Beetz, M., Cheng, G.: Transferring skills to humanoid robots by extracting semantic representations from observations of human activities. Artif. Intell. **247**, 95–118 (2017)
3. Surie, D., Pederson, T., Lagriffoul, F., Janlert, L.-E., Sjölie, D.: Activity recognition using an egocentric perspective of everyday objects. In: Indulska, J., Ma, J., Yang, L.T., Ungerer, T., Cao, J. (eds.) UIC 2007. LNCS, vol. 4611, pp. 246–257. Springer, Heidelberg (2007). https://doi.org/10.1007/978-3-540-73549-6_25
4. Bambach, S.: A survey on recent advances of computer vision algorithms for egocentric video. ArXiv, abs/1501.02825 (2015)
5. Du, Y., Wang, W., Wang, L.: Hierarchical recurrent neural network for skeleton based action recognition. In: 2015 IEEE Conference on Computer Vision and Pattern Recognition (CVPR), pp. 1110–1118 (2015)
6. Soomro, K., Shah, M.: Unsupervised action discovery and localization in videos. In: 2017 IEEE International Conference on Computer Vision (ICCV), pp. 696–705 (2017)
7. Garcia-Hernando, G., Yuan, S., Baek, S., Kim, T.-K.: First-person hand action benchmark with RGB-D videos and 3D hand pose annotations. In: 2018 IEEE/CVF Conference on Computer Vision and Pattern Recognition, pp. 409–419 (2018)
8. Liu, W., Wen, Y., Yu, Z., Li, M., Raj, B., Song, L.: SphereFace: deep hypersphere embedding for face recognition. In: 2017 IEEE Conference on Computer Vision and Pattern Recognition (CVPR), pp. 6738–6746 (2017)
9. Wang, H., et al.: CosFace: large margin cosine loss for deep face recognition. In: 2018 IEEE/CVF Conference on Computer Vision and Pattern Recognition, pp. 5265–5274 (2018)

10. Deng, J., Guo, J., Zafeiriou, S.: ArcFace: additive angular margin loss for deep face recognition. In: 2019 IEEE/CVF Conference on Computer Vision and Pattern Recognition (CVPR), pp. 4685–4694 (2019)
11. Tzeng, E., Hoffman, J., Zhang, N., Saenko, K., Darrell, T.: Deep domain confusion: maximizing for domain invariance. ArXiv, abs/1412.3474 (2014)
12. Saito, K., Ushiku, Y., Harada, T., Saenko, K.: Strong-weak distribution alignment for adaptive object detection. In: 2019 IEEE/CVF Conference on Computer Vision and Pattern Recognition (CVPR), pp. 6949–6958 (2019)
13. Sankaranarayanan, S., Balaji, Y., Jain, A., Lim, S.-N., Chellappa, R.: Learning from synthetic data: addressing domain shift for semantic segmentation. In: 2018 IEEE/CVF Conference on Computer Vision and Pattern Recognition, pp. 3752–3761 (2018)
14. Pan, S.J., Tsang, I.W.-H., Kwok, J.T., Yang, Q.: Domain adaptation via transfer component analysis. IEEE Trans. Neural Netw. **22**, 199–210 (2011)
15. Tzeng, E., Hoffman, J., Saenko, K., Darrell, T.: Adversarial discriminative domain adaptation. In: 2017 IEEE Conference on Computer Vision and Pattern Recognition (CVPR), pp. 2962–2971 (2017)
16. Pinheiro, P.H.O.: Unsupervised domain adaptation with similarity learning. In: 2018 IEEE/CVF Conference on Computer Vision and Pattern Recognition, pp. 8004–8013 (2018)
17. Laradji, I.H., Babanezhad, R.: M-ADDA: unsupervised domain adaptation with deep metric learning. ArXiv, abs/1807.02552 (2020)
18. Long, M., Cao, Y., Wang, J., Jordan, M.I.: Learning transferable features with deep adaptation networks. ArXiv, abs/1502.02791 (2015)
19. Li, L., Wang, M., Ni, B., Wang, H., Yang, J., Zhang, W.: 3D human action representation learning via cross-view consistency pursuit. In: 2021 IEEE/CVF Conference on Computer Vision and Pattern Recognition (CVPR), pp. 4739–4748 (2021)
20. Rao, H., Shihao, X., Xiping, H., Cheng, J., Bin, H.: Augmented skeleton based contrastive action learning with momentum LSTM for unsupervised action recognition. Inf. Sci. **569**, 90–109 (2021)
21. Bhatnagar, B.L., Singh, S., Arora, C., Jawahar, C.V.: Unsupervised learning of deep feature representation for clustering egocentric actions. In: IJCAI (2017)
22. Boutaleb, Y., Soladié, C., Duong, N.-D., Kacete, A., Royan, J., Séguier, R.: Efficient multi-stream temporal learning and post-fusion strategy for 3D skeleton-based hand activity recognition. In: VISIGRAPP (2021)
23. Hadsell, R., Chopra, S., LeCun, Y.: Dimensionality reduction by learning an invariant mapping. In: 2006 IEEE Computer Society Conference on Computer Vision and Pattern Recognition (CVPR 2006), vol. 2, pp. 1735–1742 (2006)
24. Wen, Y., Zhang, K., Li, Z., Qiao, Yu.: A discriminative feature learning approach for deep face recognition. In: Leibe, B., Matas, J., Sebe, N., Welling, M. (eds.) ECCV 2016. LNCS, vol. 9911, pp. 499–515. Springer, Cham (2016). https://doi.org/10.1007/978-3-319-46478-7_31
25. Hu, L.P., Bao, X.L., Wang, Q.: The repetition principle in scientific research. Zhong xi yi jie he xue bao = J. Chin. Integr. Med. **9**(9), 937–940 (2011)
26. Strehl, A., Ghosh, J.: Cluster ensembles – a knowledge reuse framework for combining multiple partitions. J. Mach. Learn. Res. **3**, 583–617 (2002)
27. Chalamalla, A.: A survey on consensus clustering techniques (2010)
28. Pereyra, G., Tucker, G., Chorowski, J., Kaiser, L., Hinton, G.E.: Regularizing neural networks by penalizing confident output distributions. ArXiv, abs/1701.06548 (2017)

29. van der Maaten, L., Hinton, G.E.: Visualizing data using t-SNE. J. Mach. Learn. Res. **9**, 2579–2605 (2008)
30. Frey, B.J., Dueck, D.: Clustering by passing messages between data points. Science **315**, 972–976 (2007)
31. Rosenberg, A., Hirschberg, J.: V-measure: a conditional entropy-based external cluster evaluation measure. In: EMNLP (2007)
32. Romano, S., Bailey, J., Nguyen, X.V., Verspoor, K.M.: Standardized mutual information for clustering comparisons: one step further in adjustment for chance. In: ICML (2014)
33. Zhang, W., Zhao, D., Wang, X.: Agglomerative clustering via maximum incremental path integral. Pattern Recogn. **46**, 3056–3065 (2013)
34. Yang, Y., Dong, X., Nie, F., Yan, S., Zhuang, Y.: Image clustering using local discriminant models and global integration. IEEE Trans. Image Process. **19**, 2761–2773 (2010)

Using Random Forest Distances for Outlier Detection

Antonella Mensi[(✉)], Ferdinando Cicalese, and Manuele Bicego

Department of Computer Science, University of Verona, Verona, Italy
{antonella.mensi,ferdinando.cicalese,manuele.bicego}@univr.it

Abstract. In recent years, a great variety of outlier detectors have been proposed in the literature, many of which are based on pairwise distances or derived concepts. However, in such methods, most of the efforts have been devoted to the outlier detection mechanisms, not paying attention to the distance measure – in most cases the basic Euclidean distance is used. Instead, in the clustering field, data-dependent measures have shown to be very useful, especially those based on Random Forests: actually, Random Forests are partitioners of the space able to naturally encode the relation between two objects. In the outlier detection field, these informative distances have received scarce attention. This manuscript is aimed at filling this gap, studying the suitability of these measures in the identification of outliers. In our scheme, we build an unsupervised Random Forest model, from which we extract pairwise distances; these distances are then input to an outlier detector. In particular, we study the impact of several Random Forest-based distances, including advanced and recent ones, on different outlier detectors. We evaluate thoroughly our methodology on nine benchmark datasets for outlier detection, focusing on different aspects of the pipeline, such as the parametrization of the forest, the type of distance-based outlier detector, and most importantly, the impact of the adopted distance.

Keywords: Outlier detection · Random forest distances · Data-dependent distances

1 Introduction

Outlier detection is the task of finding abnormal objects in a dataset [12]. These abnormal objects, called outliers, are often considered to be *dissimilar* from the remainder of the data, the inliers. In the literature, many approaches are based on this principle, and solve the outlier detection problem by finding those objects that are distant from the rest of the data, according to a predefined distance [6,17,20]. While many efforts have been put on the derivation of the distance-based method, there is a lack of studies focusing on the proper choice of the distance measure: often the Euclidean distance or other geometric distances are used. These distances may not always be the most suitable choice for an accurate identification of outliers: actually, even though very intuitive from a mathematical perspective, their intrinsic nature also hinders some problems [2] which

S. Sclaroff et al. (Eds.): ICIAP 2022, LNCS 13233, pp. 75–86, 2022.
https://doi.org/10.1007/978-3-031-06433-3_7

may lead to an inaccurate representation of the relation between two objects. An exemplar problem of geometric distances is that they do not consider how data are distributed, i.e., the context: the only factor taken into account are the objects themselves. Another issue is linked to high-dimensionality and data sparsity: often, in such cases, several pairs of objects are equally similar, according to a geometric distance, which may be severely inaccurate. Several studies [1,2,4,9,14,19,21–23] have instead tried to propose a more inclusive definition of similarity. The core concept of all these studies is that similarities should be *data-dependent*, i.e., they should take into account also the context. An example is to consider the density of the space where the objects are, either implicitly or explicitly: given two pairs of objects which are equally similar according to the Euclidean distance, if the first pair is surrounded by fewer objects than the latter, then the former pair of objects should have a higher similarity.

Among the different data-dependent similarity measures that have been proposed in literature, a relevant class is represented by *Random Forests (RF) distances* [4,9,19,21,23], i.e., distances which exploit Random Forests [5] flexibility in describing data. More in detail, a tree in a RF contains several binary tests, each one partitioning the data based on the answer to the test. A RF implicitly encodes the relationship between two objects based on how they answer to the tests, i.e., based on which nodes of the tree the two objects traverse. In general, if the paths of two objects in a tree are highly similar it means that they are near in the space of the problem, since they answer in the same way to the different questions encountered along the path. These characteristics make RFs a valid and flexible distance extractor tool, as confirmed by the variety of successful clustering methodologies that compute the clusters starting from the pairwise distances extracted from a RF [4,13,19,21,23].

Since, as previously mentioned, many outlier-based methodologies work with distances to detect outliers, it would be interesting to investigate the use of refined and informative measures, such as RF-distances, and this represents the main goal of this paper. In our approach, we build a RF based on the extreme version of the *Extremely Randomized Tree* (ERT) structure proposed in [10]: this permits to have an unsupervised model, suitable for the outlier detection task. Once trained the forest, pairwise data-dependent dissimilarities between objects are computed and input to an outlier detector that works with distances. Please note that there exist two very preliminary studies that apply this pipeline [18,21], however they have some limitations: first, they both focus only on one outlier detector method; further, they employ only old RF-distance measures, disregarding recent advances in the field [2,4,21] focuses on only one measure. The scope of the present study is instead much wider: we use 6 RF-distances [2,4,19,21,23], including very recent ones [2,4], making a fairly extended and thorough experimental analysis involving many different outlier detection methods. In detail, the evaluation is made on 9 benchmark outlier detection datasets and we test and compare the extracted RF-distances on 7 outlier detectors; results confirm a general robustness of the approach with an improved detection when using more advanced measures.

The paper is organized into four sections, including the current one. In the following section, Sect. 2, we present in detail the proposed methodology, which is divided into three steps, each thoroughly described. In Sect. 3 we make a thorough experimental evaluation, and lastly in Sect. 4 we make some conclusions.

2 Methodology

This section presents the proposed technique, which can be divided into three steps:

1. Train an ERF $\mathcal{F}$ on a vectorial dataset.
2. From the trained $\mathcal{F}$ extract a dissimilarity matrix $\mathbf{D}$. The entry (i, j) of the matrix contains the pairwise distance between the ith and j^{th} object of the dataset.
3. Input $\mathbf{D}$ to any outlier detector that works with distances and then classify the objects of the dataset as inliers and outliers.

In the next three subsections, we describe each step in detail.

2.1 Step 1: Building an ERF

Formally, an Extremely Randomized Forest $\mathcal{F}$ is composed by T Extremely Randomized Trees [10]. This tree structure is characterized by a high degree of randomness in the building procedure: in its extreme version, called *Totally Randomized Trees*, there is no optimization procedure, and the test of each node is defined completely at random. Since in outlier detection we do not have labels, here we adopt this variant: in detail, first we randomly select a feature, and then we randomly pick a cut-point in the domain of the chosen feature. Please note that in [3] several unsupervised learning strategies for RF have been proposed, aimed at extracting distances for clustering purposes: they show that building trees via completely random splits is beneficial, thus supporting our choice of adopting Totally Randomized Trees.

In detail, each ERT t is built independently of the other trees on a subsample of size S of the training set drawn randomly without replacement. The tree building procedure is recursive: at every node the training objects arrived there are split into two groups according to the chosen random test; in detail, one group follows the left branch and the other follows the right branch. We start from the root with all training objects, and end the splitting procedure when a stopping criterion is met. In particular, a node is labelled as leaf if its depth is greater than a pre-established maximum depth D.

2.2 Step 2: Extracting the Distance Matrix D

The second step consists in the computation of the matrix $\mathbf{D}$: given $\mathcal{F}$ we make all pairs of objects in the dataset traverse an ERT t and compute their distance; the procedure is repeated for all trees; finally the tree distances are aggregated at forest level. We compute **D** using six different proposals of RF-distances [1,2,4,19,21,23] which have shown to be successful in the clustering scenario. In the following, we briefly recall the main principles behind each RF-distance:

- **Shi** [19] defines two objects to be similar only if they end up in the same leaf, i.e., they traverse the same path in a tree. Therefore, at tree level, *Shi* is a binary similarity measure, aggregated at forest level via the average.
- **Zhu2** [23] is a generalization of *Shi* and it assumes that also objects that partially share their paths are similar. The degree of similarity is related to the length of the common path, i.e., to the number of nodes that a pair of objects traverse together before they go separate ways: objects which share a greater portion of their paths are more similar.
- **Zhu3** [23] is a weighted version of *Zhu2*, which considers that different nodes in the tree convey different information; in particular, each node is given a weight which is inversely proportional to the number of training objects that reached such node (nodes with few training objects describe the space in a more refined way).
- **RatioRF** [4] implements the Tversky's ratio model of similarity [22] in the context of RF. *RatioRF* computes the similarity between two objects by considering all nodes in the tree that are in the traversed path of at least one of the two objects, and not only nodes that are in both paths. In detail, the similarity between two objects increases if, given an internal node present only in one path, the two objects would traverse the same edge.
- **Ting** [21] implements in the context of RF the m_p*–dissimilarity*, a mass-based distance function proposed by [1]. The rationale is that the similarity between two objects is inversely proportional to the number of training objects contained in the minimum region enclosing both objects. In case of multidimensional data, the minimum enclosing region, and therefore the similarity, is defined independently for each dimension and then results are aggregated via the arithmetic mean. In the context of RF, as defined by [21], the m_p*–dissimilarity* corresponds to the number of training objects contained in the Lowest Common Ancestor (LCA) of the two objects since it represents the last node containing both of them. The aggregation at forest level is computed via the arithmetic mean, as defined by [1].
- **Aryal** [4] implements the mass-based distance function proposed by [2], the m_0*-dissimilarity*. At tree level, it is equivalent to *Ting*, but the aggregation at forest level is computed using the geometric mean. Using the multiplication increases the impact of those trees in which the similarity between two objects is either very high or very low.

2.3 Step 3: Distance-Based Outlier Detection

The extracted distance matrix **D** contains the pairwise distances between each pair of objects in a dataset, to be fed to a distance-based outlier detection methodology. A great variety of methodologies that work with distances have been designed; in the following, we briefly describe the main principles behind those we employed in our proposal:

- **Nearest Neighbor (NN)-based techniques** [20]. This represents a group of very simple yet often well performing methodologies. We employed four different variants: i) **KNNd** evaluates the ratio of the distance between an object and its Kth nearest neighbor to the distance between the latter and its Kth nearest neighbor. If the former distance is much bigger than the latter, then the object under analysis has an increased probability of being an outlier; ii) **KNNDist** is simply the distance to the Kth nearest neighbor: it assumes that an outlier will be distant from its own K^{th} nearest neighbor independently of the density of the latter; iii) **KNNd-Av** is analogous to *KNNd* but instead of considering just the Kth nearest neighbor it takes into account all of them via computing the average; and iv) **KNNDist-Av** is the average distance of the first K^{th} neighbors. The main drawback of all these methods is setting K.
- **Local Outlier Factor (LOF)** [6]. This represents a more complex technique based on the estimation of the relative density, computed using the distances. In particular, *LOF* compares the density of $\mathbf{x}$'s neighborhood to that of each of its neighbors. If there is at least one neighbor which has a much denser neighborhood, then the probability of $\mathbf{x}$ being an outlier increases, since the difference in density is a hint that $\mathbf{x}$ is distributed quite differently from at least one of its neighbors. Analogously to NN-based techniques, there is the drawback of setting the size of the neighborhood K.
- **K-Centers** [20]. This is a clustering-based technique, which, after computing the clusters, calculates the distance of each object to the center of the cluster to which it belongs: if an object is far away from said center then it is more likely to be an outlier. The main problem of *K-Centers* consists of setting an appropriate number of clusters K.
- **ProxIF** [17]. This represents a RF-based methodology for outlier detection that works with pairwise distances, being inspired by the well known Isolation Forest [15]. An object is more likely to be an outlier if, on average, it traverses a shorter path in the forest. Unlike the other detectors, there exists a default parametrization to build the forest which works well in most cases – the parameters to set consists of a training criterion, and S, T and D (the same ones of an ERF).

3 Experimental Evaluation

This section is dedicated to the experimental evaluation of the methodology. In the first subsection, Subsect. 3.1, we describe the datasets and other experimental details. Subsequently, in Subsect. 3.2 we study the impact of the ERF parametrization, whereas in Subsect. 3.3 we assess whether there is a best outlier detector to work with. Lastly, in Subsect. 3.4 we present several analyses concerning the behaviour of different distance measures.

3.1 Experimental Details

We perform the evaluation of the methodology on 9 different benchmark datasets for outlier detection [7,11,15,16], 6 of which are UCI-ML datasets[1] preprocessed according to [11]. Instead, *Cardiotocography, Hepatitis* and *Stamps* were taken from [7] and preprocessed accordingly. In Table 1, we report the number of objects and features and the percentage of outliers for each dataset. We can observe that the datasets cover a large range of cases, differing greatly in dimensionality (from 5 up to 164), in the outlier percentage (from 5.39% up to 45.80%) and in the number of samples (the smallest one having 80 objects whereas the biggest 7200). Many different experiments were performed by varying the value of several parameters:

- Number of trees T: $50, 100, 200$. 3 options.
- Number of samples used to train each tree S: $64, 128, 256$. 3 options.
- Maximum depth D each tree can reach: $\log_2(S), S-1$. 2 options.
- Distance measure: *Shi, Zhu2, Zhu3, RF-Ratio, Aryal, Ting*. 6 options.
- Outlier Detector: *KNNd*, *KNNDist*, *KNNd-Av*, *KNNDist-Av*, *K-Centers*, *LOF*, *ProxIF*. 7 options.

Each experiment, i.e., parametrization setting, was iterated 10 times. The iterations have been created by splitting randomly the dataset in half, one half for the training set and the other for the testing set, with the only constraint related to the absence of outliers in the former. In addition, given a dataset, the 10 partitions of training and testing sets are identical across all parametrizations. As usually done in the outlier detection field, all experiments have been evaluated by measuring the performance in terms of Area Under the ROC Curve (AUC). Further, for all statistical analyses, we set the significance level to $\alpha = 0.05$.

3.2 ERF Parametrization

The first analysis aims at finding the best ERF parametrization for each distance measure, to be used in the subsequent analyses. Indeed, different distance measures may benefit from a different parametrization setting: for example, as to the depth D, whereas some distances may suffer, others may benefit from the additional information contained in deeper trees. We made three different analyses, one for each parameter (T, S, and D): however, due to a lack of space, we report here only the analysis concerning the maximum reachable depth D, being the most interesting one, whereas for S and T we simply summarize the obtained results at the end of the section. For what concerns the analysis of the depth, given a value of D and a distance measure, we compute the average AUC for each dataset and for every parametrization (obtained by varying S and T) across the related iterations – we used as outlier detector only *NNd* (analogous to *KNNd* with $K = 1$), one of the simplest techniques for which no parameter must be set. Then, for each distance measure, we perform a Wilcoxon signed-rank test to compare the two sets of results for the two values of D. We depict

[1] Available at https://archive.ics.uci.edu/ml/index.php.

Table 1. Overview of the 9 datasets used for the experimental evaluation.

Datasets	Nr. of objects	Nr. of features	Outlier %
Annthyroid	7200	6	7.42%
Arrhythmia	452	164	45.80%
Cardiotocography	2126	21	22.15%
Hepatitis	80	19	16.25%
Ionosphere	351	32	35.90%
Pima	768	8	34.90%
Spambase	4601	57	39.40%
Stamps	340	9	9.12%
Wilt	4839	5	5.39%

Table 2. Statistical analysis for D.

Distance	Rank		p-value
	$D = \log_2(S)$	$D = S - 1$	
Shi	1.28	1.72	**1.7e−5**
Zhu2	1.21	1.79	**2.76e−10**
Zhu3	1.26	1.74	**1.36e−7**
Ting	1.62	1.38	**0.014**
RatioRF	1.52	1.48	0.701
Aryal	1.63	1.37	**6.04e−4**

the results of such analysis in Table 2: we report for each distance measure the mean rank for each value of D and the p-value output by the test. We highlight in **bold** the p-value if the difference between $D = \log_2(S)$ and $D = S - 1$ is statistically significant. From the table different observations can be derived: first, we can infer that it is significantly better to extract *Shi, Zhu2* and *Zhu3* from trees built until $D = \log_2(S)$ rather than from trees built to the end. Indeed, the probability of two objects ending up in the same leaf (and in general of having a longer common path) decreases as D increases. In other words, if we extract *Shi* from a tree where $D = S - 1$ we may obtain a sparse similarity matrix and probably less discriminative in the detection of outliers[2]. A second observation we can infer from Table 2 is that *RatioRF* is independent of the used D, i.e., it is very robust independently of the tree structure; in the subsequent analyses we therefore set $D = \log_2(S)$ for *RatioRF* – smaller trees are less expensive from a computational point of view. As to *Ting* and *Aryal*, the two mass-based RF-distances, we observe an opposite behaviour with respect to *Shi, Zhu2* and

[2] Please note that we can extend this reasoning to *Zhu2* and *Zhu3*: the resulting matrix may not be sparse, but it may contain many low similar values, thus impacting on the final outlier detection step.

Zhu3: $D = S - 1$ is ranked higher than $D = \log_2(S)$ and the difference is statistically significant. Indeed, it is likely that when growing trees to their maximum depth, we have an increased variability in these kinds of distance matrices. In other words, in trees where $D = \log_2(S)$ there may be several pairs of objects ending up in the same LCA and therefore getting the same distance value even though their *true* distance may be different.

In conclusion, we set $D = \log_2(S)$ for all RF-distances except *Aryal* and *Ting*, for which we set $D = S - 1$. As to the other parameters of an ERF, S and T, from the related analyses, we discovered a common behaviour among the different distances: in all cases the most suitable choice is to set $S = 64$ and $T = 50$.

3.3 Comparison of Outlier Detectors

In this section we assess whether, independently of the used RF-distance measure, there is a best outlier detector to use. First, after a thorough preliminary analysis not shown here, we set the parameters of each outlier detector as follows: *KNNd*, $K = 8$; *KNNDist*, $K = 4$; *KNNd-Av*, $K = 16$; *KNNDist-Av*, $K = 8$; *K-Centers*, $K = 5$; *LOF*, $K = 9$. As for *ProxIF* we set the parameters according to the guidelines presented in [17]: $T = 200, S = 256, D = \log_2(S)$ and $O-2PS_D$ as training criterion. Then we compute for each outlier detector the AUC values for each dataset and distance measure, averaging them across the 10 iterations. Subsequently, we make a non-parametric statistical analysis to compare the seven outlier detectors. In detail, we carried out a Friedman test followed by a post-hoc Nemenyi test: the former is needed to uncover whether there is a global statistically significant difference among all models, whereas the latter is used to discover the pairs of statistically different outlier detectors. In Fig. 1 the results of the statistical analysis are visualized as a critical difference (CD) diagram [8]: each outlier detector is represented via its rank on a single line, with the best rank represented on the right. Whenever two (or more) models are comparable, i.e., there is no significant difference, they are connected by a line. From Fig. 1 we can observe that the first-ranked outlier detector is *KNNDist-Av*; nevertheless, its performances are comparable to other three detectors in the ranking, confirming the robustness of the methodology across several outlier detectors based on different core concepts.

3.4 Comparison of Distance Measures

In this section we make three analyses: we compare the distances on the best outlier detector, then we try to infer whether some RF-distances are more suitable than others, and lastly we assess whether there is a best combination of distance measure and outlier detector.

The first analysis compares the performances of the 6 distance measures when using *KNNDist-Av*, the best outlier detector for this task according to the analysis made in the previous section. In Table 3, given a dataset and a distance measure, we report the average AUC across the iterations and indicate

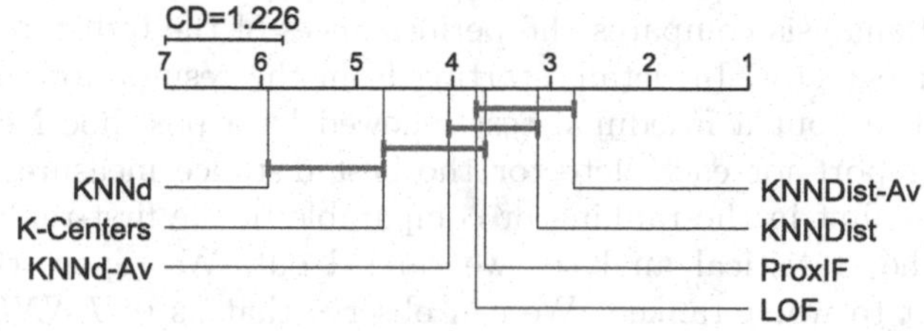

Fig. 1. Comparing outlier detectors independently of the distance measure via a CD diagram.

Table 3. Evaluation of the 6 RF-distances using KNNDist-Av.

Dataset	Shi	Zhu2	Zhu3	RatioRF	Ting	Aryal
Annthyroid	0.8560	0.8861	0.8647	0.8854	0.9044	**0.9147**
Arrhythmia	0.8131	0.8137	0.8129	**0.8210**	0.8179	0.7880
Cardiotocography	0.4564	0.4716	0.4654	**0.4935**	0.4116	0.4421
Hepatitis	0.6815	0.6989	0.6758	0.6906	0.7271	**0.7863**
Ionosphere	0.9492	0.9521	0.9466	0.9415	0.9701	**0.9786**
Pima	0.7189	0.7279	0.7184	0.7255	0.7389	**0.7538**
Spambase	0.8550	0.8589	0.8517	0.8580	0.8690	**0.8800**
Stamps	0.9236	0.9362	0.9310	**0.9395**	0.8525	0.7634
Wilt	0.7131	0.6735	0.6805	0.6827	0.7928	**0.8296**

Table 4. Best RF-distances for each outlier detector.

Outlier detector	Best gistances			
KNNDist-Av	Aryal	Ting	RatioRF	
KNNd	RatioRF	Aryal	Ting	Zhu2
KNNDist	Aryal	Zhu2	RatioRF	Ting
KNNd-Av	RatioRF	Zhu2	Aryal	Ting
K-Centers	RatioRF	Zhu2	Shi	Zhu3
LOF	RatioRF	Aryal	Zhu2	
ProxIF	Aryal	Ting		

in **bold** the best result for each dataset. We can observe that all RF-distances seem to have good and consistent performances across all datasets. The only exceptions are: *Wilt* for which using a non-mass-based distance measure leads to much poorer performances and *Stamps* for which the converse holds, i.e., using either *Aryal* or *Ting* relevantly decreases the performances. From Table 3 we can also observe that for each dataset, the best distance measure is either *Aryal* or *RatioRF*.

The second analysis compares the performances of the 6 distance measures on each outlier detector. In detail, starting from the results obtained on each dataset, we carried out a Friedman test followed by a post-hoc Nemenyi test. In Table 4 we report for each detector the best distance measures, i.e., those that were either first in the ranking or comparable to the first-ranked measure according to the statistical analyses we carried out. We report them left to right, from best to worst ranked. We can observe that as to *KNNDist-Av*, the best outlier detector for this task, *Aryal* is the best choice, even though it is comparable to both *RatioRF* and *Ting*. Overall, we can observe that *Aryal* and *RatioRF* are the only distance measures that are either the first-ranked or comparable to such measure for six out of seven outlier detectors. From Table 4 we can also conclude that using *Shi* and *Zhu3* is overall a bad choice.

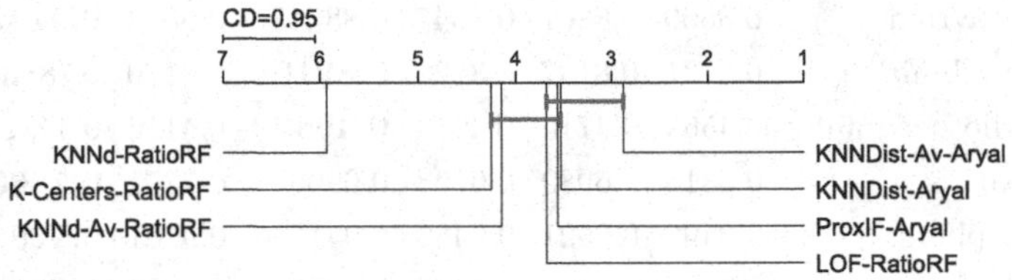

Fig. 2. CD diagram that compares each classifier combined with its best distance measure.

The last analysis consists of comparing the first-ranked distances for each outlier detector via the Friedman and Nemenyi tests, which results are depicted in the CD diagram in Fig. 2. The analysis confirms that *KNNDist-Av* seems to be the best outlier detector even though it is comparable to *KNNDist*, *LOF* and *ProxIF*, analogously to the observation made in the previous section. Three of the best ranked detectors work with the mass-based RF-distance *Aryal* confirming its suitability in detecting outliers.

Summarizing the observations made in this section, there are two distance measures which seem to be the most suitable for outlier detection: **Aryal** and **RatioRF**. Nevertheless, it is difficult to establish which between the two is the best one, since their performance also depends on the dataset and on the outlier detector.

4 Conclusions

In this paper we presented a thorough study on using RF-distances, i.e., informative data-dependent measures extracted from a RF model, to detect outliers. Indeed, even though there exist several outlier detectors based on distances, there is a lack of research on distance measures. Our manuscript, with respect to previous works, thoroughly studies the impact of several RF-distances, including quite recent and refined ones, on the identification of outliers using different outlier detectors.

The suitability of this contribution has been tested on a total of 9 datasets. It is shown how more refined distance measures, such as Aryal and RatioRF, are overall a better choice than more simplistic ones, such as Shi. Further, even though the best results are reached with a rather simple NN-based outlier detector, the methodology has shown to be robust across several classifiers based on different principles.

References

1. Aryal, S., Ting, K.M., Washio, T., Haffari, G.: Data-dependent dissimilarity measure: an effective alternative to geometric distance measures. Knowl. Inf. Syst. **53**(2), 479–506 (2017). https://doi.org/10.1007/s10115-017-1046-0
2. Aryal, S., Ting, K.M., Washio, T., Haffari, G.: A comparative study of data-dependent approaches without learning in measuring similarities of data objects. Data Min. Knowl. Discov. **34**(1), 124–162 (2019). https://doi.org/10.1007/s10618-019-00660-0
3. Bicego, M., Escolano, F.: On learning random forests for random forest clustering. In: Proceedings of the International Conference on Pattern Recognition, pp. 3451–3458 (2020). https://doi.org/10.1109/ICPR48806.2021.9412014
4. Bicego, M., Cicalese, F., Mensi, A.: RatioRF: a novel measure for random forest clustering based on the Tversky's ratio model. IEEE Trans. Knowl. Data Eng. 1 (2021). https://doi.org/10.1109/TKDE.2021.3086147
5. Breiman, L.: Random forests. Mach. Learn. **45**, 5–32 (2001). https://doi.org/10.1023/A:1010933404324
6. Breunig, M.M., Kriegel, H.P., Ng, R.T., Sander, J.: LOF: identifying density-based local outliers. In: Proceedings of the 2000 ACM SIGMOD International Conference on Management of Data, pp. 93–104 (2000). https://doi.org/10.1145/335191.335388
7. Campos, G.O., et al.: On the evaluation of unsupervised outlier detection: measures, datasets, and an empirical study. Data Min. Knowl. Discov. **30**(4), 891–927 (2016). https://doi.org/10.1007/s10618-015-0444-8
8. Demšar, J.: Statistical comparisons of classifiers over multiple data sets. J. Mach. Learn. Res. **7**, 1–30 (2006)
9. Englund, C., Verikas, A.: A novel approach to estimate proximity in a random forest: an exploratory study. Expert Syst. Appl. **39**(17), 13046–13050 (2012). https://doi.org/10.1016/j.eswa.2012.05.094
10. Geurts, P., Ernst, D., Wehenkel, L.: Extremely randomized trees. Mach. Learn. **63**(1), 3–42 (2006). https://doi.org/10.1007/s10994-006-6226-1
11. Goix, N., Drougard, N., Brault, R., Chiapino, M.: One class splitting criteria for random forests. In: Zhang, M.L., Noh, Y.K. (eds.) Proceedings of 9th Asian Conference on Machine Learning. Proceedings of Machine Learning Research, vol. 77, pp. 343–358 (2017)
12. Hawkins, D.M.: Identification of Outliers, vol. 11. Springer, Dordrecht (1980). https://doi.org/10.1007/978-94-015-3994-4
13. Kaufman, L., Rousseeuw, P.J.: Finding Groups in Data: An Introduction to Cluster Analysis, vol. 344. Wiley, Hoboken (2009). https://doi.org/10.1002/9780470316801
14. Lin, D.: An information-theoretic definition of similarity. In: Proceedings of the International Conference on Machine Learning, vol. 98, pp. 296–304 (1998)

15. Liu, F.T., Ting, K.M., Zhou, Z.H.: Isolation forest. In: IEEE International Conference on Data Mining, pp. 413–422 (2008). https://doi.org/10.1109/ICDM.2008.17
16. Liu, F.T., Ting, K.M., Zhou, Z.H.: Isolation-based anomaly detection. ACM Trans. Knowl. Discov. Data **6**(1), 3:1–3:39 (2012). https://doi.org/10.1145/2133360.2133363
17. Mensi, A., Bicego, M., Tax, D.M.: Proximity isolation forests. In: 2020 25th International Conference on Pattern Recognition (ICPR), pp. 8021–8028. IEEE (2021). https://doi.org/10.1109/ICPR48806.2021.9412322
18. Mensi, A., Franzoni, A., Tax, D.M.J., Bicego, M.: An alternative exploitation of isolation forests for outlier detection. In: Torsello, A., Rossi, L., Pelillo, M., Biggio, B., Robles-Kelly, A. (eds.) S+SSPR 2021. LNCS, vol. 12644, pp. 34–44. Springer, Cham (2021). https://doi.org/10.1007/978-3-030-73973-7_4
19. Shi, T., Horvath, S.: Unsupervised learning with random forest predictors. J. Comput. Graph. Stat. **15** (2005). https://doi.org/10.1002/sam.11498
20. Tax, D.: One-class classification; concept-learning in the absence of counter-examples. Ph.D. thesis, Delft University of Technology (2001)
21. Ting, K., Zhu, Y., Carman, M., Zhu, Y., Zhou, Z.H.: Overcoming key weaknesses of distance-based neighbourhood methods using a data dependent dissimilarity measure. In: Proceedings of the International Conference on Knowledge Discovery and Data Mining, pp. 1205–1214 (2016). https://doi.org/10.1145/2939672.2939779
22. Tversky, A.: Features of similarity. Psychol. Rev. **84**(4), 327 (1977). https://doi.org/10.1037/0033-295X.84.4.327
23. Zhu, X., Loy, C., Gong, S.: Constructing robust affinity graphs for spectral clustering. In: Proceedings of the International Conference on Computer Vision and Pattern Recognition, pp. 1450–1457 (2014). https://doi.org/10.1109/CVPR.2014.188

Case Study on the Use of the SafeML Approach in Training Autonomous Driving Vehicles

Matthias Bergler[1(✉)], Ramin Tavakoli Kolagari[1], and Kristina Lundqvist[2]

[1] Technische Hochschule Nürnberg, Keßlerplatz 12, 90489 Nürnberg, Germany
matthias.bergler@th-nuernberg.de

[2] Mälardalen University Sweden, Kögskoleplan 1, 722 20 Västeras, Sweden

Abstract. The development quality for the control software for autonomous vehicles is rapidly progressing, so that the control units in the field generally perform very reliably. Nevertheless, fatal misjudgments occasionally occur putting people at risk: such as the recent accident in which a Tesla vehicle in Autopilot mode rammed a police vehicle. Since the object recognition software which is a part of the control software is based on machine learning (ML) algorithms at its core, one can distinguish a training phase from a deployment phase of the software. In this paper we investigate to what extent the deployment phase has an impact on the robustness and reliability of the software; because just as traditional, software based on ML degrades with time. A widely known effect is the so-called concept drift: in this case, one finds that the deployment conditions in the field have changed and the software, based on the outdated training data, no longer responds adequately to the current field situation. In a previous research paper, we developed the SafeML approach with colleagues from the University of Hull, where datasets are compared for their statistical distance measures. In doing so, we detected that for simple, benchmark data, the statistical distance correlates with the classification accuracy in the field. The contribution of this paper is to analyze the applicability of the SafeML approach to complex, multidimensional data used in autonomous driving. In our analysis, we found that the SafeML approach can be used for this data as well. In practice, this would mean that a vehicle could constantly check itself and detect concept drift situation early.

Keywords: Automotive · Safety · SafeML · Machine learning · Autonomous driving

1 Introduction

In movie and television, especially in the science fiction genre, people's futuristic dreams, desires and fears become reality. But these stories do not only provide

Supported by organization MDU.

S. Sclaroff et al. (Eds.): ICIAP 2022, LNCS 13233, pp. 87–97, 2022.
https://doi.org/10.1007/978-3-031-06433-3_8

entertainment, but they also provide research incentives for science. Devices such as smartphones or medical devices such as CT or MRI are based on devices from the Star Trek or other Sci-Fi series [8]. Many of these stories also show futuristic cities with self-driving modes of transport in which passengers can pursue whatever activities they want. More and more vehicle manufacturers, such as Tesla, BMW, Mercedes etc., are therefore also dealing with the topic of autonomous driving. Tesla in particular is known for its advanced technology. Some successes and failures have already been recorded in the media. Many of the results have shown that autonomous vehicles of the highest level, i.e., completely without a driver, are possible, and semi-autonomous vehicles are already being tested in traffic. To make this possible, data from several cameras and LIDAR radar systems are used to observe the vehicle environment. Internal processing units evaluate these images of the surroundings in the form of images and sensor data and use pre-trained ML algorithms to calculate the necessary actions that have to be taken in order to get through the traffic accident-free. Unfortunately, these algorithms are not yet 100 % reliable and the tests in traffic repeatedly lead to malfunctions or failures. For example, obstacles in the form of a truck standing sideways were overlooked [16], or, as recently (as of April 2021 [4]), the vehicle started by itself without a driver. These incidents illustrate how big the technological leap to a perfectly autonomous vehicle is. These technical deficiencies are partly software-related and partly hardware-related and must therefore be viewed differently in terms of their robustness. But not only the robustness plays a major role, but also the security against the manipulation of such systems in order to consciously cause accidents or, for example, people kidnapping [5,6]. Examples worldwide show how control over vehicles can be taken over by the simplest means, for example by manipulating the infotainment system [11]. In combination with an autonomous vehicle, this can lead to devastating results. In this paper we analyse whether the Aslansefat et al. [1] approach based on distance metrics is still practicable even with a complex data set from the field of autonomous driving and is suitable for detecting a concept drift at an early stage.

2 Robustness and Safety of Machine Learning Algorithms in Autonomous Driving

The robustness of the functionality of an ML algorithm is essential for autonomous driving. Since there is no longer a driver at the highest level of autonomy, the algorithm must work reliably and without errors. But even with partial automation, it is important to have a robust algorithm so that the driver can be warned in a acceptable timeframe as soon as he has to take the steering wheel again. Since the systems used up to now are trained offline, it is not possible to relearn the algorithms based on the situation. It can very well happen that an algorithm makes the wrong decisions out of uncertainty. When analyzing the robustness of systems based on machine learning processes, the phenomenon of "concept drift" has increasingly occurred, especially in systems that have been in the field for a long time without retraining. This occurs increasingly when the input data differ too much from the original training data over time (see Fig. 1). This mostly happens due to unforeseen circumstances and cannot be prevented as seen in "An Overview of Concept Drift Applications" by Žliobaitė I. et al. [13]. An example of this would be if an algorithm for autonomous driving was only trained with data from city traffic but is used without additional training in more rural areas and must now recognize not only people, but also animals and fallen trees as obstacles. An additional example of this would be estimates and analyzes of customer buying behavior, which can change unexpectedly because of an economic crisis. Therefore, in our research we are concerned with how this phenomenon can be recognized at an early stage and warned about it. To achieve this, statistical distance measurements based on probability density functions and the characterization of data sets are dealt with in more detail. The literature shows that there is a connection between the classification results of neural networks and probability distributions of data sets as shown in Aslansefat K. et al. in "SafeML: Safety Monitoring of Machine Learning Classifiers through Statistical Difference Measure" [1], in contrast to other methods, such as support vector machines or density estimates as seen in "Detecting Concept Drift with Support Vector Machines" by Klinkenberg R. and Joachims T. [9] or "Adaptive concept drift detection" by Dries A. and Rückert U. [3].

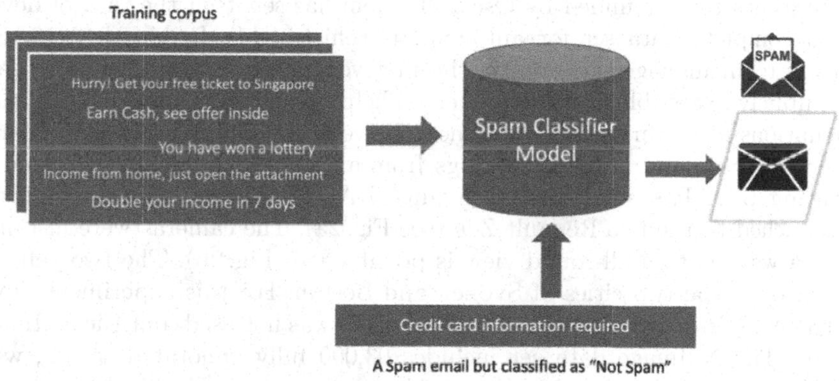

Fig. 1. A concept drift has the consequence that the model trained on historical data incorrectly interprets and processes new input data. For example, a spam filter is tricked out with new mail subjects [2].

3 SafeML Approach

When researching the early detection of a concept drift, we encountered so-called SafeML Approach of Aslansefat et al. [1]. Based on the preliminary work by Aslansefat et al. we have recreated the SafeML method. For this we started with the training phase and pre-trained our Alexnet-based CNN (see Sect. 5.2) with the help of a training and validation data set. Since we want to check the correlation between the accuracy and the distance measurements on more complex image data sets and not just data sets designed for benchmarks, we decided on the NuImages data set from the NuScenes group (see Sect. 4). This data set was specially developed for training autonomous vehicles. After the training, we determined and saved the results of the probability density function (PDF) as well as the statistical parameters of the Empirical Cumulative Distance Functions (ECDF) for each class of the training data set. In the application phase, we then classified our generated test data set using our model and compared the accuracy with the result from the training. Here, too, we determine the PDF and the statistical parameters based on the ECDFs of each class. The results from both phases are then used to determine the statistical distance. For this we used the algorithms recommended from the original paper to measure the statistical distance (see Sect. 5.3). If the statistical distance turns out to be small, the classification results are to be classified as reliable and the system can theoretically continue to work autonomously. Since we have our data labeled, we can also check whether the correlation between the accuracy of the classifier and the statistical distance is still guaranteed even with more complex data sets.

4 NuImage Data Set

The NuScenes team founded by Oscar Beijbom has set itself the task of developing a complete data set for autonomous vehicles [14]. In March 2019, the 26-strong team managed to publish the first version of the data set. The data set is publicly accessible and is intended to help researchers cope with challenging situations in urban areas for autonomous vehicles. The data set uses a large number of sensor and image recordings from an autonomously driving vehicle. A rotating LIDAR system, five long-range RADAR systems and six cameras were attached to the two Renault Zoe (see Fig. 2a). The cameras were installed in such a way that an all-round view is possible (see Fig. 2b). The two vehicles drove through the two cities of Sydney and Boston. For this experiment, however, the NuScenes data set with its 1,000 scenes was not used, but the NuImage data set. The NuImage data set includes 93,000 fully annotated scenes, with over 800,000 two-dimensional foreground annotations (see Fig. 3). There are 23 classes (see full list at [14]) that include both pedestrians and vehicles, as well as roadblocks. In addition, the annotations were supplemented by further metadata. These include information such as whether a driver is sitting in the vehicle or parking, or whether a passer-by is currently moving or standing waiting at a bus stop.

5 Experimental Setup

To carry out the experiment, we first needed to organize the data. Therefore the data set had to be loaded and the required annotations in the scene had to be detected and cut out. The tiles thus obtained were then re-saved to folders with their respective class names. The recordings obtained in this way were then

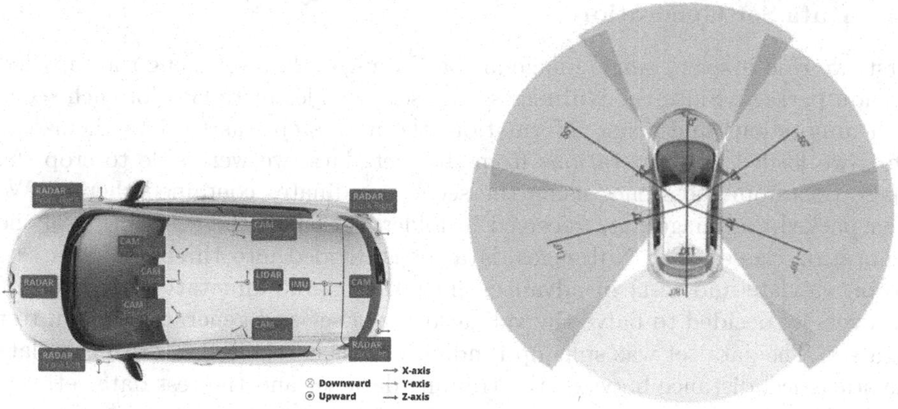

(a) This figure shows the setup of the sensors and cameras of the cars.

(b) By attaching the cameras, an all-round view is possible.

Fig. 2. Technical structure of the recording vehicles.

Fig. 3. Example image of a NuImages scene with annotated objects.

divided into three subsets for training a classifier. Half of the tiles were combined into a training data set and the other half again divided into a validation and training data set. With the help of the received training and validation subsets, a CNN based on Alexnet [12] was then trained for classification. The model obtained in this way is then used to classify the test data set. If the classification of the test data set shows sufficient accuracy, the distances between the training and the test data set are calculated using the distance metrics. If the classification accuracy turns out to be too low, the hyperparameters were optimized so that a higher accuracy could be achieved. A high performance computer was used to conduct the experiments. This is with two Intel (R) Xeon (R) Silver 4114 (2.20 GHz, 2,195 MHz, 10 cores) processors, 192 GB DDR4-2666 rg ECC memory, a SAMSUNG SSD MZVLB512HAJQ-00000 500 GB, as well as 2 NVIDIA Quadro RTX 5000 (3,072 CUDA processing units, 384 Tensor processing units).

5.1 Data Set Generation

First, we read in every single image in the NuImages data set using the supplied Python package. Since the NuImages data set provides metadata for each scene with annotations and other information, the next step was to read them out. After we loaded the annotations from the metadata, we were able to crop the image tiles from the scenes using the scene coordinates contained therein. We then packed the image data received in folders based on the description of the annotation classes. The NuImages data set is divided into three sub-data sets (train, validate and test) in advance. Since there is no annotation for the test data set, we decided to halve the validation data set and generate our own test data set. The data set was split up randomly. This is required later to calculate the statistical distance between the training data set and the test data set.

5.2 Alexnet with Spartial Pyramid Pooling Layer

To classify the image details obtained from the scenes, we first used the classic AlexNet by Krizhevsky et al. [12] used. However, the classic version of AlexNet only allows a uniform tile size of 227 × 227 pixels. This is because the fully connected layer is at the end of the network. Since the annotated tiles from the scenes have different dimensions and we wanted to keep the original resolution of the images in order not to cause scaling problems, we decided to use a spatial pyramid pooling layer as in He et al. [7] to be inserted after the last convolutional layer. This additional layer enables us to use image tiles in different dimensions instead of the 227 × 277 dimensioned images used in the paper. The spatial pyramid pooling layer generates a fixed output size (see Fig. 4) at the end of the convolutional layer, which can then be processed by the fully connected layers. However, this step only allows a batch size of 1, as the individual image tiles cannot be grouped together uniformly. This is done at the expense of running time.

5.3 Distance Measurements

After our model had achieved a sufficiently high classification rate of 99.7%, the statistical distance between the respective classes in the training and test data set was then calculated. For this purpose we have the original by Aslansefat, K. et al. recommended calculation metrics (see Fig. 5). Here we have mainly focused on the empirical cumulative distribution functions. These include:

- **Kolmogorov-Smirnov Distance:** Calculation of the maximum distance between the two ECDFs of the compared classes.
- **Kuiper Distance:** calculation and addition of the two maximum distances of the compared classes.
- **Cramer-Von Mises Distances:** Calculation and summation of the difference between several points in an interval between two ECDFs.

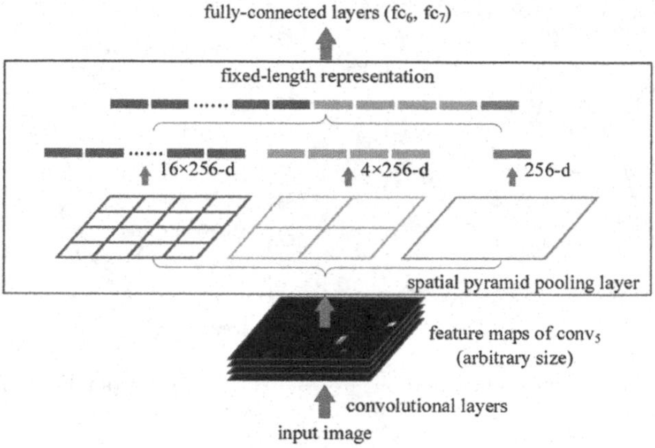

Fig. 4. Example of a spatial pyramid pooling layer. 256 describes the number of filters in the last convolutional layer conv5 [7].

- **Anderson-Darling Distance:** Calculation as for Cramer-Von Misses, but the individual differences are normalized beforehand with the standard deviation.
- **Wasserstein Distance:** Calculation of the area between two ECDFs.

This calculations were made for each class.

6 Results

In this section we present the results obtained after our application phase. The respective classification results of the application phase are listed in Table 1. The table shows the class-specific accuracy that was obtained during the application phase. The classes were taken from [14] and mapped to the numbers 1–23 in the order listed. For better visualization we decided not to include the class based results for the statistical distance measure, but visualize the trend in a separate Fig. 6. This representation allows us to better visualize the relationship between the classification accuracy and the statistical distance. In Fig. 6 we see the associated correlation with the associated distance measures for each class and for the specific distance measurement.

7 Discussion

The results of our study have shown that there is indeed a correlation between the classification accuracy and the calculated statistical distance based on ECDFs (see Fig. 6). It is particularly noticeable that a similar relationship can be determined for the Anderson-Darling (see Fig. 6e) and Wasserstein distance (see Fig. 6d), as well as for the Kuiper (see Fig. 6b) and Kolmogorov-Smirnov

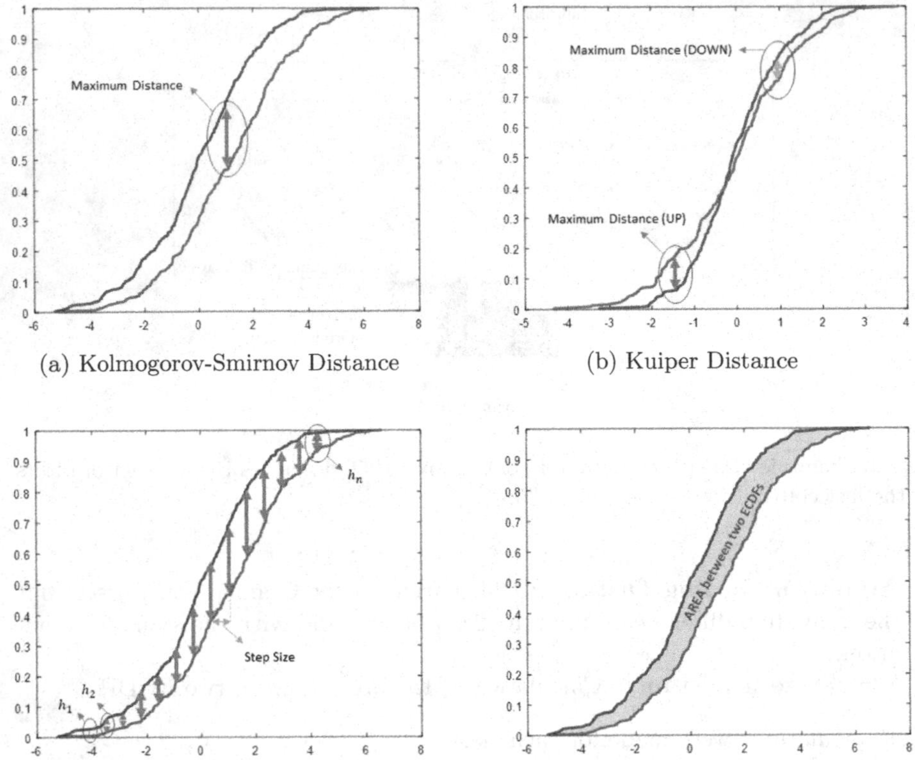

(a) Kolmogorov-Smirnov Distance (b) Kuiper Distance

(c) Cramer-Von Mises Distance (d) Wasserstein Distance

Fig. 5. Illustrations of the 5 ECDFs used. The Anderson-Darling Distance was not shown because it is illustrated identically to the Cramer-Von Mises Distance and only differs due to the normalization [10].

distance (see Fig. 6a) between the statistical distance and the classification accuracy. Class 9 is also noticeable in the tables, as it drifts far from the other classification results and distance measurements. Since this class is the "movable_object.barrier" class, we took a closer look at it. It is noticeable here that the tiles contained therein contain different types of a barrier and therefore a high degree of variance in the image content. This also explains the lower classification result and the higher statistical distance. Here one could consider dividing the class again, since a logical assignment for humans is not necessarily logical for ML algorithms. In all five distance measurements, it can be seen that the statistical distance between the training and test data set is greater for each class, even if the accuracy of the classification is better. Thus, the initial thesis that there is a correlation between the classification accuracy and the statistical distance between two data sets could also be demonstrated on a complex data set. When training the classifier, we made a conscious decision to modify AlexNet using a spatial pyramid pooling layer in order to retain the natural scaling of

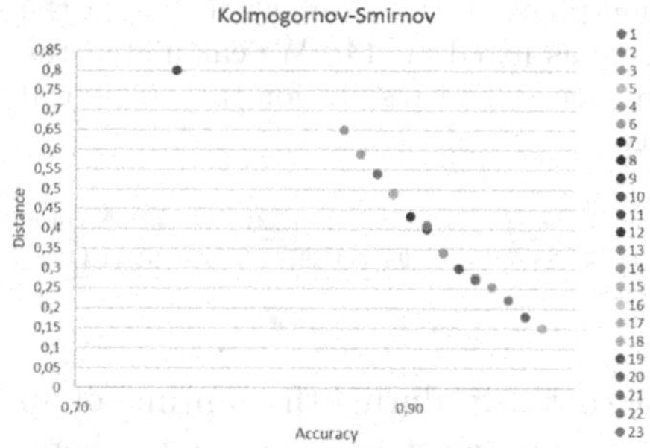

(a) Kolmogorov-Smirnov Results

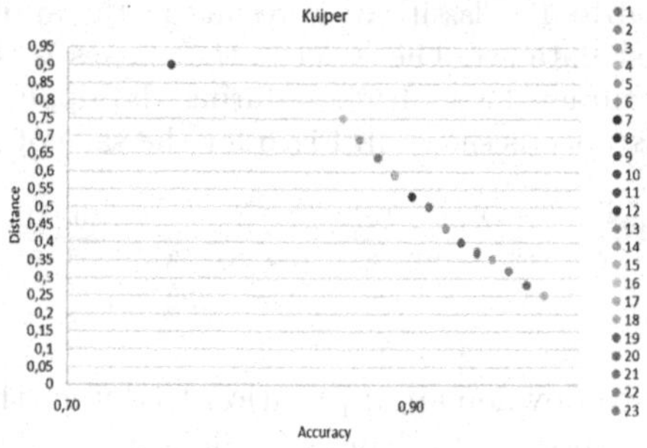

(b) Kuiper Results

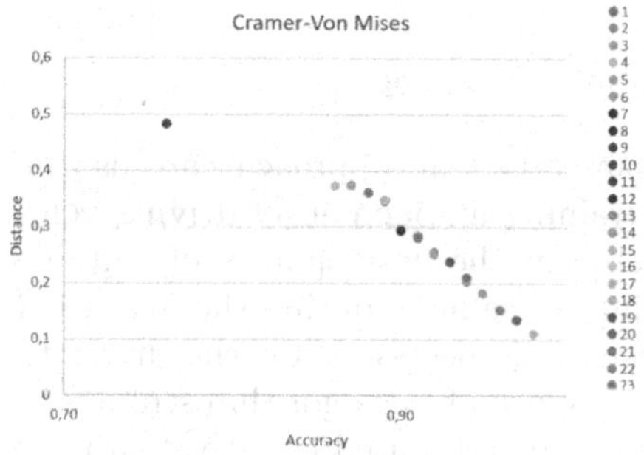

(c) Cramer-Von Mises Results

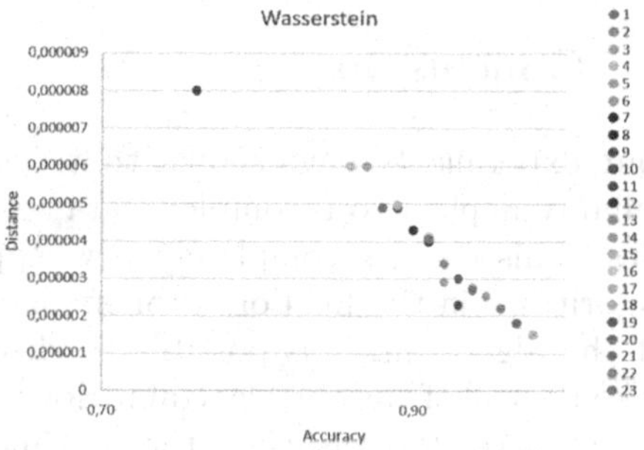

(d) Wasserstein Results

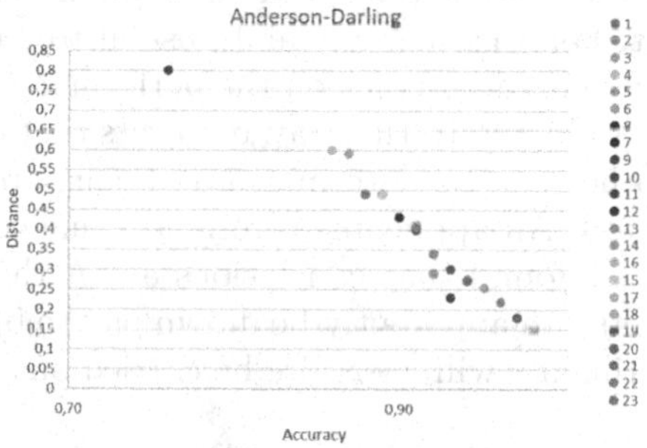

(e) Anderson-Darling Results

Fig. 6. Results for the 5 ECDFs based on classes. We calculated the statistical distance from the training with the test data set in comparision with the classification accuracy.

the image tiles obtained from the scenes. However, this had the negative consequence that both the training in the training phase and the classification in the application phase had a long calculation time. This was due to the fact that the image tiles had to be read in individually, as the different dimensions of the tiles meant that efficient batch processing was not possible. Since the long-term goal

Table 1. Classification results in the application phase of the different classes in the test data set. The order of the classes is the same as listed at [14]. We calculated the accuracy for each class during the application phase. The results for the statistical distance is shown in Fig. 6 for the sake of clarity.

Class	1	2	3	4	5	6	7	8	9	10	11	12	13	14	15	16	17	18	19	20	21	22	23	Average
Accuracy	.88	.92	.95	.86	.91	.92	.90	.89	.76	.89	.93	.91	.94	.87	.89	.91	.94	.98	.93	.97	.94	.96	.91	**.9113**

is to develop an application to detect the concept drift during the runtime of an autonomously driving vehicle, the calculation time must be shortened. Therefore we will look for another possibility to maintain the dimension for further experiments with the NuImages data set.

8 Conclusion

Our experiments have shown that the Aslansefat et al. approach can also be reliably applied to a complex data set for training autonomously driving vehicles. This enables completely new approaches in the development of suitable algorithms in the field of autonomous driving, since here, during the execution of the algorithm, it is possible to check when it is necessary for the driver to take control of the vehicle again. However, the speed of the algorithm still has to be improved for this, since the runtime of the hardware used is not yet capable of real-time. However, there was no focus on this in the current implementation. Furthermore, it will be necessary to determine suitable threshold values and to evaluate when the distance between the training and application data sets becomes too great. For this it is certainly useful to characterize data sets as in "Effects of data set characteristics on the performance of feature selection techniques" by Oreski et al. [15]. In this way, data sets that differ greatly or only partially from one another can be determined and compared with one another using the SafeML method. An applicable threshold value could then be derived from this. In addition, we found that conclusions about the appropriate modularization of the data sets can apparently be drawn on the basis of the statistical distance. Further experiments will therefore be carried out in this direction.

References

1. Aslansefat, K., Sorokos, I., Whiting, D., Tavakoli Kolagari, R., Papadopoulos, Y.: SafeML: safety monitoring of machine learning classifiers through statistical difference measures. In: Zeller, M., Höfig, K. (eds.) IMBSA 2020. LNCS, vol. 12297, pp. 197–211. Springer, Cham (2020). https://doi.org/10.1007/978-3-030-58920-2_13
2. Das, S.: Best practices for dealing with concept drift (2021). https://neptune.ai/blog/concept-drift-best-practices
3. Dries, A., Rückert, U.: Adaptive concept drift detection. Stat. Anal. Data Min. **2**, 311–327 (2009). https://doi.org/10.1002/sam.10054
4. Godwin,, C.: Tesla's autopilot 'tricked' to operate without driver (2021). https://www.bbc.com/news/technology-56854417

5. Greenberg, A.: Hackers remotely kill a jeep on the highway-with me in it (2015). https://www.wired.com/2015/07/hackers-remotely-kill-jeep-highway/
6. Greenberg, A.: The jeep hackers are back to prove car hacking can get much worse (2016). https://www.wired.com/2016/08/jeep-hackers-return-high-speed-steering-acceleration-hacks/
7. He, K., Zhang, X., Ren, S., Sun, J.: Spatial pyramid pooling in deep convolutional networks for visual recognition. In: Fleet, D., Pajdla, T., Schiele, B., Tuytelaars, T. (eds.) ECCV 2014. LNCS, vol. 8691, pp. 346–361. Springer, Cham (2014). https://doi.org/10.1007/978-3-319-10578-9_23
8. Health, U.: 5 real-life medical devices inspired by science fiction (2020). https://www.usfhealthonline.com/resources/healthcare/5-real-life-medical-devices-inspired-by-science-fiction/
9. Klinkenberg, R., Joachims, T.: Detecting concept drift with support vector machines. In: Proceedings of ICML, May 2000
10. Koorosh, A.: How to make your classifier safe (2020). https://towardsdatascience.com/how-to-make-your-classifier-safe-46d55f39f1ad
11. Kovacs, E.: Tesla car hacked remotely from drone via zero-click exploit (2021). https://www.securityweek.com/tesla-car-hacked-remotely-drone-zero-click-exploit
12. Krizhevsky, A., Sutskever, I., Hinton, G.E.: ImageNet classification with deep convolutional neural networks. Advances in Neural Information Processing Systems, vol. 25, pp. 1097–1105 (2012)
13. Žliobaitė, I., Pechenizkiy, M., Gama, J.: An overview of concept drift applications. In: Japkowicz, N., Stefanowski, J. (eds.) Big Data Analysis: New Algorithms for a New Society. SBD, vol. 16, pp. 91–114. Springer, Cham (2016). https://doi.org/10.1007/978-3-319-26989-4_4
14. NuScenes: Nuscenes by motional (2020). https://www.nuscenes.org/
15. Oreski, D., Oreski, S., Klicek, B.: Effects of dataset characteristics on the performance of feature selection techniques. Appl. Soft Comput. **52**, 109–119 (2016). https://doi.org/10.1016/j.asoc.2016.12.023
16. Templeton, B.: Tesla in Taiwan crashes directly into overturned truck, ignores pedestrian, with autopilot on (2020). https://www.forbes.com/sites/bradtempleton/2020/06/02/tesla-in-taiwan-crashes-directly-into-overturned-truck-ignores-pedestrian-with-autopilot-on/?sh=3ec11c5758e5

User-Biased Food Recognition for Health Monitoring

Mazhar Hussain[1], Alessandro Ortis[1,2(✉)], Riccardo Polosa[2,3,4], and Sebastiano Battiato[1,2]

[1] Department of Mathematics and Computer Science, University of Catania, Viale A. Doria, 6, 95125 Catania, Italy
mazhar.hussain@phd.unict.it, {ortis,battiato}@unict.it

[2] Center of Excellence for the Acceleration of HArm Reduction (CoEHAR), University of Catania, Catania, Italy
polosa@unict.it

[3] Department of Clinical and Experimental Medicine, University of Catania, Catania, Italy

[4] ECLAT Srl, Spin-off of the University of Catania, Catania, Italy

Abstract. This paper presents a user-biased food recognition system. The presented approach has been developed in the context of the FoodRec project, which aims to define an automatic framework for the monitoring of people's health and habits, during their smoke quitting program. The goal of food recognition is to extract and infer semantic information from the food images to classify diverse foods present in the image. We propose a novel Deep Convolutional Neural Network able to recognize food items of specific users and monitor their habits. It consists of a food branch to learn visual representation for the input food items and a user branch to take into account the specific user's eating habits. Furthermore, we introduce a new FoodRec-50 dataset with 2000 images and 50 food categories collected by the iOS and Android smartphone applications, taken by 164 users during their smoking cessation therapy. The information inferred from the users' eating habits is then exploited to track and monitor the dietary habits of people involved in a smoke quitting protocol. Experimental results show that the proposed food recognition method outperforms the baseline model results on the FoodRec-50 dataset. We also performed an ablation study which demonstrated that the proposed architecture is able to tune the prediction based on the users' eating habits.

Keywords: Dietary monitoring · Food recognition · Food dataset · Artificial intelligence for health

1 Introduction

Recognizing food from images is an extremely useful task for a variety of use cases. For example, it would allow people to track their food intake of what they

S. Sclaroff et al. (Eds.): ICIAP 2022, LNCS 13233, pp. 98–108, 2022.
https://doi.org/10.1007/978-3-031-06433-3_9

consume by simply taking a picture, to increase awareness of their daily diet by monitoring their eating habits, kind and amount of taken food, how much time the user spends eating during the day, how many and what times the user has a meal, analysis on user's habits changes, bad habits, and other inferences related to user's behavior and mood [1]. It can help a doctor to have a better opinion with respect to the patient's behaviour, in the applications on quitting treatment response, smoke monitoring technology [2], dietary monitoring during smoke quitting [3] and smoking cessation system [4]. Food monitoring plays a vital role in human health that is directly affected by diet [5]. Humans life is strictly affected by the food, this encourages researchers to introduce new methods for food logging and automatic food dietary monitoring [6], food retrieval and classification [7]. This paper presents a novel food recognition method that takes into account the specific user to systematically analyse and infer his/her eating habits. The idea is to introduce a bias related to the user in the food classification pipeline. In particular, inspired by deep learning approaches applied on text representation learning [8], the proposed architecture learns a user's eating habits feature representation space. We also collected a new FoodRec-50 dataset that will be used for evaluation of the food recognition technology for dietary monitoring during smoke quitting. The rest of this paper is structured as follows. Section 2 describes the related works. Section 3 presents the details about the proposed food recognition method. Section 4 discusses the proposed method comparison and evaluation experiments. Finally, Sect. 5 describes our conclusions.

2 Related Work

Recently, computer vision and deep learning techniques have gained a lot of attention due to high level of performances in various research fields and applications as well as in food recognition. Computer vision research devoted to the analysis of food images including previous works on food detection, classification, and segmentation. The paper in [9] covers food computing including acquisition, analysis, perception, recognition, retrieval, recommendation, prediction, and its applications in health, culture, agriculture, medicine, and biology. Different computer vision and machine learning techniques have been used for single-label food recognition, multi-label food recognition, food portion estimation, and personalized food recognition along with existing benchmark food datasets. The work in [10] presented a review on food recognition technology and its applications especially in the health department for dietary and calorific monitoring. Computer vision techniques for food understanding have been addressed in the areas such as food detection and recognition for automatic harvesting, food quality assessment for industry aims, dietary management, food logging and food intake monitoring, food retrieval, and classification with publicly available food datasets. Fakhrou et al. [11] proposed a smartphone application that utilizes a trained deep convolutional neural network (CNN) model for food and fruits recognition to assist children with visual impairments. Moreover, food recognition is improved using the ensemble learning approach with fusion of multiple deep CNN architectures on a customized food dataset where soft voting method is used to ensemble multiple models results. A system [12] is proposed to effectively estimate nutrient intake by using RGB depth image pairs that are captured before and after meal consumption. This system consists of a novel multi-task contextual network for food item segmentation, classification with few-shot learning-based algorithms built by limited training samples for food recognition. Pfisterer et al. [13] developed an automatic semantic food segmentation method using multi-scale encoder-decoder network architecture for food intake tracking and estimation in long-term care homes. For the encoder, ResNet architecture trained on the imagenet dataset is used because of its discriminative feature learning ability. For the decoder, a pyramid scene parsing network is chosen. The proposed method achieved comparable results to semi-automatic graph cuts. The paper in [14] designed an automatic framework for tray food analysis to find the region of interest of the input image then predict the food class for each region. Different visual descriptors have been used including opponent gabor features, chromaticity moments, color histogram, local color contrast, gabor features, complex wavelet features, and convolutional features. A semisupervised generative adversarial network is used for food recognition [15] using partially labeled data. Network architecture consists of generator and discriminator. The generator produces dataset fake samples and discriminator learns the nature of the problem and further recognizes different food items with partially labeled training data. The author claimed outperformed results on the ETH Food-101 datasets and indian food dataset as compared to the AlexNet, GoogleNet, and Ensemble Net. The ResNet deep residual learning architecture [16] is proposed for image recognition with powerful representational capability for learning discriminative features from complex scenes. ResNet network architecture designed for the classification task, trained on the imageNet dataset of natural scenes that consists of 1000 classes. Evaluation has been performed with the residual network with depths of 18-layers, 34-layers, 50-layers, 101-layers, and 152-layers. The ResNeXt [17] architecture is a combination of ResNet and InceptionNet that contains a stack of residual blocks with a split-transform-merge structure in each block. This design introduced a new dimension that is cardinality or the size of the set of transformations and further Hu et al. [18] introduced a new architectural unit squeeze-and-excitation block that comprises the squeeze operation and excitation operation with the aim to enhance the quality of representations produced by a network. Related studies described above presents traditional food image recognition without taking into account the user habits. Our proposed study differs from the recognition that happens in the development of general purposes food recognition systems as proposed approach considers the specific user that uploaded the food image to learn and monitor its eating habits.

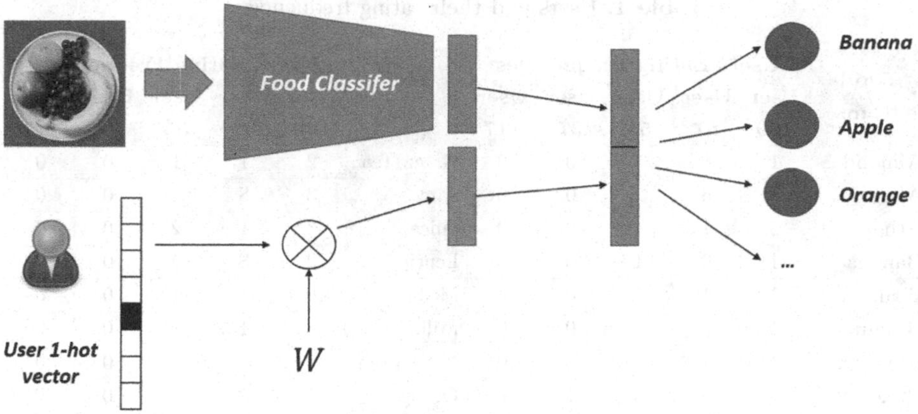

Fig. 1. Food recognition proposed architecture. Blue circles depict independent logistic activations for specific classes, which are activated by the presence of the food item in the visual content taking into account the bias given by the user. (Color figure online)

3 Proposed Method

3.1 Data Acquisition and Annotation

The objective is to build a new unique robust dataset useful for the food recognition technologies development and evaluation stages. Our dataset is specific to the users who are involved in the smoke quitting process to monitor their dietary habits. The dataset is produced to study the correlation between eating information with smoking habits. In the future such data will be used to find correlations with respect to the smoking activity of the subjects during the period of observation. To collect the food data, the iOS and Android smartphone apps are used. Users can upload a meal intake image by taking a picture what they eat and can assign labels to the food image what it contains. Annotations are required during the supervised training of the network and also to test the examples during the evaluation phase for the food recognition method. To perform the experiments, we have first extracted food images for 50 classes from the FoodRec data. Although, some of the classes (beans, breadstick, carrot, chickpeas, corn, popcorn, grape, peas, zucchini, etc.) still have few images. Further, data is annotated manually for training and evaluation of the model which contains around 1100 images.

Table 1. Users and their eating frequencies

Food Items	Users Eating Frequencies					Food Items	Users Eating Frequencies				
	User 109	User 87	User 55	User 27	User 117		User 109	User 87	User 55	User 27	User 117
Almond	1	1	1	0	0	Green Tea	2	1	1	0	0
Apple	2	6	2	0	0	Jam	3	8	1	0	0
Arugula	3	1	4	0	0	Juice	3	1	2	0	0
Banana	1	3	1	0	0	Lentil	2	3	2	0	2
Bean	1	1	1	0	0	Meat	6	1	1	0	0
Biscuit	5	1	3	0	0	Milk	8	1	1	0	0
Blueberry	1	1	1	0	0	Mushroom	1	1	1	0	0
Bread	1	8	3	1	0	Orange	1	2	2	0	0
BreadStick	1	1	1	0	0	Pasta	6	1	3	1	0
Cake	2	3	2	0	0	Peas	1	3	1	0	0
Carrot	1	1	1	0	2	Pizza	5	2	2	0	1
Cereal	3	1	1	0	1	Popcorn	1	1	3	0	0
Cheese	2	2	2	0	0	Pork	2	1	1	1	0
Chicken	1	1	2	1	0	Potato	1	1	2	0	0
Chickpeas	1	1	2	0	0	Rice	2	2	1	0	0
Chips	2	1	1	0	0	Salad	1	2	1	0	0
Chocolate	1	2	2	1	0	Soup	2	1	1	0	0
Coffee	2	2	11	2	0	Spaghetti	2	2	1	0	0
Corn	1	1	1	2	0	Strawberry	1	1	1	0	0
Cracker	1	1	3	0	2	Tea	2	2	3	1	1
Croissant	2	1	1	0	0	Tomato	2	1	1	0	1
Doughnut	1	1	1	0	0	Tortellini	2	1	1	0	0
Egg	2	1	3	0	1	Vegetable	1	1	1	0	0
Fish	2	3	1	0	0	Yogurt	2	1	1	0	0
Grape	1	3	1	0	0	Zucchini	1	1	1	0	0

3.2 Proposed FoodRec Architecture

We proposed the FoodRec architecture for data coming from the FoodRec app [3], which is specific with respect to our purposes. The proposed system aims to recognize food items of specific users and monitor their habits. This task significantly differs from the recognition of any food instance depicted by a picture, such as happens in the development of general purposes food recognition systems. Figure 1 shows the proposed FoodRec architecture. In particular, a common multi-label food classifier is composed by a Convolutional Neural Network which defines a meaningful feature representation for the input images, based on the training task. Then, the representation is fed to multiple logistic units (i.e., blue circles in the Fig. 1) which are activated if the associated food item is present in the picture. The proposed architecture will take into account the specific user that uploaded the picture. Indeed, since the proposed system is

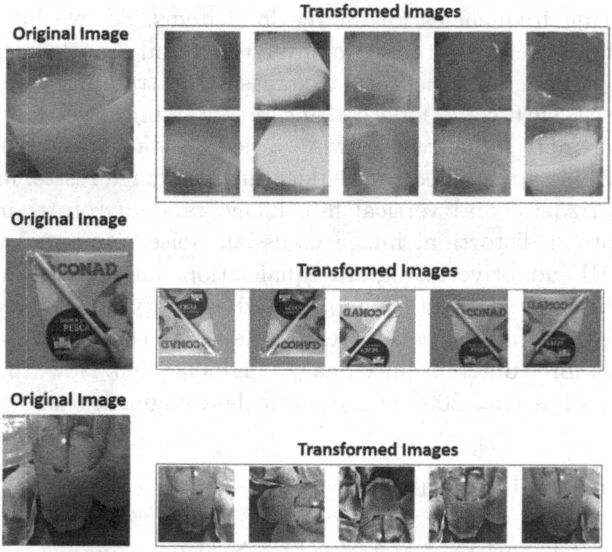

Fig. 2. Three different juices original images with transformed images.

aimed to systematically analyse and infer user habits, our objective is to add to the food classification pipeline a bias related to the user. As consequence, the individual logistic activations will be fed with a feature that is obtained by concatenating the image and user feature. The latter one, is represented by the weight matrix W in Fig. 1, which will be learned from the users' habits during the training stage.

3.3 User Data Annotation

As previously, we have extracted data for 50 classes but with only food items annotations. The proposed FoodRec architecture requires users annotation as well along with the food items eaten by them to train and test the model. So, data is annotated to embed the user information with image label so that we can feed the data simultaneously to the network. Now, the FoodRec images names contain the user ID and Image label. The idea is to extract the user ID from the image name while reading the images so that we implement the dataloader with the simultaneous user input and corresponding image data for both branches for the experiments.

3.4 Data Augmentation

Although FoodRec dataset consists of 1100 food images for 50 classes but some of the classes like beans, breadstick, carrot, chickpeas, corn, popcorn, grape, peas, zucchini, etc. still have very few images. Here we go with the data augmentation technique to deal with the lack of data. We have selected the top-20 users with the highest eating frequencies for all the food items. So, we have augmented the data for these users to produce many altered and transformed versions of the same image. Image augmentation increases the training data as we don't have enough data with some food categories containing fewer food images and make a classifier more robust with a wide variety of transformed images. Different transformations are applied to the data such as image resize, image random crop, image horizontal and vertical flip, image random rotate, image motion blur, image optical distortion, image Gaussian noise, random brightness and contrast, CLAHE adaptive histogram equalization, hue, and saturation value. Three images have been taken from each food category and augmented for the top-20 users with the highest eating frequencies for all the food items. For example, Fig. 2 shows three different juice images augmentation. Now, the FoodRec-50 dataset consists of around 2000 images after data augmentation.

4 Experimental Results

The baseline model consists of pretrained ResNet101 [16] model trained on ImageNet that is finetuned to extract a 1024-dimensional features vector to perform the food items classification. This model is trained using only the food images like the traditional classification algorithm without taking into account the user bias into the final decision making to classify the food items.

The proposed food recognition model consists of two branches, the food branch and the user branch to extract a 1024-dimensional concatenated feature map from these branches to recognize the food items. Food branch extracts 1024-dimensional feature map of food image using ResNet101 architecture with transferred weights from ImageNet dataset containing 1000 image categories, and further averaging pooling layer, flatten layer and fully connected layer are applied to get a 512-dimensional feature vector. The user branch extracts a 512-dimensional feature vector from the user bias using a fully connected layer. Finally, the output 1024-dimensional feature vector is obtained by concatenating the features extracted from both branches. User branch of the proposed network is learning 164 user eating weight vectors with 512 features. So, user weight matrix can be represented with 164 rows (one for each user) and 512 columns. The proposed network with one-hot user vector is effective because it learns specific user eating habits with the food image features as compared to the tradition food recognition systems. This approach is inspired by the document representation approach known as doc2vec presented by Le et al. [8]. Indeed, the model presented in [8] implements a document representation architecture in which the word/sentence features are affected by the document from which they have been extracted. In this way, the same word/sentence is represented differently depending on the source document which acts as a context for the encoded words and is represented as a one-hot input vector. The document branch is combined with another network branch devoted to represent single words, as we do combining the branch representing the food image with the one representing the user will act as a bias for the image representation.

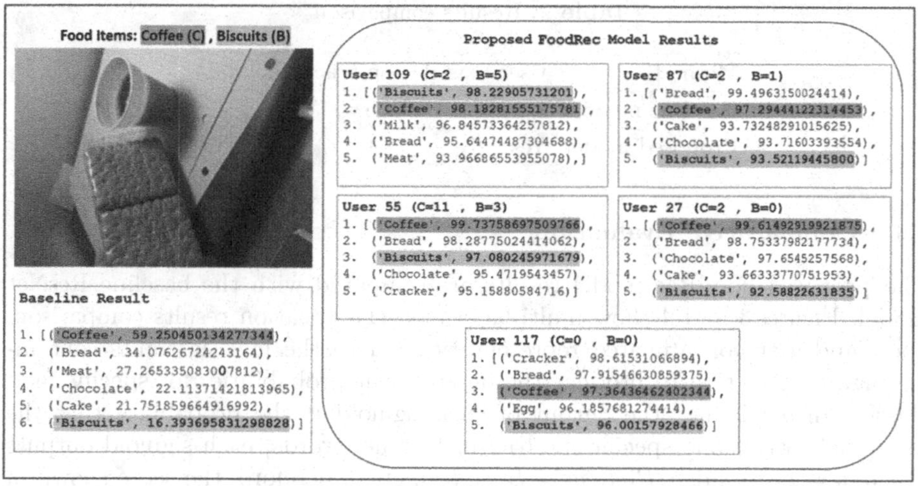

Fig. 3. Results comparison of test image containing Coffee and Biscuits.

The proposed and baseline networks are trained using same settings. Networks are trained using the Adam optimizer and the cross-entropy loss function. Initially, learning rate is set to 0.001 and it decays by a factor of 0.1 every 10 epochs. The batch size is set as 32 and networks are trained for 200 epochs. The FoodRec-50 data consists of 164 users eating different food items. Eating habits only for users 87, 109, 55, 27, and 117 for each food item is listed below in Table 1 with an individual food item and its eating frequency for that user. These users are chosen using the Euclidean Distance function to observe the difference in the decision making. The distance between user eating frequencies tells how much one user eating habits are different from other. Therefore, the distances between user eating frequencies have been used to select the users with different habits and perform specific tests aimed to assess the efficacy of our approach and its capability to encode the user eating habits. The effect can be observed in the Figs. 3 and 4 showing the results. While the baseline method finds difficulties in the recognition of multiple food items in cluttered scenarios, the proposed method shows better performances, especially for users that have high frequencies for the food items present in the test image. We defined an "eating matrix" U × N, where U is the number of users (164 rows) and N is the number of considered food items (50 columns). Each row corresponds to the eating frequencies for each food item of a specific user. Then, we computed the distance for each user to find two users (87, 109) with a maximum distance between their eating vectors. Further, we calculated the sum of eating frequencies for each user and selected one user (55) with average and two users (27, 117) with the lowest sum of eating frequencies.

Table 2. Results comparison

Method	User-biased	Top-5 accuracy (%)
Baseline model	No	59.6
Proposed model	Yes	71.1

4.1 Comparative Evaluation

The proposed FoodRec model results are compared with the baseline ResNet model. Figures 3 and 4 show multi-label food classification results comparison. Food and user concatenated representation fed to the logistics units containing sigmoid function to produce the independent probabilities for specific food classes. In particular, the output of each sigmoid is the probability that the input belongs to one specific food item. In other words, each sigmoid outputs $P(class = Banana|x), P(class = Bread|x)$, etc. Therefore, the score shown in Figs. 3 and 4 for the food types is the percentage of the sigmoid output probabilities for each food item.

For example, Fig. 3 contains a test food image with two food items coffee and biscuits with multi-label predictions. Baseline represents the ResNet model trained only with food images and results are shown below the test image in the figures. Then, the proposed FoodRec model results with five different users are shown next to the test image in the figures. We can observe from Fig. 3 that the baseline model recognizes the coffee with a score 59.25 at the very first place but biscuits with a score 16.39 occur at the 6th place in the prediction order. On the other hand, the proposed FoodRec recognizes same two food items with improved score and occur in top five predictions for all five users. For the user 109, the top two predictions are biscuits and coffee with scores 98.22 and 98.18 respectively. This happens because the user with ID 109 has a relatively higher number of instances for these kinds of food in the eating matrix. For the user 117, although it did not drink coffee (C = 0) or eat biscuits (B = 0) but the model recognizes these food items at 3rd and 5th places respectively because the model learns both image and user features. Similarly, you can also observe the difference between the FoodRec and the baseline models in the Fig. 4 with another test food image. The FoodRec model improves the score as well as learns the user dietary habits because the model is learning weight matrix for the users. As we change the user bias input, the result in the prediction is being changed according to the dietary habits for that user as you can obverse in the given figures. So, adding a bias related to the user to the food classification pipeline is effective to systematically analyse and infer user's habits. Users and their eating habits can be observed in Table 1. Moreover, the proposed model improves the general food recognition task with respect to the baseline model as shown in Table 2.

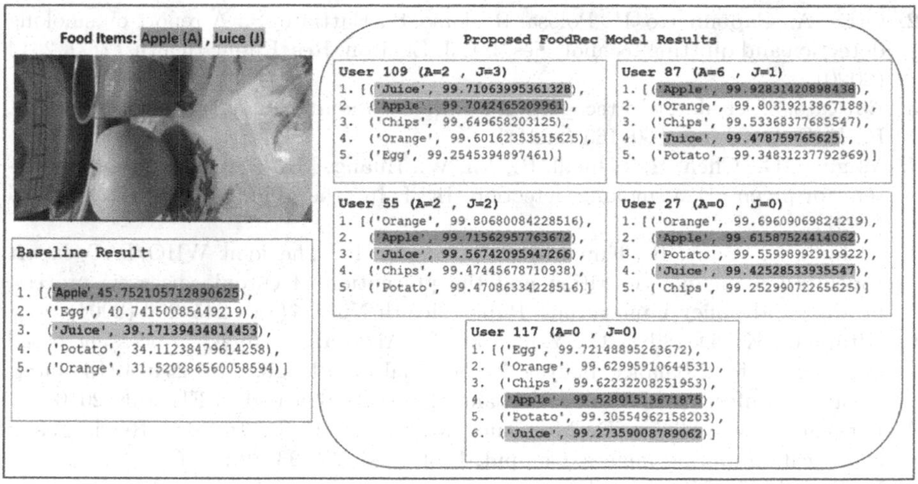

Fig. 4. Results comparison of test image containing Apple and Juice.

5 Conclusion

This paper aims to develop an automatic framework for food recognition using computer vision and deep learning techniques that plays a significant role to the health and food intake of people. The developed system acquires images of the food eaten by the user or subject over time which will then be processed by the proposed food recognition model to extract and infer semantic information from the food images. Experiments show that the proposed user-biased food recognition is effective and achieves higher results as compared to the baseline. The proposed model is hence able to influence the prediction by encoding the user bias as a result it also improves the task performance.

Acknowledgments. This investigator initiated study was sponsored by ECLAT srl, a spin-off of the University of Catania, with the help of a grant from the Foundation for a Smoke-Free World Inc., a US nonprofit 501(c)(3) private foundation with a mission to end smoking in this generation. The contents, selection, and presentation of facts, as well as any opinions expressed herein are the sole responsibility of the authors and under no circumstances shall be regarded as reflecting the positions of the Foundation for a Smoke-Free World, Inc. ECLAT srl. is a research based company from the University of Catania that delivers solutions to global health problems with special emphasis on harm minimization and technological innovation.

References

1. Ortis, A., Farinella, G.M., Battiato, S.: Survey on visual sentiment analysis. IET Image Proc. **14**(8), 1440–1456 (2020)

2. Ortis, A., Caponnetto, P., Polosa, R., Urso, S., Battiato, S.: A report on smoking detection and quitting technologies. Int. J. Environ. Res. Public Health **17**(7), 2614 (2020)
3. Battiato, S., et al.: Food recognition for dietary monitoring during smoke quitting. In: IMPROVE, pp. 160–165 (2021)
4. Maguire, G., Chen, H., Schnall, R., Xu, W., Huang, M.C.: Smoking cessation system for preemptive smoking detection. IEEE Internet Things J. **9**(5), 3204–3214 (2021)
5. Nishida, C., Uauy, R., Kumanyika, S., Shetty, P.: The joint WHO/FAO expert consultation on diet, nutrition and the prevention of chronic diseases: process, product and policy implications. Public Health Nutr. **7**(1a), 245–250 (2004)
6. Kitamura, K., De Silva, C., Yamasaki, T., Aizawa, K.: Image processing based approach to food balance analysis for personal food logging. In: 2010 IEEE International Conference on Multimedia and Expo, pp. 625–630. IEEE, July 2010
7. Farinella, G.M., Allegra, D., Moltisanti, M., Stanco, F., Battiato, S.: Retrieval and classification of food images. Comput. Biol. Med. **77**, 23–39 (2016)
8. Le, Q., Mikolov, T.: Distributed representations of sentences and documents. In: International Conference on Machine Learning, pp. 1188–1196. PMLR, June 2014
9. Min, W., Jiang, S., Liu, L., Rui, Y., Jain, R.: A survey on food computing. ACM Comput. Surv. (CSUR) **52**(5), 1–36 (2019)
10. Allegra, D., Battiato, S., Ortis, A., Urso, S., Polosa, R.: A review on food recognition technology for health applications. Health Psychol. Res. **8**(3), 9297 (2020)
11. Fakhrou, A., Kunhoth, J., Al Maadeed, S.: Smartphone-based food recognition system using multiple deep CNN models. Multimed. Tools Appl. **80**, 33011–33032 (2021). https://doi.org/10.1007/s11042-021-11329-6
12. Lu, Y., Stathopoulou, T., Vasiloglou, M.F., Christodoulidis, S., Stanga, Z., Mougiakakou, S.: An artificial intelligence-based system to assess nutrient intake for hospitalised patients. IEEE Trans. Multimed. **23**, 1136–1147 (2020)
13. Pfisterer, K.J., Amelard, R., Chung, A.G., Syrnyk, B., MacLean, A., Wong, A.: Fully-automatic semantic segmentation for food intake tracking in long-term care homes. arXiv e-prints, arXiv-1910 (2019)
14. Ciocca, G., Napoletano, P., Schettini, R.: Food recognition: a new dataset, experiments, and results. IEEE J. Biomed. Health Inform. **21**(3), 588–598 (2016)
15. Mandal, B., Puhan, N.B., Verma, A.: Deep convolutional generative adversarial network-based food recognition using partially labeled data. IEEE Sens. Lett. **3**(2), 1–4 (2018)
16. He, K., Zhang, X., Ren, S., Sun, J.: Deep residual learning for image recognition. In: Proceedings of the IEEE Conference on Computer Vision and Pattern Recognition, pp. 770–778 (2016)
17. Xie, S., Girshick, R., Dollár, P., Tu, Z., He, K.: Aggregated residual transformations for deep neural networks. In: Proceedings of the IEEE Conference on Computer Vision and Pattern Recognition, pp. 1492–1500 (2017)
18. Hu, J., Shen, L., Sun, G.: Squeeze-and-excitation networks. In: Proceedings of the IEEE Conference on Computer Vision and Pattern Recognition, pp. 7132–7141 (2018)

Multi-view Spectral Clustering via Integrating Label and Data Graph Learning

Sally El Hajjar[1], Fadi Dornaika[1,2(✉)], Fahed Abdallah[3,4], and Hichem Omrani[4]

[1] University of the Basque Country UPV/EHU, San Sebastian, Spain
fadi.dornaika@ehu.eus
[2] IKERBASQUE, Basque Foundation for Science, Bilbao, Spain
[3] Lebanese University, Beirut, Lebanon
[4] Urban Development and Mobility Department, Luxembourg Institute of Socio-Economic Research (LISER), 11 Porte des Sciences, 4366 Esch-sur-Alzette, Luxembourg

Abstract. Nowadays, one-step multi-view clustering algorithms attract many interests. The main issue of multi-view clustering approaches is how to combine the information extracted from the available views. A popular approach is to use view-based graphs and/or a consensus graph to describe the different views. We introduce a novel one-step graph-based multi-view clustering approach in this study. Our suggested method, in contrast to existing graph-based one-step clustering methods, provides two major novelties to the method called Nonnegative Embedding and Spectral Embedding (NESE) proposed in the recent paper [1]. To begin, we use the cluster label correlation to create an additional graph in addition to the graphs associated with the data space. Second, the cluster-label matrix is constrained by adopting some restrictions to make it more consistent. The effectiveness of the proposed method is demonstrated by experimental results on many public datasets.

Keywords: Multi-view clustering · Graph construction · Similarity graph · Spectral projection matrix · Cluster label space

1 Introduction

Multi-view clustering methods have recently received a lot of attention in the field of data analysis. These algorithms divide samples into different groups by using different views of the dataset [13,15,24]. Graph based methods are one of the popular methods adopted for multi-view clustering [18]. To group the data into different clusters, these approaches incorporate the similarity graph of each view into their models. The pairwise similarities between the data points are represented by the graph of each view. Most of the multi-view clustering

Supported by F.R.S-FNRS.

S. Sclaroff et al. (Eds.): ICIAP 2022, LNCS 13233, pp. 109–120, 2022.
https://doi.org/10.1007/978-3-031-06433-3_10

approaches only consider the graph similarities in the data space. Different from previous graph-based techniques, we provide a novel methodology in this study that can take advantage of both the cluster-label space and the data space described by numerous graphs. Furthermore, the additional cluster label graph can give additional information that represents the similarities between cluster labels, which can be included into the clustering model. This can contribute to enhance the clustering performance. A similar semi-supervised idea is presented in [19]. Inspired by this idea proposed for semi-supervised problem, we develop a new multi-view clustering approach called Multi-view Spectral Clustering via integrating Label and Data Graph Learning (MSLDGL). First, this approach inherits the advantages of the NESE method proposed in [1]. Thus, the final clustering assignment can be directly obtained from the nonnegative embedding matrix, without any post-processing step. In addition, our method integrates two main novelties to the NESE method. First, it integrates the cluster label graph in addition to the graphs associated with the data space. Second, it imposes a step of normalization and orthogonalization of the obtained nonnegative embedding matrix, making it more precis. The following is a summary of the main contributions of the paper.

1. Different from other multi-view clustering methods, we proposed a new approach that integrates an additional graph based on the soft cluster labels.
2. This method introduces normalization and orthogonalization constraints over the nonnegative embedding matrix. The constrained nonnegative embedding matrix can lead to better clustering results than the NESE approach.
3. The proposed approach does not require any post-processing like K-means to give the final clustering result.
4. The weight of each view is computed automatically.
5. An efficient optimization framework is provided to solve the given criterion, and its effectiveness is proved by testing different datasets.

The rest of this paper is organized as follows. Section 2 presents some relevant works on multi-view clustering and outlines the approach NESE of [1]. The proposed approach and the corresponding optimization scheme are described in detail in Sect. 3. Experimental results are presented in Sect. 4 along with a comparison of our method with several state-of-the-art methods. The paper concludes with Sect. 5.

2 Preliminaries and Related Work

2.1 Notations

In this study, matrices are written in bold upper case letters, vectors in bold lower case letter, and scalars in non-bold characters. Table 1 summarizes the notations used in this paper. Given $\mathbf{X}_v = (\mathbf{x}_v^1, \mathbf{x}_v^2, ..., \mathbf{x}_v^n) \in \mathbb{R}^{n\times d_v}$ the data matrix of each view, where n is equal to the number of data points, d_v is the total number of features in a given view v, where $v = 1, ..., V$. Given a matrix $\mathbf{Z}$, z^{ij} is its element of i-th row and j-th column. The trace of this matrix is denoted by $Tr(\mathbf{Z})$ and its transpose by $\mathbf{Z}^T$.

Table 1. Notations used in the paper.

Notation	Description
V	Number of views
n	Number of data points
d_v	Data dimensionality in view v
C	Number of clusters
$\mathbf{X}_v = (\mathbf{x}_v^1, \mathbf{x}_v^2, ..., \mathbf{x}_v^n)$	Data matrix of view v
$\mathbf{x}_v^i$	The i-th sample of $\mathbf{X}_v$
$\lVert\mathbf{Z}\rVert_F$ or $\lVert\mathbf{Z}\rVert_2$	Frobenius norm of the matrix $\mathbf{Z}$
Tr(.)	Trace operator
$\mathbf{S}_v$	Similarity matrix of view v
$\mathbf{H}$	Cluster label matrix
$\mathbf{P}_v$	Spectral representation matrix of view v
$\mathbf{I}$	Identity matrix

2.2 Related Work

In this section, we provide significant works on multi-view clustering. The co-training approach proposed in [3] is a popular multi-view clustering method, based on the assumption that the same data point should be allocated to the same cluster in different views. The spectral clustering of each view is performed to obtain the appropriate eigenvectors, after that the eigenvectors of a given view are used to give the different clusters. Thus the obtained clustering result is used to modify the graph of the other view. Another popular method is the co-regularized spectral clustering presented in [4]. This method adaptively combines different similarity matrices from different views to get more precise results.

Besides, another categories of multi-view clustering methods is presented on [11,12]. These methods allocate a weight parameter to each view, to take into consideration its contribution on the clustering assignment. Although these approaches are efficient, they require additional weight parameters for each view which can add additional cost. To remedy this issue, automatic weight learning are presented, which eliminates the necessity for hyperparameters, as seen in other methods like those in [2,6,9].

Moreover, different spectral clustering methods have been used for multi-view clustering [10,20]. These approaches create a graph that depicts the similarity between each two data points in each view. The samples are represented by the nodes, and the similarity between these data points is represented by the edges of the graph. Then, the spectral representation matrix associated to each similarity matrix is constructed. To obtain the final clustering result, a post-processing step such as k-means clustering is applied. For instance, the method described in [7] and called "Adaptively Weighted Procrustes" (AWP) is a multi-view clustering method which uses spectral rotation to give the labels for all data points. Moreover, Multi-View Subspace based clustering methods (MVSC) are a well-known multi-view clustering method [21,22]. These approaches either learn a latent space and then deliver the unified representation of the data or create a unified representation of the data from different subspaces of all views of the data. Furthermore, a new Multi-view Learning with Adaptive Neighbors algorithm (MLAN) is presented in [5], to jointly estimate the similarity matrix and the clustering assignment. Besides, this method can learn the weights of each view automatically, without any extra-parameters. However, the significant computational cost of these approaches is due to the matrix inversion and eigenvalue decomposition steps.

Numerous kernel-based approaches have been developed. Two kernel-based graph learning multi-view clustering algorithms with automatic weight learning are discussed in [2]. These two methods map the data into a linearly separable space. The first method proposed in this paper uses a single kernel per each view, whereas the second approach employs a combination of multiple kernel matrices to enhance the use of the input kernel matrix. Furthermore, without any additional parameters, these algorithms calculate the unified similarity matrix, the spectral projection matrix, and the weight of each view simultaneously.

Furthermore, as explained previously, the suggested method is an enhanced variant of the Nonnegative Embedding and Spectral Embedding (NESE) method in [1]. The main concept of NESE is to directly give the nonnegative embedding matrix $\mathbf{H}$ from the individual graph matrices $\mathbf{S}_v$ and the individual spectral representations matrices $\mathbf{P}_v$ using $\mathbf{S}_v \approx \mathbf{H}\mathbf{P}_v^T, v = 1, ..., V$. The produced matrix $\mathbf{H}$ gives the final clustering result directly without any additional parameters or post-processing step. Also, this method simultaneously offers both the consensus nonnegative embedding matrix and the spectral representation matrices.

Several existing algorithms employ two different processes to obtain the clustering result. The first phase involves learning the joint affinity matrix, and the second involves performing a hard clustering algorithm such as k-means clustering to produce the clustering result. A novel technique is proposed in [23] and called One-step Multi-view Spectral Clustering (OMSC). This method is used to solve the problem of inconsistency due to the fact that the purpose of the first step is not to achieve optimal clustering performance. It combines, in one framework, the step of learning the affinity matrix of each view and the joint affinity matrix learned from the low-dimensional space of the data, and the k-means clustering step. The joint affinity matrix is used to produce the final clustering assignment. Furthermore, the weight of each view is determined automatically to minimize the effect of noisy views.

In addition, the authors of [14] introduces a novel approach called multi-view spectral clustering via Constrained Nonnegative Embedding (CNESE). This method can address many of the problems that prior methods, particularly the NESE method, have. CNESE achieves improved performance by adding additional constraints to the nonnegative embedding matrix generated by NESE, while keeping the major advantages of the NESE approach.

All of the preceding methods have the disadvantage of ignoring the label space of the data and just extracting information from the data space. This weakness provides us with the opportunity to create a new method, which we will discuss in the following section.

2.3 Review of the (NESE) Method

The "Nonnegative Embedding and Spectral Embedding method" (NESE) is proposed in [1]. This technique can offer the clustering result without the requirement for any extra clustering steps or additional parameters by simultaneously finding the nonnegative embedding and spectral embedding matrices. The authors proposed a novel objective function in [1] to give a consistent nonnegative embedding matrix **H**, which was motivated by the symmetric nonnegative matrix factorization and the relaxed continuous Ncut. The main goal of NESE is:

$$\min_{\mathbf{H},\,\mathbf{P}_v} \sum_{v=1}^{V} \|\mathbf{S}_v - \mathbf{H}\mathbf{P}_v^T\|_2 s.t.\ \mathbf{H} \geq 0,\ \ \mathbf{P}_v^T\mathbf{P}_v = \mathbf{I}, \tag{1}$$

where $\mathbf{S}_v$ is the similarity matrix of the view v, $\mathbf{P}_v$ is the spectral projection matrix, and $\mathbf{H}$ is the unified nonnegative embedding matrix used as cluster membership matrix. This matrix eliminates the usage of extra-parameters or extra-steps, such as k-means clustering, to obtain the clustering assignment.

The authors employ an iterative optimization strategy to calculate the spectral projection matrix and the unified nonnegative embedding matrix, which are the outputs of their method.

3 Proposed Approach

In this article, we develop a new method, which is an improvement of the NESE method. This method is named "Multi-view Spectral Clustering via integrating Label and Data Graph Learning" (MSLDGL). This technique adds some constraints to the nonnegative embedding matrix **H** to enhance clustering performance. The main difference between our method and the NESE method is that it adds some constraints on the matrix **H**. First, an additional graph built from the label space is created in the form of an additional view to add richness to the data and enhance the performance. Second, a normalization and orthogonalization steps are imposed over the nonnegative embedding matrix.

For a given view v, $\mathbf{S}_v \in \mathbb{R}^{n\times n}$ is the similarity matrix, $\mathbf{P}_v \in \mathbb{R}^{n\times C}$ is the spectral representation matrix associated with the similarity matrix $\mathbf{S}_v$, and $\mathbf{H} \in \mathbb{R}^{n\times C}$ is the nonnegative embedding matrix used for clustering. Note that C is the total number of clusters. Since our new technique is an improvement of the method in [1], its first term is similar as the NESE method and is defined by:

$$\min_{\mathbf{H},\,\mathbf{P}_v} \sum_{v=1}^{V} \|\mathbf{S}_v - \mathbf{H}\mathbf{P}_v^T\|_2\ \ s.t.\ \ \mathbf{H} \geq 0,\ \ \mathbf{P}_v^T\mathbf{P}_v = \mathbf{I}. \tag{2}$$

We employ automatic view weighting, which has been used in the literature [2], to determine the importance of each view. Thus, a weight parameter is added to our objective function, and it is updated automatically during the iterative process of the proposed method.

The set of weights is given by:

$$\delta_v = \frac{1}{2 * ||\mathbf{S}_v - \mathbf{H}\mathbf{P}_v^T||_2} \qquad v = 1, ..., V. \tag{3}$$

Thus, the objective function of our method will be:

$$\min_{\mathbf{P}_v,\, \mathbf{H}} \sum_{v=1}^{V} \delta_v\, ||\mathbf{S}_v - \mathbf{H}\mathbf{P}_v^T||_2^2\, s.t.\;\; \mathbf{H} \geq 0,\;\; \mathbf{P}_v^T\, \mathbf{P}_v = \mathbf{I}. \tag{4}$$

The majority of multi-view clustering techniques collect information from the data space while disregarding cluster labels. A feature descriptor, often known as a data space, retrieves data information. For a clustering task, relying solely on the graphs created from the features and ignoring the cluster memberships (in the form of predictions) would not be the best solution. The majority of current graph-based algorithms generate graphs from the original data space, which are susceptible to noise and outliers. Relying on the cluster label space, we present a new similarity metric. Likewise, the labeling information can be utilized to generate a new graph that can be incorporated into the multiview clustering criteria. This new approach works by creating a label graph and assigning it to a different view.

The total number of views will be equal to $V + 1$. $\mathbf{S}_{V+1}$ represents the additional graph. The nodes of this graph are the data points and the weight of each edge is the Pearson correlation coefficient, given in Eq. (5). Suppose we have two samples $\mathbf{x}_i$ and $\mathbf{x}_j$, and $\mathbf{h}_i$ and $\mathbf{h}_j$ are their corresponding labels, the similarity between these data points will be computed by Eq. (5).

$$S_{V+1}(i,j) = correlation\,(\mathbf{h}_i\,, \mathbf{h}_j) = \frac{\sum_{k=1}^{C} (h_{ik} - m_i)\,(h_{jk} - m_j)}{\sqrt{\sum_{k=1}^{C} (h_{ik} - m_i)^2}\sqrt{\sum_{k=1}^{C} (h_{jk} - m_j)^2}}, \tag{5}$$

where m_i and m_j are the mean values of row vectors $\mathbf{h}_i$ and $\mathbf{h}_j$, respectively. The correlation coefficient's value ranges from -1 to $+1$. It is equal to $+1$ when the two vectors $\mathbf{h}_i$ and $\mathbf{h}_j$ are equal. A value of $+1$ means a positive perfect correlation and a value of -1 means a negative perfect correlation. Besides, the correlation coefficient is zero when the two vectors $\mathbf{h}_i$ and $\mathbf{h}_j$ are completely uncorrelated. After estimating the n^2 correlation coefficients, we remove the negative values. We generate a p-nearest neighbor graph matrix since the obtained matrix is too dense to highlight the clustering results. For every instance, we only include the first p greatest similarities and put the others to 0. This graph creates a new view $V + 1$ that is merged with the graphs of the other views.

To satisfy the non-negativity, the normalization and the orthogonality constraints, first, the negative values of $\mathbf{H}$ are set to zero, then, an orthogonalization step is applied to the obtained $\mathbf{H}$, and finally a step of normalization is applied to each row of $\mathbf{H}$, such that each row of $\mathbf{H}$ will have a sum equal to 1. Then the final mathematical formulation of our method is shown below.

$$\min_{\mathbf{P}_v,\, \mathbf{H}} \sum_{v=1}^{V+1} \delta_v \, ||\mathbf{S}_v - \mathbf{H}\,\mathbf{P}_v^T||_2^2 \; s.t. \;\; \mathbf{H}\mathbf{1} = \mathbf{1},\; \mathbf{H}^T\mathbf{H} = \mathbf{I}\,, \mathbf{H} \geq 0,\;\; \mathbf{P}_v^T\mathbf{P}_v = \mathbf{I}. \quad (6)$$

3.1 Optimization

In this section, we present the procedure for optimizing the objective function in (6). We use an effective algorithm based on an alternating minimization scheme to solve the final objective function and update the matrices $\mathbf{H}$ and $\mathbf{P}_v$.

To initialize the matrices $\mathbf{S}_v$ and their corresponding spectral representations $\mathbf{P}_v$, the same method as in [8] is used. Moreover, as mentioned earlier, to initialize the matrix $\mathbf{S}_{V+1}$, the $\mathbf{H}$ step (see Eq. (7)) of our algorithm is used with a number of views equal to V, then $\mathbf{S}_{V+1}$ is computed using the correlation coefficients of the rows of $\mathbf{H}$ as explained in the previous section. The initial spectral representation, $\mathbf{P}_{V+1}$, is set to the C eigenvectors of $\mathbf{S}_{V+1}$ associated with the C largest eigenvalues.

Then, the algorithm iteratively performs the following two steps alternately.

Update H: Fixing $\mathbf{P}_v$ and δ_v, we compute the derivative of the functional in (6) with respect to $\mathbf{H}$:

$$\frac{\partial f}{\partial \mathbf{H}} = \sum_{v=1}^{V+1} 2\,\delta_v\,(\mathbf{H} - \mathbf{S}_v\,\mathbf{P}_v\,).$$

The optimal solution $\mathbf{H}$ is obtained by vanishing this derivative. Consequently, $\mathbf{H}$ is given by:

$$\mathbf{H} = \left(\sum_{v=1}^{V+1} \delta_v\,\mathbf{I}\right)^{-1} \left(\sum_{v=1}^{V+1} \delta_v\,\mathbf{S}_v\,\mathbf{P}_v\right). \quad (7)$$

Thus, to get the matrix $\mathbf{H}$, we apply the aforementioned constraints to the element of the matrix $\mathbf{H}$ obtained by Eq. (7). Then, the element-wise ReLU (Rectified Linear Unit) operator is applied to the elements of the obtained matrix $\mathbf{H}$.

Update $\mathbf{P}_v$: Fixing $\mathbf{H}$ and δ_v, the objective function of our method will be equivalent to:

$$\min_{\mathbf{P}_v} \sum_{v=1}^{V+1} \delta_v \, ||\mathbf{S}_v - \mathbf{H}\,\mathbf{P}_v^T||_2^2 \tag{8}$$

Given that $\mathbf{P}_v^T\,\mathbf{P}_v = \mathbf{I}$, the above problem is the famous orthogonal Procrustes problem. The solution of this problem can be obtained using the singular value decomposition of $\mathbf{S}_v^T\,\mathbf{H}$. Let $\mathbf{Y}\Sigma\mathbf{T}^T = \mathrm{SVD}\,(\mathbf{S}_v^T\,\mathbf{H})$. Thus, the solution of the equation (8) is given by:

$$\mathbf{P}_v = \mathbf{Y}\,\mathbf{T}^T \quad \text{with} \quad \mathbf{Y}\Sigma\mathbf{T}^T = SVD\,(\mathbf{S}_v^T\,\mathbf{H}). \tag{9}$$

Update δ_v:

After updating the two matrices $\mathbf{H}$ and $\mathbf{P}_v$, the weight parameter δ_v of each view is updated using Eq. (3) with $v = 1, ..., V+1$.

After the estimation of the matrix $\mathbf{H}$, the cluster index of the sample $\mathbf{x}_i$ is given by the column index defined as the maximum value in the i-th row of $\mathbf{H}$.

The procedure of our MSLDGL method is summarized in Table 2.

Table 2. Algorithm 1 (MSLDGL).

Algorithm 1	(MSLDGL)
Input:	Data samples $\mathbf{X}_v \in \mathbb{R}^{n \times d_v}$, $v = 1, ..., V$
	The similarity matrix $\mathbf{S}_v$ for each view
	Parameter p
Output:	The consistent non negative embedding matrix $\mathbf{H}$
	The spectral embedding matrix $\mathbf{P}_v$ for each view
Initialization:	The weights $\delta_v = 1$
	Initialize $\mathbf{P}_v$ and $\mathbf{H}$ as mentioned in Sect. 3.1
	Initialize $\mathbf{S}_{V+1}$ and $\mathbf{P}_{V+1}$:
	Update $\mathbf{H}$ using Eq. (7)
	Estimate $\mathbf{S}_{V+1}$ using Eq. (5)
	Set $\mathbf{P}_{V+1}$ to the C eigenvectors of $\mathbf{S}_{V+1}$
	Repeat
	Update $\mathbf{P}_v$, $v = 1, ..., V+1$ using (9)
	Update $\mathbf{H}$ using (7)
	Update $\mathbf{S}_{V+1}$ using (5)
	Update δ_v, $v = 1, ..., V+1$ using (3)
	End

4 Performance Evaluation

4.1 Experimental Setup

To evaluate the effectiveness of our proposed approach, we used four image datasets: ORL[1], Out-Scene[2], NUS[3] and MSRCv1[4].

Our method is compared with some relevant methods: 1) Auto-weighted Multi-View Clustering via Kernelized graph learning (MVCSK) [2], 2) Multi-view spectral clustering via integrating Non-negative Embedding and Spectral Embedding (NESE) [1], 3) Sparse Multi-view Spectral Clustering (S-MVSC) [16], 4) Consistency-aware and Inconsistency-aware Graph-based Multi-View Clustering (CI-GMVC) [17] and 5) multi-view spectral clustering via Constrained Nonnegative Embedding (CNESE) [14].

One hyper-parameter is used in our optimization procedure: p, which reflects the number of the most similar samples in the label space for a particular data point. The value of p varies within the range [4 to 80]. To evaluate the performance of our method, the four commonly known cluster evaluation metrics provided in [25] are used, which are the clustering accuracy ACC, the normalized mutual information NMI, the purity indicator and the adjusted rand index ARI. It's worth mentioning that the better the performance, the higher the value of these indicators, meaning that the generated clusters are close to the ground-truth clusters.

4.2 Experimental Results

Our method is tested on real datasets. Table 3 presents the results obtained by MSLDGL and some recent methods on the datasets: ORL, Out-Scene, NUS, and MSRCv1. The best results are highlighted in bold in this table. On these datasets, the suggested approach MSLDGL performed better than other recent methods, and specifically it showed improvement over the NESE method. For certain competing methods indicated in Table 3, the related method is performed in several trials, and then a standard deviation for each indicator is indicated in parentheses.

4.3 Parameter Sensitivity

The sensitivity of the parameter p is analyzed in this section. The computed values of the ACC and NMI indicators for the MSRCv1 dataset are shown in Fig. 1 by increasing the parameter p from 4 to 80. As shown in this figure, the highest performance of MSLDGL is obtained for a value of the parameter p equal to 8.

---

1 https://cam-orl.co.uk/facedatabase.html.

2 https://scholar.googleusercontent.com/scholar?q=cache:Dxo2Hbfln2sJ:scholar.google.com/hl=enas-sdt=0,5.

3 https://lms.comp.nus.edu.sg/wp-content/uploads/2019/research/nuswide/NUS-WIDE.html.

4 https://www.researchgate.net/publication/335857675.

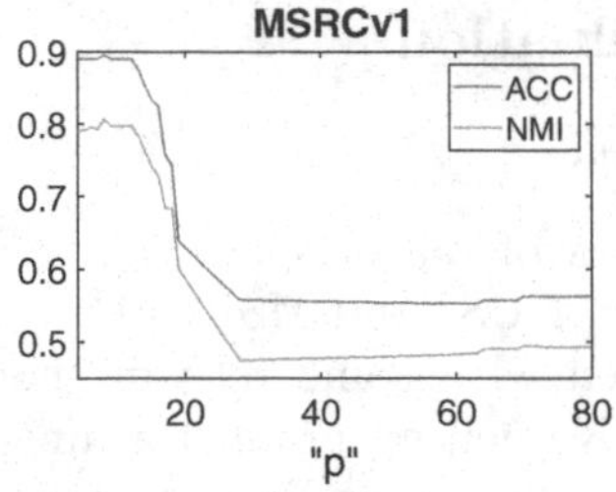

Fig. 1. Clustering performance ACC (%) and NMI (%) as a function of p on the MSRCv1 dataset.

Table 3. Clustering performance on the ORL, Out-Scene, NUS and MSRCv1 datasets.

Dataset	Method	ACC	NMI	Purity	ARI
ORL	MVCSK [2]	0.85 (±0.02)	0.94 (±0.01)	0.88 (±0.02)	0.81 (±0.02)
	NESE [1]	0.82 (±0.00)	0.91 (±0.00)	0.85 (±0.00)	0.75 (±0.00)
	S-MVSC [16]	0.80 (±0.02)	0.93 (±0.01)	0.82 (±0.02)	**0.89 (±0.01)**
	CI-GMVC [17]	0.81 (±0.00)	0.92 (±0.00)	0.85 (±0.00)	0.74 (±0.00)
	CNESE [14]	0.87 (±0.00)	**0.95 (±0.00)**	0.89 (±0.00)	0.84 (±0.00)
	MSLDGL	**0.90 (±0.00)**	**0.95 (±0.00)**	**0.91 (±0.00)**	0.86 (±0.00)
Out-Scene	MVCSK [2]	0.65 (±0.01)	0.52 (±0.00)	0.65 (±0.01)	0.42 (±0.00)
	NESE [1]	0.63 (±0.00)	0.53 (±0.00)	0.66 (±0.00)	0.46 (±0.00)
	S-MVSC [16]	0.48 (±0.01)	0.54 (±0.02)	0.65 (±0.01)	0.46 (±0.04)
	CI-GMVC [17]	0.35 (±0.01)	0.31 (±0.00)	0.35 (±0.01)	0.19 (±0.00)
	CNESE [14]	0.66 (±0.00)	**0.55 (±0.00)**	0.67 (±0.00)	**0.47 (±0.00)**
	MSLDGL	**0.70 (±0.00)**	0.53 (±0.00)	**0.70 (±0.00)**	**0.47 (±0.00)**
NUS	MVCSK [2]	0.26 (±0.01)	0.15 (±0.00)	0.28 (±0.00)	0.08 (±0.00)
	NESE [1]	0.30 (±0.00)	0.17 (±0.00)	0.32 (±0.00)	0.10 (±0.00)
	S-MVSC [16]	0.11 (±0.00)	0.01 (±0.02)	0.25 (±0.07)	0.01 (±0.01)
	CI-GMVC [17]	0.13 (±0.00)	0.05 (±0.01)	0.13 (±0.01)	0.01 (±0.00)
	CNESE [14]	0.28 (±0.00)	0.16 (±0.00)	0.31 (±0.00)	0.09 (±0.00)
	MSLDGL	**0.33 (±0.00)**	**0.18 (±0.00)**	**0.33 (±0.00)**	**0.12 (±0.00)**
MSRCv1	MVCSK [2]	0.70 (±0.02)	0.59 (±0.03)	0.70 (±0.02)	0.50 (±0.04)
	NESE [1]	0.77 (±0.00)	0.72 (±0.00)	0.80 (±0.00)	0.64 (±0.00)
	S-MVSC [16]	0.60 (±0.00)	0.69 (±0.02)	0.74 (±0.02)	**0.79 (±0.01)**
	CI-GMVC [17]	0.74 (±0.00)	0.72 (±0.00)	0.77 (±0.00)	0.59 (±0.00)
	CNESE [14]	0.86 (±0.00)	0.76 (±0.00)	0.86 (±0.00)	0.72 (±0.00)
	MSLDGL	**0.90 (±0.00)**	**0.81 (±0.00)**	**0.90 (±0.00)**	0.77 (±0.00)

4.4 Analysis of Results

It is clear from Table 3 that by applying MSLDGL, the results are always higher than NESE (the main competing method since our method is an improvement of NESE) for all the datasets. Besides, MSLDGL presents similar or higher results than other competing methods, which proves its effectiveness.

4.5 Convergence Study

Concerning the convergence of MSLDGL, Fig. 2 shows how the objective function changes as the number of iterations increases, for the MSRCv1 dataset. As shown in this figure, our approach converges very quickly, even before 20 iterations.

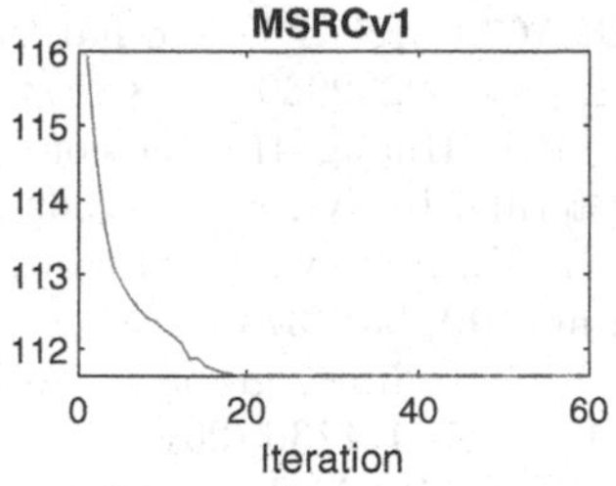

Fig. 2. Convergence of MSLDGL on the MSRCv1 dataset.

5 Conclusion

In this paper, we proposed an improved version of Nonnegative Embedding and Spectral Embedding, a novel multi-view graph-based clustering approach. We presented a new criterion that introduces two novelties over the nonnegative embedding matrix obtained by NESE to achieve better clustering performance. The initial novelty is to incorporate the label space as an additional graph into the estimating framework. The second novelty is to apply some constraints over the nonnegative embedding matrix. The usefulness and efficiency of the suggested approach are demonstrated by experimental tests on real image datasets of various types and sizes.

Acknowledgment. This research was funded by the INTER program, co-funded by the FNR (Fond National de la Recherche, Luxembourg) and the Fund for Scientific Research-FNRS, Belgium (F.R.S-FNRS), grant number 19-14016367 - 'Sustainable Residential Densification' project (SusDens, 2020–2023).

References

1. Hu, Z., Nie, F., Wang, R., Li, X.: Multi-view spectral clustering via integrating nonnegative embedding and spectral embedding. Inf. Fusion. **55**, 251–259 (2020)
2. Huang, S., Kang, Z., Tsang, I.W., Xu, Z.: Auto-weighted multi-view clustering via kernelized graph learning. Pattern Recogn. **88**, 174–184 (2019)
3. Kumar, A., Daumé, H.: A co-training approach for multi-view spectral clustering. In: Proceedings of the 28th International Conference on International Conference on Machine Learning, ICML 2011, pp. 393–400, Madison, WI, USA (2011)
4. Kumar, A., Rai, P., Daumé, H.: Co-regularized multi-view spectral clustering. In: Proceedings of the 24th International Conference on Neural Information Processing Systems, NIPS 2011, pp. 1413–1421, Red Hook, NY, USA (2011)

5. Nie, F., Cai, G., Li, J., Li, X.: Auto-weighted multi-view learning for image clustering and semi-supervised classification. IEEE Trans. Image Process. **27**(3), 1501–1511 (2017)
6. Nie, F., et al.: Parameter-free auto-weighted multiple graph learning: a framework for multi-view clustering and semi-supervised classification. In: IJCAI, pp. 1881–1887 (2016)
7. Nie, F., Tian, L., Li, X.: Multiview clustering via adaptively weighted procrustes. In: Proceedings of the 24th ACM SIGKDD international conference on knowledge discovery and data mining, pp. 2022–2030 (2018)
8. Nie, F., Wang, X., Jordan, M.I., Huang, H.: The constrained Laplacian rank algorithm for graph-based clustering. In: AAAI, pp. 1969–1976, (2016)
9. Shi, S., Nie, F., Wang, R., Li, X.: Auto-weighted multi-view clustering via spectral embedding. Neurocomputing **399**, 369–379 (2020)
10. Tang, C., et al.: Learning a joint affinity graph for multiview subspace clustering. IEEE Trans. Multimed. **21**(7), 1724–1736 (2018)
11. Xia, T., Tao, D., Mei, T., Zhang, Y.: Multiview spectral embedding. IEEE Trans. Syst. Man Cybern. Part B (Cybern.), **40**(6), 1438–1446 (2010)
12. Xu, C., Tao, D., Xu, C.: Multi-view self-paced learning for clustering. In: Proceedings of the 24th International Conference on Artificial Intelligence, IJCAI 2015, pp. 3974–3980. AAAI Press (2015)
13. Xu, Y.-M., Wang, C.-D., Lai, J.-H.: Weighted multi-view clustering with feature selection. Pattern Recogn. **53**, 25–35 (2016)
14. El Hajjar, S., Dornaika, F., Abdallah, F.: Multi-view spectral clustering via constrained nonnegative embedding. Inf. Fusion **78**, 209–217 (2021)
15. El Hajjar, S., Dornaika, F., Abdallah, F.: One-step multi-view spectral clustering with cluster label correlation graph. Inf. Sci. **592**, 97–111 (2022)
16. Hu, Z., Nie, F., Chang, W., Hao, S., Wang, R., Li, X.: Multi-view spectral clustering via sparse graph learning. Neurocomputing **384**, 1–10 (2020)
17. Horie, M., Kasai, H.: Consistency-aware and Inconsistency-aware Graph-based Multi-view Clustering. In: 2020 28th European Signal Processing Conference, pp. 1472–1476 (2021)
18. Zhan, K., Zhang, C., Guan, J., Wang, J.: Graph learning for multiview clustering. IEEE Trans. Cybern. **48**, 2887–2895 (2017)
19. Bahrami, S., Bosaghzadeh, A., Dornaika, F.: Multi similarity metric fusion in graph-based semi-supervised learning. Computation **7**, 15 (2019)
20. Sharma, K., Seal, A.: Multi-view spectral clustering for uncertain objects. Inf. Sci. **547**, 723–745 (2021)
21. Lv, J., Kang, Z., Wang, B., Ji, L., Xu, Z.: Multi-view subspace clustering via partition fusion. Inf. Sci. **560**, 410–423 (2021)
22. Zhang, G., Zhou, Y., Wang, C., Huang, D., He, X.: Joint representation learning for multi-view subspace clustering. Expert Syst. App. **166**, 113913 (2021)
23. Zhu, X., Zhang, S., He, W., Hu, R., Lei, C., Zhu, P.: One-step multi-view spectral clustering. IEEE Trans. Knowl. Data Eng. **31**, 2022–2034 (2018)
24. Ren, Z., Lei, H., Sun, Q., Yang, C.: Simultaneous learning coefficient matrix and affinity graph for multiple kernel clustering. Inf. Sci. **547**, 289–306 (2021)
25. Yang, C., Ren, Z., Sun, Q., Wu, M., Yin, M., Sun, Y.: Joint correntropy metric weighting and block diagonal regularizer for robust multiple kernel subspace clustering. Inf. Sci. **500**, 48–66 (2019)

Distance-Based Random Forest Clustering with Missing Data

Matteo Raniero, Manuele Bicego(✉), and Ferdinando Cicalese

Computer Science Department, University of Verona, Verona, Italy
{manuele.bicego,ferdinando.cicalese}@univr.it

Abstract. In recent years there has been an increased interest in clustering methods based on Random Forests, due to their flexibility and their capability in describing data. One problem of current RF-clustering approaches is that they are not able to directly deal with missing data, a common scenario in many application fields (e.g. Bioinformatics): the usual solution in this case is to pre-impute incomplete data before running standard clustering methods. In this paper we present the first Random Forest clustering approach able to *directly* deal with missing data. We start from the very recent RatioRF distance for clustering [3], which has shown to outperform all other distance-based RF clustering schemes, extending the framework in two directions, which allow the integration of missing data mechanisms directly inside the clustering pipeline. Experimental results, based on 6 standard UCI ML datasets, are promising, also in comparison with some literature alternatives.

Keywords: Random Forest clustering · Missing data · Ratio RF distance

1 Introduction

Random Forests (RFs) [6,8] represent a widely and successfully applied model for Pattern Recognition and Machine Learning. RFs are ensembles of decision trees [19], models which define, in their basic version, a hierarchical splitting of the feature space. Generally speaking, Random Forests have been mostly studied for regression and classification, whereas in alternative scenarios, such as clustering, their potentialities have not been fully exploited yet. When considering clustering, methods based on Random Forest can be broadly divided into two classes: in the first RFs (or RF-like schemes) are directly used to perform clustering [2,16,17,23,32]; in the second class [1,3,6,22,26,33] RFs are employed to derive a meaningful dissimilarity measure, to be used with a standard distance-based clustering method, such as Hierarchical clustering or Spectral clustering. In this second line, which we call *distance-based RF-clustering*, different measures have been proposed, ranging from the simplest and most employed one defined by Breiman [6,22] up to more recent and complex dissimilarities [1,3,26,33].

One problem of all these RF-clustering approaches is that they are not able to deal with missing data [18], i.e. problems where some variables do not have

S. Sclaroff et al. (Eds.): ICIAP 2022, LNCS 13233, pp. 121–132, 2022.
https://doi.org/10.1007/978-3-031-06433-3_11

a value. These scenarios are very common, especially in the biomedical field [25], in which subjects involved in a clinical study may skip some exams [15], or high-throughput sequencing technologies may return incomplete data [27]. In general, in the clustering case, the typical solutions to this problem are [31]: i) to ignore objects with missing values, or, better, ii) to complete the data with imputation methods [18]. Imputation methods, to be performed before the analysis, replace a missing value with a new one. The simplest example, called Strawman imputation, replaces a missing value in a variable with the median of all non missing values for the same variable. Since these approaches do not explicitly consider the final task (clustering), they can have some limitations, as shown in some scenarios (see e.g. [5]). Therefore a more sophisticated and recent trend appeared, which proposes to face the missing data problem *directly* inside the clustering process. In this perspective, some methods have been proposed which extend known clustering techniques (e.g. K-means) [5,7,9,13,30]. However, to the best of our knowledge, such extensions for Random Forest clustering are completely missing. In this paper we make one contribution to fill this gap, proposing the first RF-clustering method able to directly deal with incomplete data. It is important to observe that, even if RFs have been employed in the missing data context (e.g. Missforests [24]), there are no RF-clustering methods able to directly work with incomplete data, and this represents the main contribution of this paper.

We start from the very recent RatioRF distance for clustering [3], which has shown to outperform all other distance-based RF clustering schemes [1,6,22,26, 33]: this measure, defined on a set-based interpretation of the Tversky definition of similarity [28], determines the similarity between two objects by comparing their answers to a carefully selected subset of tests of the decision trees composing the forest. In this paper we propose two extensions of this framework to deal with missing data, both starting from the following observation: in the RatioRF framework a missing value represents a problem only when it is implied in a test of a node of the decision tree; in such case it is not possible to provide an answer to the binary test. To cope with this we can i) use a random decision (yes or no) or ii) keep both answers (yes/no) in an agnostic way. Due to the set-based formulation of the RatioRF distance both options can be easily integrated in the framework, as detailed in the paper. We evaluate the proposed scheme with some clustering experiments involving 6 UCI ML datasets, showing that: i) performances of RF-clustering did not degrade too much with moderate levels of missingness; ii) the two Ratio-RF modifications are equally reasonable, having different behaviours in different datasets; iii) the proposed distances compare very favourably with alternative classic distances for missing data.

The rest of the paper is organized as follows: in Sect. 2 we review the RatioRF approach, fixing the notation and introducing the basic concepts. The proposed approach is then fully presented in Sect. 3, and evaluated in Sect. 4. Finally, Sect. 5 concludes the paper.

2 Random Forest Clustering with RatioRF

In this section we will briefly introduce the starting point of our work, i.e. the very recent Random Forest clustering scheme using the RatioRF dissimilarity measure [3]. After introducing the RatioRF distance, we will briefly summarize the complete clustering scheme.

2.1 The RatioRF Distance

Assume we have set of objects/points U and a set of binary tests A (for attributes) defined over the whole set U, i.e., for each object $x \in U$ and each test $\theta \in T$ there is a unique value $\theta(x) \in \{yes, no\}$. A decision tree on a ground set of objects/points U and test set A is a binary tree T where: (i) each internal node ν is associated to a binary test $\theta_\nu \in A$; (ii) the two edges connecting the node to its children are associated with the two possible results—denoted Y for *yes* and N for *no*—of performing test θ_ν on an object from U. Further, ν_Y (resp. ν_N) denotes the child of ν connected to ν via the edge associated with Y (resp. N); $r(T)$ denotes the root of T. Let ν be a node of T at level $h+1$ and $\theta_1, b_1, \theta_2, b_2, \ldots, \theta_h, b_h$ be the sequence of nodes (tests) and edges (results), encountered on the unique path from $r(T)$ to ν. Then, it is possible to associate to ν the set of objects $S_\nu = \{x \in U \mid \theta_i(x) = b_i, i = 1, \ldots, h\}$. In words, a node ν is representative of (or it *contains*) all the objects that, when tested according to the adaptive strategy represented by the decision tree T, follow the path from the root to ν.

For each object x there is a single leaf containing it denoted as by $\ell(x)$. Let $P_T(x)$ be the set of pairs (*test*, *result*) associated to x by the strategy/tree T

$$P_T(x) = \{(\theta, b_x^\theta) \mid \theta \text{ is a test on the path from the root } r(T) \text{ to the leaf } \ell(x) \text{ and } b_x^\theta = \theta(x)\}.$$

Let θ be a test and $b \in \{Y, N\}$. It is possible to say that x agrees with (θ, b) if $\theta(x) = b$. Similarly, objects x and y agree on test θ if $\theta(x) = \theta(y)$.

A decision tree T can be used to select the set of features Φ relevant for the assessment of similarity between pairs of objects from the universe U. In particular, in [3] authors define $\Phi = \{(\theta_\nu, b) \mid \nu \text{ is a node of } T, b \in \{Y, N\}\}$, as the set of possible outcomes of the tests used by the decision tree. For an object x its feature set $X = P_T(x)$ is defined as a set of test results on the path from $r(T)$ to the leaf $\ell(x)$ associated to x by the decision tree. These are the features from Φ that are most relevant for x, in the sense of being sufficient to identify x.

Now, assume we want to compare objects x, y represented by the set of features $X = P_T(x)$, and $Y = P_T(y)$, respectively. In [3] authors define

$$X \dot{-} Y = \{(\theta, b) \mid (\theta, b) \in X \text{ and } \theta(y) \neq b\} \tag{1}$$

to be the set of features that are relevant for x and on which y disagrees. Symmetrically the set of features relevant for y and on which x disagrees are given by the set

$$Y \dot{-} X = \{(\theta, b) \mid (\theta, b) \in Y \text{ and } \theta(x) \neq b\} \tag{2}$$

They also define

$$X \Cap Y = \{(\theta, b) \in X \cup Y \mid \theta(x) = \theta(y)\} \tag{3}$$

to be the set of features on which x and y agree, among the features in $P_T(x) \cup P_T(y)$, which are those relevant for describing them (i.e., for identifying one or the other). The *Ratio-DecisionTree* similarity measure RatioDT$(\cdot, \cdot)$ is defined by [3]

$$\text{RatioDT}(x, y) = \frac{|X \Cap Y|}{|X \Cap Y| + |X \dot{-} Y| + |Y \dot{-} X|}, \tag{4}$$

As observed in [3], this similarity measure is symmetric and the corresponding dissimilarity obtained as $\sqrt{1 - \text{RatioDT}(x, y)}$ is a metric.

Remark 1. In [3], the approach above was derived following an axiomatic definition of similarity measures given by Tversky [28]. An alternative perspective on such similarity measure computation is the following: let $\Phi(XY) = P_T(x) \cup P_T(y)$ denote the set of features restricted to those employed by the tree to describe x and y. Then let $X_{\Phi(XY)}$ (resp. $Y_{\Phi(XY)}$) be the element of $\Phi(XY)$ on which x (resp. y) agrees. Then, $\text{RatioDT}(x, y) = |X_{\Phi(XY)} \cap Y_{\Phi(XY)}| / |X_{\Phi(XY)} \cup Y_{\Phi(XY)}|$, i.e., the Jaccard distance computed on the restricted set of features, that the tree selected for x and y.

The RatioDT similarity measure is straightforwardly generalized to Random Forests by averaging the decision tree distance in Eq. (4) over all the trees in the forest. More precisely, given a trained RF whose trees are $T_1, \ldots, T_m$, fix a pair of points $x, y \in U$ and let $\text{RatioDT}_\text{t}(x, y)$ be the similarity computed according to (4) from the decision tree T_t. Then, the Random Forest similarity measure RatioRF(x, y) is defined by averaging over all decision trees, i.e.

$$\text{RatioRF}(x, y) = \frac{1}{m} \sum_{t=1}^{m} \text{RatioDT}_\text{t}(x, y). \tag{5}$$

If the clustering algorithm needs in input a dissimilarity, it is possible to transform the similarity into a dissimilarity using $\sqrt{1 - \text{RatioRF}(x, y)}$, as done in [22].

2.2 The Complete Random Forest Clustering Procedure

The clustering is obtained with the following procedure:

1. **RF training**. In this step a Random Forest is trained on the data to be clustered. The main issue is that labels are not available: to face this issue it is possible to use Extremely Randomized Trees [12], i.e. trees in which the split feature and the threshold are chosen randomly – this representing a common and reasonably good solution for RF-clustering [4].

2. **Distance computation**. In this second step the RatioRF distance is computed from the trained forest, as explained in Sect. 2.1.
3. **Clustering**. Starting from the similarity, the final clustering is then obtained via any distance-based clustering algorithm, such as Hierarchical Clustering or Spectral Clustering [29].

3 Dealing with Missing Data

The RF-clustering scheme defined in the previous section requires all values for all features of the objects involved in the clustering. The presence of missing data impacts the first (RF training) and the second step (RatioRF distance computation). In the following we will introduce first how to derive the RatioRF distance with missing data, since this represents the most problematic part of the approach.

3.1 Computing RatioRF with Missing Data

The above definition of RatioDT assumes that all tests are defined on every objects. The presence of missing data in a data set is equivalent to the situation in which for some object x and test θ the value $\theta(x)$ is not defined, which is typically indicated by $\theta(x) = NAN$. There are two issues that need to be addressed if we want to employ the RatioDT(x, y) also in the presence of missing data. When an object x reaches a node ν such that $\theta_\nu(x) = NAN$:

1. should the pair (θ_ν, NAN) be part of the set of features X describing x?
2. what is the next node/test to consider for x between ν_Y and ν_N, i.e., how should we complete the partial root-to-leaf path for x beyond ν? How should we decide, considering that the test result $\theta_\nu(x)$ doesn't say which of the edges Y or N to follow?

Regarding point 1, our choice is not to consider such a node as part of the set X, as this would imply *unfounded* dissimilarity of x with any other objects y for which test θ_ν is defined. By *unfounded* we mean that we do not know whether the missing value of x on test θ_ν agrees or not with $\theta_\nu(y)$, hence it would not be fair to assume it is different.

Regarding point 2, we actually analyse two possibilities: (i) choosing at random whether to continue the root-to-leaf path for x on ν_Y or ν_N; (ii) extending the path in both directions, i.e., having $P_T(x)$ be a collection of root-to-leaf paths parting from one another at some node associated to a test where x is not defined. We call (i) the SINGLEPATH approach and (ii) the MULTIPATH approach. Accordingly, we denote by $X^{SP} = P_T^{SP}(x)$ (resp. $X^{MP} = P_T^{MP}(x)$) the set of features (pairs of test and results) selected by the SINGLEPATH (resp. MULTIPATH) approach. A simple pseudocode describing a recursive construction of such sets is given in Algorithms 1 2. Employing such procedures we assign $P_T^{SP}(x)$ =SINGLEPATH$(x, T, root(T))$ and $P_T^{MP}(x)$ =MULTIPATH$(x, T, root(T))$.

Therefore, in the presence of missing data, we can compute RatioDT(x, y) like in (4) by substituting X, Y with X^{SP}, Y^{SP} (resp. X^{MP}, Y^{MP}).

Algorithm 1: SINGLEPATH(x, T, v)

Input: A decision tree T; an object x; and a node ν of T
Output: a set X of relevant feature (pairs $(\theta_w, \theta_w(x))$ over some path from ν to a leaf of T.
if ν is a leaf **return** $\emptyset$;
if $\theta_\nu(x) = Y$ **then**
  return SINGLEPATH$(x, T, \nu_Y) \cup \{(\theta_\nu, Y)\}$
if $\theta_\nu(x) = N$ **then**
  return SINGLEPATH$(x, T, \nu_N) \cup \{(\theta_\nu, N)\}$
if $\theta_\nu(x) = NAN$ **then**
  choose ν_{next} randomly between ν_Y and ν_N;
  return SINGLEPATH(x, T, ν_{next})

Algorithm 2: MULTIPATH(x, T, v)

Input: A decision tree T; an object x; and a node ν of T
Output: a set X of relevant feature (pairs $(\theta_w, \theta_w(x))$ over some collection of paths T starting at ν and reaching a leaf.
if ν is a leaf **return** $\emptyset$;
if $\theta_\nu(x) = Y$ **then**
  return MULTIPATH$(x, T, \nu_Y) \cup \{(\theta_\nu, Y)\}$
if $\theta_\nu(x) = N$ **then**
  return MULTIPATH$(x, T, \nu_N) \cup \{(\theta_\nu, N)\}$
if $\theta_\nu(x) = NAN$ **then**
  return MULTIPATH$(x, T, \nu_Y) \cup$ MULTIPATH(x, T, ν_N)

3.2 Training Trees with Missing Data

In the training phase we also need to deal with the presence of missing data: in particular, we need to decide how an object x used in the procedure for building a tree is moved down (to the right or the left child?) after a split associated to a test for which x's value is missing/not known. Our choice is to have x continue on both child nodes. This appears to be well in the spirit of not assigning an arbitrarily imputed value to x for the test (since any choice would be unfounded as observed in the description of the testing phase)[1].

4 Experimental Evaluation

This section contains the empirical evaluation of the proposed approach. First we introduce the experimental details, then we present the results and discussion. A comparative analysis with literature alternatives concludes the section.

[1] Some experiments, not reported here, showed that empirical results would not change too much if we randomly choose one of the two paths.

4.1 Experimental Details

In order to evaluate our methods we consider some public datasets from the UCI ML Repository [10], whose details are reported in Table 1. As commonly done in clustering, we use supervised problems, remove labels, compute the clustering result and then compare it with the original labelling. In particular, we quantify the performance results for clustering quality considering the classic adjusted Rand index (ARI) [14].

Table 1. Details of the datasets used in the analysis.

Dataset	#objects	#features	#of clusters
Iris	150	4	3
Btissue	106	9	6
Wine	178	13	3
Glass	214	9	4
Leaf	340	15	30
Libras	390	90	15

To simulate missingness, data were artificially removed from these datasets using the MCAR (Missing Completely At Random) protocol [20]. This protocol consists in removing data completely at random, without taking into account any relationship between features. We only considered one constraint: no objects with all missing features can be considered. We considered 4 levels of missingness, i.e. removing 5%, 10%, 20% and 30% of the data. For every problem and each level of missingness we generated 20 datasets.

For what concerns the proposed RF-clustering scheme, in all experiments we trained RFs using the strategy described in Sect. 3.2: we used 100 trees in each forest, with $\log(n)$ for the maximum depth of each tree (with n the number of objects in the dataset). Once the RF distance is computed, the clustering is obtained using three classic approaches: spectral clustering, using the Ng-Jordan-Weiss normalized version [29], repeating the inner k-means 20 times, Affinity Propagation [11] and Hierarchical clustering, in the Ward-Link version.

4.2 Results and Discussion

In this section we compare the result of the proposed approach with the complete case, i.e. with the result obtained with the original RatioRF scheme on the complete matrix. The main goal of this analysis is to measure the impact of the missingness on the performance results for clustering quality. The results are reported in Table 2, for the different clustering methods and missingness values. In detail, the column "No Missing" contains the results with the original RF-Ratio scheme (we averaged the ARI among 20 repetitions); the other columns contain the mean ARI of the two approaches (Single Path and Multi Path), averaged over the 20 generated datasets with missing data. In order to have a statistically significant comparison, for every missing level, we perform a paired t-test ($\alpha = 0.05$) with the complete case. Bold values in table represent those cases for which there is no statistical difference between the results with and without missing data (i.e. situations in which missing data does not impact the clustering performances). From the table it is evident that the proposed approach is robust in dealing with missing data; both versions are very robust with moderate levels of missingness: in fact, the performance results for clustering quality do not degrade too much with respect to the complete case, especially for 5–10 and in some cases 20% of missing data. Please note that, when using Affinity Propagation, we have robustness also in three datasets for a remarkable 30% of missing data.

Table 2. Results for the proposed approach, in comparison with the No Missing case.

Spectral clustering

Dataset	No Missing	Missing 5%		Missing 10%		Missing 20%		Missing 30%	
		SP	MP	SP	MP	SP	MP	SP	MP
Iris	0.6903	**0.7002**	**0.6936**	**0.6914**	**0.6781**	**0.6791**	**0.6757**	0.6633	0.5855
Breast	0.4169	**0.4109**	**0.4090**	0.3860	0.3893	0.3411	0.3696	0.3138	0.3566
Libras	0.2969	**0.2890**	**0.2882**	**0.2907**	**0.2910**	0.2774	0.2833	0.2527	0.2859
Wine	0.8623	**0.8666**	**0.8689**	**0.8669**	**0.8784**	0.8339	0.8365	0.7967	0.8060
Leaf	0.3891	0.3583	0.3723	0.3227	0.3607	0.2294	0.3082	0.1640	0.2396
Glass	0.1976	**0.1904**	**0.1928**	0.1782	0.1815	0.1767	0.1785	0.1683	0.1672

Affinity Propagation

Dataset	No Missing	Missing 5%		Missing 10%		Missing 20%		Missing 30%	
		SP	MP	SP	MP	SP	MP	SP	MP
Iris	0.7021	**0.7150**	**0.6827**	**0.7216**	0.6284	**0.6942**	0.4522	**0.6698**	0.3875
Breast	0.3800	0.3483	**0.3625**	0.3552	**0.3694**	0.3382	0.3281	0.3012	0.2933
Libras	0.2125	**0.2151**	0.2235	**0.2073**	**0.2232**	0.1876	0.2287	0.1613	**0.2098**
Wine	0.6882	**0.6983**	**0.6950**	**0.7231**	**0.7207**	**0.6600**	**0.6561**	0.6047	0.4972
Leaf	0.3540	**0.3472**	**0.3612**	0.3265	**0.3488**	0.2656	0.3291	0.2173	0.2812
Glass	0.1592	**0.1553**	**0.1526**	**0.1496**	**0.1480**	**0.1460**	0.1335	**0.1449**	0.1022

Hierarchical Clustering

Dataset	No Missing	Missing 5%		Missing 10%		Missing 20%		Missing 30%	
		SP	MP	SP	MP	SP	MP	SP	MP
Iris	0.7374	0.6951	**0.7132**	0.6808	0.6701	0.6558	0.6229	0.6147	0.5576
Breast	0.3673	**0.3562**	**0.3562**	**0.3551**	**0.3539**	**0.3448**	**0.3624**	0.3198	0.3419
Libras	0.2880	**0.2849**	**0.2905**	**0.2893**	**0.2876**	**0.2846**	**0.2876**	0.2697	0.2885
Wine	0.8868	**0.8617**	**0.8583**	**0.8370**	**0.8567**	0.8010	0.8018	0.6962	0.7610
Leaf	0.3931	0.3730	**0.3888**	0.3650	**0.3872**	**0.3207**	0.3731	0.2619	0.3293
Glass	0.2250	**0.2447**	**0.2483**	**0.2325**	**0.2399**	**0.2128**	**0.1941**	0.1641	0.1862

In order to have a better comparison between the SinglePath and the MultiPath approach, we present in Fig. 1 the performance results for clustering quality of the two methods for the different datasets, averaging them among the three different clustering methodologies. In the figure, a filled mark denoted a situation in which one of the two approaches outperforms the other with a statistically significant difference (again according to a paired t-test with $\alpha = 0.05$). From the plots we can notice that there is not a clear best strategy, and the optimal solution highly depends on the datasets: it seems that MultiPath is preferable with datasets with several features (as Leaf and Libras), whereas SinglePath is more appropriate for low dimensional problems (as Iris).

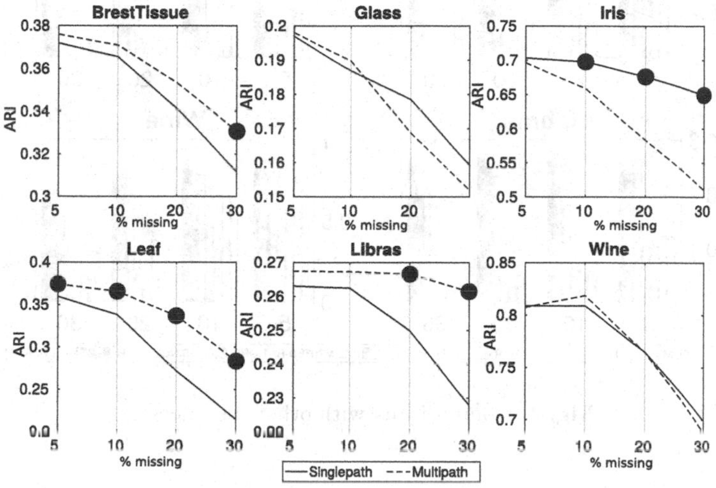

Fig. 1. SinglePath vs MultiPath

Finally, to check the overall validity of our approach, here we present a comparative analysis with some alternative distances used to deal with missing data, as described in the recent [21]. In particular we employed the Heterogeneous Euclidean-Overlap Metric (HEOM), the Heterogeneous Value Difference Metric (HVDM) and two redefinitions of these two, namely HEOM-REDEF and HVDM-REDEF. These distances can be computed with missing data (see [21] for a comparison between them), and are used in our framework as input for the three clustering procedures described before (Spectral clustering, Affinity Propagation and Hierarchical Clustering). As a further comparison, we also compute the results with the standard RatioRF pipeline on data pre-imputed with the simple Strawman method. This last comparison would permit to measure the benefits of including the management of the missing data inside the clustering procedure with respect to the pre-imputation solution.

In Fig. 2 results for the tree clustering schemes are averaged and presented using bar plots. From results, it is clear that our proposal largely outperforms

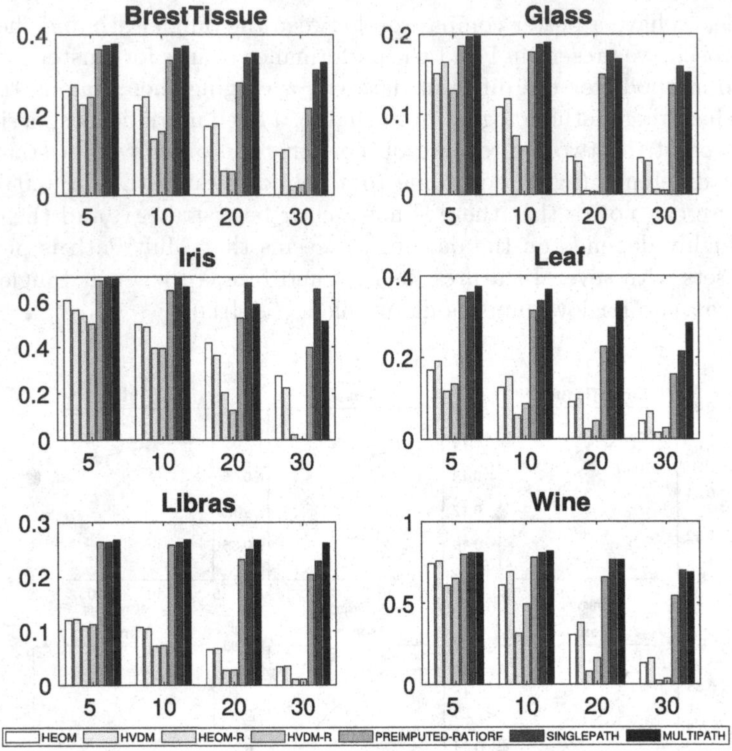

Fig. 2. Comparisons with other distances.

the alternative distances in almost all the cases, with improvements which are very relevant for large levels of missingness. Interestingly, the two proposed approaches also improve over the standard RatioRF pipeline applied on pre-imputed data, thus providing a further confirmation that it is more beneficial to deal with the missing data directly inside the clustering scheme, as shown for other clustering strategies in [5,7,9,13,30].

5 Conclusions

In this paper we presented an extension of the Random Forest clustering approach able to deal with missing data, based on two extensions of the recent RatioRF framework. An empirical evaluation confirms the robustness of the proposed strategies, both with respect to the results obtained with the complete data as well as in comparison with literature alternatives. In our future work we plan to add more empirical comparisons, in particular following two different directions: from one side we will enalrge the number of analysed datasets, in order to determine if there exists a correlation between the accuracies and the different aspects of a given dataset (its missingness nature, number of features/objects, number of clusters); on the other side we will include in the analysis more comparisons with classic as well advanced approaches to deal with missing data, like imputation (using more sophisticated approaches like MICE or knn-imputation) or marginalization.

Acknowledgements. Authors would like to thank the anonymous reviewers for providing helpful comments and suggestions.

References

1. Aryal, S., Ting, K.M., Washio, T., Haffari, G.: A comparative study of data-dependent approaches without learning in measuring similarities of data objects. Data Min. Knowl. Disc. **34**(1), 124–162 (2019). https://doi.org/10.1007/s10618-019-00660-0
2. Bicego, M.: K-random forests: a K-means style algorithm for random forest clustering. In: Proceedings of International Joint Conference on Neural Networks (IJCNN 2019) (2019)
3. Bicego, M., Cicalese, F., Mensi, A.: RatioRF: a novel measure for random forest clustering based on the Tversky's ratio model. IEEE Trans. Knowl. Data Eng. (2022, in press). https://doi.org/10.1109/TKDE.2021.3086147, https://ieeexplore.ieee.org/document/9446631
4. Bicego, M., Escolano, F.: On learning random forests for random forest clustering. In: Proceedings of International Conference on Pattern Recognition, pp. 3451–3458 (2020)
5. Boluki, S., Dadaneh, S., Qian, X., Dougherty, E.: Optimal clustering with missing values. BMC Bioinform. **20**(Suppl. 12), 321 (2019)
6. Breiman, L.: Random forests. Mach. Learn. **45**, 5–32 (2001)
7. Chi, J., Chi, E., Baraniuk, R.: k-POD: a method for k-means clustering of missing data. Am. Stat. **70**(1), 91–99 (2016)
8. Criminisi, A., Shotton, J., Konukoglu, E.: Decision forests: a unified framework for classification, regression, density estimation, manifold learning and semi-supervised learning. Found. Trends Comput. Graph. Vis. **7**(2–3), 81–227 (2012)
9. Datta, S., Bhattacharjee, S., Das, S.: Clustering with missing features: a penalized dissimilarity measure based approach. Mach. Learn. **107**(12), 1987–2025 (2018). https://doi.org/10.1007/s10994-018-5722-4
10. Dua, D., Graff, C.: UCI machine learning repository (2017). http://archive.ics.uci.edu/ml
11. Frey, B.J., Dueck, D.: Clustering by passing messages between data points. Science **315**(5814), 972–976 (2007)
12. Geurts, P., Ernst, D., Wehenkel, L.: Extremely randomized trees. Mach. Learn. **63**(1), 3–42 (2006)
13. Hathaway, R., Bezdek, J.: Fuzzy c-means clustering of incomplete data. IEEE Trans. Syst. Man Cybern. B (Cybern.) **31**(5), 735–44 (2001)
14. Hubert, L., Arabie, P.: Comparing partitions. J. Classif. **2**(1), 193–218 (1985)
15. Jakobsen, J., Gluud, C., Wetterslev, J., Winkel, P.: When and how should multiple imputation be used for handling missing data in randomised clinical trials - a practical guide with flowcharts. BMC Med. Res. Methodol. **17**, 162 (2017)
16. Moosmann, F., Triggs, B., Jurie, F.: Fast discriminative visual codebooks using randomized clustering forests. In: Advances in Neural Information Processing Systems 19, pp. 985–992 (2006)

17. Perbet, F., Stenger, B., Maki, A.: Random forest clustering and application to video segmentation. In: Proceedings of British Machine Vision Conference, BMVC 2009, pp. 1–10 (2009)
18. Pigott, T.: A review of methods for missing data. Educ. Res. Eval. **7**(4), 353–383 (2001)
19. Quinlan, J.: C4.5: Programs for Machine Learning. Morgan Kaufmann Publishers Inc., Burlington (1993)
20. Rubin, D.B.: Inference and missing data. Biometrika **63**(3), 581–592 (1976)
21. Santos, M., Abreu, P., Wilk, S., Santos, J.: How distance metrics influence missing data imputation with k-nearest neighbours. Pattern Recogn. Lett. **136**, 111–119 (2020)
22. Shi, T., Horvath, S.: Unsupervised learning with random forest predictors. J. Comput. Graph. Stat. **15**(1), 118–138 (2006)
23. Shotton, J., Johnson, M., Cipolla, R.: Semantic texton forests for image categorization and segmentation. In: Proceedings of International Conference on Computer Vision and Pattern Recognition (CVPR 2008) (2008)
24. Stekhoven, D., Buhlmann, P.: Missforest: non-parametric missing value imputation for mixed-type data. Bioinformatics **28**(1), 112–118 (2011)
25. Sterne, J., et al.: Multiple imputation for missing data in epidemiological and clinical research: potential and pitfalls. BMJ **338**, b2393 (2009)
26. Ting, K., Zhu, Y., Carman, M., Zhu, Y., Zhou, Z.H.: Overcoming key weaknesses of distance-based neighbourhood methods using a data dependent dissimilarity measure. In: Proceedings of International Conference on Knowledge Discovery and Data Mining, pp. 1205–1214 (2016)
27. Troyanskaya, O., et al.: Missing value estimation methods for DNA microarrays. Bioinformatics **17**(6), 520–525 (2001)
28. Tversky, A.: Features of similarity. Psychol. Rev. **84**(4), 327 (1977)
29. von Luxburg, U.: A tutorial on spectral clustering. Stat. Comput. **17**(4), 395–416 (2007)
30. Wagstaff, K.: Clustering with missing values: no imputation required. In: Classification, Clustering, and Data Mining Applications, pp. 649–658 (2004)
31. Wagstaff, K.: Clustering with missing values: no imputation required. In: Banks, D., McMorris, F.R., Arabie, P., Gaul, W. (eds.) Classification, Clustering, and Data Mining Applications, pp. 649–658. Springer, Heidelberg (2004). https://doi.org/10.1007/978-3-642-17103-1_61
32. Yan, D., Chen, A., Jordan, M.: Cluster forests. Comput. Stat. Data Anal. **66**, 178–192 (2013)
33. Zhu, X., Loy, C., Gong, S.: Constructing robust affinity graphs for spectral clustering. In: Proceedings of International Conference on Computer Vision and Pattern Recognition, CVPR 2014, pp. 1450–1457 (2014)

Video Analysis and Understanding

Unsupervised Person Re-identification Based on Skeleton Joints Using Graph Convolutional Networks

Khadija Khaldi(✉), Pranav Mantini, and Shishir K. Shah

University of Houston, Houston, USA
KhadijaaKhaldi@gmail.com

Abstract. With the remarkable progress of deep learning methods, person re-identification has received a lot of attention from researchers. However, the majority of previous work mainly focus on supervised learning setting, which requires expensive data annotations. In this paper, we address this problem by proposing a purely unsupervised learning model. Inspired by the effectiveness of modeling the spatio-temporal information of pedestrian video, we mine the relationships between human body joints. Specifically, we propose a novel framework by learning inter-frame and intra-frame relationships for discriminative feature learning via two Graph Convolutional Networks (GCN) modules: spatial and temporal. The spatial module captures the structural information of the human body and the temporal module propagates information across adjacent frames. At the end, we perform hierarchical clustering by selecting P identities and K instances (PK sampling) to generate pseudo-labels for the unlabeled data. By iteratively optimizing these modules, our model extracts robust spatial-temporal information that can alleviate the occlusion problem. We conduct experiments on two benchmarks: MARS and DukeMTMC-VideoReID datasets, where we demonstrate the effectiveness of our proposed method.

Keywords: Person re-identification · Unsupervised learning · Graph neural network

1 Introduction

Person re-identification (ReID) is a crucial problem, which aims to match pedestrians across non-overlapping camera views. Currently, ReID plays a significant role in the field of video surveillance and security. However, a multitude of factors such as change of illumination, occlusion, viewpoints, make person re-identification an extremely challenging task. Supervised ReID methods have achieved significant success attributing to the rapid progress in the ability of convolutional neural networks (CNN) to learn complex features. However, these methods remain impractical due to the need for large-scale annotated data in order to learn a robust embedding subspace. To relieve the requirement for expensive data annotation, in this paper, we focus on the unsupervised ReID task.

S. Sclaroff et al. (Eds.): ICIAP 2022, LNCS 13233, pp. 135–146, 2022.
https://doi.org/10.1007/978-3-031-06433-3_12

Unsupervised person ReID aims to learn discriminative features from unlabeled person images hence show better potential to apply person ReID in real applications. There have been a series of unsupervised methods commonly categorized under: 1) domain adaptation and 2) fully unsupervised learning. Domain adaptation methods [14,29] learn robust features by pre-training the model on labeled source data and then transferring the knowledge to unlabeled target data. However, due to the large gap between source and target dataset, the performance significantly degrades. Recently, fully unsupervised methods gained attention, as they achieve good performance without the need for an annotated dataset. For instance, one promising work demonstrates an iterative clustering framework [6] that learns representations by learning the similarity between images and then apply clustering to generate pseudo-labels.

Fig. 1. An illustration of three image sequences from Mars dataset. It is clear that pedestrian legs are occluded in the first and second frames but reappear in the third frame, which motivates the use of temporal GCN to propagate information to occluded nodes.

The previous person ReID methods rely heavily on appearance features without capturing the temporal information. Considering that frames suffer from occlusion, noise, or low illumination; appearance-based features in conjunction with temporal features can improve our ability to perform ReID. To this end, video-based methods can provide us with richer information. Most of the video-based ReID methods employ RNN/LSTM [30] to effectively extract high-order information between frames. Nevertheless, many of these approaches are unable to exploit the intrinsic relationships between body part within the same frame and across different frames. As shown in Fig. 1, in the first frame, some body parts of the person are occluded, however, these parts reappear in the next frames. Appearance-based ReID methods tend to fail in such cases, however, encoding the relationship between the body parts within and across frame can overcome such shortcomings.

Considering Graph Convolutional Networks (GCN) have achieved state-of-the-art performance in many tasks [19,24] and inspired by their strong ability of generalizing neural networks for data with graph structures, we propose to use GCN to model the intra and inter relationships of human body parts. In this regard, we propose an unsupervised video-based framework optimized in two steps. First, we learn discriminative features by exploiting the similarity between images using GCN, then we apply hierarchical clustering to generate pseudo labels. Specifically, given a pedestrian image we use a pose alignment model to localize body parts, then we connect them together into an intra-frame graph. This graph propagates contextual information between body parts of the same person. Then, to take into consideration the temporal information, we create an inter-graph connecting the same body part across adjacent frames of the same tracklet. Finally, a unified framework, namely Unsupervised Person Re-identification based on Skeleton Joints using Graph Convolutional Networks is proposed to simultaneously model the spatial and temporal relations of inter and intra frames regions in a person tracklet.

In summary, the contributions of this work are:

1. We propose a novel method to encode human body structure information to accomplish unsupervised ReID.
2. We employ GCN to model the inter and intra frame relations between pedestrian body joints. By combining these two relation connections, our framework is capable of alleviating the occlusion problem.

Extensive experiments show that our proposed method outperforms existing unsupervised state-of-the-art methods on two large-scale video-based person ReID datasets

2 Related Work

Unsupervised Person ReID

Unsupervised Person ReID methods can be categorized into Unsupervised Domain Adaptation (UDA) [3,18] and Fully Unsupervised ReID (FU) [6,12,29] methods. The former tries to transfer the knowledge from a labeled source domain data to an unlabeled target domain data, where these two domains differ in data distribution.

Among the existing works to improve target domain performance, methods employ a generative adversarial network (GAN) to transfer the source domain images to target-domain style. To handle the domain gap problem, SPGAN [3] used CycleGAN [33] as domain style transfer-backbone. Similarly, Wang *et al.* developed a framework TJ-AIDL [18], which learns jointly an attribute semantic and identity discriminative feature representation space for the unlabeled target domain dataset. Despite their effectiveness, these methods largely ignore the intra-domain variations such as pose and background variations in the target domain.

On the other hand, recently a lot of studies have been focusing on solving person ReID problem using a fully unsupervised setting. The general idea of these methods is to explore unlabeled data progressively by alternately assigning pseudo labels to unlabeled data and fine-tuning the model according to these pseudo-labeled data. Unlike UDA, these methods do not have access to annotations and hence are more challenging to design.

In [12], Lin *et al.* iteratively train their model using pseudo-labels generated by a bottom-up clustering approach then update their classifier in each iteration. Similarly, Khadija *et al.* [6] propose to learn a strong adaptative feature embedding using contrastive learning, before generating pseudo-labels. Different from image-based unsupervised ReID, [20] learns an effective ReID model by fully exploring the discriminative structural information of unlabeled tracklets.

Graph Neural Network

In recent years, researchers have paid more attention to graph convolutional networks (GCNs) and their variants due to their strong ability to model relations and their capacity for reasoning. GCN has been successfully applied to numerous computer vision tasks [19,24]. It performs representation learning with a neighborhood aggregation framework that combines structure information and node features in the learning process. Several recent works [1,25] employed graph models for person ReID.

Wu *et al.* [22] introduced a graph neural network to model the contextual interactions between the relevant regional features by combining pose alignment and feature affinity connections. This method focuses on modeling the inter-frame body part relationships but ignores the intra-frame structural information. To alleviate this problem, Yang *et al.* [26] proposed a Spatial-Temporal Graph Convolutional Network (STGCN) which consists of two GCN branches. The spatial branch learns the structural information of the human body. The temporal branch provides discriminant information from adjacent frames to address the occlusion problem. Wu *et al.* [21] introduced an unsupervised graph association approach UGA to mine the cross-view relationships.

Different from existing ReID graph-based methods which only use body patches as nodes, we hypothesize that modeling spatial and temporal relationships between key-points for unsupervised feature learning, can improve the ability to perform unsupervised ReID.

3 Proposed Method

First, we extract semantic local features using a CNN backbone and pose estimation model. Concretely, we consider the local features of an image as nodes of a graph. Then we apply two GCN modules: spatial and temporal, to propagate information of key-points across current frame and adjacent frames. At the end, we perform hierarchical clustering with PK sampling to generate pseudo-labels for the unlabeled data. An overview of the proposed method is shown in Fig. 2.

3.1 Preliminary

Inspired by the recent success of part-based ReID methods [26] in dealing with the occlusion problem, the goal of this module is to extract semantic features of key-point regions. Specifically, given a tracklet of person sequence V, we randomly sample T frames from the input video. We denote the tracklet as $V = \{I_1, I_2, ..., I_T\}$ where T is the number of frames sampled from the video.

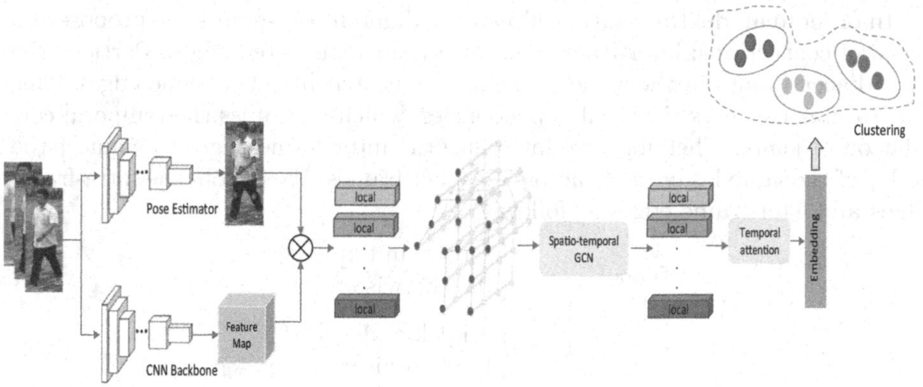

Fig. 2. An illustration of our model architecture. First, our framework extracts semantic local features of key-points using a CNN backbone and pose estimator. We view the local features of an image as nodes of a graph and propose a spatio-temporal GCN to propagate information between nodes. At the end, we generate new pseudo labels using hierarchical clustering. In an iterative end-to-end fashion, our model is trained using three optimization functions.

We first utilize a pre-trained CNN model to extract a feature map for each of the frames:

$$F_{cnn} = \{F_1, F_2, ..., F_T\}, \quad (1)$$

where F_i is the feature map of i-th frame in the video, $F_i \in \mathbb{R}^{h \times w \times c}$, in which h, w, c denotes the height, width, and channels, respectively.

We feed the same tracklet to a pose estimator model to get a heatmap of key-points M_{kp}. Through an outer product ($\otimes$) and a global average pooling operations ($g(.)$), we can get a group of appearance features $x_k^S \in \mathbb{R}^c$ for each of the key-points regions K:

$$X^S = \{x_1^S, ..., x_K^S\} = g(F_{cnn} \otimes M_{kp}), \quad (2)$$

where $K = 14$.

Thus, for a video with T frames, the total number of keypoints regions is $KP = T \times K$.

3.2 Joint Graph Construction

Inspired by the benefits of modeling the structural relationships among human body parts, we construct a spatio-temporal graph $G = (X, E)$ on a person sequence with K joints and T frames featuring both intra-body and inter-frame connection. Firstly, we utilize a pose estimation model [17] to locate 17 key-points (joints), then we fuse all key-points on the head region and get final $K = 14$ key-points. We define $X = \{x_i\}_{i=1}^{T.K}$ as the set of nodes in our constructed graph for a pedestrian video, where each node v_i corresponds to a key-point in the frame. The terms nodes and key-points are interchangeable in our work.

In order to model the relationship among different key-points, we propose two types of connections: inter-frame and intra-frame connected edges. Particularly, the adjacent joints in the same frame are connected by intra-frame edges. This natural link captures the local dependencies, which may miss the temporal correlation of joints. Therefore, we introduce the inter-frame edges to connect the nodes of the same body joints across adjacent frames. We denote the intra-frame edges and inter-frame edges as follows:

$$\varepsilon_{\text{intra}}(i,j) = \begin{cases} 1 & \mathrm{i} \neq \mathrm{j} \text{ and } \mathrm{x_j} \in \mathcal{A}_\mathrm{i} \\ 0 & \text{otherwise} \end{cases} \tag{3}$$

$$\varepsilon_{\text{inter}}(i,k) = \begin{cases} 1 & \mathrm{i} \neq \mathrm{k} \text{ and } \mathrm{x_j} \in \mathcal{T}_\mathrm{i} \\ 0 & \text{otherwise} \end{cases} \tag{4}$$

where $\mathcal{A}_i$ denotes the collection of adjacent nodes in same frame for node x_i, and $\mathcal{T}_i$ represents the node set of same part in adjacent frames for node x_i.

3.3 GCN Module

After obtaining the graph, we adopt a GCN to leverage the contextual information from the neighborhood and then update the original features iteratively. To implement graph convolutional on the human skeleton, we define the adjacency matrix A as a stack of $\{A_1, ..., A_T\}$, where A_i represents the adjacency matrix within a single frame and T is the number of frames. We symmetrically normalize each A_t using the following form [7]:

$$A_t = D_t^{-\frac{1}{2}} \hat{A}_t D_t^{-\frac{1}{2}}, \tag{5}$$

where $\hat{A}_t = A_t + I$ and D_t is the diagonal degree matrix of $\hat{A}_t$. We denote the stack of A_t and D_t as $\hat{A}$ and D, respectively.

As outlined in [7], for the GCN to work properly, we need to normalize nodes with large degree in the adjacency matrix. With the above definition, we define the graph propagation rule from layer l to layer $l+1$ as:

$$X^{(l+1)} = \sigma\left(D^{-\frac{1}{2}} \hat{A} D^{-\frac{1}{2}} X^{(l)} W^{(l)}\right), \tag{6}$$

where $W^{(l)}$ is the weight matrix at layer l.

After applying the spatio-temporal GCN, each of the node's features will get updated according to its adjacent nodes.

3.4 Temporal Attention Module

Given the updated key-point features, we use spatio-temporal attention to reflect the importance of each region in the final tracklet representation. Following [4] we compute this representation as:

$$\mathbf{x}_g = \sum_{i=1}^{T \cdot K} \frac{\|\hat{\mathbf{x}}_i\|_1}{\sum_j \|\hat{\mathbf{x}}_j\|_1} \hat{\mathbf{x}}_i \tag{7}$$

As a result, each spatial region $\hat{\mathbf{x}}_i$ from different frames is assigned with a specific attention score based on the spatio-temporal attention information.

3.5 Clustering Module

In order to train our model in a fully unsupervised manner, it is essential to design a supervision signal. To achieve this, we need to generate pseudo labels using a clustering approach. Following [31], we perform hierarchical clustering with randomly selecting P identities and K instances (PK sampling) to obtain reliable clusters at each iteration. Since we do not have ground truth labels, we assign each tracklet to a distinct cluster. Initially, the number of clusters $C = N$, where N is the number of training samples. Concretely, we regard samples in the same cluster to have the same pseudo labels. Then, at each iteration, we merge a specific number of clusters using the distance between clusters.

Formally, we formulate the proposed distance in UPGMA [16] (unweighted pair group method with arithmetic mean) as follows:

$$D_{ab} = \frac{1}{n_a n_b} \sum_{i \in C_a, j \in C_b} D\left(C_{a_i}, C_{b_j}\right), \tag{8}$$

Where $D(.)$ is the euclidean distance. C_{a_i}, C_{b_j} are two samples in the clusters C_a, C_b, respectively. n_a, n_b represent the number of samples in C_a, C_b.

Each time, the clusters with the shortest distance are merged, hence we define a merging rate $mp \in (0, 1)$ to measure the speed of cluster merging. At each step, we aim to merge $m = N \times mp$ clusters and after t iterations, the number of clusters is dynamically decreased to $C = N - t \times m$. Then we use PK sampling to generate a new dataset with P identities and K images for each identity, and use it as the input of CNN in the next iteration. We consider fine-tuning, hierarchical clustering, PK sampling and evaluation as one iteration. At the end, we iterate the model until the performance no longer improves.

3.6 Loss Functions

We employ two kinds of losses to jointly optimize our model: hard-batch triplet loss proposed in [5] and the softmax loss.

The main idea of the hard-triplet loss is mining the relationships between anchor with positive sample and negative sample. Thus, to meet the requirement for the loss function, we randomly sample P identities and K images for each mini-batch. Typically, our loss function is defined as follows:

$$\mathcal{L}_{triplet}^{kp} = \sum_{i=1}^{P} \sum_{a=1}^{K} \sum_{k=1}^{KP} \beta_k [\alpha + \overbrace{\max_{p=1\dots K} \left\| x_{a,k}^{(i)} - x_{p,k}^{(j)} \right\|_2}^{\text{hardest positive}} - \overbrace{\min_{\substack{n=1\dots N \\ j=1\dots P}} \left\| x_{a,k}^{(i)} - x_{n,k}^{(j)} \right\|_2}^{\text{hardest negative}}]_+, \tag{9}$$

Where $x_{a,k}$, $x_{p,k}$, $x_{n,k}$ are key-points features extracted from the anchor, positive and negative samples, respectively, α is the hyperparameter margin in hard-batch triplet loss and $\beta_k = \max(m_{kp}[k]) \in [0, 1]$ is the kth key-point confidence.

As a result, key-points belonging to the same identity will be closer to each other than key-points belonging to different identities.

We also employ the triplet loss to enforce the tracklet representations x_g of the same identity, defined in Eq. 7, to be close to each others in the metric space:

$$\mathcal{L}^{g}_{triplet} = \sum_{i=1}^{P}\sum_{a=1}^{K}[\alpha + \overbrace{\max_{p=1...K}\left\|x^{(i)}_{a,g} - x^{(j)}_{p,g}\right\|_2}^{\text{hardest positive}} - \overbrace{\min_{\substack{n=1...N\\ j=1...P}}\left\|x^{(i)}_{a,g} - x^{(j)}_{n,g}\right\|_2}^{\text{hardest negative}}]_+ \quad (10)$$

In addition, we employ the softmax cross-entropy loss. We can formulate it as follows:

$$\mathcal{L}_{\text{softmax}} = -\sum_{i=1}^{P}\sum_{a=1}^{K}\log\frac{\exp(V^{T}_{c,i}x^{(i)}_{a,g}/\tau)}{\sum_{j=1}^{C}\exp(V^{T}_{c,j}x^{(i)}_{a,g}/\tau)}, \quad (11)$$

where C is the number of clusters, τ is a temperature parameter [5] that controls the softness of probability distribution over classes and V_c is an external lookup table maintaining the feature vector of each cluster updated as follows:

$$V_{y_i}(t) \leftarrow \frac{1}{2}(V_{y_i} + V_{y_i}(t-1)) \quad (12)$$

During the training stage, the overall objective function of our framework is formulated as:

$$\mathcal{L} = \mathcal{L}^{kp}_{triplet} + \mathcal{L}^{g}_{triplet} + \mathcal{L}_{\text{softmax}} \quad (13)$$

4 Experimental Results

4.1 Datasets

MARS dataset [32] is the largest video-based person reidentification benchmark captured by six cameras in a university campus, where each identity is captured by at least two camera angles. It contains 1,261 identities and nearly 20,000 video sequences. These videos are generated by the DPM detector and GMMCP tracker.

DukeMTMC-VideoReID [15] is another large-scale video-based ReID dataset, derived from the DukeMTMC dataset [15]. The dataset contains a total of 4,832 tracklets and 1,812 identities, where 702 identities are used for training and testing, and 408 identities are used as distractors. The tracklets are characterized with varying viewpoints illumination and background.

4.2 Experimental Settings

Implementation Details. We adopt ResNet-50 as the CNN backbone by removing the last classification layer. For the human pose estimation model, we employ the state-of-the-art model HR-Net [17] pre-trained on the COCO dataset [11]. The model estimates 2D coordinates (X, Y) of 17 key-points, and we fuse all key-points on the head region and get final $K = 14$ key-points, including head, shoulders, elbows, wrists, hips, knees, and ankles. In the training stage, we adopt a restricted random sampling startegy [10] to randomly sample $T = 8$ frames from each video. Then, each of the frames is resized to 256×128. To ensure the quality of the pseudo labels, we use 0.6% of the labeled data to jump start the GCN training (CNN is trained only with unlabeled data). After clustering, to meet the requirement for the hard-batch triplet loss we randomly select $P = 4$ identities and randomly sample $K = 4$ images so that the mini-batch size is 16. We use Adam optimizer with an initial learning rate of 6×10^{-5} and a weight decay of 5×10^{-4}. We set the number of iterations to 20, and the margin to 0.5 in the hard-batch triplet loss. We train our model for 70 epochs for the first step and 20 epochs for the remaining steps.

4.3 Comparison with the State-of-the-Art Methods

To validate the effectiveness of our proposed method on the video-based person ReID problem, we compare our proposed method with several recent state-of-the-art methods that can be classified in two groups: (1) one-shot ReID models: EUG [15], DGM [28], Stepwise [13] and RACE [27] and (2) pure unsupervised ReID models: TAUDL [8], OIM [23], BUC [12], TSSL [20] and CUPR [6].

As shown in Table 1, we have the following observations: (1) On Mars, our model achieves 34.1% in mAP and 58.5% in rank-1, improving the state-of-the-art TSSL [20] performance by 3.6% in mAP and 2.2% in rank 1. (2) On DukeMTMC-VideoReID, our model achieves 65.3% in mAP and 76.5% in rank-1, improving the state-of-the-art BUC [12] performance by 1.7% in rank 1.

Without any human supervising, our model outperforms the state-of-the-art methods, which indicates that it is not only effective in generating pseudo labels, but it is also simple and require no complex architecture.

4.4 Algorithm Analysis

Analysis over Cluster Speed Merging. The merging rate defines the number of clusters to merge at each step. We evaluate our model on Mars dataset using different mp values. We show the performance results in Table 2. At $mp = 0.06$ we get the highest performance with $mAP = 34.1\%$ and $rank - 1 = 58.5\%$. Then, the performance decreases when we use a higher or lower merging speed.

Analysis over Model Components. In Fig. 3, we show the impact of three learning component we used to train our model. With the clustering loss our model achieves a marginal performance of 27.1% mAP and 52% in rank-1. The addition of the global hard triplet loss significantly improves the performance by 5% in mAP and 4.5% in rank-1. This demonstrates the importance of hard-batch triplet loss which can reduce the influence of hard examples and improve model performance. Finally, we use the key-points hard triplet loss which further improves the performance to 34.4% in mAP and 58.6% in rank-1. These results illustrate the effectiveness of the three self-supervised signals leading to a robust model.

Table 1. Comparisons with the state-of-the-art person ReID methods on Mars and DukeMTMC-VideoReID. The 1st and 2nd highest scores are marked by red and blue, respectively.

Methods	Labels	Mars				DukeMTMC-VideoReID			
		mAP	rank-1	rank-5	rank-10	mAP	rank-1	rank-5	rank-10
EUG [15]	One labeled image per identity	42.5	62.6	74.9	-	63.2	72.7	-	
DGM [28]		16.8	36.8	54.0	-	33.6	42.3	57.9	69.3
Stepwise [13]		19.7	41.2	55.5	-	46.7	56.2	70.3	79.2
RACE [27]		24.5	43.2	57.1	62.1	-	-	-	-
TAUDL [8]	Pure unsupervised: Camera label	29.1	43.8	-	-	-	-		
DAL [2]		23.0	49.3	65.9	72.2	-	-	-	-
UTAL [9]		35.2	49.9	66.4	-	-	-		
OIM [23]	Pure unsupervised: No label	13.5	33.7	48.1	54.8	43.8	51.1	70.5	76.2
BUC [12]		29.4	55.1	68.3	72.8	66.7	74.8	86.8	89.7
TSSL [20]		30.5	56.3	-	-	64.6	73.9	-	-
Ours		34.1	58.5	74.4	77.2	65.3	76.5	88.1	90.8

Table 2. Performance comparison of different merging speed on Mars.

Merging speed	Mars			
	mAP	rank-1	rank-5	rank-10
mp = 0.1	30.1	53.4	70.1	72.5
mp = 0.09	30.3	54.2	70.3	74.1
mp = 0.08	31.4	54.9	71.6	74.7
mp = 0.07	33.7	57.6	73.9	76.8
mp = 0.06	**34.1**	**58.5**	**74.4**	**77.2**
mp = 0.05	32.6	56.1	72.6	75.7
mp = 0.04	32.4	55.1	71.2	75.2

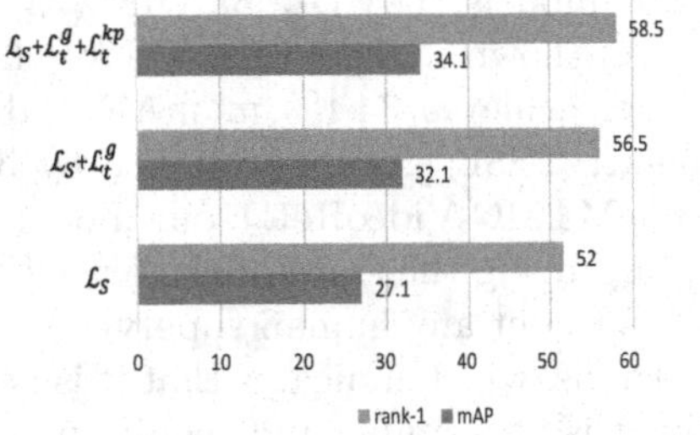

Fig. 3. Performance comparison of self-supervised learning components on Mars.

5 Conclusion

We propose a fully unsupervised ReID framework to learn inter-frame and intra-frame relation information for discriminative feature learning. By jointly optimizing the GCN spatial and temporal branches, our framework is capable of well capturing the semantic relations between key-points regions across frames and alleviating the occlusion problem. By combining our findings, we improve considerably over the previous state-of-the-art methods.

References

1. Chen, D., Xu, D., Li, H., Sebe, N., Wang, X.: Group consistent similarity learning via deep CRF for person re-identification. In: Proceedings of the IEEE Conference on Computer Vision and Pattern Recognition, pp. 8649–8658 (2018)

2. Chen, Y., Zhu, X., Gong, S.: Deep association learning for unsupervised video person re-identification. arXiv preprint arXiv:1808.07301 (2018)
3. Deng, W., Zheng, L., Ye, Q., Kang, G., Yang, Y., Jiao, J.: Image-image domain adaptation with preserved self-similarity and domain-dissimilarity for person re-identification. In: Proceedings of the IEEE Conference on Computer Vision and Pattern Recognition, pp. 994–1003 (2018)
4. Fu, Y., Wang, X., Wei, Y., Huang, T.: STA: spatial-temporal attention for large-scale video-based person re-identification. In: Proceedings of the AAAI Conference on Artificial Intelligence, vol. 33, pp. 8287–8294 (2019)
5. Hermans, A., Beyer, L., Leibe, B.: In defense of the triplet loss for person re-identification. arXiv preprint arXiv:1703.07737 (2017)
6. Khaldi, K., Shah, S.K.: CUPR: contrastive unsupervised learning for person re-identification. In: Proceedings of the 16th International Joint Conference on Computer Vision, Imaging and Computer Graphics Theory and Applications (VISI-GRAPP 2021) (2021)
7. Kipf, T.N., Welling, M.: Semi-supervised classification with graph convolutional networks. arXiv preprint arXiv:1609.02907 (2016)
8. Li, M., Zhu, X., Gong, S.: Unsupervised person re-identification by deep learning Tracklet association. In: Proceedings of the European Conference on Computer Vision (ECCV), pp. 737–753 (2018)
9. Li, M., Zhu, X., Gong, S.: Unsupervised tracklet person re-identification. IEEE Trans. Pattern Anal. Mach. Intell. **42**(7), 1770–1782 (2019)
10. Li, S., Bak, S., Carr, P., Wang, X.: Diversity regularized spatiotemporal attention for video-based person re-identification. In: Proceedings of the IEEE Conference on Computer Vision and Pattern Recognition, pp. 369–378 (2018)
11. Lin, T.-Y., et al.: Microsoft coco: common objects in context. In: Fleet, D., Pajdla, T., Schiele, B., Tuytelaars, T. (eds.) ECCV 2014. LNCS, vol. 8693, pp. 740–755. Springer, Cham (2014). https://doi.org/10.1007/978-3-319-10602-1_48
12. Lin, Y., Dong, X., Zheng, L., Yan, Y., Yang, Y.: A bottom-up clustering approach to unsupervised person re-identification. In: Proceedings of the AAAI Conference on Artificial Intelligence, vol. 33, pp. 8738–8745 (2019)
13. Liu, Z., Wang, D., Lu, H.: Stepwise metric promotion for unsupervised video person re-identification. In: Proceedings of the IEEE International Conference on Computer Vision, pp. 2429–2438 (2017)
14. Lv, J., Chen, W., Li, Q., Yang, C.: Unsupervised cross-dataset person re-identification by transfer learning of spatial-temporal patterns. In: Proceedings of the IEEE Conference on Computer Vision and Pattern Recognition, pp. 7948–7956 (2018)
15. Ristani, E., Solera, F., Zou, R., Cucchiara, R., Tomasi, C.: Performance measures and a data set for multi-target, multi-camera tracking. In: Hua, G., Jégou, H. (eds.) ECCV 2016. LNCS, vol. 9914, pp. 17–35. Springer, Cham (2016). https://doi.org/10.1007/978-3-319-48881-3_2
16. Sneath, P.H., Sokal, R.R.: Unweighted pair group method with arithmetic mean. Numer. Taxonomy 230–234 (1973)
17. Sun, K., Xiao, B., Liu, D., Wang, J.: Deep high-resolution representation learning for human pose estimation. In: Proceedings of the IEEE/CVF Conference on Computer Vision and Pattern Recognition, pp. 5693–5703 (2019)
18. Wang, J., Zhu, X., Gong, S., Li, W.: Transferable joint attribute-identity deep learning for unsupervised person re-identification. In: Proceedings of the IEEE Conference on Computer Vision and Pattern Recognition, pp. 2275–2284 (2018)

19. Wang, X., Gupta, A.: Videos as space-time region graphs. In: Proceedings of the European Conference on Computer Vision (ECCV), pp. 399–417 (2018)
20. Wu, G., Zhu, X., Gong, S.: Tracklet self-supervised learning for unsupervised person re-identification. In: Proceedings of the AAAI Conference on Artificial Intelligence, vol. 34, pp. 12362–12369 (2020)
21. Wu, J., Yang, Y., Liu, H., Liao, S., Lei, Z., Li, S.Z.: Unsupervised graph association for person re-identification. In: Proceedings of the IEEE/CVF International Conference on Computer Vision, pp. 8321–8330 (2019)
22. Wu, Y., Bourahla, O.E.F., Li, X., Wu, F., Tian, Q., Zhou, X.: Adaptive graph representation learning for video person re-identification. IEEE Trans. Image Process. **29**, 8821–8830 (2020)
23. Xiao, T., Li, S., Wang, B., Lin, L., Wang, X.: Joint detection and identification feature learning for person search. In: Proceedings of the IEEE Conference on Computer Vision and Pattern Recognition, pp. 3415–3424 (2017)
24. Yan, S., Xiong, Y., Lin, D.: Spatial temporal graph convolutional networks for skeleton-based action recognition. In: Thirty-Second AAAI Conference on Artificial Intelligence (2018)
25. Yan, Y., Zhang, Q., Ni, B., Zhang, W., Xu, M., Yang, X.: Learning context graph for person search. In: Proceedings of the IEEE/CVF Conference on Computer Vision and Pattern Recognition, pp. 2158–2167 (2019)
26. Yang, J., Zheng, W.S., Yang, Q., Chen, Y.C., Tian, Q.: Spatial-temporal graph convolutional network for video-based person re-identification. In: Proceedings of the IEEE/CVF Conference on Computer Vision and Pattern Recognition, pp. 3289–3299 (2020)
27. Ye, M., Lan, X., Yuen, P.C.: Robust anchor embedding for unsupervised video person re-identification in the wild. In: Proceedings of the European Conference on Computer Vision (ECCV), pp. 170–186 (2018)
28. Ye, M., Ma, A.J., Zheng, L., Li, J., Yuen, P.C.: Dynamic label graph matching for unsupervised video re-identification. In: Proceedings of the IEEE International Conference on Computer Vision, pp. 5142–5150 (2017)
29. Yu, H.X., Wu, A., Zheng, W.S.: Cross-view asymmetric metric learning for unsupervised person re-identification. In: Proceedings of the IEEE International Conference on Computer Vision, pp. 994–1002 (2017)
30. Yue-Hei Ng, J., Hausknecht, M., Vijayanarasimhan, S., Vinyals, O., Monga, R., Toderici, G.: Beyond short snippets: deep networks for video classification. In: Proceedings of the IEEE Conference on Computer Vision and Pattern Recognition, pp. 4694–4702 (2015)
31. Zeng, K., Ning, M., Wang, Y., Guo, Y.: Hierarchical clustering with hard-batch triplet loss for person re-identification. In: Proceedings of the IEEE/CVF Conference on Computer Vision and Pattern Recognition, pp. 13657–13665 (2020)
32. Zheng, L., et al.: MARS: a video benchmark for large-scale person re-identification. In: Leibe, B., Matas, J., Sebe, N., Welling, M. (eds.) ECCV 2016. LNCS, vol. 9910, pp. 868–884. Springer, Cham (2016). https://doi.org/10.1007/978-3-319-46466-4_52
33. Zhu, J.Y., Park, T., Isola, P., Efros, A.A.: Unpaired image-to-image translation using cycle-consistent adversarial networks. In: Proceedings of the IEEE International Conference on Computer Vision, pp. 2223–2232 (2017)

Keyframe Insights into Real-Time Video Tagging of Compressed UHD Content

Dominic Rüfenacht(✉)

Mobius Labs GmbH, Berlin, Germany
dominic.ruefenacht@gmail.com

Abstract. We present a method that can analyze coded ultra-high resolution (UHD) video content an order of magnitude faster than real-time. We observe that the larger the resolution of a video, the larger the fraction of the overall processing time is spent on decoding frames from the video. In this paper, we exploit the way video is coded to significantly speed up the frame decoding process. More precisely, we only decode keyframes, which can be decoded significantly faster than 'random' frames in the video. A key insight is that in modern video codecs, keyframes are often placed around scene changes (shot boundaries), and hence form a very representative subset of frames of the video. We show on the example of video genre tagging that keyframes nicely lend themselves to video analysis tasks. Unlike previous genre prediction methods which include a multitude of signals, we train a per-frame genre classification system using a CNN that solely takes (key-)frames as input. We show that the aggregated genre predictions are very competitive to much more involved methods at predicting the video genre(s), and even outperform state-of-the-art genre tagging that solely rely on video frames as input. The proposed system can reliably tag video genres of a compressed video between 12× (8K content) and 96× (1080p content) faster than real-time.

Keywords: Movie genre tagging · Real-time

1 Introduction

The increasing demand for ultra high resolution video content (e.g., 4K, 8K) puts a large burden on video analysis pipelines, as frame decoding times become an increasing fraction of the overall processing time. In a modern video tagging pipeline, 99% of the overall processing time for tagging compressed 8K content is spent extracting frames. This is primarily due to the fact that typical convolutional neural networks (CNNs) used for classification tasks are trained on inputs that are smaller than 300 × 300 pixels, irrespective of the original resolution of the video. To the best of our knowledge, existing work on video classification does not take into account the frame decoding time, and only report processing speeds on already decoded frames. In this work, we show how the coding structure of compressed video content can be taken advantage of and propose a system that can analyze compressed 8K content an order of magnitude faster than real-time.

S. Sclaroff et al. (Eds.): ICIAP 2022, LNCS 13233, pp. 147–157, 2022.
https://doi.org/10.1007/978-3-031-06433-3_13

The significant speedup is achieved by only decoding keyframes, which drastically reduces the bottleneck of decoding frames as they can be decoded significantly faster than frames that are predicted from other frames (i.e., so-called P- or B-frames). We observe that in modern video codecs, the rate-distortion (R-D) optimal choice is to code keyframes at (or near) shot boundaries, as frames across shots cannot be well predicted and hence shot boundaries provide a "natural" place for keyframes. This means that the sparse subset of keyframes we extract from the video are highly representative of the video and hence lend themselves nicely for video tagging applications.

The idea of only considering keyframes for video analysis tasks can be applied to a wide range of applications – essentially anything that does not require high and/or fixed frame-rate input. In this work, we validate the use of keyframes on the practical application of video genre tagging. More specifically, given an input video, the task is to assign a set of genres to the video (e.g., 'action', 'thriller', 'comedy'). The ability to automatically classify movies by genre has several applications, including (a) the indexing of video databases, (b) automatically recommending movies based on user preference modelling, and (c) automating content filtering and video summarization.

Existing genre tagging methods typically rely on a multitude of signals (e.g., densely sampled video frames, audio, subtitles, synopses) to tag the genres of a movie [5]. Apart from frames and audio, all other information is likely not available in a real scenario. Furthermore, a common belief of existing (frame-based) genre tagging methods is that temporal reasoning is crucial to a well-performing genre tagging system ([5,9]).

In this paper, we propose a system that is trained to tag genres on individual frames. While not all genres are apparent in all frames, we show that if we temporally pool those frame-level genre tags, we can obtain state-of-the-art genre tagging performance. Furthermore, the proposed system only considers keyframes at inference time. We leverage the 14K movie trailers from [5], and train a lightweight multi-label genre tagging convolutional neural network (CNN) which can tag 19 different genres (the 18 classes from [5], plus an additional 'Text' class). This additional 'Text' class is used to filter out frames that are dominated by text (as commonly seen in trailers). The resulting method can accurately tag video genres of 8K content over 10 times faster than real-time, while outperforming state-of-the-art frame-based genre tagging methods by a large margin.

2 Related Work

Existing genre tagging methods often involve a multitude of signals. Pre-deep learning works consisted of cleverly combining handcrafted features to classify video content into a small set of genres. [2] investigates the usefulness of several hand-crafted audio-visual features for very specific genres (e.g., sport versus non-sport, cartoon versus non-cartoon). [1] use a combination of low-level visual features such as colour and texture descriptors as well as audio features and perform classification into seven distinct genres using a support vector machine (SVM). In [5], the authors conduct an extensive study on the usefulness of different modalities. They consider a total of 22 feature representations, extracted from five modalities (trailer frames, trailer audio, movie synopsis, subtitles, and movie posters) using amongst others CNNs and long-short term memory (LSTM) for feature extraction. When evaluated on a dataset of 14'000 trailers, they show that movie synopsis and trailer frames are most useful as individual modalities. They report the best performance by a fusion of movie synopses and trailer frames. While it is interesting from an academic point of view to research how different modalities (and their fusion) can benefit genre tagging, most of the input will likely not be available in a practical scenario.

In this work, our goal is to provide a practical solution to video genre tagging, and only use video frames as input. A related work that also only relies on frame data is [9]. They train a very deep CNN to extract features, and propose a convolution through time module (CTT) that applies a temporal convolution on all frame features. The output is then pooled to form video-level genre tag predictions.

To the best of our knowledge, all existing video genre prediction methods that use frame information (i.e., images) pool the genre predictions across all frames during training ([9,10]). This results in the fact that they can only predict genres at video-level. In contrast, the method proposed in this paper is trained to predict the video genre from individual frames, which are then pooled only at the inference stage to form video-level predictions. This can lead to more insights (i.e., we can tell *where* in movie certain genres are present), and add additional insights even when the original video-level genre tags are available (see Fig. 5 for examples).

3 The Impact of Increasing Video Resolutions on Video Analysis Tasks

The increase in video resolution (4K and beyond) significantly slows down video analysis pipelines, which struggle dealing with an ever increasing amount of information; one frame of an 8K video contains over 33 million pixels. Raw, uncompressed 8K video recorded at 24 frames per second (fps) requires a lot of bandwidth (around 2.6 Gigabytes per second of video), which makes the use of sophisticated video compression a necessity for most practical applications.

A typical pipeline used for analyzing compressed video content consists of two main steps. First, the compressed frames have to be decoded, after which they are resized and input to the video analysis model (e.g., a CNN trained for genre tagging). One observation is that the input resolution for many tasks can be very low and still provides high prediction accuracy; typical architectures for image and video tagging use input resolutions that are lower than 300×300 pixels, irrespective of video resolution.

To the best of our knowledge, the time it takes to decode frames from compressed video has been ignored in existing literature on video analysis tasks.

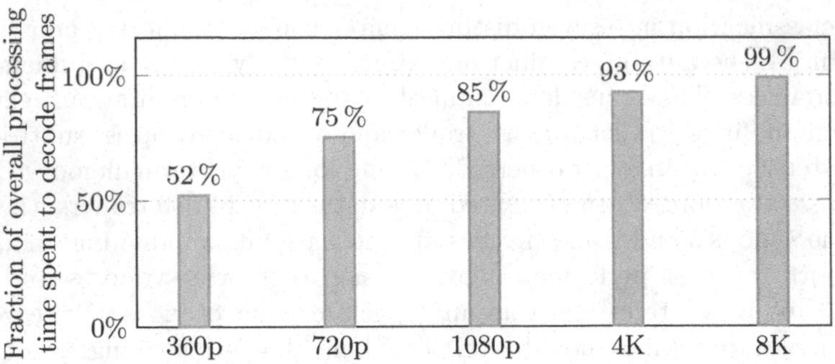

Fig. 1. Fraction of time a modern video tagging pipeline spends decoding frames. As video resolution increases, the majority of time to obtain the tags is spent decoding frames.

As it turns out, as video resolution increases, frame decoding time more and more dominates the overall processing time of the video analysis pipeline. The fraction of time spent decoding frames amounts between 50% (360p resolution) up to 99% (8K resolution) of the overall time to analyze the video.

3.1 Coding Structure of a Video

The reason compression 'works' is that there exists a lot of redundancy in a video, meaning that nearby frames of a video often look very similar. Modern video codecs exploit this fact by only coding certain frames – so-called keyframes or I-frames – as full images (they are typically placed every few seconds); all other frames are either unidirectionally predicted from previous frames (P-frames), or bidirectionally predicted from previous and future frames (B-frames) in the video, using what is known as 'motion-compensated prediction' [4].

3.2 Keyframes to the Rescue

While temporal prediction is a great help in compressing a video, the dependency of P- and B-frames on other frames makes the process of decoding one specific frame (i.e., random-access mode) slow; for example, in the case of hierarchical B-frames structure [6], $2^N + 1$ frames have to be decoded for accessing one B frame at hierarchy level N. Keyframes, on the other hand, can be much faster decoded as they are not predicted from any other frames. In this paper, we present a framework that exploits the fact that keyframes can be extracted much faster than other frames in order to severely speed up the time it takes to analyze compressed videos.

4 Keyframe-Based Video Genre Tagging

This section presents the proposed method for genre tagging in videos. As mentioned earlier, our goal here is to design a framework that provides an efficient

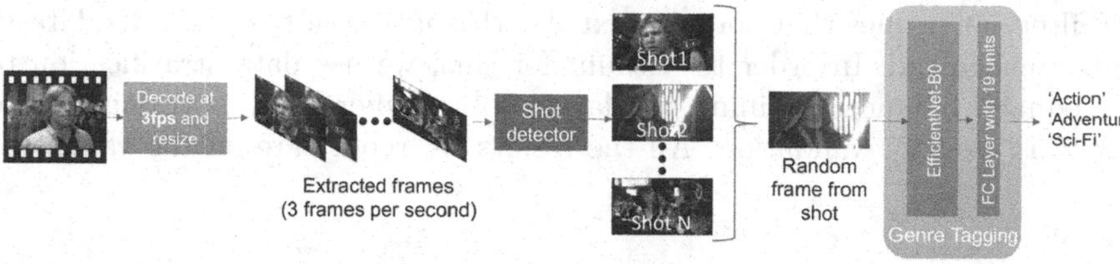

(a) Training data pipeline.

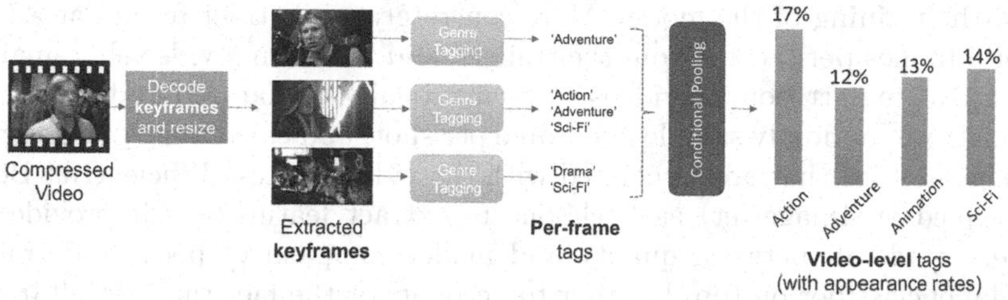

(b) Keyframe-based genre tagging at inference time.

Fig. 2. Overview of the proposed keyframe-based video genre tagging framework. (a) To train the genre tagging, we sample random frames from each shot in the trailer. (b) At inference time, we only decode keyframes and pass them through the genre tagging network. The per-frame genre tags are then passed through a pooling layer that generates the video-level predictions.

and practical solution for the task of tagging video genres that runs faster than real-time for UHD content. We achieve this by only considering keyframes extracted from the videos.

In the following, we describe the dataset we use, and show how we train a CNN to perform video genre classification. Lastly, we present how we use the trained model at inference time to tag very efficiently tag video genres.

4.1 Training Video Genre Tagging

Dataset. Learning the genres of movies can be a prohibitive task in terms of volumes of data. Observing that trailers usually provide a good summary of a movie, researchers have resorted to using trailers for training video genres. In this work, we use the dataset provided in Mangolin et al. [5], which consists of around 14'000 trailers downloaded from YouTube at 360p resolution. Each trailer is annotated with genre tags obtained from the TMDb database; in total, there are 18 distinct genre tags, and every trailer can have one or more genre tags. We add an additional tag called 'Text', which are frames that contain predominantly text and are not very informative or even misleading for genre tagging. We find this useful for two reasons: (a) text shots typically are only present in trailers and not in full-length movies, and (b) they lead to worse model performance if included. At inference time, we can use the 'Text' tag to easily filter out frames that contain text. By the nature of the task, the dataset is quite unbalanced. In order to account for this, we use data stratification to create equally balanced training/validation splits, where 80% is training data, and 20% is used for validation. All the results we report are on the validation split.

Video Genre Tagging Architecture. In order to maximize the diversity of the training data, we make use of more than just the keyframes in the trailers during the training of the model. More concretely, we extract frames at a fixed rate of 3 frames per second from the trailers, and then run a video shot analysis similar to [3] to partition the videos into semantically similar parts. At each training epoch, we randomly sample one frame per shot, and each frame gets assigned all the genre tags that are associated with the video. We use EfficientNet-B0 [8] (pre-trained on ImageNet) as backbone to extract features, as it provides an excellent trade-off between quality and model complexity/speed, and train a multi-label classifier on top. In order to account for the fact that not all frames of the trailer portray all (if any) of the genres attached to the trailer, we use label smoothing [7].

4.2 Predicting Video Genres from Keyframes at Inference Time

Figure 2(b) shows the proposed keyframe-based video genre tagging pipeline. In order to significantly speed up the frame extraction process, we only decode the keyframes from the compressed video, and subsequently resize them to 224×224 pixels. We then pass each keyframe individually through the trained video genre tagging CNN to obtain per-frame genre tag predictions. Finally, these tags are pooled together to obtain video-level tags. During this stage, we filter out all frames that contain the tag 'Text' (see previous section). A notable difference to existing works is that using the proposed approach, we can quantify how often each genre tag appears across all keyframes, which leads to more insights as to which genre(s) are present in a video. We believe that this information can even benefit databases that already contain genre tag metadata.

5 Experimental Validation

In this section, we present results that validate the proposed keyframe-based genre tagging method. We first show results on keyframe extraction speeds, which is a key ingredient that makes the proposed method run very fast. We then draw our attention to quantitative results, where we compare the keyframe-based tagging with state-of-the-art video genre tagging models. Lastly, we show qualitative results that highlight the usefulness of tagging genres at a frame-level.

5.1 Frame Extraction Speed Comparison

In this section, we compare the time it takes to extract frames at different resolutions using the commonly used approach of extracting a fixed rate of frames,

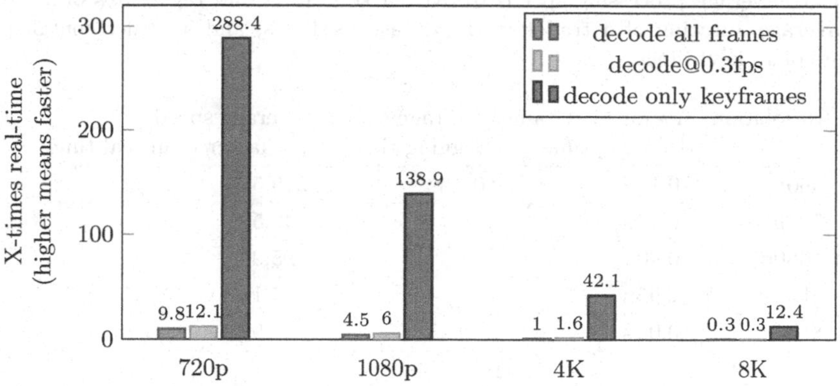

Fig. 3. Decoding speeds (measured in terms of real-time performance, see text for more details) of different frame extraction 'modes'. The bars show how fast the frame extraction is with respect to the original duration of the video. Note how across video resolutions, extracting key-frames only is over 20 times faster than decoding at a fixed frame-rate.

with the proposed way of only decoding keyframes. At the time of this writing, only a very limited number of 8K videos exist. For this experiment, we transcode five 8K reference videos using HEVC to different resolutions ranging from 720p (1280×720) up to 8K (7680×4320). We then measure the time it takes to decode the frames, including resizing them to 224×224 resolution (the input resolution of our genre tagging model).[1] Since the videos have different durations, we measure the speed with respect to real-time performance. That is, $10\times$ means that it takes 1 min to extract the frames of a 10 min video.

We compare the time it takes to decode all frames from the video with the time when only keyframes are decoded. In order to make the comparison as fair as possible, we further set a fixed frame extraction rate (0.3 fps, i.e. 1 frame every three seconds), which roughly matches the number of keyframes in the videos. As can be observed in Fig. 3, across all tested resolutions, extracting only keyframes is around 20 times faster than using a fixed frame extraction rate. Perhaps surprisingly, changing the frame extraction rate from 24 fps (i.e., decoding all frames) to 0.3 fps, does not significantly speed up the frame extraction process, despite there being 72 times fewer frames decoded. As mentioned before, this is due to the way of how frames are coded, which results in the fact that many more frames have to be decoded to reconstruct the requested frames.

5.2 Frame Genre Tagging Speed

We now turn our attention to the overall processing time of the proposed video genre tagging framework. To the best of our knowledge, we are the first to report

[1] The resizing to the same resolution guarantees that writing out the frames to disk does not further penalize larger resolution videos.

Table 1. Average processing speeds to extract keyframes and tag genres of a video. The average duration of a trailer in the dataset is 124 s, and a trailer contains an average of 71 keyframes.

Resolution	Average keyframe decoding time	Average genre tagging time	Average speed (× faster than real-time)
360p	0.17 s	0.4 s	219.5×
720p	0.43 s		149.5×
1080p	0.90 s		95.9×
4K	2.95 s		37.1×
8K	10.05 s		11.9×

processing speeds, and hence we cannot compare with existing work. However, we find it useful to show the processing speeds of our approach, as it is perhaps the most compelling feature of this work, and note that existing methods would run significantly slower. Table 1 reports the average time it takes to process the trailers at different resolutions. The average duration of the trailers in the database is 124 s, and each trailer contains 71 keyframes on average. This means there is on average a keyframe every 1.74 s, which is quite a lot and can be attributed to the fact that a trailer contains more scene shots than a typical movie. Nonetheless, one can see that the proposed framework can tag genres of a movie at 8K resolution 11.9× faster than real-time.

5.3 Quantitative Evaluation of the Proposed Genre Tagging

We compare the proposed method to the recently proposed multi-modal genre tagging system proposed in Mangolin et al. [5]. Their work provides an excellent reference point since the authors performed an extensive analysis of a range of multimodal signals (trailer frames, trailer audio, synopses, subtitles, and posters) which allows us to see how the proposed method compares to different combinations of signals. Figure 4 compares the best obtained results per single modality from [5], as well as their overall best result obtained by fusing different modalities ('BestFusion' in the figure).

We would like to draw particular attention to the two orange bars in the figure (Trailer-CTT-MMC [9] and Trailer-C3D [5]), which are the most closely related to the proposed method in that they also involve only video frame data. In contrast to our proposed method which only relies on keyframes (which can be extracted very fast), the Trailer-C3D model involves 120 frames extracted from each trailer, and a much heavier architecture involving 3D convolutions. Despite the more involved architectures, the proposed approach significantly outperforms both methods (+0.27 and +0.12 f1-score improvement, respectively). We attribute this perhaps surprising result to the fact that keyframes do indeed

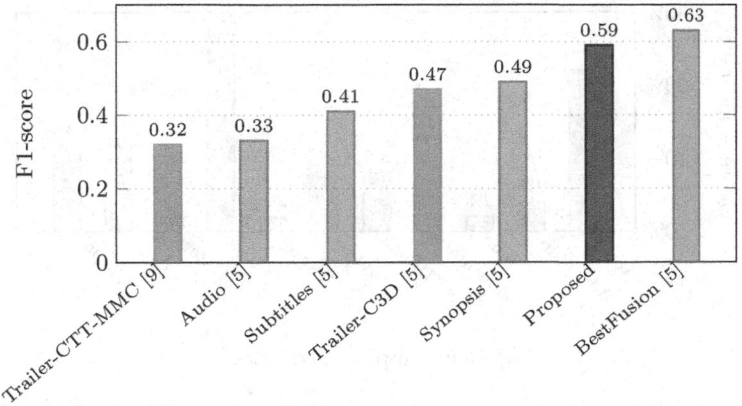

Fig. 4. Quantitative comparison between the proposed keyframe-based genre tagging and different configurations from [5], as well as CTT-MCC [9].

provide a useful subset of frames, as well as to our careful dataset preparation where we remove text shots which can adversely affect the classification performance. Furthermore, the results put in question the common belief that some form of temporal reasoning is required to reliably tag genres.

Another interesting comparison is with respect to the best performing method presented in [5], which is obtained by fusing the classifications from the Trailer-C3D and movie synopsis models. While the proposed method approach performs slightly worse, we believe that in a practical scenario, it is unlikely that a database has a movie synopsis available but no genre tags (which are much quicker to add manually), which would render genre tagging redundant anyways.

5.4 Qualitative Results

While the quantitative evaluation provides a good way of comparing with state-of-the-art, we find it important to further provide some qualitative results to provide more insight into the performance of the proposed method. Figure 5(a) shows genre prediction results on a movie trailer tagged as 'music'. We show a plot of genre appearance rates, which shows the fraction of keyframes that have a particular genre tag predicted. As can be seen, many genres other than 'music' appear in this movie (e.g., 22% 'Action'). Figure 5(b) provides the genre predictions of a few consecutive keyframes, which exemplify the quality of the individual frame genre predictions.

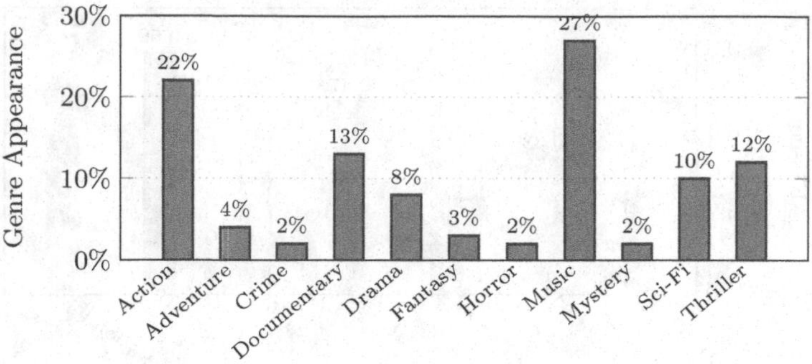

(a) Genre appearance rates

(b) Tags for keyframes

Fig. 5. Genre prediction results of the proposed method on the trailer for the movie 'Metallica – Through the Never'. The official genre tag for this movie is music. a) Our predictions reveal that there are other genres appearing frequently in the trailer. b) shows individual frame predictions, where we see that this is actually quite a dark movie, which is not what one typically expects to see in a 'music' movie.

6 Conclusions and Future Work

We have presented a very efficient approach to tag genre(s) of a compressed UHD video content. We observe that at ultra-high resolutions of 4K and beyond, more than 90% of the overall processing time to tag the genres of a movie is spend decoding the frames. In this work, we stipulate that *keyframes*, which can be extracted much faster than other frames from a compressed video, form an useful subset of all video frames for video analysis tasks. We validate this hypothesis on the task of genre tagging. The proposed keyframe-based genre tagging compares favourably to genre tagging methods that rely on multimodal inputs, and runs significantly faster. The proposed method can tag genres of 8K video content over 11 times faster than real-time, opening the door to faster-than-real-time genre tagging of UHD videos.

While this work focused on video genre tagging, we would like to end this paper by reiterating that the proposed approach can also be used for other video analysis tasks that do not rely on high framerate input.

References

1. Ekenel, H.K., Semela, T., Stiefelhagen, R.: Content-based video genre classification using multiple cues. In: ACM Workshop on Automated Information Extraction in Media Production, pp. 21–26 (2010). https://doi.org/10.1145/1877850.1877858
2. Glasberg, R., Schmiedeke, S., Mocigemba, M., Sikora, T.: New real-time approaches for video-genre-classification using high-level descriptors and a set of classifiers. In: IEEE International Conference on Semantic Computing, pp. 120–127 (2008). https://doi.org/10.1109/ICSC.2008.92
3. Gygli, M.: Ridiculously fast shot boundary detection with fully convolutional neural networks. arXiv (2017)
4. Jain, J.R., Jain, A.K.: Displacement measurement and its application in interframe image coding. IEEE Trans. Commun. **29**(12), 1799–1808 (1981). https://doi.org/10.1109/TCOM.1981.1094950
5. Mangolin, R.B., et al.: A multimodal approach for multi-label movie genre classification. Multim. Tools Appl. (2020). https://doi.org/10.1007/s11042-020-10086-2
6. Schwarz, H., Marpe, D., Wiegand, T.: Analysis of hierarchical b pictures and MCTF. In: IEEE International Conference on Multimedia and Expo, pp. 1929–1932 (2006). https://doi.org/10.1109/ICME.2006.262934
7. Szegedy, C., Vanhoucke, V., Ioffe, S., Shlens, J., Wojna, Z.: Rethinking the inception architecture for computer vision. In: IEEE Conference on Computer Vision and Pattern Recognition (2015). https://doi.org/10.1109/CVPR.2016.308
8. Tan, M., Le, Q.V.: EfficientNet: rethinking model scaling for convolutional neural networks. In: 36th International Conference on Machine Learning 2019-June, pp. 10691–10700 (2019)
9. Wehrmann, J., Barros, R.C.: Convolutions through time for multi-label movie genre classification. In: ACM Symposium on Applied Computing, pp. 114–119 (2017). https://doi.org/10.1145/3019612.3019641
10. Zhou, H., Hermans, T., Karandikar, A.V., Rehg, J.M.: Movie genre classification via scene categorization. In: Proceedings of the 18th ACM International Conference on Multimedia, January 2016, pp. 747–750 (2010). https://doi.org/10.1145/1873951.1874068

Exploring the Use of Efficient Projection Kernels for Motion Saliency Estimation

Elena Nicora(✉) and Nicoletta Noceti

MaLGa-DIBRIS, Università degli Studi di Genova, Genoa, Italy
elena.nicora@dibris.unige.it

Abstract. In this paper we investigate the potential of a family of efficient filters – the Gray-Code Kernels – for addressing visual saliency estimation guided by motion. Our implementation relies on the use of 3D kernels applied to overlapping blocks of frames and is able to gather meaningful spatio-temporal information with a very light computation. We introduce an attention module that reasons on the use of pooling strategies, combined in an unsupervised way to derive a saliency map highlighting the presence of motion in the scene. In the experiments we show that our method is able to effectively and efficiently identify the portion of the image where the motion is occurring, providing tolerance to a variety of scene conditions.

Keywords: Gray-Code Kernels · Motion saliency estimation · Motion detection

1 Introduction

Biological perception systems are very skilled in processing efficiently and effectively a huge amount of visual information, bringing the attention on attractive regions or objects for further processing. The same ability in artificial systems is related to the notion of visual saliency estimation [4], often the first step of more complex pipelines for higher level visual analysis. For this reason, it is convenient to address the task very efficiently, to avoid an eccessive burden in the computation. Motion, in particular, is known to be one of the most attractive visual features for human attention [10,27].

In this work we explicitly focus on bottom-up motion-based visual saliency, and we explore the use of Gray-Code Kernels (GCKs) as a tool to efficiently obtain a projection of a video content on top of which a new unsupervised motion-based attention module is devised, to coarsely but very efficiently detecting the moving objects in the scene.

The task of motion-based saliency estimation is tightly intertwined with motion detection, that in literature may take different shapes. Video Object Segmentation (VOS), also referred to as Salient Object Segmentation, aims at separating moving objects from the background, to obtain a very precise segmentation mask.

Classical approaches shape the problem as a change detection, in which the main idea is to segment moving objects comparing the current scene with a model of the background, that may include indeed dynamic elements [3,26]. These type of methods may be computationally feasible to the price of limited generalization capabilities, as

S. Sclaroff et al. (Eds.): ICIAP 2022, LNCS 13233, pp. 158–169, 2022.
https://doi.org/10.1007/978-3-031-06433-3_14

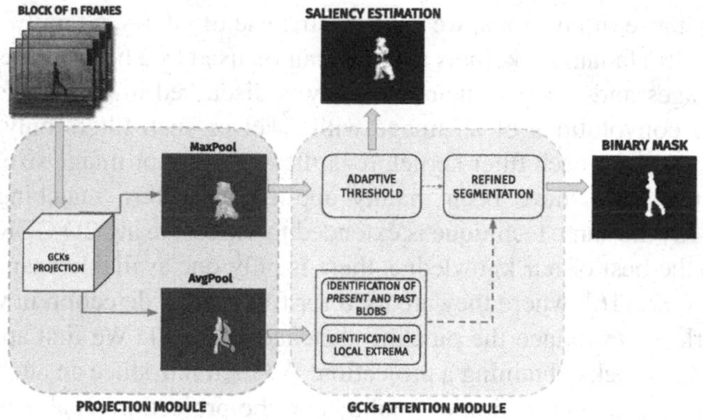

Fig. 1. A visual representation of the processing flow of our method.

their use is usually constrained to settings where the camera is fixed and the lighting conditions are kept under control.

Motion detection and segmentation methods based on optical flow provide a higher degree of flexibility to scene conditions and acquisition settings [7, 18, 25]. On the other hand, they typically are either very efficient but inaccurate, or very effective to the price of a high computational demand, especially when involving global optimization steps [30] or if based on deep architectures [29].

VOS methods aim at obtaining a mask of the entire salient object, even when only part of it is actually moving. Hence information coming only from motion cues might not be enough. In order to overcome this problem, different approaches [17, 22–24] combine motion-based cues with additional sources of information related with the probability of finding an object in some regions of the image [19, 31, 32] to refine the map. Similarly, motion and appearance information have been fused in methods leveraging deep architectures [11, 28]. This refinement step allows to reach considerable segmentation accuracy, but to the price of a significant computational cost, and the need of customized solutions whose use is limited to specific detection tasks.

Unlike these approaches, we target scenarios where motion-based saliency estimation triggers the attention towards regions of the image where the presence of relevant information can be detected. We do not target a very precise segmentation of the entire object, while it is necessary to gather very quickly hints on where and how the relevant motion is occurring. Application domains as robotics and video-surveillance are just two examples where this ability may be highly beneficial. Similarly to change detection methods, we are mainly interested to an efficient way of detecting the presence of motion in the scene, but as with optical flow we foresee a representation that can sustain different levels of motion analysis. Also, a desired property for the method is the tolerance to different viewing and scene conditions.

Inspired by these motivations, we explored the use of GCKs, a family of filters that include the Walsh-Hadamard kernels and that can be used as a highly efficient filtering scheme for images and videos. Their efficiency is discussed in [2], where it is shown that successive convolutions of an image with a set of such filters only require two operations per pixel for each filter kernel, regardless of filter or image size.

In literature, GCKs have been mainly applied to pattern matching on images [9,12,21]. In [15] the same technique is extended to videos, using 2D GCKs for motion estimation. To the best of our knowledge, there is only one available example of application of 3D GCKs [16], where they are used for foreground detection in videos.

In this work we introduce the pipeline sketched in Fig. 1. We first apply the bank of GCKs to video blocks obtaining a projection. We then introduce an attention module that, making use of a combination of poolings on the projection, is able to estimate in an unsupervised way a saliency map that reflects the presence of motion in the scene, and from which a coarse segmentation map can be derived.

We compare our method performances, in terms of segmentation quality and computational time, to basic motion detection approaches and perform an evaluation on benchmark datasets with different characteristics and complexity.

The remainder of the paper is organized as follows: in Sect. 2 we introduce the basic concepts behind the GCKs. The proposed method is reported in Sect. 3, and its experimental evaluation is discussed in Sect. 4. Section 5 is left to a final discussion.

2 Gray-Code Filter Kernels

In this section we report the basic definitions of Gray Code Kernels considering, for the sake of clarity, the simple case where both signal and kernels are one-dimensional vectors. We refer the interested reader to [2] for a more detailed discussion.

2.1 Efficient Filtering with GCK

The family of kernels is built according to a recursive definition, starting from a d-dimensional vector s of length t – the *seed* – and a parameter k. The final length of each kernel is $2^k t$, while the total number of kernels in the family is $n = 2^{kd}$. The family of kernels can be recursively defined as follows

$$\begin{aligned} V_s^{(0)} &= [s] \\ V_s^{(k)} &= [v_s^{(k-1)} \alpha_k \times v_s^{(k-1)}] \end{aligned} \tag{1}$$

with $v_s^{(k-1)} \in V_s^{(k-1)}$ and $\alpha_k \in \{+1, -1\}$. The family of kernels can be seen as leaves of a binary tree of k levels, with root $V_s^{(0)}$, and where the branches are labeled with the values of α used to create the kernels. The efficiency of filtering with the family of GCKs is related to the ordering in which these filters are applied in sequence. An optimal ordering is defined based on the *α-index*, a code composed by the labels of the edges of the path going from the root of the tree to the leaf (the kernel) of the tree. Two kernels at a certain level k of the binary tree are *α-related* iff the hamming distance between their α-indices is one.

An ordered sequence of α-related kernels is called a *Gray-Code Sequence* (GCS). The following relationship holds [2]: given two α-related kernels v_+, v_- in $V_s^{(k)}$, then

$$\begin{aligned} v_+(i) &= v_-(i) + v_-(i-\Delta) + v_+(i-\Delta) \\ v_-(i) &= v_+(i) - v_+(i-\Delta) - v_-(i-\Delta) \end{aligned} \tag{2}$$

where Δ is the length of the common portion of the two α-related kernels of v_+ and v_-. Let b_+ and b_- be the results of convolving a signal x respectively with α-related filter kernels v_+ and v_-, then, by linearity of convolution we have the following:

$$\begin{aligned} b_+(i) &= b_-(i) + b_-(i-\Delta) + b_+(i-\Delta) \\ b_-(i) &= b_+(i) - b_+(i-\Delta) - b_-(i-\Delta) \end{aligned} \tag{3}$$

From the above formulas it can be derived that, given the result of convolving the signal x with one of the two kernels, only two operations per pixel are needed to obtain the result of convolving the signal with the other kernel.

2.2 3D Gray-Code Kernels

As our interest is to explore the use of GCKs on videos, a natural choice is adopting 3D filters, able to capture spatio-temporal properties in the data. A family of 3D GCKs can be built by combining triplets of 1D GCKs. Considering the notation we used in Sect. 2.1, if V^1, V^2 and V^3 are three 1D kernels in the leaves level of the binary tree (we omit the index k of the tree level, and s, the seed, to simplify the notation), a 3D GCK $\mathcal{F}$ can be computed as $\mathcal{F} = V^1 \times V^2 \times V^3$ where $'\times'$ denotes the outer product between vectors. In this way we may derive a bank of filters $\{\mathcal{F}^h\}_{h=0}^{F-1}$, where F is the total number of filters 2^{kD} with $D = 3$, of size $2^k t \times 2^k t \times 2^k t$ (where $2^k t$ is the length of the 1D filter, t the length of the seed s).

In [2] the authors showed that the efficient properties of 1D GCKs can be generalized to higher dimensions. In general, GCKs in a d dimensional space can be computed as inner products between d 1-dimensional filters in the sequence. However, this general approach does not necessarily mean that the resulting multi-dimensional filters are already in the right order, and further processing is needed to obtain a GCS. It is worth noting that there is no a single solution for building a GCS, and different strategies may be exploited depending on the task.

Implementation. In our implementation[1], we derive the GCS by processing filters with coherent temporal sequency i.e. filters that have the same number of changes along the time axis using the snake ordering and minimizing the jumps between filters with different time component. This is due to the fact that the transition from 2D to 3D kernels brings non-trivial problems in what concerns memory efficiency, as in order to obtain the current projection, the previous Δ projections need to be stored in memory and available, with $\Delta = 2^{(k-1)}t$ corresponding to the length of the prefix that two α-related kernels have in common. According to the original work, we build a binary tree starting from an initial seed $s = [1]$ (notice this corresponds to derive a family of Walsh-Hadamard kernels) and we fix the number of levels of the tree at $k = 2$. The final GCS will be thus composed by 64 filters of size $4 \times 4 \times 4$ of ± 1 values.

[1] The implementation of our method in Python will be made soon publicly available.

3 Proposed Method for Motion Saliency Estimation

In this section we introduce our unsupervised motion-based attention module derived from 3D GCKs projections. The proposed method – a visual sketch is provided in Fig. 1 – aims at the coarse identification of regions with salient motion in a video.

The method starts from the assumption that the variability of the responses of the bank of filters may depend on (i) the type of information found in different locations of the image sequence, and (ii) the subsets of kernels considered. Supposedly, if nothing interesting is happening in a certain area of the image, then filters belonging to the same class should react coherently (or not react at all). Indeed, we can distinguish three main subgroups of filters, depending on their structure, and main properties that they are *predominantly* able to highlight: spatial ($\mathcal{S}$), temporal ($\mathcal{T}$), and spatio-temporal ($\mathcal{ST}$).

Considering our interest on motion information, we focused our analysis on temporal and spatio-temporal filters, and designed a motion-based attention module with pooling operations able to highlight oscillations in the filter responses (by reporting the maximum and the average value over time). In a pre-processing step, each frame is normalized in $[-1, +1]$ so to have values that are comparable with the kernels.

In the remainder of the section we discuss in details the role of each block.

3.1 Projection Module: Filtering and Pooling

GCKs Projections. Given an input video we split the sequence in overlapping clips $\mathcal{C}$ of size $W \times H \times n$, where W and H are, respectively, the width and height of the video frames, while $n = 2^k t$ (see previous section). We define a projection Φ that maps the original frame block in a new space obtained by filtering it with the bank of GCKs. Thus $\Phi : \mathbb{R}^{W \times H \times n} \rightarrow \mathbb{R}^{W \times H \times F}$ is mathematically defined as

$$\Phi(\mathcal{C}) = \{\mathcal{C} \circledast \mathcal{F}^h\}_{h=0}^{F-1} \tag{4}$$

where $\circledast$ denotes the convolution operator (the first full convolution is followed by F – 1 efficient convolution). We consider the central slice of the resulting cube ($n/2$) as the one maximizing the amount of information about the whole block i.e. how that filter responded in a span of n frames in time. We end up having blocks of 2D projections $\hat{\mathcal{C}} = \Phi(\mathcal{C})$, normalized between 0 and 1, for each processed block.

Pooling GCKs Responses. To analyse more globally the behaviour of the bank of projections and, as a consequence, derive a compact (and easier to manage) representation for each frame block, we apply pooling operations, considering two different strategies. More in details, we applied average and max pooling defined as follows:

$$AP(\hat{\mathcal{C}})(i,j) = \frac{1}{F_T} \sum_{h=0}^{F_T-1} \Phi^h(\mathcal{C})(i,j) \tag{5}$$

$$MP(\hat{\mathcal{C}})(i,j) = \max_{h=0}^{F_{ST}-1} \Phi^h(\mathcal{C})(i,j) \tag{6}$$

where with Φ^h we refer to the projection of the input block with the h-th kernel in the bank, while F_T and F_{ST} are the numbers of temporal and spatio-temporal features, respectively.

We give an intuition about the effects of pooling. With the max pooling on the spatio-temporal kernels we derive a visual impression of *where* the motion occurred in the time span covered by the filtering. Maximizing the contributions from the different instants we gather a representation that, in spirit, is similar to the Motion History [1]: since filtering acts on blocks of frames contiguous in time, rather than single images, our final projections provide a 2D impression of the movement evolution, with peaks in the regions where a significant amount of motion is occurring in the reference time instant. With the average pooling instead we gather cues about *how* the motion evolves in time. Averaging the values, contributions in regions covered by the motion in past instants tend to smoothly decrease, while the values tend to increase in locations where the motion is mainly evolving in the current time instant. Considering that the input frames are normalized in the interval $[-1, +1]$ and the kernels have values in $\{-1, +1\}$, values in the average pooling close to zero correspond to "neutral" responses, where nothing happens, values close to 1 correspond to salient current motion information, while lower values (i.e. negative) correspond to information from the past.

In Fig. 2 we report an example of max pooling (Fig. 2b) and average pooling (Fig. 2c) for the block with central frame the one in Fig. 2a.

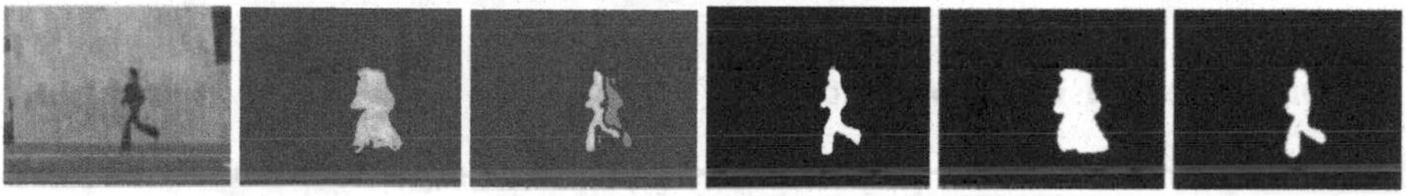

(a) (b) (c) (d) (e) (f)

Fig. 2. Example of pooling and segmentation maps obtained by our method: (a) Frame (b) MaxPool (c) AvgPool (d) Ground truth (e) MaxPool adaptive segmentation (f) Refined final map.

3.2 GCKs Attention Module

Adaptive Segmentation Based on Max Pooling. A first round of segmentation is performed on each max pooling map by means of Otsu's adaptive threshold method [20]. Initial result is refined by discarding connected components with an area smaller than a threshold τ and by applying morphological operations (opening and closing) to attenuate the effects of noise.

The maps obtained after the first threshold on max pooling only (see Fig. 2e), provide a more comprehensive view on the movement that goes beyond the instantaneous variation and still it includes a significant amount of false positives. The second step of our segmentation aims at attenuating this effect.

Avg Pooling Based Refinement and Saliency Estimation. As observed before, average pooling over temporal projections is able to convey information about the evolution of the scene in a limited amount of time frames. Roughly speaking, spatio-temporal locations correspoding to past and present phases on the motion can be identified by detecting the local minima and maxima of the pooling map using a pair of threshold values σ_1 and σ_2. We then exploit this knowledge to refine the segmentation map, as follows (i) "present" blobs are exploited to consolidate the masks obtained from max pooling, (ii) positions belonging to the "past" are discarded from the refined map. An additional step of refinement is implemented by discarding blobs with no local extrema in the average pooling. The segmentation map can finally be composed as $B = B_{max} - B_{avg}^{past} + B_{avg}^{present}$ Fig. 2f provide an example of the map obtained at the final stage. The final motion saliency map is also derived by selecting the values of the max pooling in the locations covered by B.

4 Experimental Results

In this section we discuss the quantitative and qualitative evaluation we performed to show the potentials of our method, while discussing its limitations.

Unless otherwise stated, we fix $\tau = 25$ (approximately a 5×5 patch of pixels). The thresholds σ_1 and σ_2 have been empirically set to 0.4 and 0.6 respectively, in order to discard the majority of background values.

4.1 Datasets and Evaluation Metrics

To test the potential of GCKs projections for motion saliency estimation in videos we evaluated our method on two different datasets. A first explorative analysis has been conducted on a classic action recognition dataset, the *Weizmann dataset* [8], a low resolution dataset including full-body actions of different granularity. Although rather aged, the dataset is an ideal test bed for assessing our method in a highly controlled scenario (no dynamic background, fixed camera).

The second dataset is a Video Object Segmentation benchmark, *SegTrack v2* [14], which consists in 14 high resolution videos that includes challenging conditions such as motion blur, complex deformations, occlusions and slow motion. Available annotations for both datasets include ground truth binary masks of every video frame.

As objectively evaluating the quality of motion saliency for detection is not trivial, we directly evaluate our method in terms of quality of the motion-based detection masks, according to the experimental protocol commonly used for VOS and change detection [5,32], as they are the closest tasks in literature that allow us to provide comparisons with alternative approaches on recent datasets. The metrics that we used are *mean Intersection over Union* of the estimated segmentation with respect to the ground truth, *Precision* (given by the number of true positives among all the predicted positives) and *Recall* (given by the fraction of true positives over the union of true positives and false positives). Notice this is a very challenging evaluation for us, as our method is not designed to provide very precise segmentation masks.

4.2 Method Assessment

To evaluate our method we first considered controlled scenarios with the Weizmann dataset, that includes sequences recorded with a fixed camera. In each sequence only one salient object is present and the background is static.

In Table 1, we report mean IoU, Precision and Recall for the segmentation based on max pooling (left) and after the refinement with the average pooling (right) for each one of the ten actions belonging to the dataset.

Table 1. Evaluation on the Weizmann dataset.

	Adaptive segmentation				Refinement step			
	mIoU	PR	RE	BBox	mIoU	PR	RE	BBox
bend	0.21	0.41	0.42	0.36	**0.26**	**0.47**	**0.44**	**0.41**
jack	0.32	0.50	0.48	**0.71**	**0.36**	**0.52**	**0.59**	0.63
jump	0.41	0.50	0.70	0.49	**0.59**	**0.66**	0.85	**0.69**
pjump	**0.47**	**0.54**	**0.79**	**0.65**	0.28	0.44	0.63	0.41
run	0.39	0.41	**0.90**	0.53	**0.66**	**0.74**	0.87	**0.78**
side	0.46	0.48	**0.91**	0.55	**0.56**	**0.65**	0.80	**0.73**
skip	0.44	0.45	**0.95**	0.58	**0.61**	**0.73**	0.80	**0.74**
walk	0.46	0.48	**0.94**	0.69	**0.56**	**0.69**	0.74	**0.78**
wave1	0.08	0.35	0.17	**0.21**	**0.11**	**0.44**	**0.20**	0.18
wave2	0.11	0.37	0.15	**0.30**	**0.15**	**0.42**	**0.21**	0.26
Mean	0.34	0.45	**0.64**	0.51	**0.41**	**0.58**	0.61	**0.56**

Fig. 3. Examples of obtained (even partial) detections (green: ground truth; red: our detection). (Color figure online)

Some considerations are in order. Refinement based on information about the "past" is particularly useful when the movement is wider and affecting more globally the object (as for running, skipping and walking), while it is less stable analysing movements more "compact" in space, like jumping in the same spot. Also notice that for actions like waving, bending, or doing jumping jacks, characterized by the movement of only a part of the body, our method provides less accurate (from the point of view of the metric) results, as it only detects the moving part (which is precisely its aim) instead of the whole body (annotated in the ground truth, see Fig. 3).

Motivated by the observation that, to our purposes, the results at the pixel level are not particularly important as long as the region corresponding to the moving object can be correctly detected, we also included in the analysis the computation of the mean IoU between bounding boxes, obtaining very promising results in fact of faithfulness of the detected area. As it can be appreciated on the right part of the table, the refinement provides, for the majority of actions, an improvement to the results.

4.3 Evaluations on VOS Datasets

We now discuss the performance of our method on the SegTrack dataset. Our aim is to highlight the potential of our method to generalize to more complex situations. Sequences in the dataset provide very challenging object segmentation scenarios, in

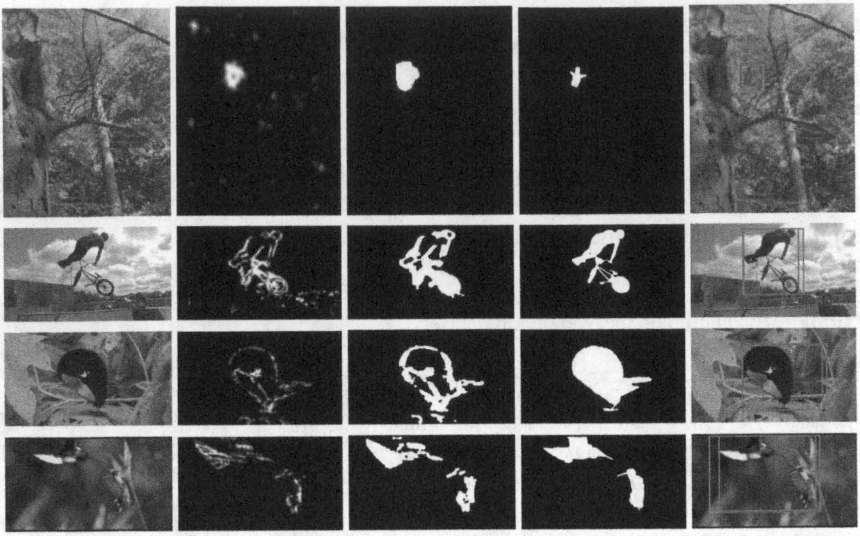

Fig. 4. Examples of output from the SegTrack dataset. From left: frame, our saliency map, our binary map, ground truth, bounding boxes (green: ground truth, red: our result). (Color figure online)

conditions of fast motion or motion blur, moving cameras and complex deformation. The subset of sequences included in Fig. 4 provides some prototypical examples.

We report in Table 2 a quantitative analysis, providing a comparison with alternative strategies for motion saliency estimation for detection of comparable complexity (thus we do not consider complex, effective but less efficient solutions based on deep architectures, as [29]). We considered in particular the use of optical flow as a starting point for motion saliency estimation, followed by the thresholding strategy we use in our approach for a fair comparison. Among the plethora of alternatives offered by the literature, we adopted the classical Gunnar-Farneback algorithm for dense optical flow estimation [6], and the approach of Sift Flow, used in a recent VOS work [32]. For the first we relied on the OpenCV implementation in Python, while for the latter an implementation is available from [32]. An additional approach considered for comparison is Background Subtraction for which we employed the Python function available in OpenCV that implements [33].

We further compare our results with a foreground segmentation method based on Gaussian Mixture Models [13], again exploiting OpenCV implementation. Our method provides competitive results in a number of videos, and, on average, a higher stability against different challenging conditions with lower standard deviation. While in videos where the camera is fixed we pay our lower attention to the fine-grained details, in more complex scenarios with camera motion, we produce consistently higher results.

4.4 Computational Analysis

We close the section with an analysis of the computation. We tested the efficiency of GCKs filtering scheme with respect to classical 3D convolution. Here we report the

Table 2. Evaluation on the Seg Track dataset.

		Segmentation IoU				Bounding Box IoU			
		[6]	[32]	[33]	gck	[6]	[32]	[33]	gck
Fixed camera	Birdfall	0.469	0.403	0.459	0.222	0.319	0.503	0.288	0.297
	Worm	0.036	0.195	0.165	0.146	0.073	0.166	0.064	0.105
	Hummingbird	0.419	0.437	0.643	0.341	0.759	0.756	0.870	0.669
	Frog	0.361	0.506	0.53	0.243	0.469	0.578	0.549	0.48
	Mean IoU	0.321	0.385	**0.449**	0.238	0.405	**0.500**	0.442	0.387
Handheld camera	Bird of paradise	0.293	0.406	0.193	0.286	0.494	0.560	0.551	0.633
	bmx	0.327	0.286	0.271	0.374	0.33	0.514	0.219	0.715
	Penguin	0.119	0.278	0.069	0.202	0.557	0.568	0.588	0.566
	Parachute	0.023	0.33	0.059	0.28	0.038	0.374	0.046	0.31
	Mean IoU	0.190	**0.325**	0.148	0.285	0.354	0.504	0.351	**0.556**
Dynamic camera	Cheetah	0.029	0.069	0.151	0.336	0.097	0.17	0.098	0.4
	Drift	0.011	0.005	0.136	0.419	0.16	0.16	0.183	0.348
	Monkey	0.035	0.01	0.07	0.063	0.087	0.086	0.087	0.086
	Monkeydog	0.048	0.069	0.068	0.114	0.165	0.188	0.142	0.199
	Soldier	0.023	0.022	0.083	0.23	0.111	0.116	0.103	0.202
	Girl	0.045	0.066	0.064	0.25	0.158	0.231	0.162	0.274
	Mean IoU	0.031	0.040	0.095	**0.235**	0.129	0.158	0.129	**0.251**
	Overall	0.159	0.220	0.211	**0.250**	0.272	0.355	0.282	**0.377**

time consumption of alternative strategies for obtaining the same 64 projections with kernels of size $4 \times 4 \times 4$, expressed in seconds per frame. All the tested solutions have been implemented in Python on a laptop with Intel(R) Core(TM) i5-8265U CPU @ 1.60 GHz and 32 GB RAM. With respect to full 3D convolution (11.45 sec/frame) or FFT-based convolution (0.7 sec/frame) our method is substantially more efficient (0.075 sec/frame). In terms of number of operations, in order to obtain M projections with a kernel of size $n \times n \times n$, using a full 3D convolution will require $M * n^3$ multiplications per pixel, while using GCKs efficient filtering scheme we are able to obtain the same result at the cost of $2(M-1)$ operations per pixel. With respect to the methods used in Table 2, the comparison is less fair as they all rely on either different languages or different frameworks. To give an idea, our implementation runs approximately with same computational time of [6,33] and is ~75 faster than [32].

5 Discussion

In this work we explored the use of the GCKs for efficient motion saliency estimation in videos. We showed that our method provides a good compromise between effectiveness and efficiency, with saliency maps that are able to reliably highlight the motion in a scene. A nice property of the GCKs is that they provide a powerful representation also for motion classification tasks, going beyond the detection: a preliminary analysis on synthetic data shows very encouraging classification performance. This may open the door to an end-to-end pipeline in which the very same low-level features could be exploited in multiple stages of the analysis, from the detection to the higher-level understanding of a dynamic event.

Acknowledgements. This work has been carried out at the Machine Learning Genoa (MaLGa) center, Università di Genova (IT). It has been supported by AFOSR with the project "Cognitively-inspired architectures for human motion understanding", grant no. FA8655-20-1-7035.

References

1. Ahad, M.A.R., Tan, J.K., Kim, H., Ishikawa, S.: Motion history image: its variants and applications. Mach. Vis. Appl. **23**(2), 255–281 (2012)
2. Ben-Artzi, G., Hel-Or, H., Hel-Or, Y.: The gray-code filter kernels. IEEE Trans. Pattern Anal. Mach. Intell. **29**(3), 382–393 (2007)
3. Bouwmans, T.: Recent advanced statistical background modeling for foreground detection-a systematic survey. Recent Pat. Comput. Sci. **4**(3), 147–176 (2011)
4. Cong, R., Lei, J., Fu, H., Cheng, M.M., Lin, W., Huang, Q.: Review of visual saliency detection with comprehensive information. IEEE Trans. Circ. Syst. Video Technol. **29**(10), 2941–2959 (2018)
5. Faktor, A., Irani, M.: Video segmentation by non-local consensus voting. In: BMVC. vol. 2, p. 8 (2014)
6. Farnebäck, G.: Two-frame motion estimation based on polynomial expansion. In: Bigun, J., Gustavsson, T. (eds.) SCIA 2003. LNCS, vol. 2749, pp. 363–370. Springer, Heidelberg (2003). https://doi.org/10.1007/3-540-45103-X_50
7. Fortun, D., Bouthemy, P., Kervrann, C.: Optical flow modeling and computation: a survey. Comput. Vis. Image Underst. **134**, 1–21 (2015)
8. Gorelick, L., Blank, M., Shechtman, E., Irani, M., Basri, R.: Actions as space-time shapes. Trans. Pattern Anal. Mach. Intell. **29**(12), 2247–2253 (2007)
9. Hel-Or, Y., Hel-Or, H.: Real-time pattern matching using projection kernels. IEEE Trans. Pattern Anal. Mach. Intell. **27**(9), 1430–1445 (2005)
10. Itti, L., Koch, C., Niebur, E.: A model of saliency-based visual attention for rapid scene analysis. IEEE Trans. Pattern Anal. Mach. Intell. **20**(11), 1254–1259 (1998)
11. Jain, S.D., Xiong, B., Grauman, K.: FusionSeg: learning to combine motion and appearance for fully automatic segmentation of generic objects in videos. In: 2017 IEEE Conference on Computer Vision and Pattern Recognition (CVPR), pp. 2117–2126. IEEE (2017)
12. Korman, S., Avidan, S.: Coherency sensitive hashing. IEEE Trans. Pattern Anal. Mach. Intell. **38**(6), 1099–1112 (2015)
13. Lee, D.S.: Effective gaussian mixture learning for video background subtraction. IEEE Trans. Pattern Anal. Mach. Intell. **27**(5), 827–832 (2005)
14. Li, F., Kim, T., Humayun, A., Tsai, D., Rehg, J.M.: Video segmentation by tracking many figure-ground segments. In: Proceedings of the IEEE International Conference on Computer Vision, pp. 2192–2199 (2013)
15. Moshe, Y., Hel-Or, H.: Video block motion estimation based on gray-code kernels. IEEE Trans. Image Process. **18**(10), 2243–2254 (2009)
16. Moshe, Y., Hel-Or, H., Hel-Or, Y.: Foreground detection using spatiotemporal projection kernels. In: 2012 IEEE Conference on Computer Vision and Pattern Recognition, pp. 3210–3217. IEEE (2012)
17. Noceti, N., Delponte, E., Odone, F.: Spatio-temporal constraints for on-line 3D object recognition in videos. Comput. Vis. Image Underst. **113**(12), 1198–1209 (2009)

18. Noceti, N., Sciutti, A., Sandini, G.: Cognition helps vision: recognizing biological motion using invariant dynamic cues. In: Murino, V., Puppo, E. (eds.) ICIAP 2015. LNCS, vol. 9280, pp. 676–686. Springer, Cham (2015). https://doi.org/10.1007/978-3-319-23234-8_62
19. Ochs, P., Malik, J., Brox, T.: Segmentation of moving objects by long term video analysis. IEEE Trans. Pattern Anal. Mach. Intell. **36**(6), 1187–1200 (2013)
20. Otsu, N.: A threshold selection method from gray-level histograms. IEEE Trans. Syst. Man Cybernet. **9**(1), 62–66 (1979)
21. Ouyang, W., Zhang, R., Cham, W.K.: Fast pattern matching using orthogonal Haar transform. In: 2010 IEEE Computer Society Conference on Computer Vision and Pattern Recognition, pp. 3050–3057. IEEE (2010)
22. Papazoglou, A., Ferrari, V.: Fast object segmentation in unconstrained video. In: Proceedings of the IEEE international conference on computer vision. pp. 1777–1784 (2013)
23. Perazzi, F., Khoreva, A., Benenson, R., Schiele, B., Sorkine-Hornung, A.: Learning video object segmentation from static images. In: Proceedings of the IEEE Conference on Computer Vision and Pattern Recognition, pp. 2663–2672 (2017)
24. Perazzi, F., Pont-Tuset, J., McWilliams, B., Van Gool, L., Gross, M., Sorkine-Hornung, A.: A benchmark dataset and evaluation methodology for video object segmentation. In: Proceedings of the IEEE Conference on Computer Vision and Pattern Recognition, pp. 724–732 (2016)
25. Rea, F., Vignolo, A., Sciutti, A., Noceti, N.: Human motion understanding for selecting action timing in collaborative human-robot interaction. Front. Robot. AI **6**, 58 (2019)
26. Stagliano, A., Noceti, N., Verri, A., Odone, F.: Online space-variant background modeling with sparse coding. IEEE Trans. Image Process. **24**(8), 2415–2428 (2015)
27. Vignolo, A., Noceti, N., Rea, F., Sciutti, A., Odone, F., Sandini, G.: Detecting biological motion for human-robot interaction: a link between perception and action. Front. Robot. AI p. 14 (2017)
28. Voigtlaender, P., Leibe, B.: Online adaptation of convolutional neural networks for video object segmentation. arXiv preprint arXiv:1706.09364 (2017)
29. Weinzaepfel, P., Revaud, J., Harchaoui, Z., Schmid, C.: Deepflow: large displacement optical flow with deep matching. In: Proceedings of the IEEE International Conference on Computer Vision, pp. 1385–1392 (2013)
30. Werlberger, M., Pock, T., Bischof, H.: Motion estimation with non-local total variation regularization. In: 2010 IEEE Computer Society Conference on Computer Vision and Pattern Recognition, pp. 2464–2471. IEEE (2010)
31. Xiao, F., Jae Lee, Y.: Track and segment: an iterative unsupervised approach for video object proposals. In: Proceedings of the IEEE Conference on Computer Vision and Pattern Recognition. pp. 933–942 (2016)
32. Zhuo, T., Cheng, Z., Zhang, P., Wong, Y., Kankanhalli, M.: Unsupervised online video object segmentation with motion property understanding. IEEE Trans. Image Process. **29**, 237–249 (2019)
33. Zivkovic, Z.: Improved adaptive Gaussian mixture model for background subtraction. In: Proceedings of the 17th International Conference on Pattern Recognition, 2004. ICPR 2004, vol. 2, pp. 28–31. IEEE (2004)

FirstPiano: A New Egocentric Hand Action Dataset Oriented Towards Augmented Reality Applications

Théo Voillemin, Hazem Wannous(✉), and Jean-Philippe Vandeborre

IMT Nord Europe, Univ. Lille, CNRS, UMR 9189 - CRIStAL, 59000 Lille, France
{theo.voillemin,hazem.wannous,
jean-philippe.vandeborre}@imt-nord-europe.fr

Abstract. Research on hand action recognition has achieved very interesting performance in recent years, notably thanks to deep learning methods. With those improvements, we can see new visions towards real applications of new Human-Machine interfaces (HMI) using this recognition. Such new interactions and interfaces need data to develop the best user experience iteratively. However, current datasets for hand action recognition in an egocentric view, even if perfectly useful for these problems of recognition, they generally lack of a limited but coherent context for the proposed actions. Indeed, these datasets tend to provide a wide range of actions, more or less in relation to each other, which does not help to create an interesting context for HMI application purposes. Thereby, we present in this paper a new dataset, FirstPiano, for hand action recognition in an egocentric view, in the context of piano training. FirstPiano provides a total of 672 video sequences directly extracted from the sensors of the Microsoft HoloLens Augmented Reality device. Each sequence is provided in depth, infrared and grayscale data, with 4 different points of view for the last one, for a total of 6 streams for each video. We also present the first benchmark of experiments using a Capsule Network over different classification problems and different stream combinations. Our dataset and experiments can therefore be interesting for research communities of action recognition and human-machine interface.

Keywords: Action recognition · Human machine interaction · Hand action dataset

1 Introduction

Augmented reality (AR) and virtual reality (VR) technologies have reached levels that were fantasized a few decades ago. However, the promises of new and natural interactions associated with the manipulation of a virtual world and the portability of these devices are still far from being fulfilled. Indeed, if VR devices have so far favored the use of joysticks, AR technologies, on the other hand, have mainly decided to use basic hand gesture recognition to permit the user to manipulate their interfaces. In both cases, hand action recognition has

S. Sclaroff et al. (Eds.): ICIAP 2022, LNCS 13233, pp. 170–181, 2022.
https://doi.org/10.1007/978-3-031-06433-3_15

yet been fully considered and implemented. Still, hand action is certainly the principal activity in everyday of a human being since we interact with the world around us with our hands. Hence, it would be interesting for AR and VR applications to draw all potential of hand action recognition features.

For such applications, algorithms for hand action recognition are needed, and for such algorithms, datasets for training them are needed. However, most of these datasets focus their efforts on providing a diversity of actions like object and environment manipulations and also interactions with people instead of focusing on a specific task. Hence, it is difficult to train such an algorithm that can be used into a real AR or VR application since these datasets do not provide a clear and useful context.

Thereby, we propose in this paper a new hand action dataset, FirstPiano, which provides data for action recognition purposes, but also for training some algorithms that can be used for useful AR applications and so helping HMI research field. The acquisition of the data of FirstPiano was thus motivated to capture an egocentric point of view of hand action within a strict and focus context for such AR purposes. This also motivated us to obtain the data from an easily reproducible setup. Hence, FirstPiano is a dataset capturing a context of piano learning acquired directly from the integrated sensors of the Microsoft HoloLens device. It contains 672 video sequences spread over up to 16 labels. Each sequence is provided with 6 different video streams, depth, infrared and grayscale images. A first benchmark using a recently developed neural network architecture is also described to propose a first estimation of results on our dataset on different data inputs and classification problem configurations.

2 Related Work

Hand Action Datasets: With the emergence and greater accessibility of RGB or depth sensors at the beginning of the 2010s, many researchers have increasingly focused on the problems of hand action recognition. We can then find a lot of datasets that offer a common base to compare different proposed algorithms. All hand action datasets are in an egocentric point of view, but we can find two different contexts. The first one is the context of only grasping an object [2,3,25], where it is interesting to analyze the exact position of the hand depending on the nature of the grasped object, such as by the fingertips when grasping a pencil, or pliers shape for holding a can. These datasets most focus on analyzing the shape of the hand, and the object grasped rather than the temporal evolution of the gesture of the hand. The second context is that of daily activities [1,11–13,17,19,22,28]. In this case, the object grasped is manipulated into fine action, such as pouring milk or cleaning glasses. First Person Hand Action dataset, presented by Hernando *et al.* [12], is certainly the current most elaborate dataset in this context with 1,175 sequences over 45 labels, but with RGB and depth data for each one of its sequences and also the exact position of the joints of the hand doing the action. EGTEA, presented by Li *et al.* [17], the evolution of GTEA [11] contains 28 h of daily activities with RGB videos, and also the gaze information and the mask of the hand. Still, these datasets, even if very interesting for hand action recognition training, lack a concrete, precise, and focused context by providing a lot of different actions, not necessarily linked to each other and whose recognition would not be of interest for an AR application. However, from a third person point of view, we can find interesting hand gesture datasets that offer precise context for applications. For example, Molchanov *et al.*, with NVGesture [20], proposed a dataset of hand gestures for designing touchless interfaces while using a car so the user can focus on driving. De Smedt *et al.* proposed DHG 14/28 [5] with a coarse and fine gesture that is executed both with the all hand opens or with only the index that can be used for interacting with computer interfaces.

Hand Action Recognition Algorithms: The last few years, deep learning algorithms have become one the most efficient method in many fields especially hand action recognition. Recently, even handcrafted approaches finally use neural network by extracting manually very precise and fine features before passing them to a network [31] or by using a neural network in parallel of handcrafted features then by merging them together [15,30]. Specialized algorithms for hand action recognition are quite rare to find in the state of the art in contrast to hand gesture recognition, where we can find neural networks that are specially designed to hand understanding by using hand skeleton with RNN architecture [4] or graph convolutional network [16]. Hand action recognition solutions are generally the same as those for human body activities [7,14,24] since both can consider a skeleton based representation as a sequence of joints [6], or 2D/3D CNN when this representation is not rich enough to capture the movement and analyze the manipulated object. Lin *et al.* [18] implemented a temporal shift module to share temporal information from a frame to the next inside a simple 2D CNN. Duarte *et al.* [8] proposed a first implementation of a 3D capsule network [26] for video understanding of human activities.

Human-Machine Interfaces on AR Device: Just before the arrival of augmented reality devices to the general public, such as the Microsoft HoloLens, some research groups have already developed their own smart glasses with hand action recognition solution integrated. For example, Schröder *et al.* [27] proposed smart glasses with RGB and depth sensors connected to an external computer to process an action made by the user to then display contextual information over the glasses. Essig *et al.* [9] proposed a similar project, but with a long-term vision to be fully personalized to the user by progressively constructing a mental representation of the user to display precise feedback over his glasses.

3 FirstPiano: Egocentric Dataset of Piano Interaction

3.1 Dataset Overview

FirstPiano is a hand action dataset focused on hand action recognition and to be used for AR applications. For this purpose, it contains a set of 672 actions videos with 6 different modalities for each of them for a total of 4,032 videos and 473,892 frames. We propose a context of piano training where a subject is asked to play a major scale in a right way or a wrong way for a total of up to 16 labels.

To understand how we get to 16 labels, we need to explain how a major scale is constructed in the musical field. The western musical language has 7 different notes (C, D, E, F, G, A, B), which are the white keys on a piano and 2 variations for each of them, flat or sharp, which are the black keys. Two musical notes are separated by what is called an interval, the smallest one being the half interval which can be observed on a piano between two consecutive keys (for example, between white A and black $B\flat$ or between white B and white C). A full interval is a succession of two half intervals (for example, between white C and white D, between black $G\sharp$ and black $B\flat$ or between white E and black $F\sharp$). To construct a major scale, one need to take a base note, which give the name of the scale, and to take the 7 next notes by following this series of intervals: 1, 1, 1/2, 1, 1, 1, 1/2. We propose examples for C and F major scales on Fig. 1.

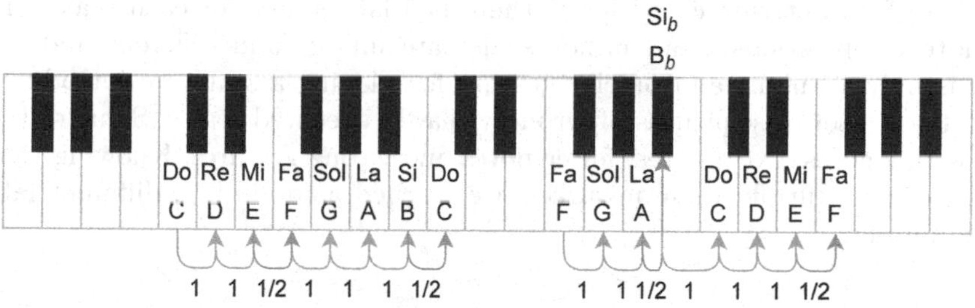

Fig. 1. Example of C and F major scale construction on a piano

Moreover, to be considered as rightly-played, we impose a precise gesture. Let's consider the fingers on a hand as 1 for the thumb, 2 for the index, 3 for the middle, 4 for the ring finger and 5 for the little finger. When played with the right hand, the 8 notes of a major scale must be played in ascending way on the following order: 1, 2, 3, 1, 2, 3, 4, 5. When played with the left hand the order is different: 5, 4, 3, 2, 1, 3, 2, 1. For descending way, the order for both hands is reversed. An example of this precise gesture with the left hand on a D major can be seen on Fig. 2.

Hence, each major scale is constructed from a start note is unique and need to have a precise gesture during its execution. We decided to limit the dataset to the

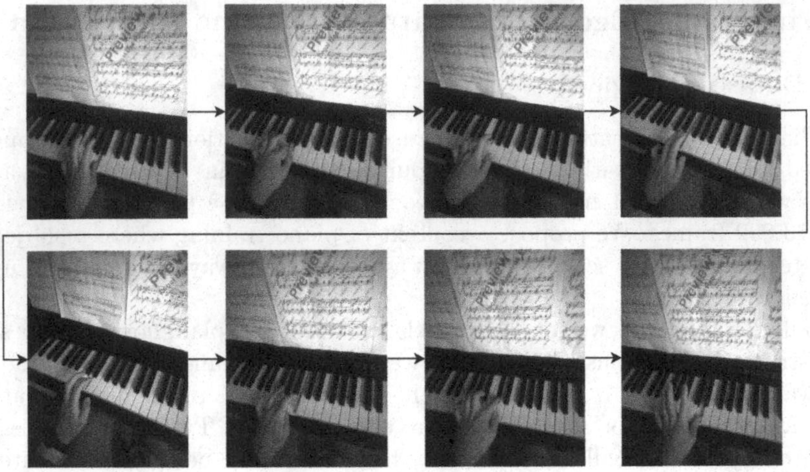

Fig. 2. Gesture imposed to right play a major scale. Example with left hand on D major

7 major scale that can be constructed from any white key. Each of these scales has multiple start emplacement in the piano and is performed with the right and the left hand, both in two other ways, crescendo and crescendo-decrescendo. We decided not to differentiate all those different playing configurations directly; therefore, for rightly-executed scale, there is 7 labels, one for each scale. Since the note compositions of all major scales are unique and different and since there is only a unique and precise gesture for playing a scale, we introduce in the dataset a set of sequences of wrongly-played executed scales. Some of them with wrong notes, extra notes, fewer notes and wrong gestures. Knowing that a sequence can multiply these mistakes, we reached a total of 16 different labels in the dataset.

3.2 Sensors and Acquisition Modalities

For the data acquisition, we decided to directly use the integrated sensors of the Microsoft HoloLens device thanks to the HoloLensForCV project[1] for data extraction. This AR headset directly provides an accessible and constant setup unlike other dataset that use a custom one. The Microsoft HoloLens is equipped with 8 different sensors. Since the frontal RGB camera has a too small viewing angle and since the long throw depth does not provide interesting data, we decided to keep the other 6 sensors to know: the short depth, long throw reflectivity, and the 4 peripheral grayscale sensors (see Fig. 3).

With a total of 66,320 of short throw depth frames, this sensor provides interesting information, especially to have precision to know if a key is pressed on the piano or else to let the possibility to extract hand pose and its skeleton

[1] https://github.com/microsoft/HoloLensForCV.

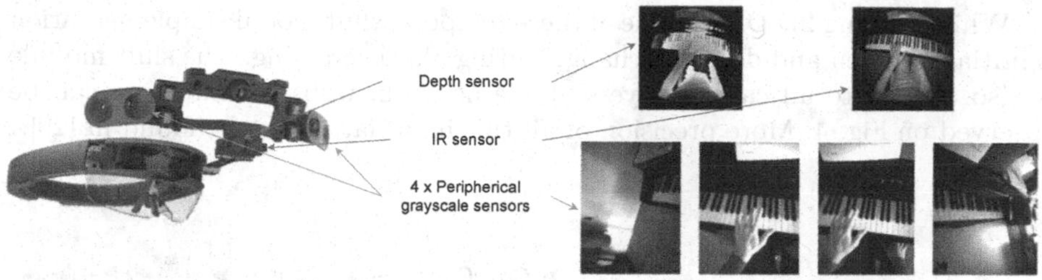

Fig. 3. The 6 sensors of the HoloLens device used and an example of each one inside the FirstPiano dataset

[10,21]. With 66,194 frames, the long throw reflectivity sensor permits to get a good central field of view of the hand of the player. The 4 peripheral grayscale sensors, with a total of 341,378 frames provide a full field of view on the entire piano, divided into 4 more precise points of view.

The acquisition was made in 2 different places and with 3 different subjects to provide a bit of diversity into the speed of execution and ease to play. Finally, each of the 7 major scales has 12 recordings, all with a rightly-played and wrongly-played recording, each in 4 different gestures, with the right hand or the left hand in a crescendo and crescendo-decrescendo way for a total of 672 sequences.

4 Benchmark Evaluation

4.1 Method: 2D Deep Video Capsule Network

We decided to use the 2D Deep Video Capsule Network approach (2D DVCN) [29] for the evaluation of FirstPiano. 2D DVCN architecture consists of the implementing of the temporal shift module [18], which permits adding temporal information into 2D neural network architecture that analyse spatial features, into a deep capsule network [23,26]. Briefly, a capsule is quite similar to a convolutional layer in its function since it captures and analyzes spatial features thanks to convolutional operations. The difference being that a capsule groups together the result of many convolutional operations to encapsulate many complex spatial features. We implemented a temporal shift module applied on the capsule by shifting part or all of the convolution operations composing the capsule of all or part of the capsule in a layer, leading us into 3 different implementations. The first one is shifting the first convolutions of the first capsules so that we share partial spatial information of some capsules to let the network to work with both past and current information. The second one is shifting first convolutions of all the capsules, since all capsules capture independent complex spatial features, it seems logical to consider all of them equally. In the last one is shifting all the convolutions of the first capsules of a layer, since all the convolutions of a capsule are linked together to represent a complex spatial feature, it also seems logical to shift all of them inside a same capsule.

While training 2D DVCN, one of these temporal shift module implementation is initially chosen and does not change during all the training. The shift module is also applied to all capsule layers of the network, whose architecture can be observed on Fig. 4. More precision of all this information can be found in [29].

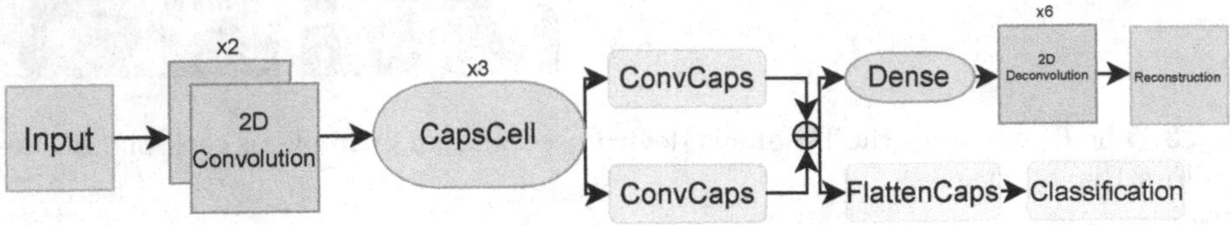

Fig. 4. Illustration of 2D DVCN [29]. ConvCaps is a standard 2d convolutional capsule layer but with temporal shift applied on it. CapsCell is a set of 3 ConvCaps with a residual branch between the first and the last

4.2 Experiments

For our experiments, we use the Keras library with Tensorflow in backend. 2D DVCN was trained on the following hardware configuration: Intel i9 9900k, Nvidia RTX Titan and 32 GB RAM. We used the Adam Optimizer during training with a starting learning rate of 0.001 and trained the network over 500 epochs.

Binary Recognition of Major Scales: For our first experiments, we considered a problem of binary classification. Even if we described that the sequences of FirstPiano are distributed over 16 labels, we firstly decided to consider only two labels, one when the major scale is rightly-played and the other if it is played wrongly-played, whatever the major scale. We justify the approach for two reasons, first one is that it permits us to test our dataset on an easy problem. The second one is that a rightly-played major scale is unique and can be deduced from the very first note. We can then justify having a network for this binary classification and a second method dedicated to identify the major scale with the first frames of a sequence classified as justify.

We start by using only two of the grayscale sensors, the central left and the central right since these two sensors are enough to get a full vision of the piano. Moreover, in this configuration, we let the network learns which hand is playing instead of giving him directly the right sensor associated with the playing hand. The two frames are simply juxtaposed as a panorama before being sent to the network. In this configuration, we reach an accuracy of 76.58% of good classification when using the shifting on the first convolutions of all capsules (see Table 1).

Table 1. Comparison of 2D DVCN results depending on the temporal shift implementation used over our FirstPiano dataset considering 2 labels and using 2 grayscale streams

Temporal shift	Accuracy (%)
1^{rst} convolutions of the 1^{rst} capsules	75.71
All convolutions of the 1^{rst} capsules	76.26
1^{rst} convolutions of all capsules	**78.56**
No shifting	76.13

We then decided to feed the network with only the depth data. This stream has a sufficient angle of view to observe the majority of the piano and since we did not acquire some playing on the extreme sides of the piano, the depth stream is sufficient alone. Moreover, in addition to being visually interpreted the same as a grayscale image, with fewer details, it also contains additional information such as the distance of each pixel to the sensor. We obtain similar results than the configuration of using as input both grayscale streams, with the temporal shift of the first convolutions of all capsules returning the best result. however, it seems that the depth stream alone does not provide as much information as the two grayscale streams since we only reached an accuracy of 73.33% of good classification in this case (see Table 2).

Table 2. Comparison of 2D DVCN results depending on the temporal shift implementation used over our FirstPiano dataset considering 2 labels and using depth stream only

Temporal shift	Accuracy (%)
1^{rst} convolutions of the 1^{rst} capsules	69.14
All convolutions of the 1^{rst} capsules	69.91
1^{rst} convolutions of all capsules	**73.33**
No shifting	70.57

Combining all the 3 previously described video streams by juxtaposing them leads us towards very similar results to those of the case using only depth data (see Table 3). However, given that we are using both grayscale video streams, we can expect to get at least similar results to the case using them only. The most logical conclusion is that the way 2D DVCN was implemented was probably not deep enough to process such high dimensional input and since capsules work with convolutional operations, the juxtaposition of depth and grayscale into a unique image lost them.

Table 3. Comparison of 2D DVCN results depending on the temporal shift implementation used over our FirstPiano dataset considering 2 labels and using 3 video streams

Temporal shift	Accuracy (%)
1^{rst} convolutions of the 1^{rst} capsules	70.14
All convolutions of the 1^{rst} capsules	**73.56**
1^{rst} convolutions of all capsules	72.54
No shifting	69.14

Multi Labelling of Major Scales Recognition: Since we obtained pretty good results in the easy configuration of binary classification, we decided to complicate the problem by multi labelling all the major scales. We then considered 8 labels, one for each different rightly-played major scale, for a total of 7 and the last one, which includes all wrongly-played scales. We can justify this choice because, even if a wrongly-played scale is associated to a note in the hierarchy of our dataset, it can be difficult, even impossible, to tell from which scale there have been wrong notes. Moreover, we think it will be far more interesting to have a specialised method to precisely identify what mistake is made in a wrongly-played scale, especially the exact frames it occurs. Since a single network cannot deal with so many particularities and special cases, we decided it was the best configuration to a multi labelling problem from a unique method.

Table 4. Comparison of 2D DVCN results depending on the temporal shift implementation used over our FirstPiano dataset considering 8 labels and using 2 grayscale streams

Temporal shift	Accuracy (%)
1^{rst} convolutions of the 1^{rst} capsules	59.09
All convolutions of the 1^{rst} capsules	**61.93**
1^{rst} convolutions of all capsules	61.36
No shifting	60.20

Since the problem has become more difficult, we can observe a degradation of the classification accuracy. Using the two grayscale video streams, we reached 61.93% of good classification when shifting all convolutional operations of the first capsules of a layer (see Table 4). Slightly worse results were obtained using all 3 depth and grayscale streams. We reached 59.45% of good classification using the same temporal shift, we also found similar degradation of results using the depth stream juxtapose next to the grayscale ones (see Table 5). These results are coherent with the complication of the classification problem. Indeed, we can explain it since the network now has to also focus exactly on the notes played and not only on the gesture or the number of notes played.

Table 5. Comparison of 2D DVCN results depending on the temporal shift implementation used over our FirstPiano dataset considering 8 labels and using 3 video streams

Temporal shift	Accuracy (%)
1^{rst} convolutions of the 1^{rst} capsules	57.23
All convolutions of the 1^{rst} capsules	**59.45**
1^{rst} convolutions of all capsules	56.46
No shifting	56.89

5 Conclusion

In this paper, we proposed a new dataset for hand action recognition oriented towards AR applications, FirstPiano. The dataset provides 672 video sequences, each one with 6 different video streams, of rightly and wrongly played major scale on a piano in various configurations. Data are extracted directly from the integrated sensors on the Microsoft HoloLens device so that the acquisition setup can be constant for everyone wanted to use the dataset for AR applications or even to complete it. We also provide the first benchmark to evaluate FirstPiano on different configurations, such as binary and multi classification using 1 to 3 video streams as input.

As future works, it could be interesting to complete the dataset even for current actions provided but also with new piano exercises such as dissociation between the left hand and the right hand during playing or even different rhythms. It would also be interesting to deepen the notation of labels, especially by including, for each sequence, the frames when a new note is being played as right as the finger used so that it could be possible to train a precise algorithm that can find the exact location of mistakes.

We believe that this work presenting a new dataset in addition to the tested benchmark can encourage new research in the hand action recognition but also in the human computer interface fields.

References

1. Bambach, S., Lee, S., Crandall, D.J., Yu, C.: Lending a hand: Detecting hands and recognizing activities in complex egocentric interactions. In: IEEE International Conference on Computer Vision (ICCV). pp. 1949–1957 (2015)
2. Bullock, I.M., Feix, T., Dollar, A.M.: The yale human grasping dataset: Grasp, object, and task data in household and machine shop environments. The International Journal of Robotics Research **34**(3), 251–255 (2015)
3. Cai, M., Kitani, K.M., Sato, Y.: A scalable approach for understanding the visual structures of hand grasps. In: IEEE International Conference on Robotics and Automation (ICRA). pp. 1360–1366 (2015)
4. Chen, X., Guo, H., Wang, G., Zhang, L.: Motion feature augmented recurrent neural network for skeleton-based dynamic hand gesture recognition. IEEE International Conference on Image Processing (ICIP), September 2017

5. De Smedt, Q., Wannous, H., Vandeborre, J.P., Guerry, J., Saux, B.L., Filliat, D.: 3D hand gesture recognition using a depth and skeletal dataset: Shrec 2017 track. In: Proceedings of the Workshop on 3D Object Retrieval. 3Dor 2017, pp. 33–38. Eurographics Association, Goslar, DEU (2017)
6. De Smedt, Q., Wannous, H., Vandeborre, J.-P.: 3D hand gesture recognition by analysing set-of-joints trajectories. In: Wannous, H., Pala, P., Daoudi, M., Flórez-Revuelta, F. (eds.) UHA3DS 2016. LNCS, vol. 10188, pp. 86–97. Springer, Cham (2018). https://doi.org/10.1007/978-3-319-91863-1_7
7. Devanne, M., Wannous, H., Daoudi, M., Berretti, S., Bimbo, A.D., Pala, P.: Learning Shape Variations of Motion Trajectories for Gait Analysis. In: International Conference on Pattern Recognition (ICPR). pp. 895–900. Càncun, Mexico (2016)
8. Duarte, K., Rawat, Y., Shah, M.: VideoCapsuleNet : a simplified network for action detection. In: Advances in Neural Information Processing Systems, pp. 7610–7619 (2018)
9. Essig, K., Strenge, B., Schack, T.: ADAMAAS: towards smart glasses for mobile and personalized action assistance.. In: 9th ACM International Conference, pp. 1–4, June 2016
10. Fang, L., Liu, X., Liu, L., Xu, H., Kang, W.: JGR-P2O: joint graph reasoning based pixel-to-offset prediction network for 3D hand pose estimation from a single depth image. In: Vedaldi, A., Bischof, H., Brox, T., Frahm, J.-M. (eds.) ECCV 2020. LNCS, vol. 12351, pp. 120–137. Springer, Cham (2020). https://doi.org/10.1007/978-3-030-58539-6_8
11. Fathi, A., Ren, X., Rehg, J.M.: Learning to recognize objects in egocentric activities. In: IEEE Conference on Computer Vision and Pattern Recognition (CVPR), pp. 3281–3288 (2011)
12. Garcia-Hernando, G., Yuan, S., Baek, S., Kim, T.K.: First-person hand action benchmark with RGB-D videos and 3D hand pose annotations. In: IEEE Conference on Computer Vision and Pattern Recognition, pp. 409–419 (2018)
13. Goyal, R., et al.: The something video database for learning and evaluating visual common sense. In: IEEE International Conference on Computer Vision (ICCV) 2017, pp. 5843–5851. Los Alamitos, CA, USA, October 2017
14. Karpathy, A., Toderici, G., Shetty, S., Leung, T., Sukthankar, R., Fei-Fei, L.: Large-scale video classification with convolutional neural networks. In: IEEE Conference on Computer Vision and Pattern Recognition, pp. 1725–1732 (2014)
15. Khan, M.A., Sharif, M., Akram, T., Raza, M., Saba, T., Rehman, A.: Hand-crafted and deep convolutional neural network features fusion and selection strategy: an application to intelligent human action recognition. Appl. Soft Comput. **87**, 105986 (2020)
16. Li, C., Li, S., Gao, Y., Zhang, X., Li, W.: A two-stream neural network for pose-based hand gesture recognition. CoRR abs/2101.08926 (2021)
17. Li, Y., Liu, M., Rehg, J.M.: In the eye of beholder: joint learning of gaze and actions in first person video. In: Proceedings of the European Conference on Computer Vision (ECCV), September 2018
18. Lin, J., Gan, C., Han, S.: Tsm: Temporal shift module for efficient video understanding. In: IEEE International Conference on Computer Vision (ICCV) (2019)
19. Moghimi, M., Azagra, P., Montesano, L., Murillo, A.C., Belongie, S.: Experiments on an RGB-D wearable vision system for egocentric activity recognition. In: 2014 IEEE Conference on Computer Vision and Pattern Recognition Workshops, pp. 611–617 (2014)

20. Molchanov, P., Yang, X., Gupta, S., Kim, K., Tyree, S., Kautz, J.: Online detection and classification of dynamic hand gestures with recurrent 3D convolutional neural network. In: IEEE Conference on Computer Vision and Pattern Recognition, pp. 4207–4215 (2016)
21. Oberweger, M., Wohlhart, P., Lepetit, V.: Hands deep in deep learning for hand pose estimation. In: Computer Vision Winter Workshop, pp. 1–10 (2015)
22. Pirsiavash, H., Ramanan, D.: Detecting activities of daily living in first-person camera views. In: IEEE Conference on Computer Vision and Pattern Recognition, pp. 2847–2854 (2012)
23. Rajasegaran, J., Jayasundara, V., Jayasekara, S., Jayasekara, H., Seneviratne, S., Rodrigo, R.: DeepCaps: going deeper with capsule networks. In: IEEE Conference on Computer Vision and Pattern Recognition (CVPR), pp. 10717–10725 (2019)
24. Rhif, M., Wannous, H., Farah, I.R.: Action recognition from 3D skeleton sequences using deep networks on lie group features. In: 24th International Conference on Pattern Recognition (ICPR), pp. 3427–3432 (2018)
25. Rogez, G., Supancic, J.S., Ramanan, D.: Understanding everyday hands in action from RGB-D images. In: IEEE International Conference on Computer Vision (ICCV), pp. 3889–3897 (2015)
26. Sabour, S., Frosst, N., Hinton, G.E.: Dynamic routing between capsules. In: Guyon, I., et al. (eds.) Advances in Neural Information Processing Systems. vol. 30. Curran Associates, Inc. Red Hook (2017)
27. Schröder, M., Ritter, H.: Deep learning for action recognition in augmented reality assistance systems. In: ACM SIGGRAPH 2017 Posters, pp. 1–2, June 2017
28. Tang, Y., Tian, Y., Lu, J., Feng, J., Zhou, J.: Action recognition in RGB-D egocentric videos. In: IEEE International Conference on Image Processing (ICIP), pp. 3410–3414 (2017)
29. Voillemin, T., Wannous, H., Vandeborre, J.P.: 2D deep video capsule network with temporal shift for action recognition. In: 25th International Conference on Pattern Recognition (ICPR), pp. 3513–3519 (2021)
30. Wang, L., Qiao, Y., Tang, X.: Action recognition with trajectory-pooled deep-convolutional descriptors. In: IEEE Conference on Computer Vision and Pattern Recognition (CVPR), pp. 4305–4314 (2015)
31. Wang, S., Hou, Y., Li, Z., Dong, J., Tang, C.: Combining convnets with handcrafted features for action recognition based on an HMM-SVM classifier. Multim. Tools Appl. **77**(15), 18983–18998 (2018)

Learning Video Retrieval Models with Relevance-Aware Online Mining

Alex Falcon[1,2](✉), Giuseppe Serra[2], and Oswald Lanz[3]

[1] Fondazione Bruno Kessler, 38123 Povo, TN, Italy
afalcon@fbk.eu

[2] University of Udine, 33100 Udine, UD, Italy
giuseppe.serra@uniud.it

[3] Free University of Bozen-Bolzano, 39100 Bolzano, BZ, Italy
lanz@inf.unibz.it

Abstract. Due to the amount of videos and related captions uploaded every hour, deep learning-based solutions for cross-modal video retrieval are attracting more and more attention. A typical approach consists in learning a joint text-video embedding space, where the similarity of a video and its associated caption is maximized, whereas a lower similarity is enforced with all the other captions, called *negatives*. This approach assumes that only the video and caption pairs in the dataset are valid, but different captions - *positives* - may also describe its visual contents, hence some of them may be wrongly penalized. To address this shortcoming, we propose the *Relevance-Aware Negatives and Positives* mining (RANP) which, based on the semantics of the negatives, improves their selection while also increasing the similarity of other valid positives. We explore the influence of these techniques on two video-text datasets: EPIC-Kitchens-100 and MSR-VTT. By using the proposed techniques, we achieve considerable improvements in terms of nDCG and mAP, leading to state-of-the-art results, e.g. +5.3% nDCG and +3.0% mAP on EPIC-Kitchens-100. We share code and pretrained models at https://github.com/aranciokov/ranp.

Keywords: Video retrieval · Cross-modal retrieval · Contrastive loss · Hard negative mining

1 Introduction

When performing a search by typing a textual query on a multimedia search engine, the user expects the retrieved contents to be semantically close to it. As one can expect, it is important for the first retrieved item to be 'exactly' what the user was looking for. Yet, the following ones should be treated as importantly as the first one, given that multiple items are likely relevant to the user query.

Supplementary Information The online version contains supplementary material available at https://doi.org/10.1007/978-3-031-06433-3_16.

S. Sclaroff et al. (Eds.): ICIAP 2022, LNCS 13233, pp. 182–194, 2022.
https://doi.org/10.1007/978-3-031-06433-3_16

In a recent work, Wray et al. [32] described this problem for the domain of video retrieval. Similar observations were also made in previous work in different domains, e.g. in image retrieval [12]. Focusing on text-video retrieval, most of the current methods learn a joint textual-visual embedding space (e.g. [3,9]). To do so, a video descriptor and a textual descriptor are computed independently for each of the pairs (video and its captions) in the dataset; then, the similarity of these descriptors is maximized. At inference time, given a textual query, the multimedia search engine would retrieve the related video as the first result. To achieve this goal, a typical choice consists in contrastive loss functions, e.g. [13,14], which contrast the similarity of the paired video and text descriptors against those of different videos and captions called 'negatives'.

Several techniques have been proposed to decide which negatives to use in order to drive the learning, e.g. 'hard' or 'semi-hard' [17,28], how many of them, e.g. one [28], two [5], or more [29]. Yet, in all these cases the pool of negatives contains the captions which are not paired in the dataset: thus, the selection of the negatives is often unaware of the overlap between the semantic content of the caption and the contents of the video. As an example, a dataset may contain a video $v^\star$ and its caption $q^\star$ about a chef preparing and baking a cheesecake; a different caption q_1 about preparing a cheesecake without baking it; and q_2 describing how to change a light bulb. By using the aforementioned strategies, during training q_1 and q_2 may be selected as negative captions, and their similarity to $v^\star$ and $q^\star$ would be lowered. While this is fine for q_2, q_1 should be treated differently considering its similarity with $q^\star$. To address this shortcoming, in this paper we propose to improve the mining by making it aware of such an overlap. In particular, we focus on 'online' mining (i.e. the negatives are picked from the batch) because it is widely used in recent works (e.g. [3,9]) and is less burdensome than 'offline' mining. Moreover, to estimate the overlap we use a relevance function [7] defined on already available captions, avoiding the need for costly annotations. Thus, we name it relevance-aware online negative mining, or RAN. Differently from previous techniques which might select less false negatives after training for some epochs, RAN improves the selection from the start thanks to the use of semantics and not only the network state.

Similarly, we extend this idea to select captions which present a considerable overlap with the video contents ('positive' captions), but are not paired to the video. In fact, video retrieval methods which perform online mining only consider the groundtruth pairs as valid positives, missing this opportunity. A few works using offline mining select them based on semantic class labels [32,33] and, in different domains, this is done by using additional data which are not available in video-text datasets, e.g. class identifiers for image retrieval [36] and person re-identification [17]. By merging this technique with RAN, we obtain RANP which carefully mines both negative and positive captions using semantics.

The main contributions of this work can be summarized as follows:

- to address the shortcoming of false negatives' selection during the online hard negative mining, we propose and formulate a relevance-aware variant that we call RAN which, differently from previous techniques, uses semantics to select better negatives from the start of the training;

- we introduce in the video retrieval field the relevance-aware online hard positive mining, which helps selecting hard positives thanks to a relevance-aware mechanism, and use it alongside RAN obtaining RANP;
- we validate the proposed techniques on two public video-and-language datasets, which are EPIC-Kitchens-100 and MSR-VTT, providing evidence of their usefulness while also achieving the new state-of-the-art on EPIC-Kitchens-100 with an improvement of +5.3% nDCG and +3.0% mAP.

2 Related Work

Video Retrieval. The approaches for cross-modal text-video retrieval usually learn a joint textual-visual embedding space [3,6,9,31]. Given the multimodal nature of videos, several authors introduced novel techniques to learn a joint representation of all the available modalities [11,22,23,25,31]. As an example, Mixture of Embedding Experts (or MoEE, by Miech et al. [23]) and T2Vlad (by Wang et al. [31]) applied techniques based on NetVLAD [1], whereas a multimodal Transformer was used in Gabeur et al. [11]. Liu et al. [22] proposed Collaborative Experts (CE), which extended previous works with a gating mechanism to modulate each feature based on the other pretrained experts. Recently, Croitoru et al. [6] shifted the attention to the textual counterpart, leveraging the availability of multiple language models. The structure of the input data was used in multiple works, e.g. by constructing embedding spaces based on the part-of-speech (Wray et al. [33]) or by employing semantic role labeling to learn global and local representations (Chen et al. [3]). All these works focus on instance-based video retrieval, where only video and caption pairs in the dataset are considered to evaluate the performance. Given that multiple descriptions can describe a video, Wray et al. [32] proposed a 'semantic' video retrieval, which considers multiple degrees of relevance when computing the evaluation metrics.

Contrastive Loss and Mining Techniques. To learn the cross-modal embedding spaces, contrastive losses [13,14,17] are often employed because they enforce a high similarity for the descriptors of (video, caption) pairs in the dataset. Hadsell et al. [14] initially computed the loss on pair of samples, and the idea has been extended to triplets [28], quadruplets [5], and 'N+1'-tuples [29]. Yet, the amount of possible tuples scales exponentially (e.g. cubically with triplets) and most of them contribute meaninglessly to the loss. Hence, mining techniques were proposed to extract less tuples, either from the dataset ('offline') or from the batch ('online'). Offline mining is often avoided because it recomputes the tuples throughout the training making it burdensome. Nonetheless, some works made use of it in several domains, e.g. deep metric learning [15,30] and video retrieval [32,33]. In 'online' mining, the positive items are given by groundtruth associations, e.g. (video, caption) pairs in the dataset, whereas the rest of the batch forms the pool of negatives. The loss is often computed on all negatives (e.g. [11,23]), but picking 'hard' or 'semi-hard' negatives (i.e. irrelevant but highly similar to the groundtruth) is often preferred, as in [3,9]. Nonetheless, recent research (e.g. Xuan et al. [35,36]) presented the usefulness of easy examples.

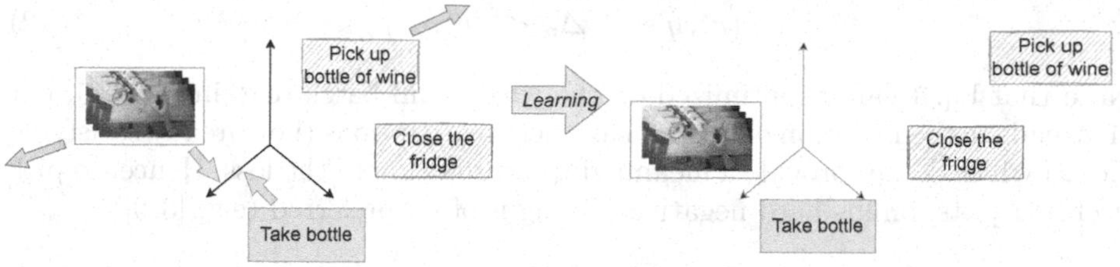

Fig. 1. By adopting the typical approach (Eq. 3), the caption 'pick up bottle of wine' is used as the hardest negative because it is not associated to the video in the dataset, and its descriptor is the most similar (closest) to the video. We visualize the change in similarity with the green (increase) and red (decrease) arrows. (Color figure online)

Mining techniques for positive items have also been proposed (e.g. in cross-modal [17,36] and near-duplicate video retrieval [19]) although they are based on the availability of groundtruth labels. In representation learning for images or videos, positive items were also artificially constructed through transformations [4,16,26,27].

We aim at introducing semantic knowledge in training by leveraging the relevance function. A similar idea is also used in [32], but we do so to improve online mining techniques, which are usually preferred. Furthermore, differently from previous video retrieval methods, we present a two-step approach to select online hard positives and show its effectiveness on two large scale datasets.

3 Training a Video Retrieval Model with Contrastive Loss and Mining

Given a video $v^\star$ and a pool of candidate textual queries, the aim of video-to-text retrieval is to return a ranked list of candidates where on top we expect the caption $q^\star$ corresponding to $v^\star$ in the dataset. 'Ranked list' implies that the output is sorted based on the similarity (computed with $s(\cdot,\cdot)$, e.g. cosine similarity) of $v^\star$ with all the candidate captions. In a common video retrieval setting, the evaluation metrics solely focus on the rank of the corresponding caption. But, multiple captions may equally describe the same video, so we focus on semantic video retrieval [32], where the evaluation is based on metrics which look at the whole ranked list. In the rest of the paper we focus on video-to-text, but text-to-video retrieval is obtained by swapping the role of $q^\star$ and $v^\star$.

To train the video retrieval model, a contrastive loss which performs online mining of triplets [28] is widely used [3,9–11] and is based on this term:

$$L_n = max(0, \Delta_n + s(v^\star, q-) - s(v^\star, q^\star)) \tag{1}$$

where Δ_n is a fixed margin, and $q-$ is a query which does not describe the video $v^\star$ (i.e. $q-$ is a negative query for $v^\star$). By optimizing with respect to Eq. 1, a margin Δ_n is enforced between the similarity of the groundtruth pair and the similarity of video and negative query, in order to satisfy the following constraint:

$$s(v^\star, q-) + \Delta_n < s(v^\star, q^\star) \tag{2}$$

Note that Eq. 1 can be optimized on the whole mini-batch (e.g. in [11,23]), but this leads to the inclusion of several easy negative captions (i.e. already satisfying Eq. 2) which do not provide a meaningful contribution to the loss. Hence, to pick useful triplets, online hard negative mining is often preferred (e.g. [3,9]).

3.1 Online Hard Negative Mining

Online hard negative mining consists in optimizing Eq. 1 only on hard negatives, which are the captions q violating Eq. 2. Formally, given $(v^\star, q^\star)$ and defining Q as the set of captions in the mini-batch, the hardest negative is identified by:

$$q- = argmax_{q \in Q \setminus \{q^\star\}} s(v^\star, q) \tag{3}$$

Since $q^\star$ describes $v^\star$ in the dataset, $q^\star$ is not considered when looking for the hardest negative. Yet, not all the queries in $Q \setminus \{q^\star\}$ should be considered negative: in fact, there may be queries which correctly (or at least partially) describe $v^\star$ although the association is not present in the dataset. As an example, let $q^\star$ be 'take bottle', q_1 'pick up bottle of wine', q_2 'close the fridge', and $s(v^\star, q_1) > s(v^\star, q_2)$ as in Fig. 1. Then, q_1 is the hardest negative because of two conditions: firstly, it is not $q^\star$ which, according to Eq. 3, makes it a possible negative; secondly, it is the closest 'negative' to $v^\star$, hence selected by $argmax$. Note that $s(\cdot, \cdot)$ is tightly bound to the network state, hence after some training it might not select some of these false negatives. Nonetheless, the techniques we propose avoid these situations from the start by using the semantics of the data.

4 Proposed Method: Relevance-Aware Online Mining

4.1 Relevance

To introduce the relevance, we start with an example. Let: (x_1) 'pick up a flowerpot and a sunflower', (x_2) 'pick an helianthus and a flowerpot', (x_3) 'pot the lily in a flowerpot', (x_4) 'put the cake in the oven'. We consider x_2 and x_1 quite similar ('helianthus' and 'sunflower' are synonyms), hence x_2 is highly relevant; x_3 is slightly relevant because of 'flowerpot', but the flowers and actions are different; and x_4 is irrelevant. Therefore, we want to capture semantic relations (e.g. synonyms) to determine how 'similar' the two captions are, i.e. the degree of relevance. In particular, we define the relevance function $\mathcal{R}(x_i, x_j)$ in terms of noun and verb classes shared among x_i and x_j, as in Damen et al. [7]. Formally:

$$\mathcal{R}(x_i, x_j) = \frac{1}{2}\left(\frac{|x_i^V \cap x_j^V|}{|x_i^V \cup x_j^V|} + \frac{|x_i^N \cap x_j^N|}{|x_i^N \cup x_j^N|}\right) \tag{4}$$

where x_i^V and x_i^N represent, respectively, the set of verb and noun classes identified in the i-th caption. We refer to 'noun class' (or 'verb class') to consider the noun (or verb) tokens which share similar semantics. When one (or both) of the inputs to $\mathcal{R}$ is a video, we consider two situations. If only one caption q_i is paired to the video v_i, we consider for v_i the noun and verb classes of q_i. Conversely, if multiple captions are available, then we construct a word set based on the classes which appear more frequently among the different captions, as also recently done in [32]. That is: $x_i^N = \{c^N | c^N \in \mathcal{D}(x_i)_{|\rho,N}\}$, where c^N is a noun class, $\mathcal{D}(x_i)$ is the set of captions available for x_i, and we define $\mathcal{D}(x_i)_{|\rho,N}$ as the reduced set of classes for the part-of-speech N which appear in at least $\rho \cdot |\mathcal{D}(x_i)|$ captions. Formally: $\mathcal{D}(x_i)_{|\rho,N} = \{c \,|\, PoS(c) = N \wedge |\{d | d \in \mathcal{D}(x_i) \wedge c \in d\}| \geq \rho \cdot |\mathcal{D}(x_i)|\}$, where $PoS(\cdot)$ determines the part-of-speech of the given class. Note that x_i^V is built equivalently. Finally, looking at the previous example, we can compute the following: $\mathcal{R}(x_1, x_2) = 1$, $\mathcal{R}(x_1, x_3) = 0.16$, and $\mathcal{R}(x_1, x_4) = 0$.

4.2 Relevance-Aware Online Hard Negative Mining

In Sect. 3.1 we intuitively describe a limitation of current online hard negative mining. Formally, we consider $\mathcal{R}$ and fix a threshold τ to determine the degree of relevance above which a caption is considered positive. Then, $\{q \,|\, \mathcal{R}(v^\star, q) \geq \tau, q \in Q \setminus \{q^\star\}\}$ may be non empty, which may consequently lead to the selection of a caption $q-$ as negative, although it is 'positive' to $v^\star$, i.e. $\mathcal{R}(v^\star, q-) \geq \tau$. Considering that the triplet loss lowers the similarity of $q-$ to $v^\star$ while increasing the similarity of $q^\star$ to $v^\star$, $q-$ would be penalized although describing it correctly. To address this shortcoming, we introduce RAN, which makes the mining process aware of the relevance of the captions to the video, in order to avoid the selection of a 'false negative'. We consider the following equation:

$$q- = argmax_{q \in Q \setminus \{q \,|\, \mathcal{R}(v^\star, q) \geq \tau\}} s(v^\star, q) \tag{5}$$

where, differently from Eq. 3, we employ $\{q \,|\, \mathcal{R}(v^\star, q) \geq \tau\}$ to capture the items which should be excluded from the pool of candidate negatives.

4.3 Relevance-Aware Online Hard Positive Mining

With the previous technique we pick high quality negative captions. Similarly, we want to select the captions which describe $v^\star$ and increase their similarity to it, to further improve the structure of the embedding space. To do so, we propose a two-steps approach for the relevance-aware online mining of positives. First of all, we compute the hardest positive $q+$ for $v^\star$, a positive caption (i.e. $\mathcal{R}(v^\star, q+) \geq \tau$) which has a far too dissimilar representation when compared to $v^\star$. By following the notation used for the negative mining, this would be:

$$q+ = argmin_{q \in Q \setminus \{q^\star\}} s(v^\star, q) \tag{6}$$

but this is not optimal, as it may select as positives the easy negative captions which were not violating Eq. 2. Hence, we propose to further employ the relevance to improve Eq. 6, by capturing the negative captions with $\{q \,|\, \mathcal{R}(v^\star, q) < \tau\}$ and excluding them from the selection of the hard positives. Therefore:

$$q+ = argmin_{q \in Q \setminus \{q \,|\, \mathcal{R}(v^\star, q) < \tau\}} s(v^\star, q) \tag{7}$$

Then, we use these positive captions in the triplet loss, in order to increase the similarity of $v^\star$ and $q+$, while at the same time decrease the similarity with $q-$. This can be formalized as:

$$L_p = max(0, \Delta_p + s(v, q-) - s(v, q+)) \tag{8}$$

Given a batch B of paired videos and captions, the final video-to-text loss is:

$$\mathcal{L}_{v-t} = \frac{1}{|B|}(\sum_{v \in B} L_p + \sum_{v \in B} L_n) \tag{9}$$

5 Results

To validate our method, we consider two large scale video-text datasets: EPIC-Kitchens-100 [7] and MSR-VTT [34]. The former contains 67217 clips for training and 9668 for testing. Each clip is annotated with a short caption describing activities in the kitchen. Moreover, for each caption, verb and noun semantic classes are available. MSR-VTT consists of 10000 clips about multiple domains, each annotated with 20 free-form captions. We follow the official split (from [34]) of 6513, 497, and 2990 clips for training, validation, and testing. To compute the semantic classes, we consider $\rho = 0.25$ (see Sect. 4.1) and employ a pipeline made of spaCy, WordNet [24], and Lesk algorithm [21] as in Wray et al. [32].

We consider both 'text-to-video' and 'video-to-text' versions of L_n (Eq. 1) and L_p (Eq. 8). We employ HGR [3] as our base model and eventually augment it with the proposed RAN and RANP. On both datasets, we perform the training for 50 epochs using a batch size of 64. For EPIC-Kitchens-100 we use TBN [20] features, provided alongside the dataset [7]. For MSR-VTT we use ImageNet-pretrained ResNet-152 features (from [3]). After training, we select the best model on the validation set to perform the evaluation on the testing set.

As recently proposed by [32], we use Normalized Discounted Cumulative Gain (nDCG) [18] and Mean Average Precision (mAP) [2] for evaluation purposes. For MSR-VTT we only use nDCG because, due to how semantic classes are computed for its videos, the relevance values of paired captions are always lower than one, making mAP unusable. More details can be found in the Supplementary.

5.1 Analysis of Hard Negatives: Relevance Distribution

In Fig. 2 we plot the distribution of relevance values of the hard negatives during one epoch of training. On EPIC-Kitchens-100 (Fig. 2 left) we observe a sizeable amount (more than 50%) of negatives which are relevant to the query. In particular, 13% of them have a relevance of 1. Moreover, we observe four modes for the relevance: 0 (45%), 50 (36%), 100 (13%), 25 (3%). Figure 2 (middle) shows how the distribution changes when we apply the proposed RAN with $\tau = 0.75$ (visualized with an orange bar) to improve the negatives' selection (see Sect. 4.2). By doing so, a lower amount of relevant items will be selected as hard negatives.

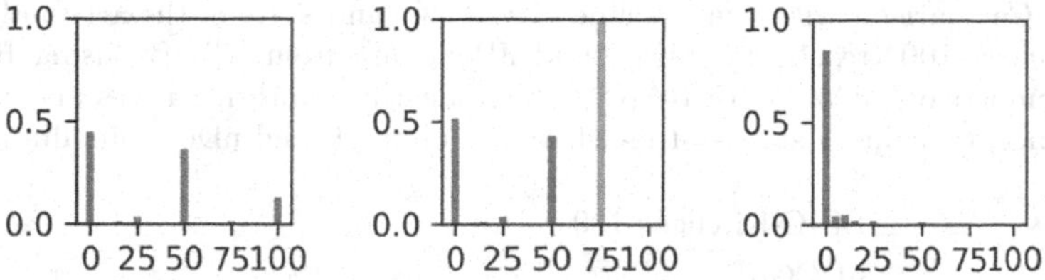

Fig. 2. Distribution of relevance values of hard negatives ($\mathcal{R}$ values from 0 to 100 on x axis, relative frequency on y axis) observed during one epoch of training (batch size 64) on EPIC-Kitchens-100 (left) and on MSR-VTT (right). In the middle, we apply the relevance-aware negative mining with $\tau = 0.75$ (visualized with the orange bar). (Color figure online)

Table 1. Performance of the baseline and the proposed RAN and RANP (improvement shown with $_{\uparrow X}$). We pick the values for the threshold τ close to the values of the modes (Fig. 2). As τ decreases, less positives are wrongly picked as negatives. Moreover, RANP pulls closer to each video more similar captions, leading to even better performance.

			RAN	RANP	RAN	RANP	RAN	RANP
	τ		0.75	0.75	0.40	0.40	0.15	0.15
EPIC	nDCG (%)	35.9	$37.4_{\uparrow 1.5}$	$40.2_{\uparrow 4.3}$	$48.8_{\uparrow 12.9}$	$\mathbf{59.0_{\uparrow 23.1}}$	$48.4_{\uparrow 12.5}$	$58.8_{\uparrow 22.9}$
	mAP (%)	39.5	$43.1_{\uparrow 4.4}$	$46.4_{\uparrow 6.9}$	$46.4_{\uparrow 6.9}$	$46.1_{\uparrow 6.6}$	$46.5_{\uparrow 7.0}$	$\mathbf{47.2_{\uparrow 7.7}}$
MSR-VTT	nDCG (%)	25.3	$25.2_{\uparrow 0.1}$	-	$26.4_{\uparrow 1.1}$	$28.0_{\uparrow 2.7}$	$28.7_{\uparrow 3.4}$	$\mathbf{31.1_{\uparrow 5.8}}$

For MSR-VTT we compute a set of classes for each video based on 'popular' classes which appear in the multiple descriptions (see Sect. 4.1). Consequently, the groundtruth captions for a given video may have a relevance lower than 1, making it harder to find relevant items within random mini-batches (Fig. 2 right). Here the top four modes are: 0 (85.2%), 10 (4.7%), 5 (4%), 15 (1.7%).

5.2 Influence of the Threshold τ on the Proposed Techniques

In Table 1 we present the results obtained by HGR when trained with the original mining technique (Eq. 3) followed by the usage of the proposed RAN and RANP mining strategies. We perform these experiments on both datasets, keeping $\Delta_n = \Delta_p = 0.2$ (as in [3]) and picking the values for τ from the modes observed in the previous section. In the Supplementary we show that minor changes can be observed by varying these values, although proper hyperparameter optimization is required. HGR achieves 35.9% nDCG and 39.5% mAP on EPIC-Kitchens-100, whereas it achieves 25.3% nDCG on MSR-VTT. By introducing the proposed strategies, we observe consistent improvements on both datasets.

First of all, lowering τ has a positive effect on both mAP and nDCG. This is likely due to the improved selection of the hard examples, which avoids several 'false negatives' and leads to a stabler training. As an example, for EPIC-Kitchens-100 $\tau = 0.75$ means that the examples with a relevance bigger than τ are no longer treated as possible negatives, and we observe 37.4% nDCG (+1.5%)

Table 2. Comparison with the baseline (HGR [3]) and state-of-the-art methods for EPIC-Kitchens-100 (results for MME and JPoSE are from [7]). By using RAN we achieve competitive mAP. With RANP, which identifies both negatives and positives via relevance, we achieve state-of-the-art results on mAP and nDCG simultaneously.

	EPIC-Kitchens-100					
	nDCG (%)			mAP (%)		
Model	t2v	v2t	avg	t2v	v2t	avg
HGR [3]	37.9	41.2	35.9	35.7	36.1	39.5
MME [33]	46.9	50.0	48.5	34.0	43.0	38.5
JPoSE [33]	51.5	55.5	53.5	38.1	49.9	44.0
Hao et al. [8]	51.8	55.3	53.5	38.5	50.0	44.2
RAN	47.1	49.7	48.4	43.1	49.9	46.5
RANP	**56.5**$_{\uparrow 4.7}$	**61.2**$_{\uparrow 5.7}$	**58.8**$_{\uparrow 5.3}$	**42.3**$_{\uparrow 3.8}$	**52.0**$_{\uparrow 2.0}$	**47.2**$_{\uparrow 3.0}$

and 43.1% mAP (+4.4%). On the other hand, for MSR-VTT the same τ leads to almost no improvements because less than 1% of the hard negatives have a relevance higher than 0.75 (see Fig. 2 right).

Secondly, by using RAN on EPIC-Kitchens-100 we observe an improvement of up to 12.9% nDCG (48.8%) and 7.0% mAP (46.5%) over the base model; on MSR-VTT the improvements measure up to 3.4% nDCG (28.7%). On the former we achieve great improvements thanks to the simplicity of the captions which makes it possible to easily find many relevant items and remove them from the pool of negatives. Conversely, the captions in MSR-VTT are multiple and free-form, making high relevance values rarer and thus showing lesser improvements.

Thirdly, the proposed RANP is also greatly useful. In fact, compared to only using the relevance-aware negative mining (i.e. RAN), on EPIC-Kitchens-100 we achieve further improvements, reaching 59.0% nDCG with $\tau = 0.40$ (+23.1% over the baseline) and 47.2% mAP with $\tau = 0.15$ (+7.7%). On MSR-VTT we also observe considerable improvements in terms of nDCG (up to +5.8%). With RANP, during training we ensure that its original caption is pulled near the video, but also other captions which describe its visual contents, further improving the quality of the ranked lists. Based on these observations, in the following experiments we use $\tau = 0.15$ for EPIC-Kitchens-100 and $\tau = 0.10$ for MSR-VTT.

5.3 Comparison with State-of-the-Art

In Tables 2 and 3 we report the results we obtain with HGR [3] augmented by the proposed techniques on both EPIC-Kitchens-100 and MSR-VTT, and propose a comparison to other popular methods.

EPIC-Kitchens-100. In Table 2 we compare to MME and JPoSE, proposed by Wray et al. [33] and used in [7] as the baselines for the challenge. We include Hao et al. [8] which is the current the state-of-the-art (53.5% nDCG and 44.2%

Table 3. For a fair comparison with MoEE [23] and CE [22] we evaluate their performance while using appearance features and the code provided by the authors. Then we evaluate the baseline (HGR) and the proposed techniques.

		Model	CE	MoEE	HGR	RAN	**RANP**
MSR-VTT	nDCG (%)	t2v	28.9	28.4	24.6	27.4	**29.1**$_{\uparrow 0.2}$
		v2t	30.0	29.5	26.1	30.1	**34.1**$_{\uparrow 4.1}$
		avg	29.4	29.0	25.3	28.7	**31.6**$_{\uparrow 2.2}$

mAP). We observe considerable improvements (+5.3% nDCG and +3.0% mAP) by using the proposed RANP, leading us to a new state-of-the-art result (58.8% nDCG and 47.2% mAP). Moreover, such an improvement is observed when looking both at the average and at task-level values (text-to-video and video-to-text), giving a clear evidence of the usefulness of the proposed techniques.

MSR-VTT. For MSR-VTT, we compare to MoEE (Miech et al. [23]) and CE (Liu et al. [22]). To have a fair comparison, in Table 3 we evaluate them using only appearance features within the open source codebase of [22]. Then, we include the results for HGR and for the proposed techniques. CE and MoEE present higher nDCG rates (respectively 29.4% and 29.0%) than the base HGR (25.3%), showing that CE and MoEE compute higher quality ranked lists than our baseline. If we augment HGR with the proposed relevance-aware hard negative mining (RAN) we observe an improvement of +3.4% nDCG (28.7%). This is due to having less positive captions wrongly selected as negatives during training. If we also introduce the proposed variant for hard positives (using the full RANP), we observe a further improvement, leading to an overall margin of +2.2% over CE (31.6% versus 29.4%). With the introduction of this second technique, less irrelevant captions are retrieved at the top of the ranked list.

6 Conclusions

Video retrieval methods are usually trained using a contrastive loss, such as the triplet loss [28]. During training, the negatives are selected among the captions or videos which are not associated in the dataset. In this paper, we showed that the typical formulation used to mine the negatives also selects captions which partially describe the input video. To emphasize the importance of this selection step, we proposed the relevance-aware online hard negative mining, which uses a relevance function to separate positive and negative items. Furthermore, in the video retrieval community positive examples hardly are mined because proper labels are usually absent. To this end, we also proposed the relevance-aware online hard positive mining. Finally, we gave empirical evidence of the strength of the proposed techniques by applying them on a deep learning model (HGR [3]) and testing it on two benchmark datasets: the recently released EPIC-Kitchens-100 [7] and MSR-VTT [34]. In both cases, the application of the proposed techniques leads to considerable improvements and state-of-the-are results.

Acknowledgements. We gratefully acknowledge the support from Amazon AWS Machine Learning Research Awards (MLRA) and NVIDIA AI Technology Centre (NVAITC), EMEA. We acknowledge the CINECA award under the ISCRA initiative, which provided computing resources for this work.

References

1. Arandjelovic, R., Gronat, P., Torii, A., Pajdla, T., Sivic, J.: Netvlad: CNN architecture for weakly supervised place recognition. In: Proceedings of the IEEE CVPR, pp. 5297–5307 (2016)
2. Baeza-Yates, R., Ribeiro-Neto, B., et al.: Modern Information Retrieval, vol. 463. ACM Press, New York (1999)
3. Chen, S., Zhao, Y., Jin, Q., Wu, Q.: Fine-grained video-text retrieval with hierarchical graph reasoning. In: Proceedings of the IEEE/CVF CVPR, June 2020
4. Chen, T., Kornblith, S., Norouzi, M., Hinton, G.: A simple framework for contrastive learning of visual representations. In: International Conference on Machine Learning, pp. 1597–1607. PMLR (2020)
5. Chen, W., Chen, X., Zhang, J., Huang, K.: Beyond triplet loss: a deep quadruplet network for person re-identification. In: Proceedings of the IEEE CVPR, pp. 403–412 (2017)
6. Croitoru, I., Bogolin, S.V., Leordeanu, M., Jin, H., Zisserman, A., Albanie, S., Liu, Y.: Teachtext: Crossmodal generalized distillation for text-video retrieval. In: Proceedings of the IEEE/CVF ICCV. pp. 11583–11593 (2021)
7. Damen, D., Doughty, H., Farinella, G.M., Furnari, A., Kazakos, E., Ma, J., Moltisanti, D., Munro, J., Perrett, T., Price, W., et al.: Rescaling egocentric vision. IJCV (2021)
8. Damen, D., Fragomeni, A., Munro, J., Perrett, T., Whettam, D., Wray, M., Furnari, A., Farinella, G.M., Moltisanti, D.: Epic-kitchens-100- 2021 challenges report. University of Bristol, Tech. rep. (2021)
9. Dong, J., Li, X., Xu, C., Yang, X., Yang, G., Wang, X., Wang, M.: Dual encoding for video retrieval by text. IEEE Transactions on Pattern Analysis and Machine Intelligence (2021)
10. Dzabraev, M., Kalashnikov, M., Komkov, S., Petiushko, A.: MDMMT: multidomain multimodal transformer for video retrieval. In: Proceedings of the IEEE/CVF CVPR, pp. 3354–3363 (2021)
11. Gabeur, V., Sun, C., Alahari, K., Schmid, C.: Multi-modal transformer for video retrieval. In: Proceedings of the IEEE ECCV. Springer (2020)
12. Gordo, A., Larlus, D.: Beyond instance-level image retrieval: leveraging captions to learn a global visual representation for semantic retrieval. In: Proceedings of the IEEE CVPR, pp. 6589–6598 (2017)
13. Gutmann, M., Hyvärinen, A.: Noise-contrastive estimation: a new estimation principle for unnormalized statistical models. In: Proceedings of the Thirteenth International Conference on Artificial Intelligence and Statistics. pp. 297–304. JMLR Workshop and Conference Proceedings (2010)
14. Hadsell, R., Chopra, S., LeCun, Y.: Dimensionality reduction by learning an invariant mapping. In: 2006 IEEE Computer Society CVPR (CVPR 2006). vol. 2, pp. 1735–1742. IEEE (2006)

15. Harwood, B., Kumar BG, V., Carneiro, G., Reid, I., Drummond, T.: Smart mining for deep metric learning. In: Proceedings of the IEEE ICCV, pp. 2821–2829 (2017)
16. He, K., Fan, H., Wu, Y., Xie, S., Girshick, R.: Momentum contrast for unsupervised visual representation learning. In: Proceedings of the IEEE/CVF CVPR, pp. 9729–9738 (2020)
17. Hermans, A., Beyer, L., Leibe, B.: In defense of the triplet loss for person re-identification. arXiv preprint arXiv:1703.07737 (2017)
18. Järvelin, K., Kekäläinen, J.: Cumulated gain-based evaluation of IR techniques. ACM Trans. Inf. Syst. **20**(4), 422–446 (2002)
19. Jiang, Q.Y., He, Y., Li, G., Lin, J., Li, L., Li, W.J.: SVD: a large-scale short video dataset for near-duplicate video retrieval. In: Proceedings of the IEEE/CVF ICCV, pp. 5281–5289 (2019)
20. Kazakos, E., Nagrani, A., Zisserman, A., Damen, D.: Epic-fusion: audio-visual temporal binding for egocentric action recognition. In: Proceedings of the IEEE/CVF ICCV, pp. 5492–5501 (2019)
21. Lesk, M.: Automatic sense disambiguation using machine readable dictionaries: how to tell a pine cone from an ice cream cone. In: Proceedings of the 5th Annual International Conference on Systems Documentation, pp. 24–26 (1986)
22. Liu, Y., Albanie, S., Nagrani, A., Zisserman, A.: Use what you have: video retrieval using representations from collaborative experts. In: BMVC (2019)
23. Miech, A., Laptev, I., Sivic, J.: Learning a text-video embedding from incomplete and heterogeneous data. arXiv preprint arXiv:1804.02516 (2018)
24. Miller, G.A.: WordNet: a lexical database for English. Commun. ACM **38**(11), 39–41 (1995)
25. Mithun, N.C., Li, J., Metze, F., Roy-Chowdhury, A.K.: Learning joint embedding with multimodal cues for cross-modal video-text retrieval. In: Proceedings of the 2018 ACM on International Conference on Multimedia Retrieval, pp. 19–27 (2018)
26. Pan, T., Song, Y., Yang, T., Jiang, W., Liu, W.: Videomoco: Contrastive video representation learning with temporally adversarial examples. In: Proceedings of the IEEE/CVF CVPR. pp. 11205–11214 (2021)
27. Qian, R., et al.: Spatiotemporal contrastive video representation learning. In: Proceedings of the IEEE/CVF CVPR, pp. 6964–6974 (2021)
28. Schroff, F., Kalenichenko, D., Philbin, J.: FaceNet: a unified embedding for face recognition and clustering. In: Proceedings of the IEEE CVPR, pp. 815–823 (2015)
29. Sohn, K.: Improved deep metric learning with multi-class n-pair loss objective. In: Michael, I., Jordan, Y.L., Solla, S.A. (eds.) Advances in Neural Information Processing Systems, pp. 1857–1865. MIT Press, London (2016)
30. Suh, Y., Han, B., Kim, W., Lee, K.M.: Stochastic class-based hard example mining for deep metric learning. In: Proceedings of the IEEE/CVF CVPR, pp. 7251–7259 (2019)
31. Wang, X., Zhu, L., Yang, Y.: T2VLAD: global-local sequence alignment for text-video retrieval. In: Proceedings of the IEEE/CVF CVPR, pp. 5079–5088 (2021)
32. Wray, M., Doughty, H., Damen, D.: On semantic similarity in video retrieval. In: Proceedings of the IEEE/CVF CVPR, pp. 3650–3660 (2021)
33. Wray, M., Larlus, D., Csurka, G., Damen, D.: Fine-grained action retrieval through multiple parts-of-speech embeddings. In: Proceedings of the IEEE ICCV, pp. 450–459 (2019)
34. Xu, J., Mei, T., Yao, T., Rui, Y.: MSR-VTT: a large video description dataset for bridging video and language. In: Proceedings of the IEEE CVPR, pp. 5288–5296 (2016)

35. Xuan, H., Stylianou, A., Liu, X., Pless, R.: Hard negative examples are hard, but useful. In: Proceedings of the IEEE ECCV, pp. 126–142 (2020)
36. Xuan, H., Stylianou, A., Pless, R.: Improved embeddings with easy positive triplet mining. In: Proceedings of the IEEE/CVF Winter Conference on Applications of Computer Vision, pp. 2474–2482 (2020)

Foreground Detection Using an Attention Module and a Video Encoding

Anthony A. Benavides-Arce(✉), Victor Flores-Benites, and Rensso Mora-Colque

Universidad Católica San Pablo, Arequipa, Peru
{anthony.benavides,victor.flores,rvhmora}@ucsp.edu.pe

Abstract. Foreground detection is the task of labelling the foreground or background pixels in the video sequence and it depends on the context of the scene. For many years, methods based on background model have been the most used approaches for detecting foreground; however, their methods are sensitive to error propagation from the first background model estimations. To address this problem, we proposed a U-net based architecture with an attention module, where the encoding of the entire video sequence is used as attention context to get features related to the background model. We tested our network on sixteen scenes from the CDnet2014 dataset, with an average F-measure of 88.42. The results also show that our model outperforms traditional and neural networks methods. Thus, we demonstrated that an attention module on a U-net based architecture can deal with the foreground detection challenges.

Keywords: Foreground Detection · U-Net · Attention · Video encoding

1 Introduction

Foreground Detection (FD) is a binary classification task which assigns each pixel in video sequences with labels belonging to either moving objects considered as foreground and the non-moving objects as background. It is used as a pre-processing step in applications like hand gesture identification [11], video compression [27], segmentation [23] and vehicle counter [8].

FD remains a challenging task due to various factors in video sequences as: (1) Illumination changes, caused by the variation of lighting in the scene due to changes in the lights or changes in the brightness of the sun. (2) Dynamic background, due to possible background changes during the scene, especially in outdoor scenes, e.g. waves or swaying tree leaves. (3) Camera jitter, due to a motion camera, the background is not static. (4) Camouflage, It is when the foreground is falsely labelled as background because they have similar colour [5]. The main challenge of FD is to be able to distinguish the dithering effect at the boundaries of foreground objects, overcoming the previous challenges.

The background subtraction technique is the most used method for FD, it determines the foreground by extracting the difference between the current frame

S. Sclaroff et al. (Eds.): ICIAP 2022, LNCS 13233, pp. 195–205, 2022.
https://doi.org/10.1007/978-3-031-06433-3_17

and the background. This technique consists of two stages: 1) an estimation of the background model from the frame sequence and 2) foreground extraction by subtracting the background model [13]. In this context, a background model consists of a reference to compare with the incoming video frames.

Background subtraction is a complex and subjective task because it can fall into the propagation of errors due to a poorly estimation of the background model. Hence, we propose a model based on U-net architecture [20] that allows the network to incorporate global and local features. Our main contribution is the use of a video encoding and an attention module to process the high encoding of U-net and compare it against the features obtained from the video encoding. As a result, our model projects the video information into a latent space, where it has rich spatial and temporal information. Our method was tested on sixteen scenes of CDnet2014 dataset, with F-measure and percentage of wrong classifications (PWC) metrics. The proposed model is obtaining better F-measure and PWC scores than the state-of-the-art.

The next sections are: Sect. 2 is devoted to a literature review on the most significant methods for FD. Our methodology and network architecture are analysed in Sect. 3. In Sect. 4 we present our experiments and discuss our results. Finally, the conclusions and future work are provided in Sect. 5.

2 Related Work

We divide the FD proposals as handcrafted methods and neural network methods. We called handcrafted methods to the most traditional FD approaches and neural network methods to Convolutional Neural Network (CNN) approaches. Handcrafted approaches employ statistical methods such as Gaussian Mixture Models [3] and Conditional Random Field [30], which typically detect the foreground at the pixel level of a frame [19]. Many of these approaches employ a background model [12] or multiple background models [22]. However, handcrafted methods perform blurry segmentation of the foreground, and they are very sensitive to the background model.

Neural networks were introduced to deal with these problems because they are robust, fast, flexible to background changes, and they can label an object accurately. Babaee et al. [5] proposed a model which employs two handcrafted methods to generate and update the background model, and a CNN to perform FD. However, this model is sensitive to updates of its background model. On the other hand, Gao et al. [9] proposed a method that uses 3D-CNNs to get temporal information, and Akilan et al. [1] presented a model that employs 3D-CNNs and an LSTM to capture long-short-term spatio-temporal features. U-net architecture [20] has been successful in segmentation tasks, hence, there are models that use it for the FD task [14,15]. These methods deal with FD using temporal information from the scene and distinguish the dithering effect at the boundaries of foreground objects.

3 Methodology

We present a U-net based model, which extracts features and generates a segmentation map of foreground objects. For this purpose, we propose an attention module that uses a context to attend to moving objects. This context is obtained from the relevant features of the entire video sequence, such that the context has a representation of the regularities in the scene. The detailed architecture of the network is shown in Fig. 1.

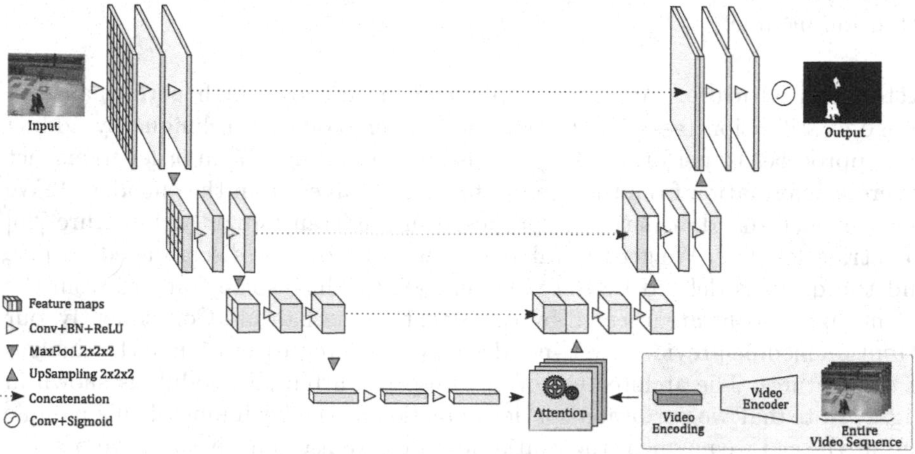

Fig. 1. Our network architecture. Our model is based on a U-net architecture, we present the encoder and decoder feature maps with green and white colour respectively. Unlike vanilla U-net, we add an attention module on the last encoder layer and a visual encoding as an attention context. Thus, the attention module is able to select features with useful information for FD. Finally, the last network layer has a sigmoid activation function because the network output is going to label each pixel with a probability between 0 and 1 in order to distinguish between background and foreground respectively.

U-net Based Architecture. Models inspired by the U-net architecture [14, 15] have reported significant results for the FD task. Hence, we use a U-net architecture [20] because its residual connections avoid degradation problems in the resulting segmented frame. See Fig. 1, our main contribution is in the last layer of the encoder, we employ an attention module using a video encoding as context, which highlights the high features of the network. Finally, we use a sigmoid activation function in the last layer of the network, which means a significant change to vanilla U-net because the network output is going to label each pixel with a probability between 0 and 1 in order to distinguish between background and foreground respectively.

Video Encoding. We require a context of the video sequence for the attention module, however, unlike other proposals that use a background model methodology [12,19,22], we use a static video encoding obtained from the entire video sequence. This allows the model to recognize the background regularities in the scene. The video encoder consists of a ResNet-34 [10] pre-trained on imageNet dataset [21] without the last fully connected layer. We employ this network to calculate codifications from each frame and for each sequence we average these features, these features are the video encoding. The ResNet-34 is fed with the entire video sequence and the obtained features are averaged.

The resulting video encoding is compared to the latent space obtained in the attention module.

Attention Module. Attention approaches have shown high performance in image classification tasks [6,24,29], hence, we employ this technique by adding two approaches to perform FD: 1) We use the disentangle [7] in order to extract different information from the general futures obtained from the encoder. 2) We perform multi-head attention as proposed in the Transformer architecture [25] to extract features. We get the value and the key from the disentangled output, and the query is defined by the video encoding. Thus, some features from the disentangle output are selected by the attention mechanism. Consequently, our attention module provides the U-net decoder with a representation of the changes of the features. The architecture of our proposed attention module is shown in Fig. 2. Note that we perform a spatial reduction at the beginning of this module and at the end we augment the spatial features, we use adaptive average pooling for both cases.

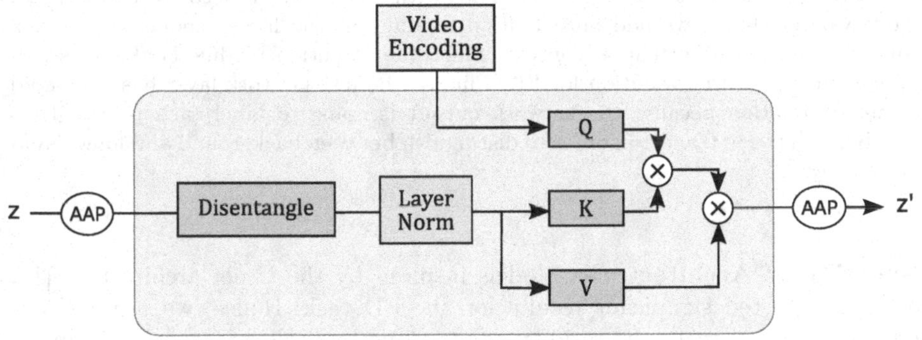

Fig. 2. Our proposed attention module takes the high representation of the features z. This module consists of two steps: i) We disentangle z to generate independent representations, in order to attend various aspects of the latent space where we apply layer normalization [4] to it. ii) Then, we generate the attention maps using linear transformations on the disentangle output to obtain the value V and key K, and we use the video encoding as input for the query value Q. This results in the feature selection from the disentangle output. In the figure, AAP means adaptive average pooling.

4 Evaluation

4.1 CDnet2014 Dataset

We perform our experiments on CDnet2014 dataset [26], because it contains a wide variety of challenging scenarios, specifically 11 categories: baseline, camera jitter, bad weather, dynamic background, intermittent object motion, low frame rate, night videos, PTZ (panning-tilting-zooming), shadow, thermal and turbulence. All FD challenges are covered with the CDnet2014 dataset. Each dataset category contains four to six sequences, and there are 53 different video sequences in total. The spatial resolutions of the video frames range from 320×240 to 720×576 pixels. In addition, video sequences can contain between 900 and 7000 frames. The environments vary from video to video, which makes this dataset suitable for measuring the robustness of a model in different scenarios. Moreover, CDnet2014 has the ground-truth of the foreground and background of the video frames, for testing and training supervised models such as ours. Sixteen scenes of the CDnet2014 dataset are described in Table 1.

Table 1. Sixteen scenes from the CDnet2014 dataset with their size and the number of frames [1].

Category	Scene name	Frame size (W × H)	N frames
Baseline	Highway	320 × 240	1229
	Office	360 × 240	1447
	Pedestrians	360 × 240	753
	PETS2006	720 × 576	900
Dynamic background	Canoe	320 × 240	342
	Boats	320 × 240	6026
	OverPass	320 × 240	440
	Fall	720 × 480	1400
Camera jitter	Boulevard	320 × 240	1004
Shadow	CopyMachine	720 × 480	1401
	PeopleInShade	380 × 244	829
	BusStation	360 × 240	832
PTZ camera	TwoPosPTZCam	570 × 340	449
Low framerate	Turnpike	320 × 240	350
Intermittent object motion	Sofa	320 × 240	2243
Night time	TramStation	480 × 295	1250

4.2 Evaluation Metrics

The evaluation metrics used are the F-measure and the percentage of wrong classifications (PWC). The F-measure combines precision and recall into one metric and [26] indicated a strong correlation of F-measure with ranks on CDnet2014 website, and in general is considered as a good indicator for comparison purposes. Given a number of true positives (TP), false positives (FP), and false negatives (FN), F-measure is defined as:

$$F - measure = \frac{2TP}{2TP + FP + FN}. \tag{1}$$

Although F-measure is widely used, it does not consider the number of true negatives (TN) and thus is sensitive to very small moving objects. At the limit, due to lack of a foreground object may lead to a TP of 0 and therefore an F-measure of 0. In order to compensate this, we used the PWC metric, which incorporates TN:

$$PWC = \frac{100 \times (FN + FP)}{TP + FN + FP + TN}, \tag{2}$$

The idea of this metrics is that the methods maximize the F-measure while minimizing the PWC.

4.3 Trainable Parameters

We trained our network on a set of 60% of the total frames in the scene, 10% of the training set was selected to validate the training, and the rest of the frames were used to test our model, rather than select only 20 frames from the test set like Yang et al. [28]. We consider it is important to evaluate the model in all frames of the dataset to obtain a better analysis. As the state-of-the-art methods, in our experiments, we used Adam optimizer, a learning rate of 0.0001, a batch size of 5, and we trained the network for 30 epochs using Dice loss. For the output, we employ a threshold of 0.75 [1] as a post-processing step to label the output pixels as foreground or background if the probability is higher or lower, respectively.

4.4 Results

Quantitative Analysis. We perform two types of analysis: i) We use a 60% of the initial frames of the video sequence for the training set and the rest for the test set. We compare our model against [1,12,22] in Tables 2 and 4. ii) On the other hand, models [5,16] use a random selection of training and test sets, probably biasing their results. We employed the same strategy for the training and test sets. This analysis is presented in Table 3. For the first analysis, we obtained an average f-measure of 88.42, outperforming state-of-the-art models. In addition, we obtained a 98.88 f-measure using random selection of training and test sets, demonstrating that our model outperforms models employing that strategy.

Table 2. F-measure performance comparison without having any selection of the training and test frames. PBAS [12], UMBS [22], 3DCNN-LSTM [1] and the proposed method were tested in sixteen scenes of CDnet2014, n/a means that the result was not available. The scores of the PBAS and UMBS methods were acquired from [1].

Category	Scene	PBAS	UMBS	3D-LSTM	Ours
Baseline	Highway	94.51	92.17	**98.76**	98.50
	Office	94.20	97.19	97.40	96.49
	Pedestrians	93.63	95.66	95.01	**97.48**
	PETS2006	87.36	86.48	92.18	94.04
Dynamic background	Canoe	71.96	93.45	75.36	88.88
	Boats	36.11	90.41	95.90	**97.97**
	OverPass	79.25	89.90	84.33	70.93
	Fall	**87.14**	56.68	34.05	38.55
Camera jitter	Boulevard	66.02	86.72	90.23	**95.13**
Shadow	CopyMachine	87.27	87.11	77.71	94.56
	PeopleInShade	89.19	90.16	85.24	89.39
	BusStation	86.09	86.95	95.08	76.29
PTZ camera	TwoPosPTZCam	n/a	79.59	88.70	**94.21**
Low framerate	Turnpike	n/a	89.01	97.84	**98.59**
IOM	Sofa	73.81	84.55	92.33	**96.14**
Night videos	TramStation	82.43	**88.56**	85.61	87.63
Overall		80.64	87.16	87.12	**88.42**

Table 3. F-measure performance comparison with random selection of training and test sets. DeepBS [5], FgSegNet [16] and the proposed method were tested in eight categories of CDnet2014. The scores of the DeepBS and FgSegNet methods were acquired from [16].

Category	DeepBS	FgSegNet	Ours
Baseline	95.80	**99.75**	99.74
Camera jitter	89.90	**99.45**	**99.45**
Dynamic background	87.61	**99.39**	99.25
IOM	60.97	99.33	**99.77**
Shadow	93.04	**99.54**	99.49
Low frame rate	59.00	95.58	**98.80**
Night videos	63.59	**97.79**	96.55
PTZ	33.06	**98.93**	98.03
Overall	72.87	98.72	**98.88**

Table 4. Performance comparison with PWC without having any selection of the training and test frames. UMBS [22], DeepBS [5], 3DCNN-LSTM [1], U-Net MDI [14] and the proposed method were tested in eight categories of CDnet2014. n/a means that the result was not available. The scores of the PBAS, UMBS and DeepBS methods were acquired from [1].

Category	UMBS	DeepBS	3D-LSTM	U-Net MDI	Ours
Baseline	0.64	0.24	0.21	**0.13**	0.30
Camera jitter	0.30	0.89	3.03	3.54	**0.28**
Dynamic background	1.17	**0.20**	0.35	0.28	1.28
IOM	2.63	4.12	**1.15**	3.07	2.44
Shadow	1.78	**0.74**	1.10	0.64	1.10
Low frame rate	0.92	1.35	**0.34**	0.90	0.46
Night videos	3.97	2.57	0.90	**0.56**	0.84
PTZ	**0.51**	7.72	0.98	na	0.77
Overall	1.49	1.99	1.01	1.30	**0.93**

Qualitative Analysis. The Fig. 3 shows a comparison of the foreground detected by our model and the ground-truth. We bound the qualitative presentation to one sample per data sequence due to space constraints. The qualitative results showed the proposed model has tightly detected the foreground and background when compared to the ground-truth segmentations. However, it is important to evaluate its performance across all the test frames of all the video sequences.

4.5 Discussion

We have trained our network without having any selection of the training and test frames to be closer to a real environment, unlike other methods like [5,16] which choose randomly selected frames. Therefore, our model exhibits a drastic drop in performance in the fall scene, as shown in the results in the Table 2, with an F-measure of 38.55, because the foreground objects are unbalanced in this video scene, thus, the test set has foreground objects never seen in the training step. We validate this with the results in Table 3, where the model uses a random selection of frames from the video sequence. Our model is able to segment the foreground with high performance in each CDnet2014 scene.

Our method has better performance than other models that use a background model methodology [12,19,22], because their methods are sensitive to error propagation from the first background model estimations. Note that we are using a video encoding as a context which is computed once, which allows a high-level abstraction of the video, leaving out small or video changes. This context in the attention module provides the regularities in the video sequence, therefore the network has additional information to segment foreground objects.

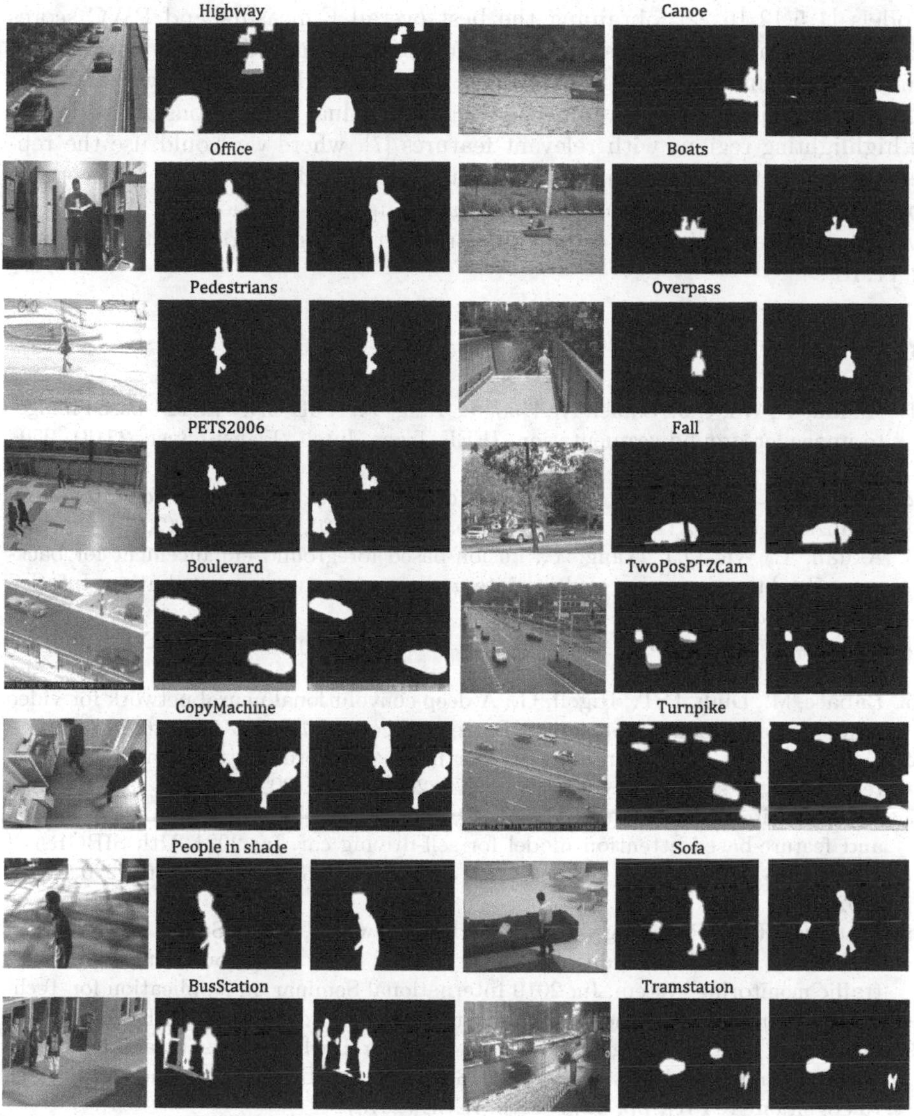

Fig. 3. Qualitative results of the proposed model. For each scene are shown the input frame, the ground-truth and the result obtained from the proposed method.

5 Conclusions

We proposed a U-net based model with an attention module using video encoding of the entire video sequence as a context for the foreground detection task. The attention module provides the U-net decoder with a representation of the common patterns in the video. The proposed model is outperforming the evaluated models [1,5,12,16,22], obtaining the best overall F-measure and PWC scores in sixteen CDnet2014 scenes, where illumination changes, dynamic background, camera jitter and camouflage challenges are exhibited. As future work, we are going to add spatial attention modules in the residual connections with the aim of highlighting regions with relevant features [7], where we would use the representation of features obtained in the attention module in order to recognize the spatial regions where these features are found. In addition, we are going to evaluate the generalization of our network by training a model with all scenes [2,17,18].

References

1. Akilan, T., Wu, Q.J., Safaei, A., Huo, J., Yang, Y.: A 3D CNN-LSTM-based image-to-image foreground segmentation. IEEE Trans. Intell. Transp. Syst. **21**(3), 959–971 (2019)
2. Akilan, T., Wu, Q.J.: sEnDec: an improved image to image CNN for foreground localization. IEEE Trans. Intell. Transp. Syst. **21**(10), 4435–4443 (2019)
3. Akilan, T., Wu, Q.J., Yang, Y.: Fusion-based foreground enhancement for background subtraction using multivariate multi-model gaussian distribution. Inf. Sci. **430**, 414–431 (2018)
4. Ba, J.L., Kiros, J.R., Hinton, G.E.: Layer normalization. arXiv preprint arXiv:1607.06450 (2016)
5. Babaee, M., Dinh, D.T., Rigoll, G.: A deep convolutional neural network for video sequence background subtraction. Pattern Recognit. **76**, 635–649 (2018)
6. Dosovitskiy, A., et al.: An image is worth 16 × 16 words: Transformers for image recognition at scale. arXiv preprint arXiv:2010.11929 (2020)
7. Flores-Benites, V., Mugruza-Vassallo, C.A., Mora-Colque, R.: TVAnet: a spatial and feature-based attention model for self-driving car. In: 2021 34th SIBGRAPI Conference on Graphics, Patterns and Images (SIBGRAPI), pp. 263–270. IEEE (2021)
8. Fratama, R.R., Partiningsih, N.D.A., Rachmawanto, E.H., Sari, C.A., Andono, P.N., et al.: Real-time multiple vehicle counter using background subtraction for traffic monitoring system. In: 2019 International Seminar on Application for Technology of Information and Communication (iSemantic), pp. 1–5. IEEE (2019)
9. Gao, Y., Cai, H., Zhang, X., Lan, L., Luo, Z.: Background subtraction via 3D convolutional neural networks. In: 2018 24th International Conference on Pattern Recognition (ICPR), pp. 1271–1276. IEEE (2018)
10. He, K., Zhang, X., Ren, S., Sun, J.: Deep residual learning for image recognition. In: Proceedings of the IEEE Conference on Computer Vision and Pattern Recognition, pp. 770–778 (2016)
11. Hema, C.: Hand gesture identification using preprocessing, background subtraction and segmentation techniques. Int. J. Appl. Eng. Res. **11**(5), 3221–3228 (2016)
12. Hofmann, M., Tiefenbacher, P., Rigoll, G.: Background segmentation with feedback: the pixel-based adaptive segmenter. In: 2012 IEEE Computer Society Conference on Computer Vision and Pattern Recognition Workshops, pp. 38–43. IEEE (2012)
13. Huynh-The, T., Banos, O., Lee, S., Kang, B.H., Kim, E.S., Le-Tien, T.: NIC: a robust background extraction algorithm for foreground detection in dynamic scenes. IEEE Trans. Circuits Syst. Video Technol. **27**(7), 1478–1490 (2016)

14. Kim, J.Y., Ha, J.E.: Foreground objects detection by U-Net with multiple difference images. Appl. Sci. **11**(4), 1807 (2021)
15. Kim, J.Y., Ha, J.E.: Spatio-temporal data augmentation for visual surveillance. arXiv preprint arXiv:2101.09895 (2021)
16. Lim, L.A., Keles, H.Y.: Foreground segmentation using convolutional neural networks for multiscale feature encoding. Pattern Recognit. Lett. **112**, 256–262 (2018)
17. Patil, P.W., Biradar, K.M., Dudhane, A., Murala, S.: An end-to-end edge aggregation network for moving object segmentation. In: proceedings of the IEEE/CVF Conference on computer Vision and Pattern Recognition, pp. 8149–8158 (2020)
18. Patil, P.W., Dudhane, A., Murala, S.: Multi-frame recurrent adversarial network for moving object segmentation. In: Proceedings of the IEEE/CVF Winter Conference on Applications of Computer Vision, pp. 2302–2311 (2021)
19. Piccardi, M.: Background subtraction techniques: a review. In: 2004 IEEE International Conference on Systems, Man and Cybernetics (IEEE Cat. No. 04CH37583), vol. 4, pp. 3099–3104. IEEE (2004)
20. Ronneberger, O., Fischer, P., Brox, T.: U-Net: convolutional networks for biomedical image segmentation. In: Navab, N., Hornegger, J., Wells, W.M., Frangi, A.F. (eds.) MICCAI 2015, Part III. LNCS, vol. 9351, pp. 234–241. Springer, Cham (2015). https://doi.org/10.1007/978-3-319-24574-4_28
21. Russakovsky, O., et al.: ImageNet large scale visual recognition challenge. Int. J. Comput. Vis. **115**(3), 211–252 (2015). https://doi.org/10.1007/s11263-015-0816-y
22. Sajid, H., Cheung, S.C.S.: Universal multimode background subtraction. IEEE Trans. Image Process. **26**(7), 3249–3260 (2017)
23. Tarafdar, A., Roy, S., Mondal, A., Sen, R., Adhikari, A.: Image segmentation using background subtraction on colored images. In: 2019 International Conference on Opto-Electronics and Applied Optics (Optronix), pp. 1–4. IEEE (2019)
24. Touvron, H., Cord, M., Douze, M., Massa, F., Sablayrolles, A., Jégou, H.: Training data-efficient image transformers and distillation through attention. In: International Conference on Machine Learning, pp. 10347–10357. PMLR (2021)
25. Vaswani, A., et al.: Attention is all you need. In: Advances in Neural Information Processing Systems, pp. 5998–6008 (2017)
26. Wang, Y., Jodoin, P.M., Porikli, F., Konrad, J., Benezeth, Y., Ishwar, P.: CDnet 2014: an expanded change detection benchmark dataset. In: Proceedings of the IEEE Conference on Computer Vision and Pattern Recognition Workshops, pp. 387–394 (2014)
27. Wu, L., Huang, K., Shen, H., Gao, L.: A foreground-background parallel compression with residual encoding for surveillance video (2020)
28. Yang, L., Li, J., Luo, Y., Zhao, Y., Cheng, H., Li, J.: Deep background modeling using fully convolutional network. IEEE Trans. Intell. Transp. Syst. **19**(1), 254–262 (2017)
29. Zhang, Y., et al.: VidTr: video transformer without convolutions. In: Proceedings of the IEEE/CVF International Conference on Computer Vision, pp. 13577–13587 (2021)
30. Zou, W., Bai, C., Kpalma, K., Ronsin, J.: Online glocal transfer for automatic figure-ground segmentation. IEEE Trans. Image Process. **23**(5), 2109–2121 (2014)

Test-Time Adaptation for Egocentric Action Recognition

Mirco Plananamente[1,2], Chiara Plizzari[1(✉)], and Barbara Caputo[1]

[1] Politecnico di Torino, Turin, Italy
{mirco.planamente,chiara.plizzari,barbara.caputo}@polito.it
[2] CINI Consortium, Venice, Italy

Abstract. Egocentric action recognition is becoming an increasingly researched topic thanks to the rising popularity of wearable cameras. Despite the numerous publications in the field, the learned representations still suffers from an intrinsic "environmental bias". To address this issue, domain adaptation and generalization approaches have been proposed, which operate by either adapting the model to target data *during training* or by learning a model able to generalize to unseen videos by exploiting the knowledge from multiple source domains. In this work, we propose to adapt a model trained on source data to novel environments *at test time*, making adaptation practical to real-world scenarios where target data are not available at training time. On the popular EPIC-Kitchens dataset, we present a new benchmark for Test-Time Adaptation (TTA) in egocentric action recognition. Moreover, we propose a new multi-modal TTA approach, which we call RNA^{++}, and combine it with a new set of losses aiming at reducing classifier's uncertainty, showing remarkable results w.r.t. existing TTA methods inherited from image classification. Code available: https://github.com/EgocentricVision/RNA-TTA.

Keywords: Egocentric action recognition · Test-time adaptation

1 Introduction

In the last years, the technological advances in the field of wearable devices led to a growing interest in egocentric vision due to the possibility to capture information about how humans perceive the world and interact with the environment, without the need of a fixed recording system. The first person perspective unlocks a variety of applications, including wearable sport cameras, human-robot interaction, and human assistance. Contrary to traditional third-person views, the recording equipment is worn by the observer and it moves with her, posing new issues such as ego-motion, occluded objects, and significantly more variations in lighting, perspective, and environment.

The recent release of the EPIC-Kitchens large-scale dataset [8], as well as the contests that accompanied it, has sparked interest in more efficient architectures

M. Plananamente and C. Plizzari—Equally contributed to this work.

S. Sclaroff et al. (Eds.): ICIAP 2022, LNCS 13233, pp. 206–218, 2022.
https://doi.org/10.1007/978-3-031-06433-3_18

capable of dealing with these issues. Despite the numerous publications in the field [32], egocentric action recognition still has one major flaw that remains unsolved, known as "environmental bias" [39]. This problem arises from the network's heavy reliance on the environment in which the activities are recorded, which inhibits the network's ability to recognize actions when they are conducted in unfamiliar (unseen) surroundings. In general, this problem is referred to in the literature as *domain shift*, meaning that a model trained on a source labeled dataset cannot generalize well on an unseen dataset, called target. Usually, it is addressed by reducing the problem to an unsupervised domain adaptation (UDA) setting [25], where an unlabeled set of samples from the target is available and used to learn and adapt the model to the target distribution.

However, the UDA scenario is not always realistic, as (i) the target domain should be known a priori and (ii) the target data should be available at training time. To overcome those limitations, authors of [29] proposed an alternative solution which simply leverages the shared knowledge from multiple sources available during training to learn a representation that is able to generalize to any unseen domain, regardless of the possibility to access target data – known as Domain Generalization (DG) setting.

Differently from previous works, in this paper we investigate a solution that focuses on performing adaptation *during testing*. The proposed approach is based on the simple assumption that the samples received by the network during testing can be considered as a hint of the target distribution. Thus, we seek to adapt a pre-trained model to new videos coming from the test set. To best of our knowledge, this approach, known as *Test-Time Adaptation* (TTA), has never been examined in an egocentric context before. Indeed, its use in this context is even more relevant, as (i) since online adaptation does not require additional parameters, it increases the portability on multiple devices and its access to diverse users and (ii) as test data is not required to be stored, it respects privacy concerns; this is of crucial importance in the case of the first person videos as anonymization is more difficult than standard third person videos or images [38].

In this work, we present a new benchmark for multi-modal TTA in egocentric action recognition on the well-known EPIC-Kitchens dataset. Moreover, we propose a new TTA approach, called RNA^{++}, which extends RNA-Net [29], a recent multi-modal DG method, to operate on different video clips at test time. We further combine it with a new set of losses meant to reduce the classifier's confusion on test data. Results show the effectiveness of multi-modal learning in enhancing the ability of the model to adapt to new data and further validate the effectiveness of the proposed methods.

2 Related Works

2.1 Egocentric Action Recognition

The community's interest for *First Person Action Recognition* (FPAR) has quickly grown in recent years. FPAR's architectures are generally inherited from third-person literature [3,23,43]. However, due to the complexity of the setting, the multi-modal approach is the most popular technique, consisting in combining traditional visual RGB data with motion data, such as optical flow [3,10,19,25,36,43]. However, as shown in [7], the use of optical flow limits the application of several methods in online scenarios, pushing the community either towards single-stream architectures [7,28,49], or to investigate alternative modalities, e.g., audio [19,20] or event data [31]. This work is the first one exploiting audio modality, jointly with its visual counterpart, in a test-time adaptation scenario.

2.2 Cross-domain Action Recognition

Under the *Unsupervised Domain Adaptation (UDA)* setting, an unlabeled set of samples from the target is available for adaptation during training. Most of the approaches have been designed for image classification tasks [11,13,22,24]. Recently, many works started to analyze UDA for video classification tasks [5,17,21,25,27,35]. Those use adversarial learning with temporal attention [5,27], multi-modal cues [25], clip order prediction [6] or contrastive losses [21,35].

The *Domain Generalization (DG)* setting, instead, aims at finding a representation able to generalize to any unseen domain, regardless of the possibility to access target data at training time. Existing approaches in DG are mostly designed for image data [2,9,40]. Only one work investigated the DG setting in third person action recognition [46]. Recently, authors of [29] proposed a solution to this problem in first person action recognition, by proposing a feature-level solution which exploits the collaboration of audio-visual signals.

In this work, we further explore the possibility to adapt the model directly on test data under a *test-time adaptation* setting. While the latter has been widely explored on image data [26,33,37,42,48], only one work explored the possibility to adopt it on videos [1]. In this work, we take a step ahead by extending the setting to the egocentric action recognition scenario.

3 Problem Formulation

Test-Time Adaptation (TTA) for Action Recognition. This setting consists in learning the target distribution using just the unlabelled videos available during test. Due the capability of wearable devices to capture data in a variety of situations and surroundings, the target distribution is extremely variable and hard to generalize to using DG techniques. Moreover, the availability of a set of target data to learn the unseen distribution from during training, as well as the continuous access to source data to re-train the model on novel environments, are both impracticable in this case, making the Source Free DA and UDA settings unfeasible. In this work, we propose TTA as an intriguing and significant setting that has yet to be investigated in the egocentric literature. Indeed, it allows to optimize the network on test data during inference by introducing an additional, but negligible w.r.t. standard training, inference cost (Table 1).

Under the video setting, two aspects have to be considered, (i) *multi-modality*, which translates in a multi-modal input $x = (x_i^v, x_i^a)$, where we denote with v and

Table 1. Adaptation settings differ by the data and losses used during train and test. The terms x_s and y_s refer to the labeled distribution, known as *source*, while x_t the unlabelled one, known as *target*. Our TTA setting only needs the target data x_t.

Setting	Source	Target	Train loss	Test loss
Unsupervised Domain Adaptation (UDA)	x_s, y_s	x_t	$\mathcal{L}(x_s, y_s) + \mathcal{L}(x_s, x_t)$	–
Domain Generalization (DG)	x_s, y_s	–	$\mathcal{L}(x_s, y_s)$	–
Source-Free DA	–	x_t	$\mathcal{L}(x_t)$	–
Test-Time Adaptation (TTA)	x_s, y_s	x_t	–	$\mathcal{L}(x_t)$

a the visual and audio modality respectively, and with i the i-th sample, and (ii) *temporality*, consisting in having an input x_i^m composed of k *clips* representing different temporal positions within the video, i.e., $x_i^m = \{x_{i1}^m, ..., x_{ik}^m\}$.

Problem Setting. We assume a model trained on different source domains $\{\mathcal{S}_1, ..., \mathcal{S}_n\}$, where each $\mathcal{S} = \{(x_{s,i}, y_{s,i})\}_{i=1}^{N_s}$ is composed of N_s source samples with label space Y_s known, and a target domain $\mathcal{T} = \{x_{t,i}\}_{i=1}^{N_t}$ of N_t target samples whose label space Y_t is unknown. The main assumptions is that the label space is shared, $\mathcal{Y}_s = \mathcal{Y}_t$. Our objective is to perform *test-time adaptation* by adapting the model trained on source data to samples available at test time. During the forward pass, each modality input (x_i^v, x_i^a) is fed to a separate feature extractor, F^v and F^a respectively (Fig. 1). The resulting features $f^v = F^v(x_i^v)$ and $f^a = F^a(x_i^a)$ are then passed to the separate classifiers G^v and G^a, whose outputs correspond to distinct score predictions (one for each modality). The final prediction results from the combination of the different modality predictions of each clip (*late fusion*), followed by the average prediction over all the clips.

4 Test-Time Adaptation for Action Recognition

In this section, we describe the proposed approach, consisting in the extension of RNA-Net to the TTA scenario (RNA^{++}) and its combination with losses aiming at reducing the classifier's uncertainty on test data (Class Relative (CR) losses).

4.1 Multi-modal Test-Time Adaptation

A very recent work [29] showed that exploiting the multi-modal nature of videos allows one to exploit the shared knowledge available from multiple sources to build a model able to generalize to unseen data. The same strategy has also been shown to be effective as an adaptation technique when using unlabeled target data [30]. In particular, authors of [29] brought to light that the discrepancy between the two modalities' mean feature norms inhibits the network from learning equally from the two during training, i.e., the network privileges the modality with greater feature norm, while penalizing the other. This causes the final model to perform sub-optimally in comparison to the uni-modal one, a problem which has also been shown in [44]. Authors of [29] address it by proposing an

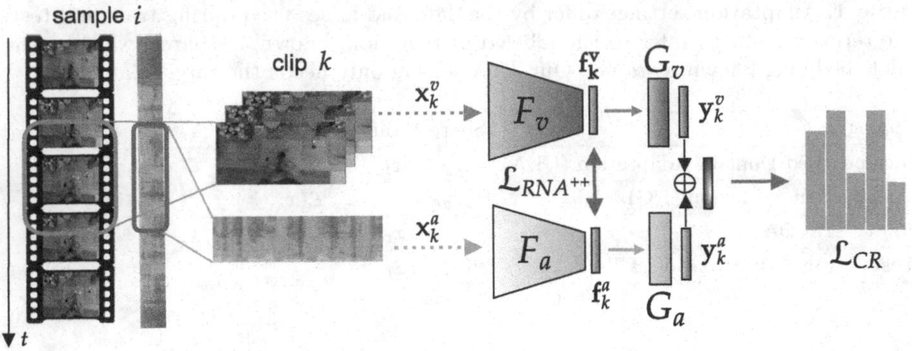

Fig. 1. Unlabeled test visual x_k^v and audio x_k^a inputs for the k-th clip are fed to the respective feature extractors F^v and F^a. The $\mathcal{L}_{RNA++}$ loss operates at feature-level by balancing the relative feature norms of the two modalities. The latter is combined with Class Relative losses $\mathcal{L}_{CR}$, namely $\mathcal{L}_{MCC}$, $\mathcal{L}_{CENT}$ or $\mathcal{L}_{IM}$ losses.

audio-visual loss which minimizes the discrepancy between the two modalities' feature norms during training, in order to better exploit multi-modal learning and thus leading to better generalization results.

This problem, referred to as the "norm-unbalance problem", still exists at test-time, negatively affecting the final prediction. In fact, the latter is biased towards the modality with greater feature norm [47], reducing the potentiality of having multiple modalities. For these reasons, in this work, we extend the proposed *Relative Norm Alignment (RNA)* loss to re-balance the mean feature norms of the two modalities during testing. This loss, which we call RNA^{++}, is designed to deal with the multi-clip nature of the test phase and it is defined as

$$\mathcal{L}_{RNA^{++}} = \left(\frac{\mathbb{E}[h(X_k^v)]}{\mathbb{E}[h(X_k^a)]} - 1 \right)^2, \tag{1}$$

where $h(X_k^m) = (\|\cdot\|_2 \circ f_k^m)(X_k^m)$ indicates the L_2-norm of the features f_k^m, $\mathbb{E}[h(X_k^m)] = \frac{1}{N}\sum_{x_{ik}^m \in \mathcal{X}_k^m} h(x_{ik}^m)$ with k the k-th clip of the m-th modality and N denotes the number of samples of the test set $\mathcal{X}_k^m = \{x_{1k}^m, ..., x_{Nk}^m\}$.

4.2 Class Relative Losses

By operating at feature level, the RNA^{++} loss promotes the cooperation between the two modalities, increasing the robustness of their final embeddings and, as a result, leading to a more robust classifier which is less affected by the domain shift. However, as the RNA^{++} loss is not backpropagated through the classifier, it focuses only on the multi-modal embeddings and ignores the classification layer's uncertainty on target data. To tackle this weakness, a natural choice might be to introduce in our multi-modal framework the standard *entropy loss* [13], which is commonly used to minimize prediction *uncertainty*. However, the entropy term alone is insufficient to provide stability, as a trivial solution is the one in which the predicted single-class samples may prevail over the others [12,45], especially when dealing with unbalanced datasets. It has also been proven that the entropy loss is not able to correctly measure the "class confusion" between correct and ambiguous classes [18]. As a result, this classifier's prediction uncertainty tends to introduce noise in the multi-clip prediction, as wrong clip predictions might dominate the correct one. Based on these considerations, minimizing the entropy is not sufficient to reduce the uncertainty of the final classifier on test samples. Thus, we re-purpose losses that bring attention to the relation between all per-class predictions in order to reduce uncertainty and we refer to them as Class Relative losses (CR losses). It follows a detailed description of these methods.

Minimum Class Confusion (MCC). This loss [18] minimizes the *inter-class confusion* on test data so that no samples are ambiguously classified into two classes at the same time. It is formalized as:

$$\mathcal{L}_{MCC} = \frac{1}{|\mathcal{C}|} \sum_{j=1}^{|\mathcal{C}|} \sum_{j' \neq j}^{|\mathcal{C}|} |\tilde{\mathbf{C}}_{jj'}| \tag{2}$$

where $\mathcal{C}$ is the number of classes and $\tilde{\mathbf{C}}_{jj'}$ measures the confusion between each class pair (j, j'). The latter is derived from the **Class Correlation** term $\mathbf{C}_{jj'}$, which is defined as:

$$\mathbf{C}_{jj'} = \hat{y}_{.j}^{T} \mathbf{W} \hat{y}_{.j'} \tag{3}$$

where we denote with $\hat{y}_{.j'}$ the j-th column of the probability matrix $\hat{Y}_{ij}$, which represents probability of the i-th samples to belong to the j-th class. $\hat{Y}_{ij}$ is obtained by summing the audio and visual probability matrices $\hat{Y}_{ij}^{a}$ and $\hat{Y}_{ij}^{v}$ respectively. The diagonal matrix $\mathbf{W}$ is used to re-weight $\mathbf{C}_{jj'}$ in order to emphasize the class with the highest class ambiguity. Finally, $\tilde{\mathbf{C}}_{jj'}$ is obtained by *category normalization* of the $\mathbf{C}_{jj'}$ value as in [41].

Information Maximization (IM). The objective of IM loss [12,14,34] is to make test-time predictions individually certain and *globally diverse* to avoid trivial solutions caused by entropy minimization alone. Indeed, it combines a conditional entropy term and a diversity term:

$$\mathcal{L}_{div} = -\mathbb{E}_{x \in \mathcal{X}} \sum_{c=1}^{\mathcal{C}} \sigma_c(h(x)) \log \sigma_c(h(x)) + \sum_{c=1}^{\mathcal{C}} \bar{p}_c \log \bar{p}_c \tag{4}$$

where $h(x) = G^v(F^v(x^v)) + G^a(F^a(x^a))$ is the C-dimensional output of each sample, summed over each modality input, and $\bar{p} = \mathbb{E}_{x \in \mathcal{X}}[\sigma(h(x))]$ is the mean of the softmax outputs for the current batch.

Complement Entropy (CENT). Considering our setting where multiple clip predictions are considered during test, the CENT loss aims at neutralizing the negative effects of incorrectly predicted clips on the final prediction. It accomplishes this by "flattening" the predicted probabilities of "complement classes", i.e., all classes except the predicted one. As a result, when several clip predictions are considered, the voting process' noise is reduced. We refer to this loss as "complement entropy" objective, as it consists in maximizing the entropy for low-confident classes rather than minimizing it for the most confident one, as standard entropy minimization does. Given the k-th clip, it is defined as:

$$\mathcal{L}_{CENT} = \frac{1}{N}\sum_{i=1}^{N}\mathcal{H}(\hat{y}_{i\bar{c}}) = -\frac{1}{N}\sum_{i=1}^{N}\sum_{j=1, j\neq p}^{C}\left(\frac{\hat{y}_{ij}}{1-\hat{y}_{ip}}\log\frac{\hat{y}_{ij}}{1-\hat{y}_{ip}}\right) \tag{5}$$

where N is the total number of samples in the batch, $\hat{y}_{ip}$ represents the predicted probability of the class p with the higher score for the i-th sample, i.e., $\hat{y}_{ip} = max_j(\hat{y}_{ij})$, and $\mathcal{H}(\cdot)$ is the entropy function computed on the prediction of complement classes $\hat{y}_{i\bar{c}}$ ($\bar{c} \neq p$). The formulation is similar to the one in [4], and we extend it to operate in an unsupervised fashion. In our multi-modal setting, $\hat{y}_{ij}$ results from the sum of audio and visual predictions $\hat{y}^a_{ij}$ and $\hat{y}^v_{ij}$ respectively.

5 Experimental Results

In this section, we first introduce the dataset and the experimental setup, followed by a brief overview of the baseline methods used (Sect. 5.1). Finally, we present the experimental results (Sect. 5.2).

5.1 Experimental Setting

Dataset. We use the EPIC-Kitchens-55 dataset [8] and we adopt the same experimental protocol of [25], where the three kitchens are handpicked from the 32 available. We refer to them here as D1, D2, and D3 respectively.

Input. For RGB, during inference, 5 equidistant clips of 16 frames are fed to the network. During adaptation, we apply random crops, scale jitters and horizontal flips for data augmentation, while at pure inference time only center crops are applied. Regarding aural information, we follow [19] and convert the audio track into a 256 × 256 matrix representing the log-spectrogram of the signal. As for visual information, 5 equidistant audio clips in correspondence to the visual ones are used during both adaptation and inference.

Implementation Details. Our network is composed of two streams, one for each modality m, with distinct feature extractor F^m and classifier G^m. The RGB stream uses I3D [3] as in [25]. The audio feature extractor uses the BN-Inception model [15] pretrained on ImageNet, which proved to be a reliable backbone for processing audio spectrograms [19]. Each F^m produces a 1024-dimensional representation f_m which is fed to the classifier G^m, consisting in a fully-connected layer that outputs the score logits. Then, the two modalities are fused by summing the outputs. During adaptation, the network is optimized with a batch size of 32, SGD optimizer with momentum 0.1, and weight decay

$1e^{-7}$. We optimized the learning rate $lr \in \{1e^{-2}, 1e^{-3}, 1e^{-4}, 1e^{-5}\}$, loss weights $\alpha, \beta, \gamma, \delta, \epsilon^1 \in \{1, 0.1, 0.5, 0.01\}$, and optimization steps $n \in \{1, ..., 10\}$ for all methods, reporting the accuracy scores averaged on three different runs. The code is implemented using Pytorch framework and the models are trained using Intel(R) Core(TM) i7-9800X CPU and two GPUs Titan RTX (24 GB).

Baseline Methods. We adapted the most popular image-based TTA methods to our video scenario. Those are:

- **Prediction-time BN**: authors of [26,33,48] proved that either updating [33,48] or replacing [26] batch normalization statistics μ and σ^2 with the ones from test data during inference achieves good adaptation results. In our experiments, we do not entirely replace the source statistics, but we rather update them with the ones from the target.
- **TENT** [42]: the adaptation is performed by optimizing the modulation parameters γ and β of the Batch Normalization (BN) layers by minimizing the entropy loss [13]. The normalization statistics μ and σ^2 are initialized on source data and updated for each layer in turn, during the forward pass, on test data batch statistics. We also tried a different variation of TENT, which we refer to as TENT-C, where we also optimize the classifier.
- **T3A** [16]: it is a backpropagation-free method which adjusts the classifier at test-time. In particular, it creates a pseudo-prototype for each class using online test data and the classifier pre-trained source, and then classifies each test sample basing on its distance to the pseudo-prototype.

5.2 Results

In this section, we evaluate TTA results by considering both (i) a network which has been trained on multiple source domains (*DeepAll*) and (ii) a network which has been trained with RNA-Net [29], a method which aims to improve generalization results by exploiting audio-visual correlations at feature level. On the two, we evaluate three different approaches: (i) **baseline methods**, which are standard image-based TTA methods which we adapted to our setting, (ii) **RNA^{++}**, the extension of RNA loss to operate *at feature level* on test data and (iii) **Class Relative (CR) losses**, which are losses operating *at prediction level.*

Baseline Methods. We show in Table 2 and Table 3 the effects of applying existing TTA methods, namely BN [33,48], TENT [42], and T3A [16]. Despite its simplicity, BN shows a consistent improvement over both the DeepAll and RNA-Net baselines. This proves that the feature distribution varies greatly from source to target domains, an thus simply updating batch normalization statistics with the ones from target data is effective in coping with the domain shift. Both TENT and TENT-C improve over the baselines, showing that methods inherited from the image-based domain scale well to our multi-modal action recognition setting. TENT-C achieves slightly better results than TENT, proving that optimizing the

[1] $\alpha, \beta, \gamma, \delta, \epsilon$ are the weights of RNA^{++}, MCC, ENT, IM and CENT losses respectively.

Table 2. Top-1 Accuracy (%) of different test-time adaptation methods in a Multi-Source DG scenario when applied to a **DeepAll baseline**. $D_i, D_j \rightarrow D_k$ indicates that we trained on D_i and D_j and we tested on D_k.

	D2, D3 → D1	D3, D1 → D2	D1, D2 → D3	Mean	Gain
DeepAll	51.34	43.22	41.07	45.21	–
BN [33,48]	50.82	44.14	43.91	46.29	▲+1.08
TENT [42]	49.86	42.99	43.96	45.60	▲+0.39
TENT-C [42]	49.83	43.07	44.00	45.63	▲+0.42
T3A [16]	40.28	36.86	39.24	38.79	▼−6.42
RNA^{++}	50.79	43.91	43.87	46.19	▲+0.98
ENT	51.81	43.60	43.33	46.25	▲+1.04
MCC	52.09	44.06	43.11	**46.42**	▲+1.21
IM	50.38	43.68	44.40	46.15	▲+0.94
CENT	51.10	43.30	44.84	46.41	▲+1.20
ENT+RNA^{++}	50.86	43.60	43.91	46.12	▲+0.91
MCC+RNA^{++}	51.88	44.06	44.00	**46.65**	▲+1.44
IM+RNA^{++}	51.95	43.52	44.44	46.64	▲+1.43
CENT+RNA^{++}	50.58	43.45	45.42	46.48	▲+1.28

Table 3. Top-1 Accuracy (%) of different test-time adaptation methods in a Multi-Source DG scenario when applied to **RNA-Net baseline**.

	D2, D3 → D1	D3, D1 → D2	D1, D2 → D3	Mean	Gain
RNA-Net [29]	55.75	46.67	50.53	50.98	–
BN [33,48]	57.56	46.90	52.04	52.17	▲+1.18
TENT [42]	54.18	47.43	53.29	51.63	▲+0.65
TENT-C [42]	54.28	47.89	53.24	51.81	▲+0.83
T3A [16]	49.69	41.15	34.58	41.81	▼−9.17
RNA^{++}	57.56	46.90	52.00	52.15	▲+1.17
ENT	57.46	46.51	52.09	52.02	▲+1.04
MCC	57.70	47.05	52.04	52.26	▲+1.28
IM	57.46	47.13	52.09	52.23	▲+1.25
CENT	57.53	47.43	52.31	**52.42**	▲+1.44
ENT+RNA^{++}	57.29	46.67	52.04	52.00	▲+1.02
MCC+RNA^{++}	57.63	46.90	52.04	52.19	▲+1.21
IM+RNA^{++}	57.56	47.20	52.18	52.31	▲+1.33
CENT+RNA^{++}	57.67	47.51	52.18	**52.45**	▲+1.47

classifier parameters using target data is effective in improving generalization. Both techniques benefit from having a model that has been pre-trained using a multi-modal DG strategy, since the improvement on RNA-Net is more consistent than the improvement on DeepAll.

However, the improvements are limited and slighty lower than the one obtained by BN, confirming the difficulties of this task in the egocentric context and the importance of this new benchmark to promote future research in this new field. We can further notice that, differently from the others approches, T3A does not scale to this setting. Indeed, this is explainable by the fact that the dataset used is strongly unbalanced, and a method which exploits a per-class pseudo-prototype representation could lead to sub-optimal results making prediction of classes with fewer samples almost impossible.

RNA^{++}. In Table 2 and Table 3 we illustrate the effects of minimizing RNA^{++} at test-time. It can be seen that RNA^{++} outperforms the baseline DeepAll and RNA-Net by 0.98% and 1.17% respectively, showing that re-balancing the mean feature norms of the two modalities on test samples further improves the adaptation ability of the network. It can be noticed also that starting from the robust initialization of RNA-Net to perform the rebalancing operation, it helps the RNA++ to be more effective at test time. However the limited improvement of RNA++ over existing techniques, particularly when compared to BN, highlights its need to be guided by a loss that acts on the final prediction.

CR Losses. We show the performance of the entropy loss w.r.t. CR losses in Table 2 and Table 3. When applying the entropy loss, we fine-tune all the network. The entropy loss surpasses both DeepAll and RNA-Net baselines and yields results comparable to all existing TTA approaches. Except for one case, the proposed CR losses (MCC, IM, and CENT losses) surpass the entropy loss. This proves the limitation of entropy loss in this context (see Sect. 4.2) and highlights the benefits of using losses which take into account all the predictions.

Combining RNA^{++} with CR Losses. When further combining those losses with RNA^{++}, performance increase on both settings in almost all configurations. This confirms the effectiveness of fine-tuning the network through a loss which operates not only on features but also on predictions. Indeed, the combination of RNA^{++} with CR losses proved to be the most effective technique.

The combination of RNA^{++} with the entropy loss does improves over RNA^{++} alone, while on the other side the combination of it with CR losses outperforms RNA^{++} in all cases. This provides additional evidence to support the mentioned limits of entropy (see Sect. 4.2) in the TTA scenario.

6 Conclusions

In this work, we investigate the test-time adaptation setting for audio-visual egocentric action recognition. We propose a new benchmark for this context, showing the performance of current image-based test-time adaptation algorithms which we adapted to the video domain. Moreover, we propose RNA^{++}, a new test-time adaptation approach which extends RNA-Net, a recent multi-modal domain generalization method. Finally, we prove the importance of combining it with a set of losses meant to further reduce classifier's uncertainty on test data. We regard our work as a starting point for future research into new settings that allow action recognition algorithms to be applied in real-world scenarios.

Acknowledgements. This work was supported by the CINI Consortium through the VIDESEC project.

References

1. Azimi, F., Palacio, S., Raue, F., Hees, J., Bertinetto, L., Dengel, A.: Self-supervised test-time adaptation on video data. In: WACV, pp. 3439–3448 (2022)
2. Bucci, S., D'Innocente, A., Liao, Y., Carlucci, F.M., Caputo, B., Tommasi, T.: Self-supervised learning across domains. TPAMI (2021)
3. Carreira, J., Zisserman, A.: Quo vadis, action recognition? A new model and the kinetics dataset. In: CVPR. pp. 6299–6308 (2017)
4. Chen, H.Y., et al.: Complement objective training. arXiv:1903.01182 (2019)
5. Chen, M.H., Kira, Z., AlRegib, G., Yoo, J., Chen, R., Zheng, J.: Temporal attentive alignment for large-scale video domain adaptation. In: CVPR, pp. 6321–6330 (2019)
6. Choi, J., Sharma, G., Schulter, S., Huang, J.-B.: Shuffle and attend: video domain adaptation. In: Vedaldi, A., Bischof, H., Brox, T., Frahm, J.-M. (eds.) ECCV 2020, Part XII. LNCS, vol. 12357, pp. 678–695. Springer, Cham (2020). https://doi.org/10.1007/978-3-030-58610-2_40
7. Crasto, N., Weinzaepfel, P., Alahari, K., Schmid, C.: Mars: motion-augmented RGB stream for action recognition. In: CVPR, June 2019
8. Damen, D., et al.: Scaling egocentric vision: The epic-kitchens dataset. In: ECCV, pp. 720–736 (2018)
9. Dou, Q., Coelho de Castro, D., Kamnitsas, K., Glocker, B.: Domain generalization via model-agnostic learning of semantic features. NIPS **32**, 6450–6461 (2019)
10. Furnari, A., Farinella, G.: Rolling-unrolling LSTMs for action anticipation from first-person video. TPAMI **43**(11), 4021–4036 (2020)
11. Ganin, Y., Lempitsky, V.: Unsupervised domain adaptation by backpropagation. In: International Conference on Machine Learning, pp. 1180–1189. PMLR (2015)
12. Gomes, R., Krause, A., Perona, P.: Discriminative clustering by regularized information maximization. In: NIPS (2010)
13. Grandvalet, Y., Bengio, Y.: Semi-supervised learning by entropy minimization. In: NIPS. vol. 367, pp. 281–296, January 2004
14. Hu, W., Miyato, T., Tokui, S., Matsumoto, E., Sugiyama, M.: Learning discrete representations via information maximizing self-augmented training. In: ICML, pp. 1558–1567 (2017)
15. Ioffe, S., Szegedy, C.: Batch normalization: accelerating deep network training by reducing internal covariate shift. In: ICML, pp. 448–456 (2015)
16. Iwasawa, Y., Matsuo, Y.: Test-time classifier adjustment module for model-agnostic domain generalization. In: Beygelzimer, A., Dauphin, Y., Liang, P., Vaughan, J.W. (eds.) NIPS (2021)
17. Jamal, A., Namboodiri, V.P., Deodhare, D., Venkatesh, K.: Deep domain adaptation in action space. In: BMVC (2018)

18. Jin, Y., Wang, X., Long, M., Wang, J.: Minimum class confusion for versatile domain adaptation. In: Vedaldi, A., Bischof, H., Brox, T., Frahm, J.-M. (eds.) ECCV 2020, Part XXI. LNCS, vol. 12366, pp. 464–480. Springer, Cham (2020). https://doi.org/10.1007/978-3-030-58589-1_28
19. Kazakos, E., Nagrani, A., Zisserman, A., Damen, D.: Epic-fusion: audio-visual temporal binding for egocentric action recognition. In: ICCV, October 2019
20. Kazakos, E., Nagrani, A., Zisserman, A., Damen, D.: Slow-fast auditory streams for audio recognition. In: ICASSP, pp. 855–859 (2021)
21. Kim, D., et al.: Learning cross-modal contrastive features for video domain adaptation. In: ICCV, pp. 13618–13627 (2021)
22. Li, Y., Wang, N., Shi, J., Hou, X., Liu, J.: Adaptive batch normalization for practical domain adaptation. Pattern Recognit. **80**, 109–117 (2018)
23. Lin, J., Gan, C., Han, S.: TSM: temporal shift module for efficient video understanding. In: ICCV, pp. 7083–7093 (2019)
24. Long, M., Cao, Y., Wang, J., Jordan, M.: Learning transferable features with deep adaptation networks. In: ICML, pp. 97–105 (2015)
25. Munro, J., Damen, D.: Multi-modal domain adaptation for fine-grained action recognition. In: CVPR, June 2020
26. Nado, Z., Padhy, S., Sculley, D., D'Amour, A., Lakshminarayanan, B., Snoek, J.: Evaluating prediction-time batch normalization for robustness under covariate shift. arXiv preprint arXiv:2006.10963 (2020)
27. Pan, B., Cao, Z., Adeli, E., Niebles, J.C.: Adversarial cross-domain action recognition with co-attention. In: AAAI, vol. 34, pp. 11815–11822 (2020)
28. Planamente, M., Bottino, A., Caputo, B.: Self-supervised joint encoding of motion and appearance for first person action recognition. In: ICPR, pp. 8751–8758 (2021)
29. Planamente, M., Plizzari, C., Alberti, E., Caputo, B.: Domain generalization through audio-visual relative norm alignment in first person action recognition. In: WACV, pp. 1807–1818, January 2022
30. Plizzari, C., Planamente, M., Alberti, E., Caputo, B.: PoliTO-IIT submission to the epic-kitchens-100 unsupervised domain adaptation challenge for action recognition. arXiv preprint arXiv:2107.00337 (2021)
31. Plizzari, C., et al.: E^2(go) motion: Motion augmented event stream for egocentric action recognition. arXiv preprint arXiv:2112.03596 (2021)
32. Rodin, I., Furnari, A., Mavroedis, D., Farinella, G.M.: Predicting the future from first person (egocentric) vision: a survey. CVIU **211**, 103252 (2021)
33. Schneider, S., Rusak, E., Eck, L., Bringmann, O., Brendel, W., Bethge, M.: Improving robustness against common corruptions by covariate shift adaptation. arXiv preprint arXiv:2006.16971 (2020)
34. Shi, Y., Sha, F.: Information-theoretical learning of discriminative clusters for unsupervised domain adaptation. arXiv preprint arXiv:1206.6438 (2012)
35. Song, X., et al.: Spatio-temporal contrastive domain adaptation for action recognition. In: CVPR, pp. 9787–9795, June 2021
36. Sudhakaran, S., Escalera, S., Lanz, O.: LSTA: long short-term attention for egocentric action recognition. In: CVPR, pp. 9954–9963 (2019)
37. Sun, Y., Wang, X., Liu, Z., Miller, J., Efros, A., Hardt, M.: Test-time training with self-supervision for generalization under distribution shifts. In: ICML, pp. 9229–9248. PMLR (2020)
38. Thapar, D., Nigam, A., Arora, C.: Anonymizing egocentric videos. In: ICCV, pp. 2320–2329 (2021)
39. Torralba, A., Efros, A.A.: Unbiased look at dataset bias. In: CVPR 2011, pp. 1521–1528. IEEE (2011)

40. Volpi, R., Namkoong, H., Sener, O., Duchi, J.C., Murino, V., Savarese, S.: Generalizing to unseen domains via adversarial data augmentation. In: NIPS, pp. 5334–5344 (2018)
41. Von Luxburg, U.: A tutorial on spectral clustering. Stat. Comput. **17**(4), 395–416 (2007). https://doi.org/10.1007/s11222-007-9033-z
42. Wang, D., Shelhamer, E., Liu, S., Olshausen, B., Darrell, T.: Tent: Fully test-time adaptation by entropy minimization. arXiv preprint arXiv:2006.10726 (2020)
43. Wang, L., et al.: Temporal segment networks: towards good practices for deep action recognition. In: Leibe, B., Matas, J., Sebe, N., Welling, M. (eds.) ECCV 2016, Part VIII. LNCS, vol. 9912, pp. 20–36. Springer, Cham (2016). https://doi.org/10.1007/978-3-319-46484-8_2
44. Wang, W., Tran, D., Feiszli, M.: What makes training multi-modal classification networks hard? In: CVPR, pp. 12695–12705 (2020)
45. Wu, X., Zhou, Q., Yang, Z., Zhao, C., Latecki, L.J., et al.: Entropy minimization vs. diversity maximization for domain adaptation. arXiv:2002.01690 (2020)
46. Yao, Z., Wang, Y., Wang, J., Yu, P., Long, M.: VideoDG: generalizing temporal relations in videos to novel domains. TPAMI (2021)
47. Ye, J., Lu, X., Lin, Z., Wang, J.Z.: Rethinking the smaller-norm-less-informative assumption in channel pruning of convolution layers. In: ICLR (2018)
48. You, F., Li, J., Zhao, Z.: Test-time batch statistics calibration for covariate shift. arXiv preprint arXiv:2110.04065 (2021)
49. Zhao, J., Snoek, C.G.: Dance with flow: two-in-one stream action detection. In: CVPR, pp. 9935–9944 (2019)

Combining EfficientNet and Vision Transformers for Video Deepfake Detection

Davide Alessandro Coccomini, Nicola Messina(✉), Claudio Gennaro, and Fabrizio Falchi

Institute of Information Science and Technologies (ISTI), Italian National Research Council (CNR), Via G. Moruzzi 1, 56124 Pisa, Italy
{davidealessandro.coccomini,nicola.messina,claudio.gennaro, fabrizio.falchi}@isti.cnr.it

Abstract. Deepfakes are the result of digital manipulation to forge realistic yet fake imagery. With the astonishing advances in deep generative models, fake images or videos are nowadays obtained using variational autoencoders (VAEs) or Generative Adversarial Networks (GANs). These technologies are becoming more accessible and accurate, resulting in fake videos that are very difficult to be detected. Traditionally, Convolutional Neural Networks (CNNs) have been used to perform video deepfake detection, with the best results obtained using methods based on EfficientNet B7. In this study, we focus on video deep fake detection on faces, given that most methods are becoming extremely accurate in the generation of realistic human faces. Specifically, we combine various types of Vision Transformers with a convolutional EfficientNet B0 used as a feature extractor, obtaining comparable results with some very recent methods that use Vision Transformers. Differently from the state-of-the-art approaches, we use neither distillation nor ensemble methods. Furthermore, we present a straightforward inference procedure based on a simple voting scheme for handling multiple faces in the same video shot. The best model achieved an AUC of 0.951 and an F1 score of 88.0%, very close to the state-of-the-art on the DeepFake Detection Challenge (DFDC). The code for reproducing our results is publicly available here: https://tinyurl.com/cnn-vit-dfd.

Keywords: Deep fake detection · Transformer networks · Deep learning

1 Introduction

With the recent advances in generative deep learning techniques, it is nowadays possible to forge highly-realistic and credible misleading videos. These methods have generated numerous fake news or revenge porn videos, becoming a severe problem in modern society [5]. These fake videos are known as *deepfakes*. Given the astonishing realism obtained by recent models in the generation of human faces, deepfakes are mainly obtained by transposing one person's face

S. Sclaroff et al. (Eds.): ICIAP 2022, LNCS 13233, pp. 219–229, 2022.
https://doi.org/10.1007/978-3-031-06433-3_19

onto another's. The results are so realistic that it is almost like the person being replaced is actually present in the video, and the replaced actors are rigged to say things they never actually said [35].

The evolution of deepfakes generation techniques and their increasing accessibility forces the research community to find effective methods to distinguish a manipulated video from a real one. Nowadays, models based on Transformer architecture are gaining ground in the field of Computer Vision, showing excellent results in image processing [19], document retrieval [25], and efficient visual-textual matching [28,29]. Unlike Vision Transformers, CNNs still maintain an important architectural prior, the spatial locality, which is very important for discovering image patch abnormalities and maintaining good data efficiency. CNNs, in fact, have a long-established success on many tasks, ranging from image classification [12,37] and object detection [1,7,32] to abstract visual reasoning [26,27].

In this paper, we use the power of convolutional and transformer models to tackle the problem of video deepfake detection. Specifically, we analyze different solutions based on the combination of convolutional networks—particularly the EfficientNet B0—with different types of Vision Transformers [9]. We compare the results with the current state-of-the-art, keeping into consideration both accuracy and network complexity. Our proposed models are frame-based, as many others in literature. Nevertheless, we also propose a method to handle multiple sequential frames at inference time. Specifically, we propose a simple yet effective voting mechanism that handles multiple face instances across multiple frames to judge the genuineness of the video shot. We show that this methodology could lead to better and more stable results.

2 Related Works

2.1 Deepfake Generation

There are mainly two generative approaches to obtain realistic faces: Generative Adversarial Networks (GANs) [14] and Variational AutoEncoders (VAEs) [21].

GANs employ two distinct networks. The discriminator, the one that must be able to identify when a video is fake or not, and the generator, the network that actually modifies the video in a sufficiently credible way to deceive its counterpart. With GANs, very credible and realistic results have been obtained, and over time, numerous approaches have been introduced such as StarGAN [6] and DiscoGAN [20]; the best results in this field have been obtained with StyleGAN-V2 [18].

VAE-based solutions, instead, make use of a system consisting of two encoder-decoder pairs, each of which is trained to deconstruct and reconstruct one of the two faces to be exchanged. Subsequently, the decoding part is switched, and this allows the reconstruction of the target person's face. The best-known uses of this technique were DeepFaceLab [31], DFaker[1], and DeepFaketf[2].

[1] https://github.com/dfaker/df.

[2] https://github.com/StromWine/DeepFake_tf.

2.2 Deepfake Detection

The problem of deepfake detection has a widespread interest not only in the visual domain. For example, the recent work in [11] analyzes deepfakes in tweets for finding and defeating false content in social networks.

In an attempt to address the problem of deepfakes detection in videos, numerous datasets have been produced over the years. These datasets are grouped into three generations, the first generation consisting of DF-TIMIT [22], UADFC [38] and FaceForensics++ [33], the second generation datasets such as Google Deepfake Detection Dataset [10], Celeb-DF [23], and finally the third generation datasets, with the DFDC dataset [8] and DeepForensics [17]. The further the generations go, the larger these datasets are, and the more frames they contain.

In particular, on the DFDC dataset, which is the largest and most complete, multiple experiments were carried out trying to obtain an effective method for deepfake detection. Very good results were obtained with EfficientNet B7 ensemble technique in [34]. Other noteworthy methods include those conducted in [30], who attempted to identify spatio-temporal anomalies by combining an EfficientNet with a Gated Recurrent Unit (GRU). Some efforts to capture spatio-temporal inconsistencies were made in [24] using 3DCNN networks and in [2], which presented a method that exploits optical flow to detect video glitches. Some more classical methods have also been proposed to perform deepfake detection. In particular, the authors in [15] proposed a method based on K-nearest neighbors, while the work in [38] exploited SVMs. Of note is the very recent work of Giudice et al. [13] in which they presented an innovative method for identifying so-called GAN Specific Frequencies (GSF) that represent a unique fingerprint of different generative architectures. By exploiting the Discrete Cosine Transform (DCT) they manage to identify anomalous frequencies.

More recently, methods based on Vision Transformers have been proposed. Notably, the method presented in [36] obtained good results by combining Transformers with a convolutional network, used to extract patches from faces detected in videos.

State of the art was then recently improved by performing distillation from the EfficientNet B7 pre-trained on the DFDC dataset to a Vision Transformer [16]. In this case, the Vision Transformer patches are combined with patches extracted from the EfficientNet B7 pre-trained via global pooling and then passed to the Transformer Encoder. A distillation token is then added to the Transformer network to transfer the knowledge acquired by the EfficientNet B7.

3 Method

The proposed methods analyze the faces extracted from the source video to determine whenever they have been manipulated. For this reason, faces are pre-extracted using a state-of-the-art face detector, MTCNN [39]. We propose two mixed convolutional-transformer architectures that take as input a pre-extracted face and output the probability that the face has been manipulated. The two presented architectures are trained in a supervised way to discern real from fake examples. For this reason, we solve the detection task by framing it as a binary classification problem. Specifically, we propose the *Efficient ViT* and the *Convolutional Cross ViT*, better explained in the following paragraphs.

The proposed models are trained on a face basis, and then they are used at inference time to draw a conclusion on the whole video shot by aggregating the inferred output both in time and across multiple faces, as explained in Sect. 4.3.

The Efficient ViT. The Efficient ViT is composed of two blocks, a convolutional module for working as a feature extractor and a Transformer Encoder, in a setup very similar to the Vision Transformer (ViT) [9]. Considering the promising results of the EfficientNet, we use an EfficientNet B0, the smallest of the EfficientNet networks, as a convolutional extractor for processing the input faces. Specifically, the EfficientNet produces a visual feature for each chunk from the input face. Each chunk is 7×7 pixels. After a linear projection, every feature from each spatial location is further processed by a Vision Transformer. The CLS token is used for producing the binary classification score. The architecture is illustrated in Fig. 1a. The EfficientNet B0 feature extractor is initialized with the pre-trained weights and fine-tuned to allow the last layers of the network to perform a more consistent and suitable extraction for this specific downstream task. The features extracted from the EfficientNet B0 convolutional network simplify the training of the Vision Transformer, as the CNN features already embed important low-level and localized information from the image.

The Convolutional Cross ViT. Limiting the architecture to the use only small patches as in the Efficient ViT may not be the ideal choice, as artifacts introduced by deepfakes generation methods may arise both locally and globally. For this reason, we also introduce the Convolutional Cross ViT architecture. The Convolutional Cross ViT builds upon both the Efficient ViT and the multi-scale Transformer architecture by [4]. More in detail, the Convolutional Cross ViT uses two distinct branches: the *S-branch*, which deals with smaller patches, and the *L-branch*, which works on larger patches for having a wider receptive field. The visual tokens output by the Transformer Encoders from the two branches are combined through cross attention, allowing direct interaction between the two paths. Finally, the CLS tokens corresponding to the outputs from the two branches are used to produce two separate logits. These logits are summed, and a final sigmoid produces the final probabilities. A detailed overview of this architecture is shown in Fig. 1b. For the Convolutional Cross ViT, we use two different CNN backbones. The former is the EfficientNet B0, which processes 7×7 image patches for the S-branch and 56×56 for the L-branch. The latter is the CNN by Wodajo et al. [36], which handles 7×7 image patches for the S-branch and 64×64 for the L-branch.

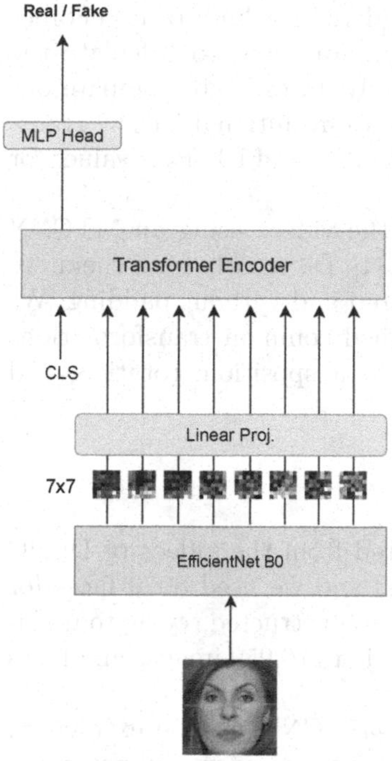

(a) Efficient ViT architecture.

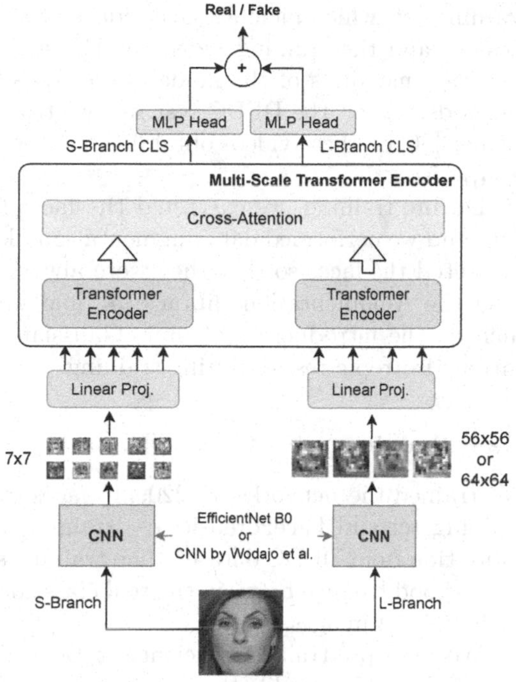

(b) Convolutional Cross ViT architecture.

Fig. 1. The proposed architectures. Notice that for the Convolutional Cross ViT in (b), we experimented both with EfficientNet B0 and with the convolutional architecture by [36] as feature extractors.

4 Experiments

We probed the presented architectures against some state-of-the-art methods on two widely-used datasets. In particular, we considered Convolutional ViT [36], ViT with distillation [16], and Selim EfficientNet B7 [34], the winner of the Deep Fake Detection Challenge (DFDC). Notice that the results for Convolutional ViT [36] are not reported in the original paper, but they are obtained executing the test code on DFDC test set using the available pre-trained model released by the authors.

4.1 Datasets and Face Extraction

We initially conducted some tests on FaceForensics++. The dataset is composed of original and fake videos generated through different deepfake generation techniques. For evaluating, we considered the videos generated in the Deepfakes, Face2Face, FaceShifter, FaceSwap and NeuralTextures sub-datasets. We also used the DFDC test set containing 5000 videos. The model trained on the entire training set, which includes fake videos of all considered methods of FaceForensics++ and the training videos of DFDC dataset, was used to calculate the accuracy measures of the model, reported separately. In order to compare our methods also on the DFDC test set, we tested the Convolutional Vision Transformer [36] on these videos obtaining the necessary AUC and F1-score values for comparison.

During training, we extracted the faces from the videos using an MTCNN [39], and we performed data augmentation like in [34]. Differently from them, we extracted the faces so that they were always squared and without padding. We used the Albumentations library [3], and we applied common transformations such as the introduction of blur, Gaussian noise, transposition, rotation, and various isotropic resizes during training.

4.2 Training

We trained the networks on 220,444 faces extracted from the videos of DFDC training set and FaceForensics++ training videos, and we used 8070 faces for validation from DFDC dataset. The training set was constructed trying to maintain a good balance between the real class composed of 116,950 images and fakes with 103,494 images.

We used pre-trained EfficientNet B0 and Wodajo CNN feature extractors. However, we observed better results when fine-tuning them, so we did not freeze the extraction layers. We used the standard binary cross-entropy loss as our objective during training. We optimized our network end-to-end, using an SGD optimizer with a learning rate of 0.01.

4.3 Inference

At inference time, we set a real/fake threshold at 0.55 as done in [16]. However, we proposed a slightly more elaborated voting procedure instead of averaging all ratings on individual faces indistinctly within the video. Specifically, we merged the scores, grouping them by the identifier of the actors. The face identifier is available as an output from the employed MTCNN face detector. The scores from different actors are averaged over time to produce a probability of the face being fake. Then, the per-actor scores are merged using hard voting. In particular, if there is at least one actor face passing the threshold, the whole video is classified as fake. The procedure is graphically explained in Fig. 2a. We claim that this approach is helpful to handle videos in which only one of the actors' faces has been manipulated.

In addition, it is interesting to evaluate how the performance changes when a varying number of faces are considered at inference time. To ensure that the tests are as light yet effective as possible, we experimented on one of our networks to see how the F1-score varies with the number of faces considered at testing time (Fig. 2b). We noticed that a plateau is reached when no more than 30 faces are used, so employing more than this number of faces seems statistically useless at inference time.

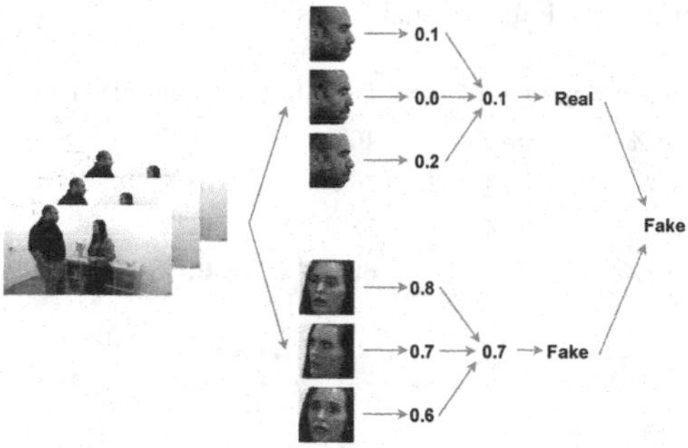

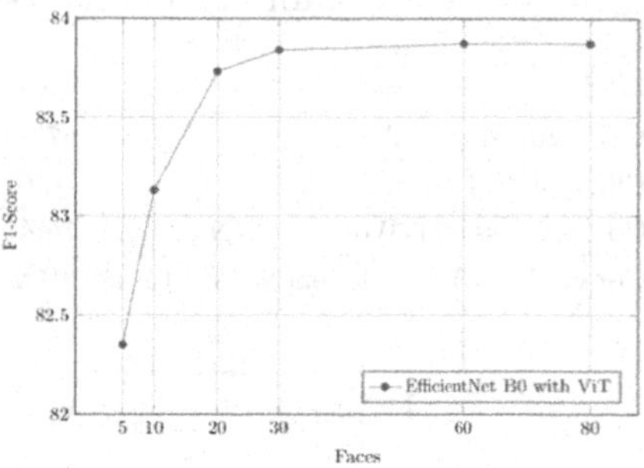

(a) Inference strategy with multiple faces in the same video.

(b) F1-score versus the number of extracted faces.

Fig. 2. Inference.

Table 1. Results on DFDC test dataset

Model	AUC	F1-score	# params
ViT with distillation [16]	0.978	91.9%	373M
Selim EfficientNet B7 [34][a]	0.972	90.6%	462M
Convolutional ViT [36]	0.843	77.0%	89M
Efficient ViT (our)	0.919	83.8%	109M
Conv. Cross ViT Wodajo CNN (our)	0.925	84.5%	142M
Conv. Cross ViT Eff.Net B0 - Avg (our)	0.947	85.6%	101M
Conv. Cross ViT Eff.Net B0 - Voting (our)	0.951	88.0%	101M

[a] Uses an ensemble of 6 networks.

4.4 Results

Table 1 shows that all models developed with EfficientNet achieve considerably higher AUC and F1-scores than the Convolutional ViT presented in [36], providing initial evidence that this specific network structure may be more suitable for this type of task. It can also be noticed that the models based on Cross Vision Transformer obtain the best results, confirming the theory that joined local and global image processing brings to better anomaly detection.

The models with Cross Vision Transformer show a particularly marked improvement when using the EfficientNet B0 as a patch extractor. Although the AUC and F1-score remain slightly below other state-of-the-art methods (in the first two rows of Table 1), these results were obtained using neither distillation nor ensemble techniques that complicate both training and inference. In fact, we can notice how the Cross Vision Transformer with the EfficientNet extractor can reach a competitive performance using less than 1/3 of the parameters of the top methods.

Furthermore, in the last two rows of Table 1 we can notice how our voting procedure used at inference time can slightly improve the results with respect to

Table 2. Models accuracy on FaceForensics++

Model	Mean	FaceSwap	DeepFakes	FaceShifter	NeuralTextures
Convolutional ViT [36]	67%	69%	**93%**	46%	60%
Efficient ViT (our)	76%	78%	83%	76%	68%
Conv. Cross ViT Wodajo CNN (our)	76%	81%	83%	73%	67%
Conv. Cross ViT EfficientNet B0 (our)	**80%**	**84%**	87%	**80%**	**69%**

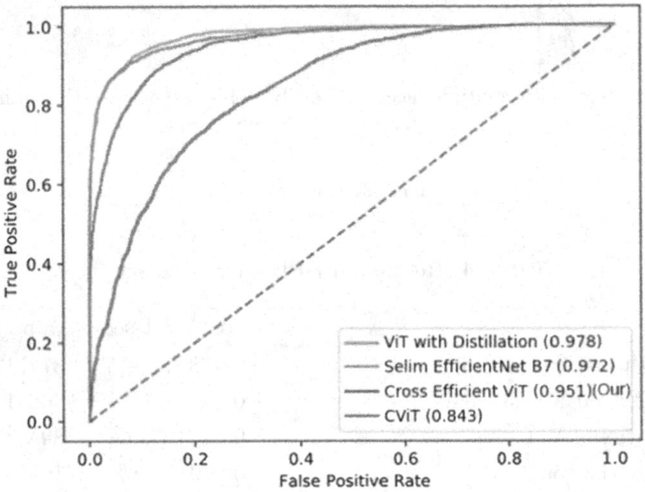

Fig. 3. ROC Curves comparison between our best model and others on DFDC test set.

a plain average of the scores from all the faces indistinctly, as done by the other methods. In Fig. 3 we report a detailed ROC plot for the architectures on the DFDC dataset.

In order to compare the developed models also on another dataset, we carried out some tests also on FaceForensics++. As shown in Table 2, our models outperform the original Convolutional ViT [36] on all sub-datasets of FaceForensics++, excluding DeepFakes. This is probably because the network could better generalize on very specific types of deepfakes. It is worth noting how the results obtained in terms of accuracy on the various sub-datasets confirm the assumption already made in [36]: some deepfakes techniques such as NeuralTextures produce videos that are more difficult to find, thus resulting in lower accuracy values than other sub-datasets. However, the average of all our three models is higher than the average obtained by the Convolutional ViT. The Convolutional Cross ViT achieves the best result with the EfficientNet B0 backbone, obtaining a mean accuracy of 80%.

5 Conclusions

In this research, we demonstrated the effectiveness of mixed convolutional-transformer networks in the Deepfake detection task. Specifically, we used pre-trained convolutional networks, such as the widely used EfficientNet B0, to extract visual features, and we relied on Vision Transformers to obtain an informative global description for the downstream task. We showed that it is possible to obtain remarkable results, very close to the state-of-the-art, without necessarily resorting to distillation techniques or ensemble networks. The use of a patch extractor based on EfficientNet proved to be particularly effective even by simply using the smallest network in this category. EfficientNet also led to better results than the generic convolutional network trained from scratch used in Wodajo et al. [36]. We then proposed a mixed architecture, the Convolutional Cross ViT, that works at two different scales to capture local and global details. The tests carried out with these models demonstrated the importance of multi-scale analysis for determining the manipulation of an image.

We also paid particular attention to the inference phase. In particular, we presented a simple yet effective voting scheme for explicitly dealing with multiple faces in a video. The scores from multiple actor faces are first averaged over time, and only then hard voting is used to decide if at least one face was manipulated. This inference mechanism yielded slightly better and stable results than the global average pooling of the scores performed by previous methods.

References

1. Amato, G., Ciampi, L., Falchi, F., Gennaro, C., Messina, N.: Learning pedestrian detection from virtual worlds. In: Ricci, E., Rota Bulò, S., Snoek, C., Lanz, O., Messelodi, S., Sebe, N. (eds.) ICIAP 2019, Part I. LNCS, vol. 11751, pp. 302–312. Springer, Cham (2019). https://doi.org/10.1007/978-3-030-30642-7_27
2. Amerini, I., Galteri, L., Caldelli, R., Del Bimbo, A.: Deepfake video detection through optical flow based CNN. In: Proceedings of the IEEE/CVF International Conference on Computer Vision Workshops (2019)
3. Buslaev, A., Parinov, A., Khvedchenya, E., Iglovikov, V.I., Kalinin, A.A.: Albumentations: fast and flexible image augmentations. ArXiv e-prints (2018)
4. Chen, C.F.R., Fan, Q., Panda, R.: CrossViT: cross-attention multi-scale vision transformer for image classification. In: Proceedings of the IEEE/CVF International Conference on Computer Vision, pp. 357–366 (2021)
5. Chesney, B., Citron, D.: Deepfakes: a looming challenge for privacy, democracy, and national security. Calif. L. Rev. **107**, 1753 (2019)
6. Choi, Y., Choi, M., Kim, M., Ha, J.W., Kim, S., Choo, J.: StarGAN: unified generative adversarial networks for multi-domain image-to-image translation. In: 2018 IEEE/CVF Conference on Computer Vision and Pattern Recognition, pp. 8789–8797 (2018). https://doi.org/10.1109/CVPR.2018.00916
7. Ciampi, L., Messina, N., Falchi, F., Gennaro, C., Amato, G.: Virtual to real adaptation of pedestrian detectors. Sensors **20**(18), 5250 (2020)
8. Dolhansky, B., et al.: The deepfake detection challenge (DFDC) dataset. arXiv preprint arXiv:2006.07397 (2020)

9. Dosovitskiy, A., et al.: An image is worth 16 × 16 words: transformers for image recognition at scale. In: International Conference on Learning Representations (2020)
10. Dufour, N., Gully, A.: Contributing data to deep-fake detection research (2019). https://ai.googleblog.com/2019/09/contributing-data-to-deepfake-detection.html
11. Fagni, T., Falchi, F., Gambini, M., Martella, A., Tesconi, M.: TweepFake: about detecting deepfake tweets. PLoS ONE **16**(5), e0251415 (2021)
12. Foret, P., Kleiner, A., Mobahi, H., Neyshabur, B.: Sharpness-aware minimization for efficiently improving generalization. In: 9th International Conference on Learning Representations, ICLR 2021, Virtual Event, Austria, 3–7 May 2021 (2021)
13. Giudice, O., Guarnera, L., Battiato, S.: Fighting deepfakes by detecting GAN DCT anomalies. J. Imaging **7**(8), 128 (2021)
14. Goodfellow, I.J., et al.: Generative adversarial networks. In: Advances in Neural Information Processing Systems, vol. 27 (2014)
15. Guarnera, L., Giudice, O., Battiato, S.: Deepfake detection by analyzing convolutional traces. In: Proceedings of the IEEE/CVF Conference on Computer Vision and Pattern Recognition Workshops (2020)
16. Heo, Y.J., Choi, Y.J., Lee, Y.W., Kim, B.G.: Deepfake detection scheme based on vision transformer and distillation. arXiv preprint arXiv:2104.01353 (2021)
17. Jiang, L., Li, R., Wu, W., Qian, C., Loy, C.C.: DeeperForensics-1.0: a large-scale dataset for real-world face forgery detection. In: Proceedings of the IEEE/CVF Conference on Computer Vision and Pattern Recognition, pp. 2889–2898 (2020)
18. Karras, T., Laine, S., Aittala, M., Hellsten, J., Lehtinen, J., Aila, T.: Analyzing and improving the image quality of styleGAN. In: Proceedings of the IEEE/CVF Conference on Computer Vision and Pattern Recognition, pp. 8110–8119 (2020)
19. Khan, S., Naseer, M., Hayat, M., Zamir, S.W., Khan, F.S., Shah, M.: Transformers in vision: a survey. ACM Comput. Surv. (CSUR) (2021). https://doi.org/10.1145/3505244
20. Kim, T., Cha, M., Kim, H., Lee, J.K., Kim, J.: Learning to discover cross-domain relations with generative adversarial networks. In: International Conference on Machine Learning, pp. 1857–1865. PMLR (2017)
21. Kingma, D.P., Welling, M.: Auto-encoding variational Bayes. In: 2nd International Conference on Learning Representations, ICLR 2014, Banff, AB, Canada, 14–16 April 2014, Conference Track Proceedings (2014)
22. Korshunov, P., Marcel, S.: Deepfakes: a new threat to face recognition? assessment and detection. arXiv preprint arXiv:1812.08685 (2018)
23. Li, Y., Yang, X., Sun, P., Qi, H., Lyu, S.: Celeb-DF: a large-scale challenging dataset for deepfake forensics. In: Proceedings of the IEEE/CVF Conference on Computer Vision and Pattern Recognition, pp. 3207–3216 (2020)
24. de Lima, O., Franklin, S., Basu, S., Karwoski, B., George, A.: Deepfake detection using spatiotemporal convolutional networks. arXiv preprint arXiv:2006.14749 (2020)
25. MacAvaney, S., Nardini, F.M., Perego, R., Tonellotto, N., Goharian, N., Frieder, O.: Efficient document re-ranking for transformers by precomputing term representations. In: Proceedings of the 43rd International ACM SIGIR Conference on Research and Development in Information Retrieval, pp. 49–58 (2020)
26. Messina, N., Amato, G., Carrara, F., Falchi, F., Gennaro, C.: Testing deep neural networks on the same-different task. In: 2019 International Conference on Content-Based Multimedia Indexing (CBMI), pp. 1–6. IEEE (2019)

27. Messina, N., Amato, G., Carrara, F., Gennaro, C., Falchi, F.: Solving the same-different task with convolutional neural networks. Pattern Recognit. Lett. **143**, 75–80 (2021)
28. Messina, N., Amato, G., Esuli, A., Falchi, F., Gennaro, C., Marchand-Maillet, S.: Fine-grained visual textual alignment for cross-modal retrieval using transformer encoders. ACM Trans. Multimedia Comput. Commun. Appl. (TOMM) **17**(4), 1–23 (2021)
29. Messina, N., Falchi, F., Esuli, A., Amato, G.: Transformer reasoning network for image-text matching and retrieval. In: 2020 25th International Conference on Pattern Recognition (ICPR), pp. 5222–5229. IEEE (2021)
30. Montserrat, D.M., et al.: Deepfakes detection with automatic face weighting. In: Proceedings of the IEEE/CVF Conference on Computer Vision and Pattern Recognition Workshops, pp. 668–669 (2020)
31. Perov, I., et al.: DeepFaceLab: A simple, flexible and extensible face swapping framework. arXiv preprint arXiv:2005.05535 (2020)
32. Ren, S., He, K., Girshick, R., Sun, J.: Faster R-CNN: towards real-time object detection with region proposal networks. Adv. Neural Inf. Process. Syst. **28**, 91–99 (2015)
33. Rossler, A., Cozzolino, D., Verdoliva, L., Riess, C., Thies, J., Nießner, M.: Face-Forensics++: learning to detect manipulated facial images. In: Proceedings of the IEEE/CVF International Conference on Computer Vision, pp. 1–11 (2019)
34. Seferbekov, S.: DFDC 1st place solution (2020). https://github.com/selimsef/dfdc_deepfake_challenge
35. Tolosana, R., Vera-Rodriguez, R., Fierrez, J., Morales, A., Ortega-Garcia, J.: Deepfakes and beyond: a survey of face manipulation and fake detection. Inf. Fusion **64**, 131–148 (2020)
36. Wodajo, D., Atnafu, S.: Deepfake video detection using convolutional vision transformer. arXiv preprint arXiv:2102.11126 (2021)
37. Xie, S., Girshick, R., Dollár, P., Tu, Z., He, K.: Aggregated residual transformations for deep neural networks. In: Proceedings of the IEEE Conference on Computer Vision and Pattern Recognition, pp. 1492–1500 (2017)
38. Yang, X., Li, Y., Lyu, S.: Exposing deep fakes using inconsistent head poses. In: ICASSP 2019 - 2019 IEEE International Conference on Acoustics, Speech and Signal Processing (ICASSP), pp. 8261–8265. IEEE (2019)
39. Zhang, K., Zhang, Z., Li, Z., Qiao, Y.: Joint face detection and alignment using multitask cascaded convolutional networks. IEEE Signal Process. Lett. **23**(10), 1499–1503 (2016)

Human Action Recognition with Transformers

Pier Luigi Mazzeo[2(✉)], Paolo Spagnolo[1,2], Matteo Fasano[1], and Cosimo Distante[2]

[1] Università del Salento, Via Monteroni sn, 73100 Lecce, Italy
[2] ISASI - CNR c/o DHITECH, Via Monteroni sn, 73100 Lecce, Italy
pierluigi.mazzeo@cnr.it

Abstract. Having a reliable tool to predict the actions performed in a video can be very useful for intelligent security systems, for many applications related to robotics and for limiting human interactions with the system. In this work we present an architecture trained to predict the action present in digital video sequences. The proposed architecture consists of two main blocks: (i) a 3D backbone that extracts features from each frame of video sequence and (ii) a temporal pooling. In this case, we use Bidirectional Encoder Representations from Transformers (BERT) as technology for temporal pooling instead of a Temporal Global Average Pooling (TGAP). The output of the proposed architecture is the prediction of the action taking place in the video sequence. We use two different backbones, ip-CSN and ir-CSN, in order to evaluate the performance of the entire architecture on two publicly available datasets: IIMDB-51, UCF-101. A comparison has been made with the most important architectures that constitute the state of the art for this task. We have obtained results that outperform the state of the art in terms of Top-1 and Top-3 accuracy.

Keywords: Human Action Recognition · BERT · 3D-CNN

1 Introduction

Video surveillance has received a lot of attention from the computer vision community in last years. The increasing demand for safety and security has resulted in more research in intelligent surveillance. It has a wide range of applications, such as observing people in large waiting rooms, shopping centers, hospitals, eldercare, home-nursing, campuses or monitoring vehicles inside/outside cities, on highways, bridges, in tunnels etc. There is also an increasing desire and need in video surveillance applications to be able to analyze human behaviors. Behavior analysis involves the analysis and the recognition of motion patterns to produce a high-level description of actions and interactions among objects. All this is designed with the hope of having more and more automated devices and structures that do not require human supervision.

The aim of this work is to study, design and develop an architecture that predicts the type of action that is seen into a video clip (HAR task).

S. Sclaroff et al. (Eds.): ICIAP 2022, LNCS 13233, pp. 230–241, 2022.
https://doi.org/10.1007/978-3-031-06433-3_20

The main contribution of this paper is to propose a novel architecture that combines a channel-separated 3D CNN backbone with Transformers methodology. In particular we have tested as transformers a BERT-based block. BERT (Bidirectional Encoder Representations from Transformers) was introduced for Natural Language Processing (NLP) applications [3]. It has been applied in several kinds of applications, but only a few works have tried to solve the issues of HAR by using BERT [12,18]. To the best of our knowledge, our work is the first to introduce channel separation in a BERT-based architecture for Human Action Recognition, with the goal of taking advantage of both spatial and spectral contents coming from video sequences. As a bridge among the two blocks we have introduced a new block which reduces the vector dimensions to be computed by the BERT-based block. The proposed architecture has been evaluated on two public available datasets (HMDB-51 [14] and UCF-101 [22]) and the obtained results demonstrated that it outperforms the referred State of the Art performance. The remainder of this paper is organized as follows. Section 2 contains a review of the state of the art. Section 3 describes the proposed architecture. Section 5 discusses extensive experimental results. Finally, Sect. 6 reports conclusions.

2 Related Works

Among different applications in computer vision, Human Action Recognition (HAR) has become one of the most attractive research fields, as the growing number of published papers testify. ConvNet Architecture for spatio-temporal Feature Learning are introduces in [23] for 3D feature extractors. It uses 3D convolutions on video frames. In [21] two-Stream CNN for Action Recognition in Videos based on two separate recognition streams (spatial and temporal) and later combined is used. The spatial stream performs action recognition from fixed video frames, instead, the temporal stream is trained to recognize action from motion in the form of dense optical flow. Beyond Short Snippets - Deep Networks for Video Classification [30] introduces feature pooling method that processes each frame independently and uses max-pooling on local information to combine frame-level information. In this case feature pooling is performed with a Long Short-Term Memory (LSTM) that is connected to the output of the underlying CNN. Temporal Segment Networks proposed in [28] improves the two stream architecture. Temporal 3D ConvNets [4] introduces a new temporal layer that models variable temporal convolution kernel depths. This temporal layer is added in the proposed 3D CNN (i.e. DenseNet). Attentional Pooling for Action Recognition [9] that proposes a modification to the networks by extending the existing architectures with attention maps that focus computation on specific parts of the input. Attention Clusters [16] introduced the idea is to focus on local features instead of trying to capture global features using attention clusters mechanisms. Video Action Transformer Network is introduced in [8] and introduces a transformer-based architecture for classifying actions. The attention model learns to focus on hands and faces which is often crucial in differentiating between actions. Late Temporal Modeling in 3D CNN Architectures with BERT that is the first approach that uses BERT as temporal pooling method has been introduced in [13], in order to better utilize the temporal information with BERT's attention mechanism. BERT is inserted after a traditional 3D CNN backbone.

3 Proposed Methodology

One of the main challenge in Human Action Recognition is to find the best combination among the 3D-CNN backbones and the temporal pooling strategies. The basic idea of this work is the integration of channel-separated convolutional network within a BERT-based framework. In Fig. 1 the proposed architecture is showed; it is composed of two main blocks: a 3D-CNN backbone block and a BERT-based temporal pooling block. A selected k video frames coming from an input sequence are propagated through the 3D CNN backbone; despite traditional approaches in the field of Human Action Recognition, in this approach we do not execute a temporal global average pooling. The features extracted are, then, passed to a temporal pooling block; in the proposed architecture we used BERT [3]. In order to correctly maintain the position information, a positional encoding is added to the extracted features. This way in order to perform action classification with BERT, an additional classification embedding x_{cls} is appended. The classification vector y_{cls} is computed by the fully connected layer (see Fig. 1), producing the predicted output label ŷ. Considering the single head self-attention model of BERT, the explicit form of y_{cls} can be written as:

$$y_{cls} = PFNN\left(\frac{1}{N(x)}\sum_{\forall j} g(x_j) f(x_{cls}, x_j)\right) \tag{1}$$

where: x_i values are the embedding vectors that consists of extracted temporal visual information and its positional encoding; i indicates the index of the target output temporal position; j denotes all possible combinations; and $N(x)$ is the normalization term. Function $g(\cdot)$ is the linear projection inside the self-attention mechanism of BERT, whereas function $f(\cdot,\cdot)$ denotes the similarity between x_{cls} and x_j: $f(x_{cls}, x_j) = softmax_j(\Theta(x_{cls})^T \Phi(x_j))$, where the functions $\Theta(\cdot)$ and $\Phi(\cdot)$ are also linear projections. The learnable functions $g(\cdot)$, $\Theta(\cdot)$ and $\Phi(\cdot)$ try to project the feature embedding vectors to a better space where the attention mechanism works more efficiently. $PFFN(\cdot)$ is Position-wise Feed-forward Network applied to all positions separately and identically: $PFFN(x) = W_2 GELU(W_1 x + b_1) + b_2$, where $GELU(\cdot)$ is the Gaussian Error Linear Unit (GELU) [11] activation function.

We use as backbone a Channel-separated convolutional networks (CSN) architecture (see Subsect. 3.1) which take as input a 224×224 tensor [24], and 64-frame length video sequence. The CSN backbone has 152 layers and produces an output features vector of 2048 elements. An additional block is added to adapt this 2048 vector to be an input of the BERT temporal pooling block. This new block simply reduces the output dimension from 2048 to 512. To evaluate the architecture performance we consider Top-N Accuracy. Top N accuracy is the standard accuracy of the true class being equal to any of the N most probable classes predicted by the classification model.

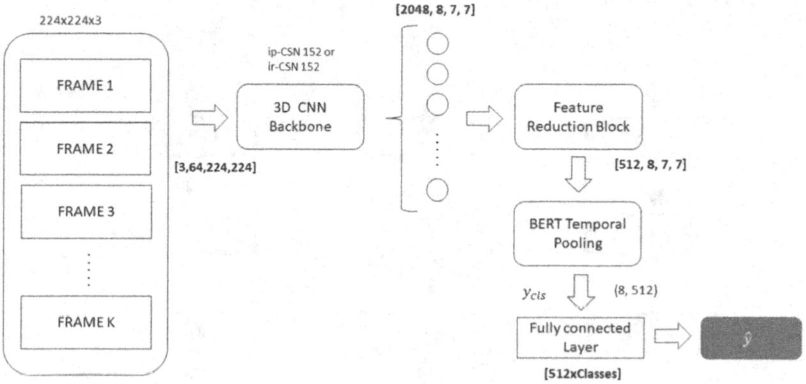

Fig. 1. Proposed architecture: K video frames are used as input of 3D CNN backbone. The tensor notation is [C, K, W, H] where C represents the number of channels, K input frames number, W, H the width and height, dimensions, respectively. A Feature reduction block reduces the features from 2048 to 512. The BERT output tensors dimension (8, 512) is given from the 8 attention heads, 512 is the number of BERT hidden layers. At the end of the architecture there is a fully connected layer that infers the predicted output label.

3.1 Channel Separated Convolutional Network

Channel-separated convolutional networks (CSN) [24] are defined as 3D CNNs in which all convolutional layers (except for conv1) are composed by $1 \times 1 \times 1$ conventional convolutions or $k \times k \times k$ depthwise convolutions (where, typically, $k = 3$). Traditional CNNs model channel interactions and local interactions (i.e., spatial or spatiotemporal) jointly in their 3D convolutions. Instead, channel-separated networks decompose these two types of interactions into two distinct layers: $1 \times 1 \times 1$ conventional convolutions for channel interaction (but no local interaction) and $k \times k \times k$ depthwise convolutions for local spatiotemporal interactions (but not channel interaction). So there are two ways to factorize a 3D bottleneck block using channel-separated convolutional networks: interaction-preserved and interaction-reduced channel-separated bottleneck blocks. This two type of block are described in Fig. 2. Interaction-preserved channel-separated bottleneck block (b) is obtained from the standard bottleneck block by replacing the $3 \times 3 \times 3$ convolution with a $1 \times 1 \times 1$ traditional convolution and a $3 \times 3 \times 3$ depthwise convolution. This block allows to reduce parameters and FLOPs of the traditional convolution significantly, but preserves all channel interactions via the addition of $1 \times 1 \times 1$ convolution. The resulting architecture is interaction-preserved channel-separated network (ip-CSN).

Interaction-reduced channel-separated bottleneck block (c) is derived from the preserved bottleneck block by removing the extra $1 \times 1 \times 1$ convolution. This implies that the complete block has a reduced number of channel interactions. This design is called interaction-reduced channel-separated bottleneck block and the resulting architecture interaction-reduced channel-separated network (ir-CSN).

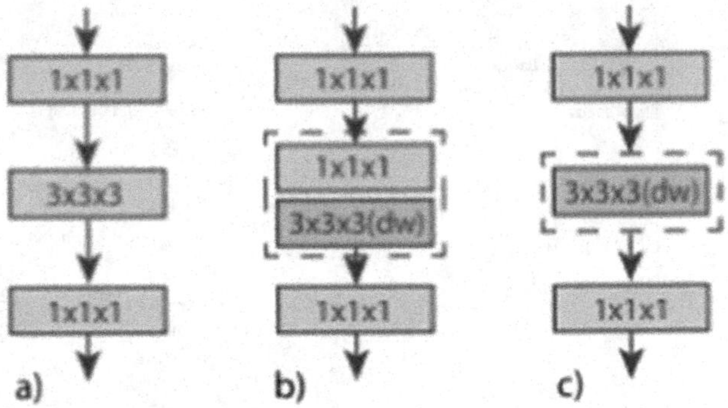

Fig. 2. Channel-separated convolutional network (CSN) block [24]

4 Datasets

4.1 HMDB-51

HMDB [14] is composed of videos from various sources, mostly from movies, and a small proportion from public databases such as the Prelinger archive, YouTube and Google videos. The dataset contains 6849 clips divided into 51 action categories, each containing a minimum of 101 clips. The authors publish also a split of the dataset that allows to divide all videos in three split. The actions categories can be grouped in five types:

- General facial actions: smile, laugh, chew, talk.
- Facial actions with object manipulation: smoke, eat, drink.
- General body movements: cartwheel, clap hands, climb, climb stairs, dive, fall on the floor, backhand flip, handstand, jump, pull up, push up, run, sit down, sit up, somersault, stand up, turn, walk, wave.
- Body movements with object interaction: brush hair, catch, draw sword, dribble, golf, hit something, kick ball, pick, pour, push something, ride bike, ride horse, shoot ball, shoot bow, shoot gun, swing baseball bat, sword exercise, throw.
- Body movements for human interaction: fencing, hug, kick someone, kiss, punch, shake hands, sword fight.

Some of the images in the dataset are shown in Fig. 3.

Fig. 3. Some HMDB 51 actions [14]

4.2 UCF-101

UCF-101 [22] is an action recognition dataset of realistic action videos, collected from YouTube, having 101 action categories. This data set is an extension of UCF-50 data set which has 50 action categories.

With 13320 videos from 101 action categories, UCF-101 gives the largest diversity in terms of actions and with the presence of large variations in camera motion, object appearance and pose, object scale, viewpoint, cluttered background, illumination conditions, etc.,. The videos in 101 action categories are grouped into 25 groups, where each group can consist of 4–7 videos of an action. The videos from the same group may share some common features, such as similar background, similar viewpoint, etc.

The action categories can be divided into five types:

- Human-Object Interaction
- Body-Motion Only
- Human-Human Interaction
- Playing Musical Instruments
- Sports

UCF-101 classes are reported in Fig. 4.

Fig. 4. UCF 101 actions [22]

4.3 IG-65M

IG-65M is firstly used in [7] where the authors leverage over 65 million public, user-generated videos from a social media website and use the associated hashtags as labels for pre-training. The label noise and temporal noise makes their training framework weakly-supervised. Unlike all existing fully-supervised video datasets which required expensive annotation, their training data is truly extensible to billion-scale without incurring any annotation costs. They demonstrate significant performance gains on various target tasks.

To construct pre-training video datasets, they use several seed action label sets and gather videos that are related to these labels. It is collected by using the Kinetics-400 class names as hashtags on Instagram. Specifically, for a given seed action label "catching a fish," they first construct all possible meaningful phrases (i.e., relevant hashtags) from it. An example, relevant hashtags ("catching a fish") = #catchingafish, #catchfish, #fishcatching, ... They then download public videos that are tagged with at least one of the hashtags from the set of relevant hashtags ("catching a fish") and associate them with the initial seed label. They use the seed labels as the final labels for videos during pre-training.

This is the most recent dataset between the presented, but it achieves very important result in architecture like R(2+1)D and CSN.

The dataset is not public for the time being but pre-trained models are available.

5 Experimental Results

We tested the proposed architecture on two public available datasets, namely HMDB-51 [14] and UCF-101 [22], see Sect. 4 for more details. We, also, used another dataset in the pretrained models as backbones, this dataset is known as IG65M [7]. All the experiments presented in this work only support RGB modality. As optimizer and learning rate we use SGD and $1e^{-2}$, respectively, when we do not include BERT. Instead when we use BERT temporal pooling we use ADAMW [17] as optimizer with a learning rate $1e^{-5}$. In order to normalize the input data we use: multi scale crop, random horizontal flip and normalization for training set, center crop and normalization for validation set. For all the training sessions we reduce the learning rate "on plateau", it means that if no improvements are seen for a 'patience' number of epochs, the learning rate is reduced. We set the 'patience' to 5. BERT is composed by eight attention heads and one transfomer block. In PFFN() (Eq. 1) the dropout ratio is set to 0.9 and the mask operation is applied with a probability of 0.2. Mask token is not used. In order to replace it, the attention weight of masked feature is set to zero. The learned positional embeddings are initialized as zero mean normal weight with 0.02 standard deviation. The classification token (x_{cls}) starts with all zero.

Firstly we use interaction-preserved version of CSN as backbone. It is composed by 152 layers. In this case training phase is performed using $1e^{-5}$ learning rate and number of epochs is set to 40. So to make the results clearer, all models in this experiments are tested on HMDB-51 and on UCF-101. In the ablation study (see Subsect. 5.1), models are tested without BERT temporal pooling in order to highlight the real benefits of the proposed approach.

CSN is also tested with its 'interaction reduced' version. In this architecture the complete block has a reduced number of channel interactions compared with those of ip-CSN. Also in this case input dimensions are applied following the paper [24] and 64-frame length is applied. The basic architecture is a simple ResNet with 152 layers. This backbone is pre-trained on IG-65M dataset. In this case training phase is performed using $1e^{-5}$ learning rate and 4 as batch size. Number of epochs is set to 60. All experiments are carried out on Pytorch, with a NVIDIA Device Geforce TITAN RTX 24 GB.

All results are summarized in Table 1 where is reported the 3-Fold accuracy obtained from testing the architectures. In particular we can see that very good results are obtained in both version on HMDB-51 and comparable with the SOTA on UCF-101.

5.1 Ablation Study

In this subsection, architectures without BERT were tested, in order to see the real benefits that it can lead. In Table 2 the results obtained using ip-CSN as

Table 1. CSN 152 architectures performance

Gray ip-CSN 152	Top-1	Top-3
HMDB-51	**0,8530**	0,9555
UCF-101	0,9851	0,9977
Gray ir-CSN 152	Top-1	Top-3
HMDB-51	**0,8517**	0,951
UCF-101	0,9832	0,9977

backbone on HMDB-51 dataset are represented. It can see how by exploiting BERT as temporal pooling, the overall performance increases by about 2%.

Furthermore, Table 2 contains a performance comparison of the proposed architecture with and without BERT as temporal pooling, on the UCF-101 dataset. It can be noticed that the use of BERT as temporal pooling improves the performance of about 0.20%.

Ablation study confirms that using BERT as temporal pooling improves the overall performance of the proposed architecture.

Table 2. ip-CSN 152 performance on split 1 with and without BERT

Dataset	BERT	Top-1	Top-3
HMDB-51	No	0.8379	0.9451
	Yes	0.8582	0.9569
UCF-101	No	0.9800	0.9944
	Yes	0.9819	0.9963

5.2 Performance Comparison

In this subsection, we compare the results obtained from the proposed architecture against the up-to-date state-of-the-art methodologies. Table 3 resumes obtained results, first column lists the methodology, second one indicates if it uses optical flow or not, third shows the dataset used for pre-trained backbones, and finally forth and fifth ones contain the Top-1 accuracy obtained, respectively, on HMDB-51 and UCF-101 datasets.

Table 3 shows the best result which is obtained with the proposed architecture using a pre-trained backbone on IG-65M, and a combination of ip-CSN 152 with BERT on HMDB51 dataset (bolded value). The same combination obtains the second best results on UCF-101 dataset. This might be owing to the fact that HMDB51 involves some actions that can be resolved only using temporal reasoning and the benefits that came from BERT temporal pooling. Also ir-CSN 152 works well on HMDB51 dataset. Furthermore, from Table 3 it can be noticed the effect of using of backbone pre-training with the IG-65M dataset. This demonstrates the importance of having a larger dataset to train the backbone with,

Table 3. Performance comparison against the State-of-art

Model	Flow	Extra training data	HMDB51	UCF101
IDT [26]	Yes		61.70	–
Two-Stream [21]	Yes	ImageNet	59.40	88.00
Two-streamFusion+IDT [6]	Yes	ImageNet	69.20	93.50
ActionVlad+IDT [10]	Yes	ImageNet	69.80	93.60
TSN [29]	Yes	ImageNet	71.00	94.90
RSTAN+IDT [5]	Yes	ImageNet	79.90	95.10
TSM [15]	No	Kinetics-400	73.50	95.90
R(2+1)D [25]	No	Kinetics-400	74.50	96.80
R(2+1)D [25]	Yes	Kinetics-400	78.70	97.30
I3D [1]	Yes	Kinetics-400	80.90	97.80
MARS+RGB+Flow [2]	Yes	Kinetics-400	80.90	98.10
FcF [20]	No	Kinetics-400	81.10	–
ResNeXt101	Yes	Kinetics-400	81.78	97.46
EvaNet [19]	Yes	Kinetics-400	82.3	–
HAF+BoW/FV halluc [27]	No	Kinetics-400	82.48	–
ResNeXt101 BERT [13]	Yes	Kinetics-400	83.55	97.87
R(2+1)D (32f) [13]	No	IG65M	80.54	98.17
R(2+1)D BERT (32f) [13]	No	IG65M	83.99	**98.65**
R(2+1)D BERT (64f) [13]	No	IG65M	85.10	**98.69**
ip-CSN 152	No	IG65M	83.79	98.00
ir-CSN 152	No	IG65M	83.46	97.82
ip-CSN 152 BERT (proposed)	No	IG65M	**85.30**	98.51
ir-CSN 152 BERT (proposed)	No	IG65M	**85.17**	98.32

even if it was collected in a weakly-supervised manner. Another consideration coming from the observation of Table 3 is given from the combination of ip-CSN and ir-CSN with BERT as temporal pooling that leads a performance increasing of about 2%. Anyway, ip-CSN 152 combined with BERT outperforms the state of the art for HMDB-51 dataset, and reaches it for the UCF-101 dataset.

6 Conclusions

This work integrated a Transformers architecture, a new technology used in NLP tasks, into Human Action Recognition (HAR) task. The goal of the task is to predict the action that someone is performing within a video sequence. A novel architecture based on the joint use of a 3D convolutional neural network and BERT for an action recognition task is proposed. BERT can be seen as a new way to perform temporal pooling, different from other methods already used such as LSTM, average pooling and concatenation pooling. Different backbones are tested such as ip-CSN and ir-CSN, comparing performances achieved with the state of the art over two public available datasets: HMDB-51 and UCF-101. Obtained results outperforms the state of the art on HMDB-51 dataset and reaches it on UCF-101. Future work will be addressed to test other 3D architectures as backbones and evaluate their performance. BERT could be replaced with more recent technologies such as RoBERTa or with new technologies that involve the use of linformer and performer rather than transformer (it should allow to reduce the complexity of self-attention from quadratic to linear in both time and space).

References

1. Carreira, J., Zisserman, A.: Quo vadis, action recognition? a new model and the kinetics dataset. In: Proceedings of the IEEE Conference on Computer Vision and Pattern Recognition, pp. 6299–6308 (2017)
2. Crasto, N., Weinzaepfel, P., Alahari, K., Schmid, C.: Mars: motion-augmented RGB stream for action recognition. In: Proceedings of the IEEE Conference On Computer Vision Aand Pattern Recognition, pp. 7882–7891 (2019)
3. Devlin, J., Chang, M.W., Lee, K., Toutanova, K.: Bert: pre-training of deep bidirectional transformers for language understanding. arXiv preprint arXiv:1810.04805 (2018)
4. Diba, A., et al.: Temporal 3D convnets: new architecture and transfer learning for video classification. arXiv preprint arXiv:1711.08200 (2017)
5. Du, W., Wang, Y., Qiao, Y.: Recurrent spatial-temporal attention network for action recognition in videos. IEEE Trans. Image Process. **27**(3), 1347–1360 (2017)
6. Feichtenhofer, C., Pinz, A., Zisserman, A.: Convolutional two-stream network fusion for video action recognition. In: Proceedings of the IEEE Conference on Computer Vision and Pattern Recognition, pp. 1933–1941 (2016)
7. Ghadiyaram, D., Tran, D., Mahajan, D.: Large-scale weakly-supervised pre-training for video action recognition. In: Proceedings of the IEEE Conference on Computer Vision and Pattern Recognition, pp. 12046–12055 (2019)
8. Girdhar, R., Carreira, J., Doersch, C., Zisserman, A.: Video action transformer network. In: Proceedings of the IEEE Conference on Computer Vision and Pattern Recognition, pp. 244–253 (2019)
9. Girdhar, R., Ramanan, D.: Attentional pooling for action recognition. In: Advances in Neural Information Processing Systems, pp. 34–45 (2017)
10. Girdhar, R., Ramanan, D., Gupta, A., Sivic, J., Russell, B.: Actionvlad: learning spatio-temporal aggregation for action classification. In: Proceedings of the IEEE Conference on Computer Vision and Pattern Recognition, pp. 971–980 (2017)
11. Hendrycks, D., Gimpel, K.: Gaussian error linear units (gelus). arXiv preprint arXiv:1606.08415 (2016)
12. Hira, S., Das, R., Modi, A., Pakhomov, D.: Delta sampling R-bert for limited data and low-light action recognition. In: Proceedings of the IEEE/CVF Conference on Computer Vision and Pattern Recognition (CVPR) Workshops, pp. 853–862 (2021)
13. Kalfaoglu, M., Kalkan, S., Alatan, A.A.: Late temporal modeling in 3D CNN architectures with bert for action recognition. arXiv preprint arXiv:2008.01232 (2020)

14. Kuehne, H., Jhuang, H., Garrote, E., Poggio, T., Serre, T.: HMDB: a large video database for human motion recognition. In: Proceedings of the International Conference on Computer Vision (ICCV) (2011)
15. Lin, J., Gan, C., Han, S.: TSM: temporal shift module for efficient video understanding. In: Proceedings of the IEEE International Conference on Computer Vision, pp. 7083–7093 (2019)
16. Long, X., Gan, C., De Melo, G., Wu, J., Liu, X., Wen, S.: Attention clusters: purely attention based local feature integration for video classification. In: Proceedings of the IEEE Conference on Computer Vision and Pattern Recognition, pp. 7834–7843 (2018)
17. Loshchilov, I., Hutter, F.: Fixing weight decay regularization in Adam (2018)
18. Nagrani, A., Sun, C., Ross, D., Sukthankar, R., Schmid, C., Zisserman, A.: Speech2action: cross-modal supervision for action recognition. In: Proceedings of the IEEE/CVF Conference on Computer Vision and Pattern Recognition, pp. 10317–10326 (2020)
19. Piergiovanni, A., Angelova, A., Toshev, A., Ryoo, M.S.: Evolving space-time neural architectures for videos. In: Proceedings of the IEEE International Conference on Computer Vision, pp. 1793–1802 (2019)
20. Piergiovanni, A., Ryoo, M.S.: Representation flow for action recognition. In: Proceedings of the IEEE Conference on Computer Vision and Pattern Recognition, pp. 9945–9953 (2019)
21. Simonyan, K., Zisserman, A.: Two-stream convolutional networks for action recognition in videos. In: Advances in neural information processing systems, pp. 568–576 (2014)
22. Soomro, K., Zamir, A.R., Shah, M.: Ucf101: a dataset of 101 human actions classes from videos in the wild. arXiv preprint arXiv:1212.0402 (2012)
23. Tran, D., Ray, J., Shou, Z., Chang, S.F., Paluri, M.: Convnet architecture search for spatiotemporal feature learning. arXiv preprint arXiv:1708.05038 (2017)
24. Tran, D., Wang, H., Torresani, L., Feiszli, M.: Video classification with channel-separated convolutional networks. In: Proceedings of the IEEE International Conference on Computer Vision, pp. 5552–5561 (2019)
25. Tran, D., Wang, H., Torresani, L., Ray, J., LeCun, Y., Paluri, M.: A closer look at spatiotemporal convolutions for action recognition. In: Proceedings of the IEEE Conference on Computer Vision and Pattern Recognition, pp. 6450–6459 (2018)
26. Wang, H., Schmid, C.: Action recognition with improved trajectories. In: Proceedings of the IEEE International Conference on Computer Vision, pp. 3551–3558 (2013)
27. Wang, L., Koniusz, P., Huynh, D.Q.: Hallucinating IDT descriptors and I3D optical flow features for action recognition with CNNS. In: Proceedings of the IEEE International Conference on Computer Vision, pp. 8698–8708 (2019)
28. Wang, L., et al.: Temporal segment networks: towards good practices for deep action recognition. In: Leibe, B., Matas, J., Sebe, N., Welling, M. (eds.) ECCV 2016. LNCS, vol. 9912, pp. 20–36. Springer, Cham (2016). https://doi.org/10.1007/978-3-319-46484-8_2
29. Wang, L., et al.: Temporal segment networks for action recognition in videos. IEEE Trans. Pattern Anal. Mach. Intell. **41**(11), 2740–2755 (2018)
30. Yue-Hei Ng, J., Hausknecht, M., Vijayanarasimhan, S., Vinyals, O., Monga, R., Toderici, G.: Beyond short snippets: deep networks for video classification. In: Proceedings of the IEEE Conference on Computer Vision and Pattern Recognition, pp. 4694–4702 (2015)

Decontextualized I3D ConvNet for Ultra-Distance Runners Performance Analysis at a Glance

David Freire-Obregón(✉), Javier Lorenzo-Navarro, and Modesto Castrillón-Santana

SIANI, Universidad de Las Palmas de Gran Canaria, Las Palmas de Gran Canaria, Spain
david.freire@ulpgc.es

Abstract. In May 2021, the site runnersworld.com published that participation in ultra-distance races has increased by 1,676% in the last 23 years. Moreover, nearly 41% of those runners participate in more than one race per year. The development of wearable devices has undoubtedly contributed to motivating participants by providing performance measures in real-time. However, we believe there is room for improvement, particularly from the organizers point of view. This work aims to determine how the runners performance can be quantified and predicted by considering a non-invasive technique focusing on the ultra-running scenario. In this sense, participants are captured when they pass through a set of locations placed along the race track. Each footage is considered an input to an I3D ConvNet to extract the participant's running gait in our work. Furthermore, weather and illumination capture conditions or occlusions may affect these footages due to the race staff and other runners. To address this challenging task, we have tracked and codified the participant's running gait at some RPs and removed the context intending to ensure a runner-of-interest proper evaluation. The evaluation suggests that the features extracted by an I3D ConvNet provide enough information to estimate the participant's performance along the different race tracks.

Keywords: Sports · I3D ConvNet · Human action evaluation

1 Introduction

Our ability to evaluate an athlete's performance depends on the sporting context. For instance, in a soccer game scenario, the players gait/pose or precision

Supported by the ULPGC under project ULPGC2018-08, by the Spanish Ministry of Economy and Competitiveness (MINECO) under project RTI2018-093337-B-I00, by the Spanish Ministry of Science and Innovation under project PID2019-107228RB-I00, and by the Gobierno de Canarias and FEDER funds under project ProID2020010024.

S. Sclaroff et al. (Eds.): ICIAP 2022, LNCS 13233, pp. 242–253, 2022.
https://doi.org/10.1007/978-3-031-06433-3_21

when kicking the ball may provide valuable insights about the players condition. Similarly, the way an ultra-distance participant is running (i.e., gait, pace, and trajectory) may also provide some intuitions about the runner performance. However, despite numerous potential applications and existing wearable technologies, this ability remains a challenge for state-of-the-art visual recognition systems.

In contrast to the purpose of traditional human action recognition (HAR) to infer the label from predefined action categories [28], the aim of human action evaluation (HAE) is to automatically quantify how well people perform actions given a particular metric. HAE has been mainly exploited in applications for healthcare and rehabilitation [20], self-learning platforms for practicing professional skills, or sports [9]. In this regard, different approaches have been considered to tackle this problem in the sporting context, i.e., virtual reality [3] and wearable sensors [27]. The former main drawback is the indoor environment restriction. In contrast, the latter can be considered an invasive proposal requiring a device calibration and may lead to some privacy concerns.

In this work, we take a step towards the athlete performance evaluation by processing short video clips recorded at different recording points (RPs) of an ultra-running track as input. First, we introduce a performance classifier built on top of a pre-trained deep neural network that reports highly competitive results in HAR. As can be seen in Fig. 1, the classifier provides an output that represents the athlete performance. The performance is discretized by categorizing the runners RP qualification time into a set of categories (i.e., excellent, very good, and so on). We have conducted several experiments to predict the runner performance, considering the runner video clip as input. Moreover, we have evaluated our model to predict the current RP performance given a video clip and the following RP performance estimation given the current RP video input. Additionally, we have developed a thoughtful context analysis to show the relevance of the environment in the proposed pipeline.

This work aims at answering some interesting questions: Can the runner performance be estimated from his/her motion? Is the context relevant? To what extent can the context be removed? Does this simplification come with a cost? We have evaluated our model in a dataset collected to evaluate Re-ID methods in complex real-world scenarios. The dataset contains 214 ultra-distance runners captured at different RPs along the track. The achieved results are remarkable (up to an 83.7% of accuracy), and they have also provided interesting insights. The first one is that increasing the number of categories for quality assessment negatively affects, as expected, the classifier performance. Another insight is related to the importance of contextual information for the pre-trained I3D ConvNet and the limitations observed during the transfer learning.

The paper is organized into five sections. The next section discusses some related work. Section 3 describes the proposed pipeline. Section 4 reports the experimental setup and the experimental results. Finally, conclusions are drawn in Sect. 5.

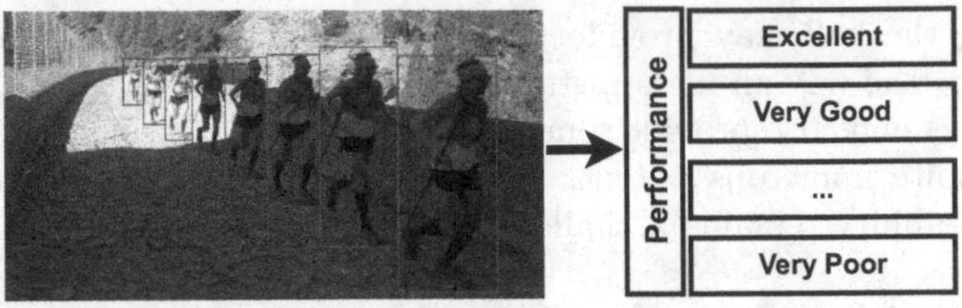

Fig. 1. Runner video sequence. We evaluate the runner performance with a model that efficiently classifies him/her motion into several categories in a single forward pass. Each category quantifies the runner's performance. The Figure shows frames of a clip present in the considered dataset [24].

2 Related Work

Contrary to traditional HAR systems [28,30], HAE systems [23] are designed not to identify a specific action but to capture the motion of the human body and to measure the completion quality of the captured motion through an evaluation technique. Both systems share common properties in terms of significant stages: First, a human detection algorithm locates the region-of-interest (ROI) in the image [29]. Next, a pose representation algorithm [4,11] is computed all over the ROI to extract a set of keypoints representing the human pose. In this regard, the skeleton-based representation approach has been widely used as a data source to solve the human pose representation [23]. Finally, a deep neural network [34] is trained on the extracted features representation for classification/regression purposes. This same primary approach has also been applied to some lower abstraction-level problems such as gait [8] or hand-action [25] recognition systems.

Athletes can measure their motion quality by judging their postures and movements through HAE techniques. The human pose representation plays a crucial role in evaluating the performed action. Lei et al. have identified three primary pose representations for the human action quality assessment [16]. The challenge relies on finding robust features from pose sequences and establishing a method to measure the similarity of pose features [31]. Second, the aforementioned skeleton-based representations encode the relationship between joints and body parts [18,23]. However, the estimated skeleton data can often be noisy in realistic scenes due to occlusions or changing lighting conditions [5]. Finally, the deep learning methods for assessing the athlete's action quality. In this representation approach, convolutional neural networks (CNN) can be combined with recurrent neural networks (RNN) due to the temporal dimension provided by the video input [21]. A typical network used for sports quality assessment is the 3D convolution network (C3D). This deep neural network that learns spatio-temporal features is increasingly being used for HAR [30]. Precisely, our work can be framed in the deep learning methods for assessing the athlete's action quality suggested by Lei et al. We make use of the Inflated 3D ConvNet (I3D), which has been used to tackle the HAR problem in the past [6,7]. This network

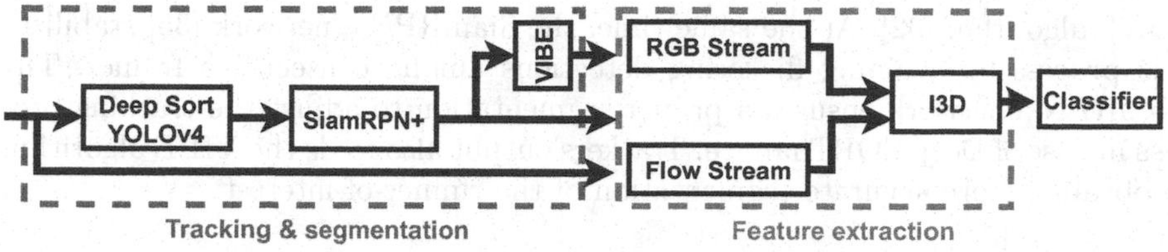

Fig. 2. The proposed pipeline for the performance quality assessment context-driven problem. The devised process comprises two major parts: the tracking and segmentation block and the features extraction block.

passes a two-stream input (RGB and flow) through a combination of 3D convolutions, Inception modules, and max-pooling layers. It uses asymmetrical filters for max-pooling, maintaining time while pooling over the spatial dimension.

Several sports datasets have been collected in the past few years. Most of them were collected from international competitions events. In this regard, some of the most notable datasets are MTL-AQA [22], UNLV AQA-7 [19] and Fis-V [33]. The sports collected in those datasets are usually practiced indoors or in a not-occlusive environment, i.e., diving, skating, skiing, snowboarding, and trampoline. Also, the sports exhibitions in those datasets take no longer than a few seconds or minutes. Our work considers an ultra-distance race collection where professional and non-professional runners compete in a 30 h race. There is a high set of variations in terms of lighting conditions, backgrounds, occlusions, and accessories due to the duration of the race. We strongly believe that our non-invasive quality assessment can provide relevant cues about the runner performance variation.

In summary, the work presented in this paper evaluates the race participant's performance considering an I3D network. The considered dataset provides a scenario in the wild. Further, we evaluate our pipeline, considering the raw video sequence as input and segmenting the runners to analyze the importance of context in I3D networks.

3 Runner Performance Pipeline

As can be seen in Fig. 2, this work proposes and evaluates a sequential pipeline divided into two major blocks. The first block is the subjects tracking and segmentation, and it provides the necessary information to locate and label the runner of interest in the scene. The output of this block can be divided into raw data, the runner bounding box (BB), and a fine-grained runner segmentation using the Video Inference for Human Body Pose and Shape Estimation (VIBE) [15] (see Fig. 3). VIBE is a video pose and shape estimation method. It predicts the parameters of SMPL body model [17] for each frame of an input video. Figure 2 also shows how the tracking and segmentation block feeds the features extraction block. In more detail, the subject is located and given the same id (the same id indicates the same subject across frames) by the Deep

SORT algorithm [32]. At the same time, the SiamRPN+ network [35] stabilizes this process by avoiding flickering detections among consecutive frames. The SiamRPN+ network ensures a proper segmentation to adjust the tracking process in case of Deep SORT failure. Trackers output also feeds the VIBE algorithm to obtain a more accurate segmentation of the runner-of-interest.

3.1 Runners Tracking and Segmentation

In the past few years, the Simple Online and Realtime Tracking (SORT) [2] has shown a remarkable performance in object tracking. Moreover, SORT with deep association metric (Deep SORT) [32] has been proposed for pedestrian detection as an extension of the SORT algorithm. Deep SORT aims to track people and correctly label the subjects in the scene. Recently, Deep SORT has reported stable tracking results in the sporting context [12]. Even though Deep SORT achieves overall good performance in tracking precision and accuracy, we have observed that illumination changes or partial occlusions can generate some tracking failures, such as detection flickering. In order to keep the Deep SORT label consistency, we have included a second tracker. The SiamRPN+ [35] plays a backup role for the Deep SORT algorithm. This neural network has been introduced as an evolution of SiamRPN. The SiamRPN+ benefits from a deeper backbone like ResNet, leading to a remarkable robustness [13].

Consequently, the previous described trackers provide a robust runner-of-interest bounding box (see Fig. 3, second column). Finally, VIBE makes use of the already detected bounding boxes to perform a runner fine-grain segmentation (see Fig. 3, third column). Then, several RNNs consider these features as input to process the sequential nature of human motion. Finally, a temporal encoder and regressor are used to predict the body parameters for the whole input sequence.

3.2 Runners Features Extraction

A few years ago, Carreira and Zisserman proposed the Inflated 3D Convnet (I3D) based on a two-stream network [6]. This deep neural network applies a two-stream structure for RGB and optical flow to the Inception-v1 [26] along with 3D CNNs. Nowadays, I3D is one of the most common feature extraction methods for video processing. The approach presented in this work exploits the pre-trained model on the Kinetics dataset as a backbone model [6]. Kinetics [14] is a large HAR dataset that includes a considerable number of action categories. Our proposal considered the backbone model trained on the Kinetics version of 400 action categories.

Consequently, the I3D acts as a feature extractor to encode the network input into a 400 vector feature representation that feeds the classifier. Here, the 400 output logits of the I3D are used as our classifier input. In this regard, five different classifiers were tested during the conducted experiments: Decision Tree, Random Forest, XGBoost, Linear SVM, and Logistic Regression. However, only the best classifier (XGBoost) results are reported in the next section.

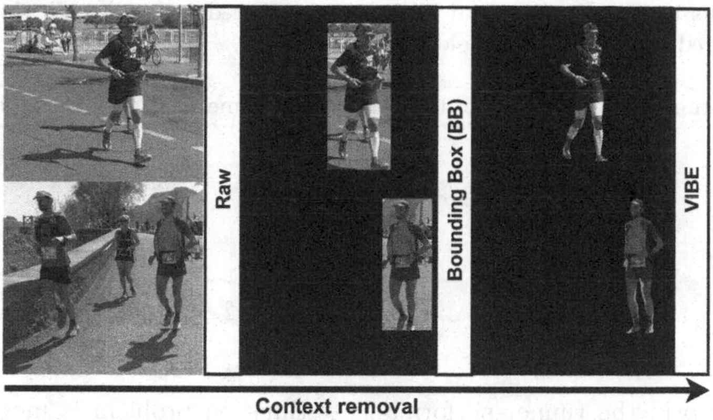

Fig. 3. Levels of context removal. We analyze the runner's performance with a model that efficiently classifies the quality assessment in a single forward pass and includes two types of context-removal processes. The first column shows original frames, while the second and third columns show frames where an increasing segmentation process is imposed.

4 Experimental Evaluation

4.1 Experimental Setup

Dataset. To evaluate below the ultra-runners performance variation, we have partially used the dataset published by Penate et al. [24]. The mentioned dataset was collected during an ultra-running competition, known as Transgrancanaria (TGC), held in March 2020. TGC comprises six running distances, but the annotated data covers just participants in the TGC Classic who must cover 128 km in 30 h, at most.

Although the TGC dataset contains annotations for almost 600 participants in six different RPs, just 214 of them were captured after km 84 with daylight, see Table 1. In our work, just the last three RPs are considered in the experiments below, when performance drops are more likely due to fatigue. Moreover, during the last RPs, the different performances among participants, the gap between leaders and last runners increases along the track. Therefore, the split time variance is higher, and performance can be analyzed more accurately. For each participant, seven seconds clips at 25 fps are fed to the tracking block described in Sect. 3.

Quality Assessment Metric. The quality of the observed movement is assessed considering the runner's RP split time. The used dataset contains runners footage at k different RPs where n^p are the runner's samples at the point p. Then, there is a footage (v_i^p) for each runner (i) that represents his/her passing through the RP (p), where t_i^p stands for the runner split time.

Table 1. RPs locations, times and kilometer (extracted from [24]). In this work we have considered the locations in bold type.

Location	Km	Start rec. time	Footage (frames)	# annot. runners
RP1	16.5	00:06	140,616	419
RP2	27.9	01:08	432,624	586
RP3	**84.2**	**07:50**	**667,872**	**203**
RP4	**110.5**	**10:20**	**1,001,208**	**139**
RP5	**124.5**	**11:20**	**1,462,056**	**114**

In this work, the runner-performance estimation problem is modeled as a classification problem rather than a regression problem. Therefore, each runner must be associated with a category. Let d be the function that maps the time of each runner into a category, then $c_i^p = d(t_i^p)$ represents the category associated with runner i at a given RP (p). As aforementioned, we are using a set of descriptors ($\mathbf{x}_i^p$) associated with each footage (see Sect. 3.2). Consequently, for each RP a dataset is defined as $D^p = \{(\mathbf{x}_i^p, c_i^p)\}$ and the complete dataset can be formulated as $D = D^1 \cup \cdots \cup D^k$.

We previously stated that this work aims to check whether it is possible to estimate a runner's performance or not by evaluating just a few seconds of footage samples from a running that may last from 13 h for the winner and 30 h for the last runner. Furthermore, this estimation can be computed at the same RP, or at a later one is given the footage-motion data (v_i^p). To achieve this task, our purpose is to find a function f such that $\hat{c}_i^{p+j} = f(\mathbf{x}_i^p)$ for $j = 0 \ldots k - p$.

The footage descriptors $\mathbf{x}_i^p$ are computed by the pre-trained I3D ConvNet. An equal-width discretization strategy is considered as a mapping function d between the split time and the different categories. For example, in Table 2, when two categories are considered, category one corresponds to those runners with a split time lower than the median, and category two corresponds to those with split time above the median. Similarly, for four categories, the runners are labeled as category one, two, three, and four as those whose split times are in the first, second, third, and fourth quartile, respectively. Finally, the runner-performance estimation is considered at the same RP ($\hat{c}_i^p = f(\mathbf{x}_i^p)$), and in the next RP ($\hat{c}_i^{p+1} = f(\mathbf{x}_i^p)$). The results presented in the next section refer to the average accuracy computed on 100 iterations. For each iteration, train and test data are chosen randomly, and the results are averaged after considering a stratified 4-fold cross-validation.

4.2 Experimental Results

We conducted a set of experiments to validate the effectiveness of the described proposal. These experiments took place through a grid search considering different classifiers (see Sect. 3.2). Only the best classifier (XGBoost) results are reported in this section, with a configuration of 200 estimators, a maximum tree

Table 2. Accuracy achieved for the current (Curr) and the next RP estimation. Each row considers a different number of categories. The last column shows the results considering a 3D convolution network instead of an I3D ConvNet.

RP-# categ.	Raw	BB	VIBE	C3D [30]	3DRes [10]
Curr-2	**83.7 ± 2.8%**	77.5 ± 2.8%	76.1 ± 3.1%	74.3 ± 4.5%	81.3 ± 2.9%
Curr-3	62.2 ± 3.4%	58.9 ± 4.1%	57.2 ± 3.8%	53.9 ± 3.8%	61.5 ± 3.6%
Curr-4	54.4 ± 3.6%	45.4 ± 3.6%	46.6 ± 3.8%	37.5 ± 2.9%	47.1 ± 3.7%
Next-2	**77.1 ± 2.4%**	73.5 ± 3.0%	73.9 ± 2.5%	72.1 ± 4.9%	75.5 ± 2.9%
Next-3	59.7 ± 3.3%	55.3 ± 3.8%	53.8 ± 3.5%	52.5 ± 3.4%	57.7 ± 2.4%
Next-4	49.1 ± 3.5%	42.5 ± 3.1%	40.0 ± 3.6%	36.5 ± 2.7%	46.3 ± 3.8%

depth of seven, and cross-entropy loss. The most relevant parameters were the number of categories, the I3D ConvNet input, and the classifier configuration.

As we argued in Sect. 4.1, the number of categories can be fixed, providing different perspectives over the data. Using a lower number of categories where the density of the underlying classes is high (i.e., two categories for good and bad performance) maximizes the number of available samples per class. Alternatively, using a higher number of categories reduces the number of available samples per class. The second relevant parameter related to our proposal is the I3D ConvNet input. In this sense, we have described three possible configurations in Sect. 3.1: original video sequence (Raw), the runner-of-interest bounding boxes (BB), and the runner-of-interest fine-grained segmentation (VIBE).

In this work, the performance of the runners has been divided into different categories. Good/Poor performance corresponds to over/under the median split time. Good/Average/Poor corresponds with the first/second/third tertiles. Excellent/Good/Poor/Bad corresponds to the quartile of the split time.

Table 2 is divided into two horizontal blocks. The first block shows the accuracy of the quality assessment for a runner at a given RP. It means that our system can predict the runner performance using the embeddings extracted from the RP video sequence as input. Results are pretty promising when considering two categories the raw input reported rates are around 84%. Observe that this classification is done using only seven seconds captured from RPs located more than 10 km away from each other. The classifier also achieves noticeable rates when BB (around 78%) and VIBE (around 76%) configurations feed the I3D ConvNet. As expected, performance drops as the number of categories increases. It happens because of the classes redistribution. In other words, having the same number of samples divided into a higher number of categories affects the model.

Table 2 second block shows the results of the estimated quality assessment for a runner on the next RP. Results are slightly worse than the rate reported in the first block of the table. It makes sense because the model predicts how the participant will perform on the next RP by observing the current RP. However, the results are quite interesting. Between the two experiments, the performance drops around 3–5%. At the same time, some runners tend to improve

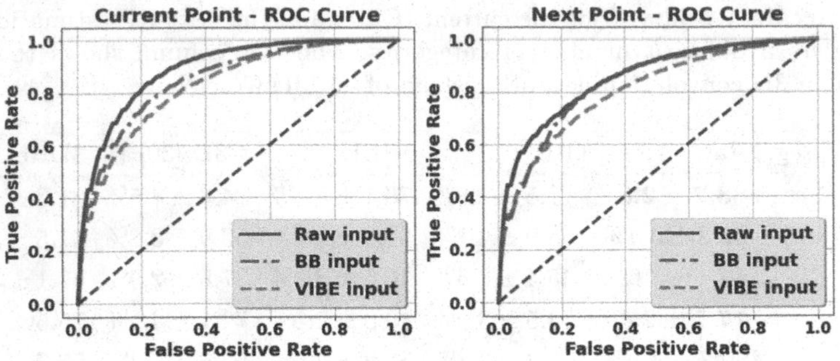

Fig. 4. ROC curves computed from the results of the devised experiments when two categories are considered. The left graph shows the ROC curves for the current RP experiment, while the right graph shows the ROC curves for the next RP experiment.

performance when they realize that they are being recorded (See Fig. 5). Table 2 shows how the model performs on different inputs under the number of categories variations, i.e., how the context reduction affects the model performance. Three interesting issues must be highlighted in this regard (Fig. 4).

First, when the raw video sequence is considered as input, the model performs best in any considered case. It can be explained because the I3D ConvNet has been pre-trained using the Kinetics 400 (see Sect. 3.2 for a further description). This dataset provides context to the model, meaning that contextual information plays a relevant role. Moreover, the I3D ConvNet may experience difficulties through the optical flow stream input when the context is removed. In this sense, we believe that the feature matching could fail for regions without context. The table shows a 3–5% accuracy loss between the Raw and the BB inputs. This loss is lower between the BB and the VIBE inputs (around 1–2%) because the amount of context information removed is not much (see Fig. 3). It seems to happen that the model is not affected by multiple runners in the scene in order to evaluate a specific runner performance. The second issue is related to the best classifier. As aforementioned, XGBoost reports the best rates. It can be explained by observing how this algorithm works. While traditional Random Forest builds trees in parallel, in boosting, trees are built sequentially, i.e., each tree is grown and boosted using information from previously grown trees. Reducing the context may come with a computational advantage, i.e., the system may be faster if it only needs to process a small fraction of the scenes [1]. Finally, we have considered a C3D and a 3D ResNet for comparison purposes (see Sect. 2). Contrary to the two streams of data (RGB and Flow) used I3D ConvNet, both C3D and 3D ResNets operate on single 3D stream input. As can be seen in Table 2, I3D ConvNet outperforms C3D and 3D ResNets by a 10%–15% and a 2%–5% respectively.

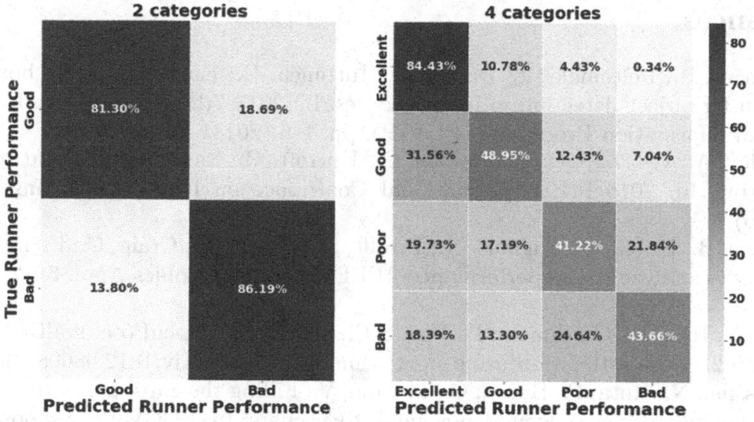

Fig. 5. Current RP confusion matrices considering raw data as input.

5 Conclusions

This paper presents an HAE approach to automatically provide an assessment for running quality and provide interpretable feedback. For this reason, we have conducted several experiments that combine an HAR pre-trained deep neural network with a quality assessment metric. The contribution represents an exciting challenge since we are unaware of any HAE research on this scenario.

We have proven that both human motion and environmental backgrounds facilitate the necessary spatio-temporal information to the I3D ConvNet to generate a set of valid embeddings. Contrary to the works detailed in Sect. 2, the quality assessment metric is not defined by a collection of body configurations (i.e., as happens in diving or skating). The assessment metric relies on categorizing the runner's split time at each RP. Consequently, our proposal is sufficiently complex to effectively distinguish between the previously categorized classes with 83.7% of accuracy at the current RP. We have also discussed how varying categories can affect the model accuracy. In this sense, scalability has been evaluated, and increasing the number of categories does not seem to improve a simpler model. This effect is caused by reducing samples per class during the training stage. Finally, we have shown that our proposal can predict the runner performance at the next RP with a 77.1% of accuracy. Of course, health surveillance must be mentioned among the most relevant uses. Athletes face physically demanding situations in ultra-distance races. When they reach RPs, the race staff has first-hand information about a runner's state by watching him. For the race organizers, it is more feasible to place a set of cameras than placing medical staff along the different race tracks due to human resources. Our proposed technique provides a categorization that can be informative for the medical staff to evaluate runners' risky situations at any given time.

References

1. Adhikari, B., Peltomaki, J., Puura, J., Huttunen, H.: Faster bounding box annotation for object detection in indoor scenes. In: 2018 7th European Workshop on Visual Information Processing (EUVIP), pp. 1–6 (2018)
2. Bewley, A., Ge, Z., Ott, L., Ramos, F., Upcroft, B.: Simple online and realtime tracking. In: 2016 IEEE International Conference on Image Processing (ICIP) (2016)
3. Bideau, B., Kulpa, R., Vignais, N., Brault, S., Multon, F., Craig, C.: Using virtual reality to analyze sports performance. IEEE Comput. Graphics Appl. **30**(2), 14–21 (2010)
4. Cao, Z., Hidalgo, G., Simon, T., Wei, S.E., Sheikh, Y.: OpenPose: realtime multi-person 2D pose estimation using Part Affinity Fields. arXiv:1812.08008 (2018)
5. Carissimi, N., Rota, P., Beyan, C., Murino, V.: Filling the gaps: predicting missing joints of human poses using denoising autoencoders. In: ECCV Workshops (2018)
6. Carreira, J., Zisserman, A.: Quo vadis, action recognition? a new model and the kinetics dataset. In: 2017 IEEE Conference on Computer Vision and Pattern Recognition (CVPR), pp. 4724–4733 (2017)
7. Freire-Obregón, D., Barra, P., Castrillón-Santana, M., Marsico, M.D.: Inflated 3D ConvNet context analysis for violence detection. Mach. Vis. Appl. **33**(1), 15 (2021)
8. Freire-Obregón, D., Castrillón-Santana, M., Barra, P., Bisogni, C., Nappi, M.: An attention recurrent model for human cooperation detection. Comput. Vis. Image Underst. **197–198**, 102991 (2020)
9. Ghasemzadeh, H., Jafari, R.: Coordination analysis of human movements with body sensor networks: a signal processing model to evaluate baseball swings. IEEE Sens. J. **11**(3), 603–610 (2011)
10. Hara, K., Kataoka, H., Satoh, Y.: Can spatiotemporal 3D CNNs retrace the history of 2D CNNs and imageNet. In: Proceedings of the IEEE Conference on Computer Vision and Pattern Recognition (CVPR), pp. 6546–6555 (2018)
11. He, Y., Yan, R., Fragkiadaki, K., Yu, S.I.: Epipolar transformers. In: Proceedings of the IEEE/CVF Conference on Computer Vision and Pattern Recognition, pp. 7779–7788 (2020)
12. Host, K., Ivašić-Kos, M., Pobar, M.: Tracking handball players with the DeepSORT algorithm. In: 9th International Conference on Pattern Recognition Applications and Methods (ICPRAM). vol. 1, pp. 593–599 (2020)
13. Huang, B., Xu, T., Jiang, S., Chen, Y., Bai, Y.: Robust visual tracking via constrained multi-kernel correlation filters. IEEE Trans. Multimedia **22**(11), 2820–2832 (2020)
14. Kay, W., et al.: The kinetics human action video dataset. CoRR abs/1705.06950 (2017)
15. Kocabas, M., Athanasiou, N., Black, M.J.: Vibe: video inference for human body pose and shape estimation. In: The IEEE Conference on Computer Vision and Pattern Recognition (CVPR) (2020)
16. Lei, Q., Du, J.X., Zhang, H.B., Ye, S., Chen, D.S.: A survey of vision-based human action evaluation methods. Sensors **19**(19) (2019)
17. Loper, M., Mahmood, N., Romero, J., Pons-Moll, G., Black, M.J.: SMPL: a skinned multi-person linear model. ACM Trans. Graph. (Proc. SIGGRAPH Asia) **34**(6), 248:1–248:16 (2015)
18. Paiement, A., Tao, L., Camplani, M., Hannuna, S., Damen, D., Mirmehdi, M.: Online quality assessment of human motion from skeleton data. In: Proceedings of the British Machine Vision Conference. BMVA Press (2014)

19. Parmar, P., Morris, B.: Action quality assessment across multiple actions. In: IEEE Winter Conference on Applications of Computer Vision, WACV 2019, 7–11 Jan 2019, pp. 1468–1476. Waikoloa Village, HI, USA. IEEE (2019)
20. Parmar, P., Morris, B.T.: Measuring the quality of exercises (2016)
21. Parmar, P., Morris, B.T.: Learning to score olympic events (2017)
22. Parmar, P., Morris, B.T.: What and how well you performed? A multitask learning approach to action quality assessment. In: IEEE Conference on Computer Vision and Pattern Recognition, CVPR 2019, 16–20 June 2019, pp. 304–313. Long Beach, CA, USA. Computer Vision Foundation/IEEE (2019)
23. Patrona, F., Chatzitofis, A., Zarpalas, D., Daras, P.: Motion analysis: action detection, recognition and evaluation based on motion capture data. Pattern Recogn. **76**, 612–622 (2018)
24. Penate-Sanchez, A., Freire-Obregón, D., Lorenzo-Melián, A., Lorenzo-Navarro, J., Castrillón-Santana, M.: Tgc20reid: a dataset for sport event re-identification in the wild. Pattern Recogn. Lett. **138**, 355–361 (2020)
25. Simon, T., Joo, H., Matthews, I., Sheikh, Y.: Hand keypoint detection in single images using multiview bootstrapping. In: CVPR (2017)
26. Szegedy, C., et al.: Going deeper with convolutions. In: 2015 IEEE Conference on Computer Vision and Pattern Recognition (CVPR), pp. 1–9 (2015)
27. Tedesco, S., et al.: Wearable motion sensors and artificial neural network for the estimation of vertical ground reaction forces in running. In: 2020 IEEE SENSORS, pp. 1–4 (2020)
28. Tran, D., Wang, H., Torresani, L., Ray, J., LeCun, Y., Paluri, M.: A closer look at spatiotemporal convolutions for action recognition. In: 2018 IEEE/CVF Conference on Computer Vision and Pattern Recognition, pp. 6450–6459 (2018)
29. Tu, Z., Li, H., Zhang, D., Dauwels, J., Li, B., Yuan, J.: Action-stage emphasized spatiotemporal VLAD for video action recognition. IEEE Trans. Image Process. **28**(6), 2799–2812 (2019)
30. Varol, G., Laptev, I., Schmid, C.: Long-term temporal convolutions for action recognition. IEEE Trans. Pattern Anal. Mach. Intell. **40**(6), 1510–1517 (2018)
31. Wnuk, K., Soatto, S.: Analyzing diving: a dataset for judging action quality. In: Koch, R., Huang, F. (eds.) ACCV 2010. LNCS, vol. 6468, pp. 266–276. Springer, Heidelberg (2011). https://doi.org/10.1007/978-3-642-22822-3_27
32. Wojke, N., Bewley, A., Paulus, D.: Simple online and realtime tracking with a deep association metric. In: 2017 IEEE International Conference on Image Processing (ICIP), pp. 3645–3649 (2017)
33. Xu, C., Fu, Y., Zhang, B., Chen, Z., Jiang, Y.G., Xue, X.: Learning to score figure skating sport videos. IEEE Trans. Circuits Syst. Video Technol. **30**(12), 4578–4590 (2020)
34. Ye, F., Tang, H., Wang, X., Liang, X.: Joints relation inference network for skeleton-based action recognition. In: 2019 IEEE International Conference on Image Processing (ICIP), pp. 16–20 (2019)
35. Zhang, Z., Peng, H.: Deeper and wider siamese networks for real-time visual tracking. In: Proceedings of the IEEE/CVF Conference on Computer Vision and Pattern Recognition, pp. 4591–4600 (2019)

Densification of Sparse Optical Flow Using Edges Information

Antonio Buemi[1(✉)], Giuseppe Spampinato[1], Arcangelo Bruna[1], and Viviana D'Alto[2]

[1] ST Microelectronics, System, Research and Application, Catania Lab, Stradale Primosole, 50, 95121 Catania, Italy
{antonio.buemi,giuseppe.spampinato,arcangelo.bruna}@st.com

[2] ST Microelectronics, System, Research and Application, Agrate Lab, Via Camillo Olivetti, 2, 20864 Agrate Bianza, Italy
viviana.dalto@st.com

Abstract. Optical flow methods, which estimate a dense motion field starting from a sparse one, are playing an important role in many visual learning and recognition applications. The proposed system is based only on sparse optical flow and line detector. It is able to densify the starting optical flow, reaching good performances in objective and subjective manner, using common applications like clustering and standard KITTI evaluation kit. In particular, an appreciable improvement has been achieved in terms of quantity of motion vectors grows (up to 540%). Since often in smart cameras both optical flow and lines are available, the proposed approach avoids overloading the Engine Control Unit to transmit the entire image flow and allows reducing the power consumption, realizing a real-time robust system.

Keywords: Optical flow · Densification · Line detector · KITTI evaluation

1 Introduction

Optical flow gives important information about camera movement and the spatial arrangement of moving objects. It is essential for many higher-level visual tasks, such as visual learning and recognition applications. Optical flow is the set of two (or three) dimensional motion vectors, represented by modules and directions of points of an image sequence. This set is considered dense if it is composed by all the points of an image sequence, otherwise it is considered sparse.

In general, dense optical flow methods generate accurate motion vectors, but they process the whole image data, so they are computationally expensive. Recent approaches take advantage of sparse-to-dense interpolation schemes to increase the motion vectors density [1].

Classical sparse-to-dense approaches [2–6] use complex interpolation filter of robust matching features like respectively: RLOF [7], SIFT [8], Stereoscan [9] and DeepMatches [10]. Even if they are very efficient methods, they are slow due to the heavy steps of feature calculation, matching and interpolation. Other methods use simpler but

S. Sclaroff et al. (Eds.): ICIAP 2022, LNCS 13233, pp. 254–264, 2022.
https://doi.org/10.1007/978-3-031-06433-3_22

limited schemes like nearest neighborhood (NN) to find correspondences of features on a grid [11, 12] and energy model optimization in the classical Total Variation [13, 14]. Other computational heavy approach consists in layered models [15, 16], in which segmentation is related to the scene geometry. Recently a deep learning-based approach to the problem has been proposed. It is based on the design of a fully convolutional neural network able to perform the interpolation from sparse to dense optical flow, taking also in input the edge map [17]. The edges information is important in optical flow estimation in order to address the issue of edge-blurring near the image and motion boundaries has remained a challenge in flow field estimation [18].

Our solution is completely different compared to the aforementioned approaches. The main strengths of the proposed method are:

- Exploiting of visual information nowadays available in most common smart cameras, *i.e.,* sparse optical flow and lines, useful to densify the optical flow.
- The system is able to densify motion vectors in the scene making use of motion vectors between adjacent frames (obtained from any sparse algorithm) and lines in the current frame.
- The resulting densification allows to achieve better quality results in further steps. In particular, to show such improvement, we choose to evaluate the performances growth achieved in a general clustering algorithm we presented in [20].
- In the context of embedded systems, with reduced memory and computational capabilities, finding application in automotive, low-cost solutions like the one proposed in this paper can be easily integrated at hardware or firmware levels.

This paper is structured as follows: the Sect. 2 presents the proposed optical flow densification system, the Sect. 3 discusses the experimental results proofing the benefit of using this approach inside a clustering system, then the Sect. 4 provides the final remarks.

2 The Proposed Approach

The proposed approach is based on the idea of exploiting the lines information in order to make the optical flow denser. The Fig. 1 shows the algorithm basic scheme. The input data are the Optical Flow (OF) at the current frame i and the detected lines (edges map) at the previous frame $i - 1$. The optical flow is described as a list of Motion Vectors (MVs), whilst the edges map is a list of points with an associated label. If two points in the edge map share the same label, they are intended to belong to the same line.

Since usually the edge map representation is quite coarse, with a lot of short segments, it can be made finer by joining the close lines, thanks to the optional step *Close Lines Merging* (*CLM*). When this step is activated, all the segments (represented by a set of points with the same label) are taken into consideration. At this point, Euclidean distances between the endpoints of each different segments are calculated. If this distance is lower or equal than a threshold, the two segments are merged in one. This step allows to reduce the total number of lines, increasing the number of motion vectors pairs, placed near the same line, to achieve a better densification.

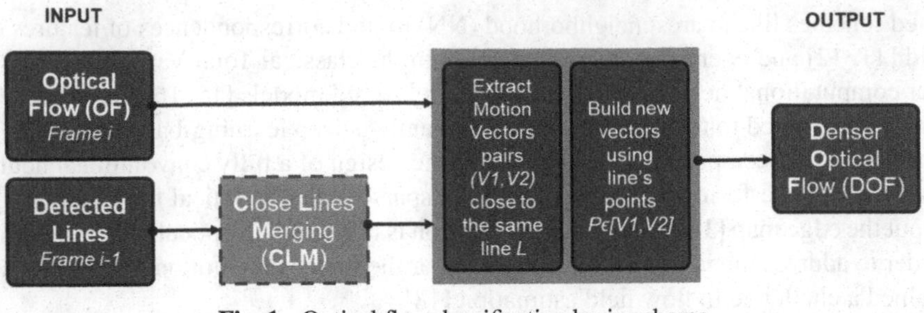

Fig. 1. Optical flow densification basic scheme

The core block of the algorithm is the *Densificator*, which consists of two modules: *Extract Motion Vectors* (*EMV*) and *Build New Vectors* (*BNV*). The first one (*EMV*) extracts the Motion Vectors endpoints pairs (V_1, V_2) of the same segment *S*, while the second one (*BNV*) builds new vectors using all the points *P* belonging to the segment V_1V_2.

To better explain the second module (*BNV*), we have to consider that each *MV* is defined by its spatial coordinates (X,Y) and its horizontal and vertical displacements (*dX*, *dY*) respect to the previous frame and each segment *S* is described by endpoints of MVs $V_1(X_1, Y_1, dX_1, dY_1)$ and $V_2(X_1, Y_1, dX_1, dY_1)$. For each point $P(X, Y)$ (see Fig. 2), laying in the segment V_1V_2, new MVs $P'(X, Y, dX', dY')$ are created by interpolation of the related endpoints horizontal and vertical displacements. In particular, the displacements dX', dY' are computed as a weighted average, depending on the Euclidean distances *dist* among *P*, V_1, V_2, as follows:

$$dX' = w1 \cdot dX_1 + w2 \cdot dX_2 \tag{1}$$

$$dY' = w1 \cdot dY1 + w2 \cdot dY_2 \tag{2}$$

$$w1 = 1.0 - dist(V_1, P)/(dist(V_1, P) + dist(V_2, P)) \tag{3}$$

$$w2 = 1.0 - (dist(V_2, P))/(dist(V_1, P) + dist(V_2, P)) \tag{4}$$

where *dist* is the Euclidean distance between two vectors.

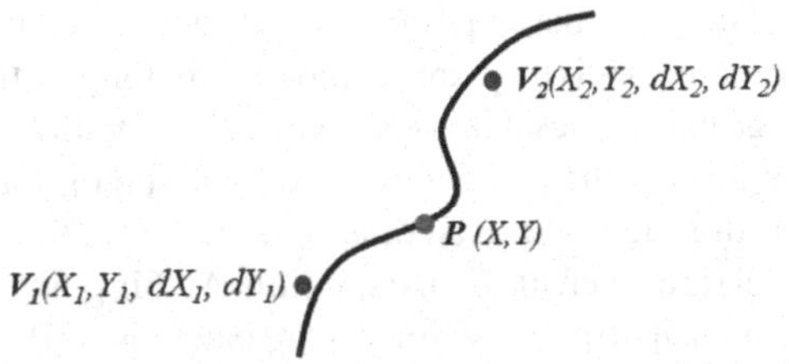

Fig. 2. Motion vector interpolation procedure

3 Experimental Results

We performed three kinds of experiments, with the purpose of measuring the impact of densification on the motion vectors amount, the benefits of densification in clustering and the densification effect on the odometry task.

The first experiment is oriented to quantify the objective increase of the motion vectors provided by the densification; the second one is designed to estimate the performances gain achieved by the clustering algorithm trough to the densification step addition. The set used for this experiment consists of five outdoor video sequences (about 2000 frames) VGA, YUV 4:2:0 format, acquired using a static camera with fisheye lens frames.

At last, the third experiment aims to evaluate the impact of the densification on the odometry algorithm implemented in the KITTI 2015 visual evaluation kit [21]. In this case, the dataset is different form the one used in the first two and it consists of 10 sequences from KITTI Flow 2015 dataset (see [21] for details).

In general, in our experiments, optical flow is generated using the chain described in [19]: FAST algorithm as feature extraction, BRIEF as descriptor generation and Hamming distance as matching algorithm. Instead, lines are obtained applying a simple Sobel filter.

Since our application target is related to embedded systems, the analysis is focused on the impact of the densification on existing algorithms rather than on the comparison with classic densification algorithms that cannot be implemented in a real time system due to their complexity cost.

The following subsections describe the results in detail.

3.1 Test 1: Objective Impact of Densification

The first class of tests allows to evaluate the performances of the method through an objective metric: the growth of the MVs number, comparing them with and without the application of the prosed method. Even such a measure doesn't give information about the reliability of the added MVs, it confirms that the proposed approach generates a meaningful number of new vectors and then the computational overload is justified. The algorithm has been tested with the "Close Lines Merging" (CLM) flag enabled and disabled, to evaluate the effect of such option, too.

The chart in Fig. 3 shows the MVs increase frame by frame. The magenta points (the lowest) represent the number of original OF for each frame, the gray histogram (the middle one) shows the number of points obtained applying the algorithm with the CLM option disabled and the blue histogram (the higher) represents the number of points generated by the algorithm with the CLM option enabled. The growth is appreciable: 420% on the average with CLM disabled and 540% on the average with CLM enabled. Note that the peaks in the histogram correspond of the frame's segments where moving objects are present, except for the higher peak that is created due a slow translation of the camera pose.

Figure 4 shows an example of output in a frame where a dark car is moving in the scene. The black lines represent the detected edges, the red segments are the original MVs

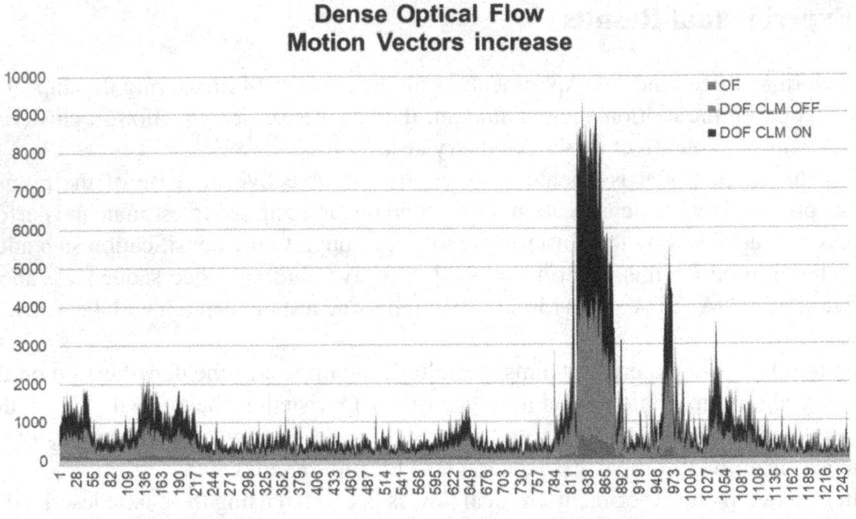

Fig. 3. Number of MVs outputted by the original OF (magenta), Dense OF with CLM disabled (gray) and Dense OF with CLM enabled (blue) algorithms (Color figure online)

and the white lines are the added motion vectors (the length of the segments represents the amount of displacement).

Figure 5 and Fig. 6 show two more examples, respectively with a motorbike and a light car moving in the scene.

Table 1 summarizes the results for the three sample frames. Such frames have been chosen as an example because they correspond to scenes with moving objects with evident edges, that is the densification algorithm main target case. Similar results are achieved in the majority of the frame due to the chosen scenario, whilst the densification impact can be less evident in the less frequent case of homogeneous scenes.

3.2 Test 2: Benefits of Densification in Clustering

The second set of experiments allows to evaluate the OF densification effect on an algorithm based on OF, *i.e.* general clustering application [20].

The clustering algorithm is common and important task, widely used in computer vision as low-level task for different applications: camera surveillance [22], Cross Traffic Alert (CTA) [23], Intelligent, Transport System (ITS) [24], object detection and recognition [25], people detection and recognition [26], people behavior analysis (*e. g.* anti-theft, falling detection, etc.) [27] and user interfaces by gesture [28].

It is able to retrieve clusters of moving objects in a scene acquired by an image sensor device starting from the motion vectors only, so the increase of the number of motion vectors should make it more reliable.

As a matter of fact, the results show an evident visual improvement in the case of large moving objects, because they are not split in different clusters (see Fig. 7 and Fig. 8). With standard size moving objects (as cars), no degradation is registered with the application of the proposed method.

Fig. 4. On the left, the frame 121 of the testing sequence with OF (red) and edges map (black). The enlarged details on the right show the original OF (top), the Dense OF obtained without the CLM step (middle) and the Dense OF obtained with the CLM option enabled (bottom). (Color figure online)

Fig. 5. On the left, the frame 641 of the testing sequence with OF (red) and edges map (black). The enlarged details on the right show the original OF (top), the Dense OF obtained without the CLM step (middle) and the Dense OF obtained with the CLM option enabled (bottom). (Color figure online)

Fig. 6. On the left, frame 1038 of the testing sequence with OF (red) and edges map (black). The enlarged details on the right show the original OF (top), the Dense OF obtained without the CLM step (middle) and the Dense OF obtained with the CLM option enabled (bottom). (Color figure online)

Table 1. Summary of the results obtained processing the frames number 121, 641 and 1038.

Number of motion vectors			
Frame	Original OF	Dense OF (CLM OFF)	Dense OF (CLM ON)
121	118	220	366
641	112	394	504
1038	137	1044	1195

Fig. 7. Clustering performances with OF (left) and dense OF (right). The densification process allows to correctly detect as a single object.

Fig. 8. Clustering performances with OF (left) and dense OF (right). The densification process allows to correctly detect as a single object.

The following objective metrics have been also computed on five annotated video sequences (about 2000 frames) to validate the results:

$$True\ Positive\ Rate = TP/(TP + FN)$$
$$Precision = TP/(TP + FP)$$
$$Accuracy = (TP + TN)/(TP + TN + FP + FN)$$

where:

$$TP = True\ Positives$$
$$TN = True\ Negatives$$
$$FP = False\ Positives$$
$$FN = False\ Negatives$$

Table 2 summarizes the results. Note that the use of the densified optical flow allows to improve the clustering algorithm performances in terms of TPR, Precision and Accuracy. Such improvement is more evident in the sequence number 3 because it contains large moving objects.

Table 2. Objective metrics highlight the improvement achieved by the clustering algorithm using the dense optical flow

seq	OF			Dense OF			Gain		
	TPR	Precision	Accuracy	TPR	Precision	Accuracy	TPR	Precision	Accuracy
1	81%	94%	91%	96%	96%	97%	15%	2%	6%
2	97%	100%	97%	98%	100%	98%	1%	0%	1%
3	94%	88%	83%	98%	100%	98%	3%	12%	14%
4	64%	93%	61%	66%	96%	65%	2%	3%	3%
5	92%	98%	91%	95%	100%	95%	3%	2%	4%
Avg	**86%**	**95%**	**84%**	**89%**	**99%**	**89%**	**3%**	**4%**	**5%**

3.3 Test 3: Densification Impact on Odometry

The last experiments class exploits the optical flow evaluation section of the KITTI 2015 visual evaluation kit [21]. Testing consists of processing the KITTI sequences using both the original sparse optical flow and the dense optical flow, evaluating the resulting error rate in the trajectory reconstruction. Table 3 summarizes the results. Note that, despite the increment of the number of motion vectors, the error rate remains almost unchanged. The main reason is that most of the interpolated vectors lie near the existing motion vectors, so global motion vectors frame by frame will be similar with and without the application of the proposed method.

Table 3. Testing results on KITTI 2015 sequences using the sparse and dense OF.

N.	Sparse OF	Dense OF	Sparse OF	Dense OF	Gain
	f_err	f_err	MVs number	MVs number	MVs number
1	16.775	16.775	189	296	1.57
2	11.384	11.389	525	947	1.80
3	13.8	13.8	115	265	2.30
4	18.882	18.882	107	107	1.00
5	16.089	16.095	194	1796	9.26
6	18.867	18.867	246	415	1.69
7	9.851	9.851	138	167	1.21
8	15.092	15.092	195	396	2.03
9	14.58	14.58	161	396	2.46
10	12.154	12.157	658	1637	2.49

4 Conclusion

A complete optical flow densification, working only with input sparse optical flow and line detector has been presented. The proposed system has been experimentally tested on a representative dataset of scenes, reaching the goal to densify optical flow, obtaining better quality results. Different kind of tests, both subjective and objective, have been executed revealing a great increasing of total number of motion vectors, 540% on the average, with increasing visual quality for specific applications, like generic clustering. Moreover, KITTI evaluation kit reveals that the growth of the number of motion vectors doesn't yield variation to the error rate. At last, the proposed system has been developed with the characteristic to be very efficient and flexible, so can be used with any algorithm which estimates motion vectors between adjacent frames and any lines detector and in contexts requiring low memory and computational consumption.

References

1. Fortun, D., Bouthemy, P., Kervrann, C.: Optical flow modeling and computation: a survey. Comput. Vis. Image Understand. **134**, 1–21 (2015)
2. Senst, T., Geistert, J., Sikora, T.: Robust local optical flow: long-range motions and varying illuminations. In: International Conference on Image Processing (2016)
3. Xu, L., Jia, J., Matsushita, Y.: Motion detail preserving optical flow estimation. IEEE Trans. Pattern Anal. Mach. Intell. **34**, 1744–1757 (2012)
4. Wulff, J., Black, M.J.: Efficient sparse-to-dense optical flow estimation using a learned basis and layers. In: Conference on Computer Vision and Pattern Recognition (2015)
5. Revaud, J., Weinzaepfel, P., Harchaoui, Z., Schmid, C.: EpicFlow: edge-preserving interpolation of correspondences for optical. In: Conference on Computer Vision and Pattern Recognition (2015)
6. Manandhar, S., Bouthemy, P., Welf, E., Roudot, P., Kervrann, C.: A sparse-to-dense method for 3D optical flow estimation in 3D light-microscopy image sequences. In: 15th IEEE International Symposium on Biomedical Imaging ISBI (2018)
7. Senst, T., Eiselein, V., Sikora, T.: Robust local optical flow for feature tracking. IEEE Trans. Circuits Syst. Video Technol. **22**(9), 1377–1387 (2012)
8. Lowe, D.G.: Distinctive image features from scale-invariant keypoints. Int. J. Comput. Vis. **6**(2), 91–110 (2004)
9. Geiger, A., Ziegler, J., Stiller, C.: Stereoscan: dense 3d reconstruction in real-time. In: IEEE Intelligent Vehicles Symposium (2011)
10. Weinzaepfel, P., Revaud, J., Harchaoui, Z., Schmid, C.: DeepFlow: large displacement optical flow with deep matching. In: International Conference on Computer Vision (2013)
11. Brox, T., Malik, J.: Large displacement optical flow: descriptor matching in variational motion estimation. Pattern Anal. Mach. Intell. **33**(3), 500–513 (2011)
12. Leordeanu, M., Zanfir, A., Sminchisescu, C.: Locally affine sparse-to-dense matching for motion and occlusion estimation. In: IEEE International Conference on Computer Vision (ICCV) (2013)
13. Liu, C., Yuen, J., Torralba, A.: Sift flow: dense correspondence across scenes and its applications. Trans. Pattern Anal. Mach. Intell. **33**(5), 978–994 (2011)
14. Xu, L., Jia, J., Matsushita, Y.: Motion detail preserving optical flow estimation. Trans. Pattern Anal. Mach. Intell **34**(9), 1744–1757 (2012)
15. Sun, D., Wulff, J., Sudderth, E., Pfister, H., Black, M.: A fully-connected layered model of foreground and background flow. In: Computer Vision and Pattern Recognition (CVPR) (2013)
16. Sun, D., Sudderth, E., Black, M.: Layered segmentation and optical flow estimation over time, In: Computer Vision and Pattern Recognition (CVPR) (2012)
17. Zweig, S., Wolf, L.: InterpoNet, a brain inspired neural network for optical flow dense interpolation. In: Conference on Computer Vision and Pattern Recognition (CVPR) (2017)
18. Zhang, C., Ge, L., Chen, Z., Li, M., Liu, W., Chen, H.: Refined TV-L1 optical flow estimation using joint filtering. IEEE Trans. Multimedia **22**(2) (2020)
19. Watroba, R., Sebyleau, Y., Spampinato, G., Motshagen, D.: Self-contained, distributed automotive Ethernet camera system. In: International Conference and Exhibition for Automotive Electronic Systems (CESA) (2016)
20. Spampinato, G., Bruna, A., D'Alto, V., Curti, S.: Advanced low cost clustering system. In: International Conference on Image Processing Theory, Tools and Applications (IPTA) (2016)
21. Menze, M., Geiger, A.: Object scene flow for autonomous vehicles. In: Conference on Computer Vision and Pattern Recognition (CVPR) (2015)

22. Coppi, D., Calderara, S., Cucchiara, R.: Transductive people tracking in unconstrained surveillance. IEEE . Circuits Syst. Video Technol. **26**(4), 762–775 (2016). https://doi.org/10.1109/TCSVT.2015.2416555
23. Balisavira, V., Pandey, V.K.: Real-time object detection by road plane segmentation technique for ADAS. In: Eighth International Conference on Signal Image Technology and Internet Based Systems (SITIS) (2012)
24. Seenouvong, N., Watchareeruetai, U., Nuthong, C., Khongsomboon, K., Ohnish, N.: A computer vision based vehicle detection and counting system. In: Eighth International Conference on Knowledge and Smart Technology (KST) (2016)
25. Kalogeiton, V., Ferrari, V., Schmid, C.: Analysing domain shift factors between videos and images for object detection. IEEE Trans. Pattern Anal. Mach. Intell. **38**(11), 2327–2334 (2016). https://doi.org/10.1109/TPAMI.2016.2551239
26. García-Martín, A., Martínez, J.M.: People detection in surveillance: classification and evaluation. IET Comput. Vis. **9**(5), 779–788 (2015). https://doi.org/10.1049/iet-cvi.2014.0148
27. Ozcan, K., Velipasalar, S.: Wearable camera- and accelerometer-based fall detection on portable devices. IEEE Embedded Syst. Lett. **8**(1), 6–9 (2016). https://doi.org/10.1109/LES.2015.2487241
28. Dang, X., Wang, W., Wang, K., Dong, M., Yin, L.: A user-independent sensor gesture interface for embedded device. IEEE Sensors (2011)

Cycle Consistency Based Method for Learning Disentangled Representation for Stochastic Video Prediction

Ujjwal Tiwari(✉), P. Aditya Sreekar, and Anoop Namboodiri

Center for Visual Information Technology, International Institute of Information Technology, Hyderabad, India
{ujjwal.t,paditya.sreekar,anoop}@iiit.ac.in

Abstract. Video frame prediction is an interesting computer vision problem of predicting the future frames of a video sequence from a given set of context frames. Video prediction models have found wide-scale perspective applications in autonomous navigation, representation learning, and healthcare. However, predicting future frames is challenging due to the high dimensional and stochastic nature of video data. This work proposes a novel cycle consistency loss to disentangle video representation into a low dimensional time-dependent pose and time-independent content latent factors in two different VAE based video prediction models. The key motivation behind cycle consistency loss is that future frame predictions are more plausible and realistic if they reconstruct the previous frames. The proposed cycle consistency loss is also generic because it can be applied to other VAE-based stochastic video prediction architectures with slight architectural modifications. We validate our disentanglement hypothesis and the quality of long-range predictions on standard synthetic and challenging real-world datasets such as Stochastic Moving MNIST and BAIR.

Keywords: Video frame prediction · Variational autoencoders · Cyclic consistency

1 Introduction

The ability to predict the possible future states of the visual world has enabled humans to plan their actions ahead of time in highly dynamic real-world settings [1]. The goal of video frame prediction is to build intelligent systems that are capable of replicating this behaviour. To accomplish this task, it is essential for these models to develop an intrinsic understanding of the laws that govern the physical world. Consequently, video frame prediction models have been studied in a variety of different contexts such as activity prediction and early recognition [2, 3], human pose and trajectory forecasting [4–6], robotics [7], and health care [8].

However, despite recent advancements in generative modeling techniques and the abundance of multimedia data, methods that try to predict future frames face two significant challenges, which are as follows:

S. Sclaroff et al. (Eds.): ICIAP 2022, LNCS 13233, pp. 265–277, 2022.
https://doi.org/10.1007/978-3-031-06433-3_23

- Learning the non-linear transformation between frames is computationally expensive and produces degraded long-range predictions.
- The task of video frame prediction is inherently multi-modal due to the stochastic nature of video data; modeling the predictive distribution of future frames is challenging in the high-dimensional pixel space.

Current video prediction methods [9–11] overcome the first challenge by decomposing video representations into low dimensional time-dependent pose and time-independent content features. Learning disentangled representation of video sequences significantly reduces the complexity of predicting future frames, which can then be modeled as predicting the low dimensional pose latent vector for each time step. DRNET [11] learns to disentangle representations into pose and content by using an adversarial loss penalising pose representations from encoding content like features. DDPAE [10], decomposes video into components with low dimensional temporal dynamics. These methods successfully demonstrate the benefits of pose/content disentanglement for video prediction on various synthetic datasets.

Stochastic video prediction models [12–14] explicitly model the inherent uncertainty in real-world videos using stochastic latent variables in order to resolve the second challenge listed above. These models predict stochastic but plausible future frames for different samples of its low-dimensional latent random variable. In the proposed work, we extend the idea of learning factorized video representations to stochastic video prediction models. Our approach is motivated by the idea that the preeminent cause of stochasticity in video sequences arises from the uncertainty in the pose of various objects in the scene. We make slight architectural modifications on two different stochastic video prediction models, SAVP [12], and SVG-LP [13] and incorporate our cycle consistency loss to their learning objective. We empirically demonstrate that our method of disentangling video representations and application of cyclic consistency loss considerably improves the quality of predicted frames when compared to SAVP-baseline model on BAIR dataset. In addition, we also evaluate the pose/content disentanglement and long-range frame prediction capabilities of our approach on both synthetic, SMNIST, and real-world, BAIR datasets.

2 Related Work

In the recent past, the task of predicting future video frames has received an increasing amount of interest from the research community. A range of deep video prediction models that assume the environment to be deterministic [5,15–17] have been proposed in the literature. Several action conditioned methods leverage the deterministic setting in video games for future frame prediction [18,19]. However, models with such naive assumptions lead to blurry frame predictions due to their inability to account for the stochasticity exhibited in real-world videos.

Transformation-based methods primarily focus on modeling the motion within a video sequence rather than explicitly reconstructing appearance. These models predict a subsequent frame by transforming the previous known frame through a constrained geometric distortion [20,21]. [7] models motion as the transformation of masked out groups of pixels by exploring the temporal consistency in videos. Other, concurrent methods [6,22] predict optical flow conditioned on context frames.

Learning to disentangle video representation for video prediction into time-dependent pose and time-independent content representations has been explored by [9–11] on synthetic datasets. In our approach, we also learn to factorize video representation into pose and content components which is done in a self-supervised manner and explicitly model the predictive distribution of the pose latent variables to account for the inherent stochasticity in videos. We validate the efficacy in learning disentangled representations on both synthetic and real-world videos.

Stochastic video prediction frameworks [14,23] use latent variables to model the ambiguous nature of future frames to capture the stochasticity in videos with real-world dynamics. These variational autoencoders (VAE) [24] based methods disentangle video into deterministic and stochastic components. While these methods are expressive in terms of producing diverse predictions, they tend to produce blurry frames. SVG-LP [13] is an example of one such VAE based video prediction model. While video prediction methods with GAN [25] formalism tend to produce sharper prediction, they are susceptible to mode collapse, and training instability, particularly in conditional settings [26]. SAVP [12] combines adversarial losses with latent variables to enable realistic stochastic video frame prediction. Our method builds upon SAVP and SVG-LP by employing a cycle consistency loss and a minor architectural change of using a content encoder to disentangle time-dependent features pose from time-independent content representations.

Another probabilistic approach to frame prediction include the autoregressive techniques that model the full joint distribution over raw pixels [27,28]. While, these methods produce crisp future frames, but training and inference are computationally expensive. Even highly parallelized approaches have also been proven to be impractical for high-resolution videos [29].

3 Our Approach

Our task of video prediction is to generate $T - C$ future frames conditioned on C context frames, where T is the total number of frames in the video sequence. That is, given context frames, $x_{1:C}$, we need to predict $\hat{x}_{C+1:T}$ future frames. As mentioned in Sect. 1, predicting future frames is challenging and computationally expensive due to the high dimensional nature of video frames. To overcome this challenge, we propose a simple description of video sequences. Two types of information are adequate to describe any scene: the list of various objects (content information) and their respective positions within the scene (pose information). For simplicity, we assume that the appearance of the objects in the video sequence does not change with time.

On top of the video description mentioned above, we also consider that the video sequences are generated from an underlying directed graphical model in which each frame x_i is generated from two disjoint latent variables representing content, z_c, and pose, $z_{i,p}$. Therefore, to predict a frame at time t, we need the pose $z_{t,p}$ and content z_c representations of the frame. As the latent content variable, z_c is assumed to be the same for all frames, the task of predicting future frames gets reduced to a much simpler task of predicting low-dimensional future pose representation, $z_{t,p}$ from the previous pose representation vector, $z_{2:t-1,p}$ of various objects in the scene.

In our approach a future frame $\hat{x}_t$ is predicted using the pose, $z_{t,p}$ and content, z_c latent factors by a generator G. To achieve proper pose/content disentanglement, we condition the generator G with time independent content information. We use a recurrent content encoder E_{con}, inspired from [10] to produce content encodings z_c which are consistent across all context frames. We implement our disentanglement approach by applying our novel cycle consistency loss on two different VAE based stochastic video prediction architectures where, the inference module consists of two recurrent encoder networks, E_{pos} and E_{prior}. While, $E_{pos}(x_t)$ models the posterior distribution, $q(z_{t,p}|x_{2:t})$, $E_{prior}(x_{t-1})$ and is used to model the learned prior distribution, $p(z_{t,p}|x_{1:t-1})$ of the pose latent variables. Pose latent variable $z_{t,p}$ is sampled from the posterior distribution during training and from the learned prior distribution at inference time. It is important to note that, In case of our VAE-GAN based SAVP-baseline, two learned discriminators D and D^{VAE} are used to discriminate the ground truth samples from posterior generated frames and prior generated frames respectively.

In Sect. 3.1, we explain the general learning objective of VAE-GAN models. Subsequent subsection Sect. 3.2, we introduce our novel *cycle consistency* loss term. The following Sect. 3.3, explains minor changes and architectures of two different stochastic video prediction models [12,13] in detail, for the purpose of learning proper disentangled representation and incorporating cycle consistency loss to their learning objective.

3.1 VAE-GAN Models

In this section, we introduce the learning objective of VAE-GAN based stochastic adversarial video prediction model [12]. The learning objective, $L_{VAE-GAN}$ given by Eq. 1, is comprised of three main loss terms which are, reconstruction loss L_{Recon}, KL Divergence L_{KL}, and two different GAN loss terms, L_{GAN} and L_{GAN}^{VAE}. Each of these terms is explained below as follows:

$$L_{VAE-GAN} = \lambda_{Recon} L_{Recon} + \beta L_{KL} + \alpha_1 L_{GAN} + \alpha_2 L_{GAN}^{VAE} \quad (1)$$

Reconstruction Loss: Reconstruction loss is minimized between the generated frames and ground truth frames. Assuming that the predictive distribution of future frames conditioned on latent variables is Gaussian, maximizing the log-likelihood leads to a pixel-wise l_2 loss. Since minimization of l_2 loss leads to blurry frame predictions [4], we use l_1 loss for the real-world BAIR dataset, and l_2 loss for the synthetic SMNIST dataset.

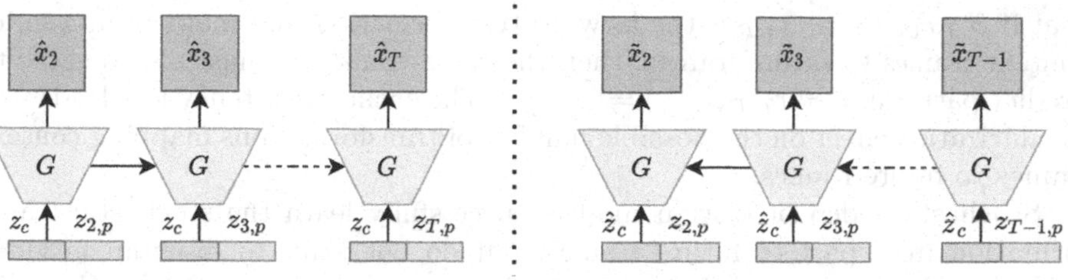

Fig. 1. Implementation of cycle consistency. Left: Forward prediction of $\hat{x}_{2:T}$. Right: Reverse prediction of $\tilde{x}_{(T-1):2}$

$$L_{Recon} = \mathbb{E}_{x_{1:T}, z_{t,p} \sim q(z_{t,p}|x_{2:t})|_2^T} \left[\frac{1}{T-1} \sum_{2}^{T} ||x_t - G(z_c, z_{t,p})||_n \right] \tag{2}$$

KL Divergence: While the pose latent variables are sampled from the posterior distribution $q(z_{t,p}|x_{2:t})$ during training, the predicted frame is produced by sampling from the learned prior distribution $p(z_{t,p}|x_{1:t-1})$ at inference time. KL divergence L_{KL} enforces that the two distributions encode similar information by making the learned prior distribution to be a strong approximation of the posterior distribution.

$$L_{KL} = -\mathbb{E}_{x_{1:T}} \left[\frac{1}{T-1} \sum_{2}^{T} D_{KL}(q(z_{t,p}|x_{2:t})||p(z_{t,p}|x_{1:t-1})) \right] \tag{3}$$

GAN Loss: In our implementation, we use two different learned discriminators D and D^{VAE} to train the frame generator G. While, D discriminates between ground truth frames and the ones generated by sampling pose $z_{t,p}$ from the posterior distribution $q(z_{t,p}|x_{2:t})$, to discriminate between the ground truth frames and the frames generated by sampling from the learned prior $p(z_{t,p}|x_{1:t-1})$ we use D^{VAE}. GAN loss terms, L_{GAN} and L_{GAN}^{VAE} force the posterior and learned prior distributions to be close to the true data generating distribution. We have mentioned the loss term, L_{GAN} in Eq. 4. The other adversarial loss L_{GAN}^{VAE} is analogous to L_{GAN}, where instead of sampling from the posterior, samples are taken from the prior distribution, and a different discriminator D^{VAE} is used. Both the discriminators have the same architecture but do not share weights.

$$L_{GAN} = \mathbb{E}_{x_{1:T}} \left[log(D(x_{2:T}) \right] + \mathbb{E}_{x_{1:T}, z_{t,p} \sim q(z_{t,p}|x_{2:t})|_2^T} \left[log(1 - G(z_c, z_{2:T,p})) \right] \tag{4}$$

3.2 Cycle Consistency Loss

In this section, we introduce our novel *Cycle Consistency* loss term. Video prediction is essentially the task of learning a transformation from the domain of context frames to a set of plausible future frames. Cycle consistency term enforces transitivity of the learned transformation; that is, if future frames are predicted from past frames, then by transforming future frames in reverse order, we should be able to recover past frames. Formally, Cycle consistency ensures that if $S : x_{1:C} \rightarrow x_{C+1:T}$ is the forward transformation that maps from a set of context frames to future frames, then the same transform must also be able to predict past frame $S(x_{T:T-C+1}) = x_{T-C:2}$. The cyclic consistency loss leads to a regularization effect on the possible number of transformations mapping context frames to future frames.

Stochastic video prediction models successfully learn the stochastic transformation from past to future frames, but do not learn to disentangle video representations into time-dependent and independent components. We make slight architectural changes along with incorporating our novel cycle consistency term to learn disentangled representations on two different VAE based models [12,13]. In SVG-LP [13] we conditioning the generator with content-encoding, which enables the model to learn the complementary pose-like information for the proper prediction of future frames. A recurrent content encoder, E_{con}, is used to produce the content-encoding z_c, which is consistent across all context frames. However, we do not make any such change to SAVP baseline, which is already a conditional video prediction model, wherein the generator is conditioned on the previous frames. Empirically, we validate our claim that the cycle consistency loss is essential in learning proper content/pose disentangled video representation in stochastic video prediction models and leads to visually stable long-range frame predictions.

To incorporate cycle consistency loss, L_{CC}, future frames $\hat{x}_{2:T}$ are predicted in forward prediction. Then a content encoding, $\hat{z}_c$, is inferred from $\hat{x}_{T:T-C+1}$ using the content encoder E_{con}. As shown in Fig. 1, in reverse prediction $\tilde{x}_{T-1:2}$ are predicted using $\hat{z}_c$ and pose latent variable in reverse order $z_{T-1:2,p}$. In the general formulation of this loss, the distance between $x_{2:T-1}$ and $\tilde{x}_{2:T-1}$ is minimized using n-norm, where both the forward predicted frames $\hat{x}_{2:T}$ and reverse frames $\tilde{x}_{2:T-1}$ are generated using the same pose $z_{2:T,p}$ encoding. Specifically, we use l_2 norm for SMNIST and l_1 norm for BAIR dataset.

$$L_{CC} = \mathbb{E}_{x_{1:T}, z_{t,p} \sim q(z_{t,p}|x_{2:t})|_2^T} \left[\frac{1}{T-2} \sum_{T-1}^{2} ||x_t - G(\hat{z}_c, z_{t,p})||_n \right] \tag{5}$$

So the total loss L_{Total} becomes as in Eq. 6

$$L_{Total} = L_{VAE-GAN} + \lambda_{CC} L_{CC} \tag{6}$$

Let $\mathbb{E}$ be the set of all encoder, E_{pos}, E_{prior} and E_{con} and let $\mathbb{D}$ be the set of all discriminators, D and D^{VAE}. Then the training objective is the optimization problem given in Eq. 7.

$$G^*, \mathbb{E}^*, \mathbb{D}^* = arg \min_{G,\mathbb{E}} \max_{\mathbb{D}} L_{Total} \tag{7}$$

3.3 Architecture Details

We test the disengagement efficacy of our novel cycle consistency loss on BAIR and SMNIST datasets. While we adopt a VAE-GAN based SAVP, [12] video

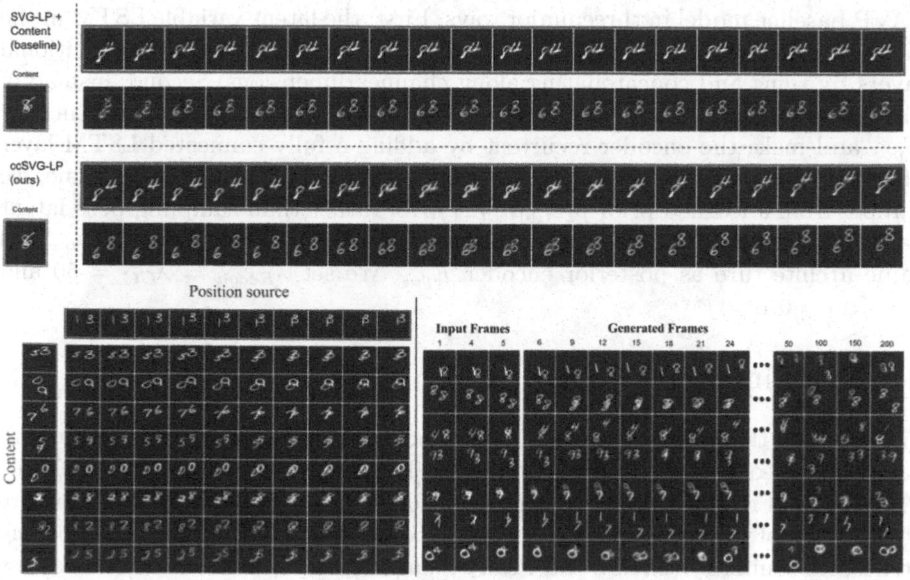

Fig. 2. Top: Comparison between our baseline and CC-SVGLP for learning disentangled. Pose latent variables from the red border sequence are used to generate a new sequence along with content information from the images in the green border. Bottom-left: We repeat the experiment from the top image on CC-SVGLP with multiple contents from images in the green border. Bottom-right: Each row is a long-range prediction of 200 frames by CC-SVGLP, given five context frames. (Color figure online)

prediction framework for the real-world BAIR dataset. In the case of, synthetic dataset, SMINST, a VAE based SVGLP [13] model with slight modification has been used in our experiments. Code for all the experiments will be released upon acceptance of this paper.

Architecture Details for SMNIST: The baseline model for SMNIST is SVG-LP. In this work, we condition the prediction model on previous frame encoding by using a recurrent DCGAN [30] content encoder E_{con}. The content encoder infers only one content encoding for all context frames [10]. We build upon this baseline by incorporating cycle consistency loss to get our cycle consistent SVGLP **CC-SVGLP** model. Weights for the gan losses L_{GAN} and L_{GAN}^{VAE}, α_1 and α_2 in Eq. 1, are set to 0. For CC-SVGLP, we set $\lambda_{Recon} = \lambda_{CC} = 0.5$ and for the baseline mentioned above we set it to $\lambda_{Recon} = 1$ & $\lambda_{CC} = 0$.

Architecture Details for BAIR: For the real-world dataset (BAIR), we adopt a VAE-GAN model, which is consistent cycle SAVP, **CC-SAVP**. SAVP is a conditional video prediction model in which the content information is passed to the generator as previous frames. As, the previous frame, x_{t-1} or $\hat{x}_{t-1}$ can be treated as the latent content representation, z_c, we do not condition the generator of SAVP with a content encoder. However, CC-SAVP differs from the

SAVP baseline model in three major ways. First, the latent variable LSTM in the generator has been removed and we pass $z_{t-1,p}$ along with $z_{t,p}$ to all convolutional layers by tiling and concatenating along channel dimensions. Second, instead of passing two frames $x_{t-1:t}$ we pass only one frame x_t to the posterior encoder E_{pos} and make the encoder recurrent by adding a fully connected LSTM layer after average pooling. Finally, for the sake of consistency, at inference time, we sample from a learned prior $p(z_{t,p}|x_{1:t-1})$ [13] rather than sampling pose latent variables $z_{t,p}$ from the standard normal distribution. Prior encoder E_{prior} has same architecture as posterior encoder E_{pos}. We set $\lambda_{Recon} = \lambda_{CC} = 50$ and $\alpha_1 = \alpha_2 = 0.1$.

4 Experiments

We evaluate the competence of our method in learning disentangled representation of video sequences and its effect on long-range predictions on synthetic, SMNIST and real-world, BAIR datasets. For BAIR, we quantitatively compare the fidelity of generated frames between SAVP and CC-SAVP by computing structural similarity (SSIM) and Peak Signal-to-Noise Ratio (PSNR). We assess the perceptual quality of the generated frames by computing the cosine similarity between VGG features [12].

4.1 Stochastic Moving MNIST

Stochastic Moving MNIST (SMNIST) dataset consists of sequences with one or more MNIST [5] digits bouncing in a frame of size 64×64. SMNIST digits move with constant velocity and change velocity and direction randomly upon collision with frame boundaries. We consider SVGLP with content encoder as the baseline architecture for this dataset. Figure 2 (Top) demonstrates the pose/content disentanglement result of both baseline architecture and the proposed CC-SVGLP. We use the pose representations $z_{2:T,p}$ from the predicted sequence, highlighted in red, and different content representation z_c from the image highlighted in green to generate 20 future frames. The digits in the generated sequence follow the pose of the source sequence, which clearly shows that our baseline is able to learn factorized video representations. However, the digits appear to be hovering around the center of the frame due to improper pose/content factorization. In the sequence generated by our model CC-SVGLP, the digits demonstrate an expressive display of movement. Thus, the cycle consistency loss is essential for proper disentanglement of video representations.

In Fig. 2 (Bottom-left) we show the pose/content disentanglement results of our approach, CC-SVGLP. The sequence on the top, highlighted in red, is the source sequence of the pose. We can see that the pose representation is agnostic to content in different frames. The digits in all the generated sequences follow the pose of digits from the source sequence irrespective of content. In Fig. 2 (Bottom-right), it can be seen that the frames generated by CC-SVGLP are sharp and depict varied video predictions up to 200 frames into the future. Note that all predictions are conditioned on the initial five ground truth frames.

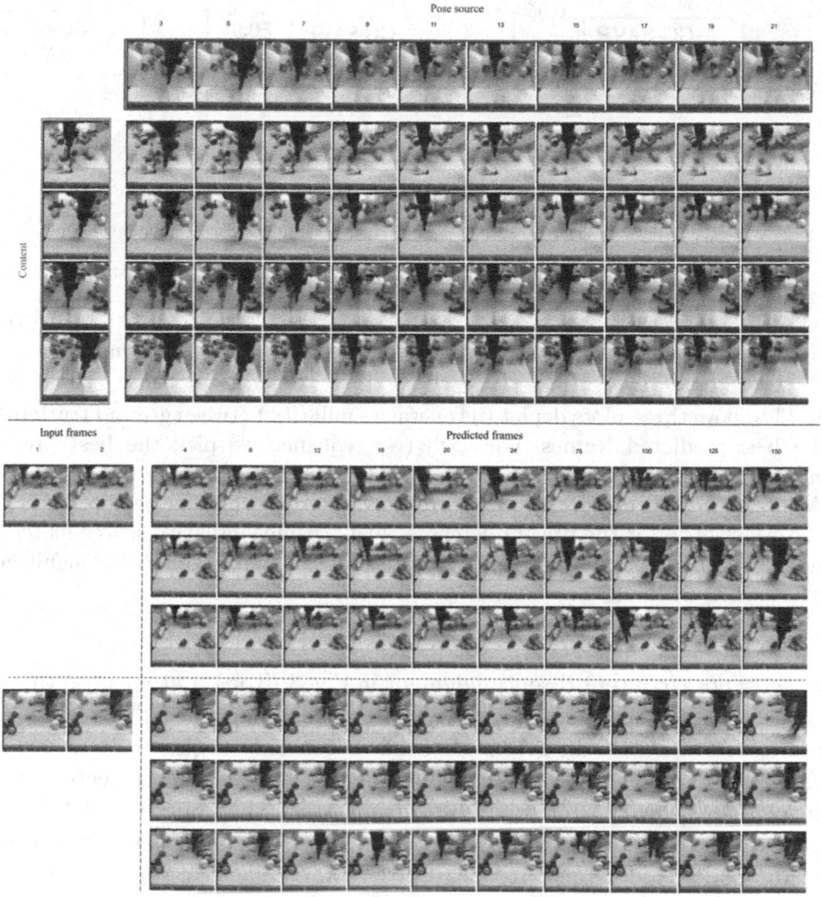

Fig. 3. Top: Demonstration of pose/content disentanglement on BAIR real-world video dataset by CC-SAVP. The pose information of the predicted sequence from the top row (in red) are applied on frames (in green) with different content information to generate subsequent sequences. It can be seen that the model has accurately learned to disentangle video representations. Bottom: Input frames on the left are used to generate multiple long-range stochastic video predictions. It can be seen that the system generates varied, sharp predictions up to 150-time steps. Note that the generator has been conditioned on only two context frames. (Color figure online)

4.2 BAIR Robot Pushing Dataset

BAIR robot pushing dataset [31] consists of video sequences of a robotic arm performing stochastic movements while interacting with various objects. The dataset contains a diverse set of objects and a large cluttered scene with 64×64 spatial resolution. The nature of the data set encapsulates a complex scene with evolving real-world dynamics, providing a challenging ground for video prediction.

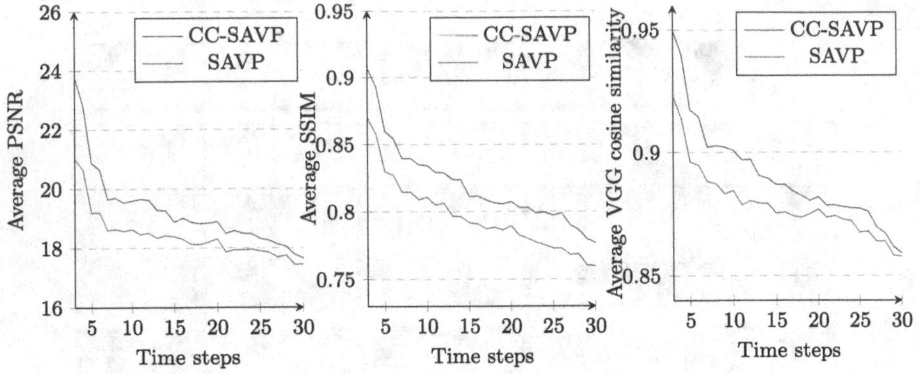

Fig. 4. The above three plots depict the average similarity between ground truth frames and the best predicted frames. For each test sequence we pick the best predicted sequence out of 100 predictions and average the similarity score over all test samples. Although, PSNR and SSIM [32] correspond poorly to human perception [33], we include these metric for the of completeness. VGG cosine similarity scores match better with human perception [33]. It can be seen that our model CC-SAVP significantly outperforms SAVP on all three evaluation metrics.

We show pose/content disentanglement in Fig. 3 (Top). Latent pose variables from the top predicted sequence in red are used along with different content information from frames in green are used to generate sequences shown in the grid. The robot arm in the generated sequences follows the trajectory of the arm in the pose source. These results demonstrate that our model CC-SAVP learns to disentangle pose from content. CC-SAVP is the first stochastic video prediction model to demonstrate clean pose/content factorization on real-world videos to the best of our knowledge.

In Fig. 3 (Bottom) long-range stochastic video predictions generated from our model are shown. We generate different possible future sequences for each input sequence by sampling from prior; it generates sharp and varied long-range predictions. We display 150 future frame predictions here, but our model can predict up to and beyond 500 future frames with graceful degradation over time. Cycle consistency loss supports long-range predictions by forcing the model to learn proper pose/content disentanglement.

In Fig. 4 we show the quantitative comparison between CC-SAVP and SAVP. These plots show the average similarity between the ground truth sequence and the closest predicted sequence. Left and middle plots are average PSNR and SSIM metric scores; these are not designed for the task of video prediction and are known to correspond to human judgment poorly. The third plot is average VGG cosine similarity, which has been shown to match human perception better. Our model CC-SAVP outperforms SAVP on all three metrics. We have used publicly available code to produce SAVP predictions.

5 Conclusion

We experimentally demonstrate the efficacy of our approach in learning pose/ content disentanglement on two different stochastic video predictions by simply conditioning the generator with time-invariant context information using a content encoder. Further, we formulate an easy to interpret and train cycle consistency loss term, which helps learn a refined predictive model of the time-variant stochastic latent variable pose. Our ablation results concretely establish the qualitative and quantitative advantages of using cycle consistency loss and our approach in learning factorized pose and content representation of long-range video prediction. By applying cycle consistency on two different stochastic video prediction models, we show that this approach is sufficiently general and can be easily incorporated in other similar latent variable video frame prediction models. Cycle consistency enables the model to learn complex distributions of the pose

References

1. Bubić, A., Cramon, D., Schubotz, R.: Prediction, cognition and the brain. Front. Hum. Neurosci. **4**, 25 (2010)
2. Lan, T., Chen, T.-C., Savarese, S.: A hierarchical representation for future action prediction. In: Fleet, D., Pajdla, T., Schiele, B., Tuytelaars, T. (eds.) ECCV 2014. LNCS, vol. 8691, pp. 689–704. Springer, Cham (2014). https://doi.org/10.1007/978-3-319-10578-9_45
3. Soran, B., Farhadi, A., Shapiro, L.: Generating notifications for missing actions: don't forget to turn the lights off! In: Proceedings of the IEEE International Conference on Computer Vision, pp. 4669–4677 (2015)
4. Mathieu, M., Couprie, C., LeCun, Y.: Deep multi-scale video prediction beyond mean square error. arXiv:1511.05440 (2015)
5. Srivastava, N., Mansimov, E., Salakhudinov, R.: Unsupervised learning of video representations using LSTMS. In: International Conference on Machine Learning, pp. 843–852 (2015)
6. Xue, T., Wu, J., Bouman, K., Freeman, B.: Visual dynamics: probabilistic future frame synthesis via cross convolutional networks. In: Advances in Neural Information Processing Systems, pp. 91–99 (2016)
7. Finn, C., Goodfellow, I., Levine, S.: Unsupervised learning for physical interaction through video prediction. In: Advances in Neural Information Processing Systems, pp. 64–72 (2016)
8. Paxton, C., Barnoy, Y., Katyal, K., Arora, R., Hager, G.D.: Visual robot task planning. arXiv:1804.00062 (2018)
9. Villegas, R., Yang, J., Hong, S., Lin, X., Lee, H.: Decomposing motion and content for natural video sequence prediction. arXiv:1706.08033 (2017)
10. Hsieh, J.-T., Liu, B., Huang, D.-A., Fei-Fei, L.F., Niebles, J.C.: Learning to decompose and disentangle representations for video prediction. In: Advances in Neural Information Processing Systems, pp. 517–526 (2018)
11. Denton, E.L., et al.: Unsupervised learning of disentangled representations from video. In: Advances in Neural Information Processing Systems, pp. 4414–4423 (2017)

12. Lee, A.X., Zhang, R., Ebert, F., Abbeel, P., Finn, C., Levine, S.: Stochastic adversarial video prediction. arXiv:1804.01523 (2018)
13. Denton, E., Fergus, R.: Stochastic video generation with a learned prior. arXiv:1802.07687 (2018)
14. Babaeizadeh, M., Finn, C., Erhan, D., Campbell, R.H., Levine, S.: Stochastic variational video prediction. arXiv:1710.11252 (2017)
15. Ranzato, M., Szlam, A., Bruna, J., Mathieu, M., Collobert, R., Chopra, S.: Video (language) modeling: a baseline for generative models of natural videos. arXiv:1412.6604 (2014)
16. Vondrick, C., Pirsiavash, H., Torralba, A.: Generating videos with scene dynamics. In: Advances in Neural Information Processing Systems, pp. 613–621 (2016)
17. Lotter, W., Kreiman, G., Cox, D.: Deep predictive coding networks for video prediction and unsupervised learning. arXiv:1605.08104 (2016)
18. Chiappa, S., Racaniere, S., Wierstra, D., Mohamed, S.: Recurrent environment simulators. arXiv:1704.02254 (2017)
19. Oh, J., Guo, X., Lee, H., Lewis, R.L., Singh, S.: Action-conditional video prediction using deep networks in atari games. In: Advances in Neural Information Processing Systems, pp. 2863–2871 (2015)
20. Jia, X., De Brabandere, B., Tuytelaars, T., Gool, L.V.: Dynamic filter networks. In: Advances in Neural Information Processing Systems, pp. 667–675 (2016)
21. Liu, W., Sharma, A., Camps, O., Sznaier, M.: Dyan: a dynamical atoms-based network for video prediction. In: European Conference on Computer Vision, pp. 175–191. Springer (2018)
22. Walker, J., Gupta, A., Hebert, M.: Dense optical flow prediction from a static image. In: Proceedings of the IEEE International Conference on Computer Vision, pp. 2443–2451 (2015)
23. Henaff, M., Zhao, J., LeCun, Y.: Prediction under uncertainty with error-encoding networks. arXiv:1711.04994 (2017)
24. Kingma, D.P., Welling, M.: Auto-encoding variational bayes. arXiv:1312.6114 (2013)
25. Goodfellow, I., et al.: Generative adversarial nets. In: Advances in Neural Information Processing Systems, pp. 2672–2680 (2014)
26. Zhou, Y., Berg, T.L.: Learning temporal transformations from time-lapse videos. In: Leibe, B., Matas, J., Sebe, N., Welling, M. (eds.) ECCV 2016. LNCS, vol. 9912, pp. 262–277. Springer, Cham (2016). https://doi.org/10.1007/978-3-319-46484-8_16
27. Kalchbrenner, N., et al.: Video pixel networks. In: Proceedings of the 34th International Conference on Machine Learning-Volume 70. JMLR.org, pp. 1771–1779 (2017)
28. Salimans, T., Karpathy, A., Chen, X., Kingma, D.P.: PixelCNN++: improving the pixelCNN with discretized logistic mixture likelihood and other modifications. arXiv:1701.05517 (2017)
29. Reed, S., et al.: Parallel multiscale autoregressive density estimation. arXiv:1703.03664 (2017)
30. Radford, A., Metz, L., Chintala, S.: Unsupervised representation learning with deep convolutional generative adversarial networks. arXiv:1511.06434 (2015)
31. Ebert, F., Finn. C., Lee, A.X., Levine, S.: Self-supervised visual planning with temporal skip connections. arXiv:1710.05268 (2017)

32. Wang, Z., Bovik, A.C., Sheikh, H.R., Simoncelli, E.P., et al.: Image quality assessment: from error visibility to structural similarity. IEEE Trans. Image Process. **13**(4), 600–612 (2004)
33. Zhang, R., Isola, P., Efros, A.A., Shechtman, E., Wang, O.: The unreasonable effectiveness of deep features as a perceptual metric. In: Proceedings of the IEEE Conference on Computer Vision and Pattern Recognition, pp. 586–595 (2018)

SeeFar: Vehicle Speed Estimation and Flow Analysis from a Moving UAV

Mang Ning[1], Xiaoliang Ma[1(✉)], Yao Lu[2], Simone Calderara[3], and Rita Cucchiara[3]

[1] KTH Royal Institute of Technology, Stockholm, Sweden
{mangn,liang}@kth.se

[2] University of Bristol, Bristol, UK
yl1220@my.bristol.ac.uk

[3] University of Modena and Reggio Emilia, Modena, Italy
{simone.calderara,rita.cucchiara}@unimore.it

Abstract. Visual perception from drones has been largely investigated for Intelligent Traffic Monitoring System (ITMS) recently. In this paper, we introduce SeeFar to achieve vehicle speed estimation and traffic flow analysis based on YOLOv5 and DeepSORT from a moving drone. SeeFar differs from previous works in three key ways: the speed estimation and flow analysis components are integrated into a unified framework; our method of predicting car speed has the least constraints while maintaining a high accuracy; our flow analysor is direction-aware and outlier-aware. Specifically, we design the speed estimator only using the camera imaging geometry, where the transformation between world space and image space is completed by the variable Ground Sampling Distance. Besides, previous papers do not evaluate their speed estimators at scale due to the difficulty of obtaining the ground truth, we therefore propose a simple yet efficient approach to estimate the true speeds of vehicles via the prior size of the road signs. We evaluate SeeFar on our ten videos that contain 929 vehicle samples. Experiments on these sequences demonstrate the effectiveness of SeeFar by achieving 98.0% accuracy of speed estimation and 99.1% accuracy of traffic volume prediction, respectively.

Keywords: Visual speed estimation · Traffic monitoring system

1 Introduction

ITMS has been widely developed in traffic light control [18], vehicle violation detection [14] and emergency control [35]. Benefiting from the progress in computer vision, visual perception has been researched in ITMS. By deploying cameras at traffic intersections, technology such as vehicle counting [4], density estimation [6], and vehicle speed estimation [44] are all developed on the basis of CNNs.

This study is carried out in the project METRIC and financially supported by the Swedish Transport Admistration. The computation was enabled by resources provided by the Swedish National Infrastructure for Computing (SNIC).

S. Sclaroff et al. (Eds.): ICIAP 2022, LNCS 13233, pp. 278–289, 2022.
https://doi.org/10.1007/978-3-031-06433-3_24

Different from the limited field of view and high deployment cost of static traffic camera, drone-based ITMS has gained extensive attention in recent years [1,5,13,22]. Relying on object detection and multi-object tracking (MOT) technology, some work [1,5,9,24] have implemented vehicle tracking and speed estimation from the perspective of Unmanned Aerial Vehicles (UAVs). But the instability of the UAV's flight state brin additional noise to visual perception tasks. Hence, previous work on vehicle speed estimation from drones imposed various constraints on the system, for example, the UAV must be stationary [9,24] or the car size must be a prior [5]. Besides, obtaining the real speed information of vehicles is very tricky and expensive, resulting in insufficient benchmarking [1] in previous papers. Finally, studies of integrating speed estimation and traffic volume analysis are still missing.

To address these three issues, we propose SeeFar, an end-to-end system that employs YOLOv5 [20] and DeepSORT [43] as the base components to perform speed estimation and traffic flow analysis (see Fig. 1). We demonstrate that a simple speed estimator grounded on the geometry of the camera imaging system can achieve state-of-the-art performance without any object prior or camera perspective preconditions. Furthermore, we design a novel method for estimating the ground truth vehicle speed via the prior size of road arrows. This allows us to benchmark our speed estimator from a large amount of video data at a low cost. Finally, our flow analysor provides rich traffic information, including the direction of the main traffic flow, the cars outside of the main flow, and the traffic volume. Overall, the contribution of this paper is three-fold:

1. SeeFar is the first end-to-end system that achieves speed estimation and traffic flow analysis for traffic monitoring from a UAV.
2. Our designed speed estimator is simple and performs accurately under little preconditions and constraints.
3. We present a novel method that addresses the difficulty of obtaining ground truth labels of car speed in public roads.

2 Related Work

2.1 Object Detection

Object detection, also known as localising and classifying the target object in images, has demonstrated remarkable performance in deep learning domain. Object detection models are often classified into two categories: two-stage detectors and one-stage detectors.

R-CNN family [16,17,19,36] proposed by Girshick and Ren et al. are the representative two-stage models. They employ a coarse-to-fine pipeline, generating a large number of candidate objects and regressing their coordinates further. In comparison, a one-stage detector predicts the object class and corresponding coordinates in an end-to-end manner. The typical one-stage detectors are YOLO series [7,15,20,32–34]. The latest YOLOv5 is optimised on YOLOv3 by incorporating Cross Stage Partial component [41], Path Aggregation Network [25]

Fig. 1. A prediction demo of SeeFar. The two main vehicle flows are identified and represented by the green boxes and blue boxes. By contrast, the red boxes are detected outliers of the main traffic flow. The cars with yellow boxes are being calculated their speed. The central blue and yellow areas are responsible for computing the traffic volume of the main flows, the results are present in the top left corner. Finally, the green and blue dots suggest the historical positions of cars. (Color figure online)

and adaptive image size. Apart from YOLO, the performance of other models (CenterNet [12] and FCOS [39]) is also noteworthy.

2.2 Multi-Object Tracking

Tracking-by-detection paradigm [2,3,8,42,43] is the mainstream in MOT due to the success of CNNs. Ciaparrone et al. [10] summarise MOT as 4 components: object detection, feature extraction, data association and track management.

A popular tracking-by-detection model is SORT [3] which utilises IOU to measure the distance of object pairs. Furthermore, Wojke et al. propose DeepSORT [43] to overcome the ID-switch issue of SORT. To better handle the unreliable detections, Chen et al. [8] design a scoring mechanism to select good candidates. Moreover, JDE algorithm [42] achieves a faster inference speed by fusing the object detector and the appearance embedding model.

In addition to the explicit distance metric, people also use RNNs to directly predict tracks [29,31]. Recently, Transformer [40] has been introduced into MOT. TransTrack [37] updates the existing tracks and predict the new tracks via two separate queries. Also, Meinhardt et al. [28] demonstrate an encoder-decoder architecture by formulating MOT as a frame-to-frame set prediction problem.

2.3 Visual Speed Estimation

Predicting the speed of an object through monocular vision is challenging because both the camera and the object itself are moving. The key step is to build the transformation from pixel coordinate space to the world coordinate system. In the early stage, the camera has to be mounted on the traffic signal pole to simplify the conditions [44], which is inapplicable for UAV scenarios.

Li et al. [24] implement the speed estimation from a moving UAV by taking the car size as a prior. However, the main limitation of this approach is that the video must be captured from an exact top-view camera. In a similar fashion, Chen et al. [9] pre-sample the transformation curve with the same condition of top view images. Likewise, Biswas et al. [5] divide all vehicles into two groups and utilise the average length as the prior. Thus, the result is a rough estimation of the actual speed.

To overcome the constraints of top view imaging and rough prior of vehicle sizes, MultEye [1] estimate the speed according to the camera imaging geometry. However, MultEye requires the target car to be located in the centre area of the image to ensure a 2D plane speed estimation. To remove this limitation, we apply car projection to extend the speed estimation to 3D space, presenting a robust speed estimator. Table 1 summarises the constraints of the aforementioned models and SeeFar has an obvious advantage of application scenarios.

Table 1. Model comparison of visual speed estimation from UAVs

	Without constrains of					
	Top view	Static drone	Prior car size	Image registration	Car position	Unavailable ground truth
Li et al. [24]	✗	✗	✗	✗	✔	✗
Chen et al. [9]	✗	✗	✗	✗	✔	✗
Biswas et al. [5]	✔	✔	✗	✗	✔	✗
MultEYE [1]	✔	✔	✔	✔	✗	✗
SeeFar (Our)	✔	✔	✔	✔	✔	✔

3 Methods

3.1 System Architecture

The whole architecture SeeFar is shown in Fig. 2. First, the detector predicts the positions of all vehicles with bounding boxes (x, y, w, h). Subsequently, the object tracker associates bounding boxes over frames and assign each car with a unique ID. After that, the speed estimator calculate the speed based on the drone's state and camera's parameters. In the meanwhile, the speed estimator outputs the displacement vector $\vec{D}_{box}$ and motion direction md for each car, providing the prerequisite for subsequent flow analysis. Finally, the vehicle flow analysor first identify the two flow directions and exclude the outliers. Then, only these cars belonging to the vehicle flow are marked as flow state $fs = in$ and are used to compute the traffic volume.

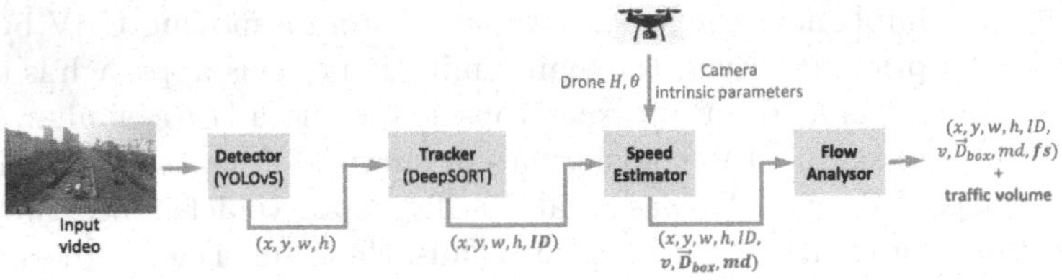

Fig. 2. The architecture of SeeFar. YOLOv5 and DeepSORT are the base modules of SeeFar. Taking extra drone states and intrinsic camera parameters as inputs, Speed Estimator predicts the speed v, motion vector $\vec{D}_{box}$ and motion direction md. Finally, Flow Analysor identifies the flow state fs for each car and calculate the traffic volume.

3.2 Detector and Tracker

In this paper, we optimise YOLOv5 [20] to detect vehicles as YOLOv5 can be configured to arbitrary size by just adjusting the depth factor and width factor. In practice, we apply grid search for these two factors and find that YOLOv5m is suitable for car detection regarding the tradeoff between accuracy and speed. Considering that most of the vehicles have relatively small sizes from the view of a flying UAV, we optimise vanilla YOLOv5m by doubling the channel size of the small object detection branch and reducing half channels of the large object detection branch. We benchmark our models on the dataset UAVDT-M [11] and our optimised model achieves a higher mean average precision (0.8%) while saving 16% of model parameters.

As for the object tracking, we employed DeepSORT [43] as the meta model. DeepSORT applies Kalman Filter [21] to do track prediction for the next time step. Then, the appearance similarity will be measured between the predicted tracks and newly arrived detections. After that, Hungarian algorithm [23] accomplishes data association and the track management updates the status of each track according to the heuristic rules. To better handle the ID switch issue in car occlusion, we train the Re-ID network of DeepSORT on the dataset VeRi-776 [26] which contains 50k images for 776 types of cars.

3.3 Speed Estimator

The object displacement $\vec{D}_{box}$ in the pixel space can be easily computed by tracking the object motions. Therefore, we can estimate the real speed of vehicles once we know the transformation from pixel space to world space. Fortunately,

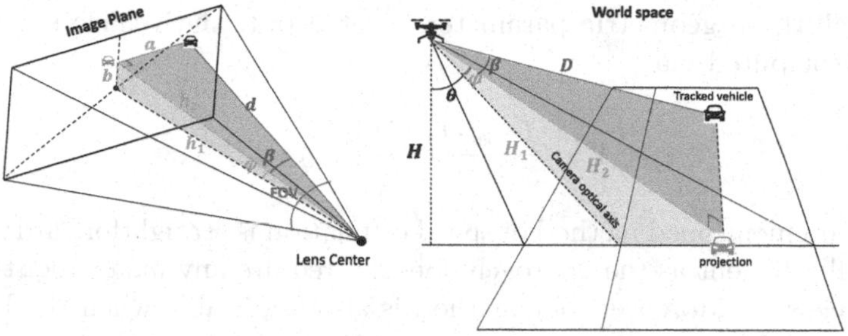

Fig. 3. The Geometry illustration of the UAV camera imaging. The green triangles in the image plane and world space share the common angel ϕ. Similarly, the red triangles have the same angle β. (Color figure online)

this transformation relationship can be directly deriv the geometry of the camera imaging system (see Fig. 3).

Assuming that the UAV moves following the direction of the traffic flow at a speed $\vec{V}_{UAV}$. The height of the UAV is H and the tilt angle of the camera is θ. We need to know the distance D between the target car and the camera to derive the Ground Sampling Distance (GSD). GSD stands for the size of a single pixel in the real world and can be computed via Eq. 1 where Sw is the actual width of the camera sensor and Pw is the amount of pixels in width.

$$GSD = \frac{D}{d} \cdot \frac{Sw}{Pw} \tag{1}$$

To compute the distance d and D, we first project the car into the vertical plane that the UAV belongs to. In this case, the distance d and D are both solvable if we know the length h_2, H_2 and the angle β.

From the lens centre to the image plane, we have two right triangles to utilise. In the green triangle, h_2 and ϕ can be derived from

$$tan\phi = \frac{b}{h_1} \quad , \quad h_2 = \sqrt{h_1^2 + b^2} \tag{2}$$

where h_1 is the focal length of the camera that we already know. b is the vertical distance from the object to the image plane centre. Given the coordinate of the object (x, y) in pixel space (w, h), a and b is calculated by

$$a = \left(x - \frac{w}{2}\right) \cdot \frac{Sw}{Pw} \quad , \quad b = \left(y - \frac{h}{2}\right) \cdot \frac{Sw}{Pw} \tag{3}$$

Therefore, we get β and d from the red triangle by the following equations

$$tan\beta = \frac{a}{h_2} \quad , \quad d = \sqrt{h_2^2 + a^2} \tag{4}$$

Also, in the world space, H_2 and D can be derived from

$$H_2 = \frac{H}{cos(\theta + \phi)} \quad , \quad D = \frac{H_2}{cos(\beta)} \tag{5}$$

With all those geometric parameters figured out, the vehicle speed V can finally be computed via

$$V = \frac{\vec{D}_{box} \cdot GSD}{t} - \vec{V}_{UAV} \tag{6}$$

The abovementioned method of speed estimation is straightforward and computationally efficient as the approach does not require any image registration or prior vehicle size. Moreover, our method is also applicable when the UAV flies above the vehicle flow and the cars move vertically in the image.

3.4 Flow Analysor

We design the flow analysor module to measure the traffic volume. However, there is often the case that many vehicles outside the main road are also captured by the drone camera. Thus it is necessary to identify and remove these 'outliers' to ensure an accurate estimation of the traffic volume. Correspondingly, we propose a rule-based system to analyse the vehicle flow.

In the first step, we sample the vehicles and maintain a history of tracked vehicles. Then we divide these cars into five bins (*up*, *down*, $left$, $right$, *static*) according to the direction and magnitude of the displacement vector $\vec{D}_{box}$. The two bins with the largest number of vehicles indicate the main directions of the traffic flow. We also calculate and preserve the average displacement vector ($\vec{D}_{flow1}$, $\vec{D}_{flow2}$) for both vehicle flows.

In the second step, we aim at identifying all outliers. To this end, the motion angle γ is calculated between each new car and the two flow directions ($\vec{D}_{flow1}$ and $\vec{D}_{flow2}$). Then we measure the distance s between each new vehicle and the centre of the historical vehicle cluster. Since the vehicle cluster roughly forms a linear distribution along the flow direction, it is optimal to apply Mahalanobis Distance [27] to remove the correlation and scale influence. Therefore, the distance s is computed by Eq. 7 where S is the covariance matrix of the vehicle cluster and $\bar{x}$ is the centre of the vehicle cluster. If both γ and s are under the predefined thresholds, the vehicle is considered as a valid sample and its flow state fs is assigned with in.

$$s = \sqrt{(x - \bar{x})^T S^{-1} (x - \bar{x})} \tag{7}$$

In the final step, the traffic volume is computed in the 'counter region' which is perpendicular to the motion direction of the vehicle flow. To construct this region, the mean of $\vec{D}_{flow1}$ and $\vec{D}_{flow2}$ is utilised to represent the overall motion direction $\vec{D}_{avg}$. Eventually, the car moving across the blue and yellow region contributes to the traffic volume.

Fig. 4. The failure scenario of our speed estimator

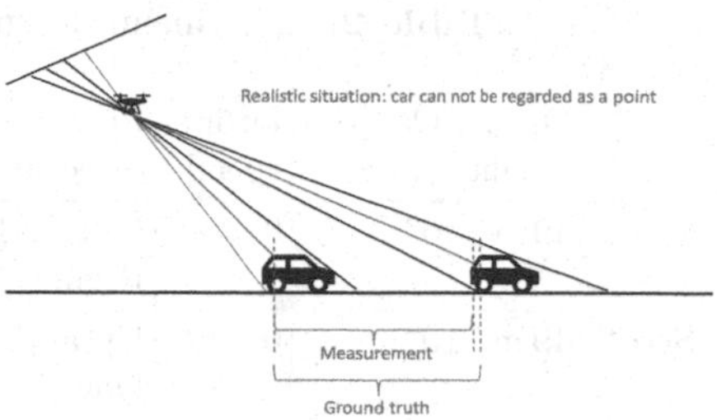

Fig. 5. The centroid shift of the bounding box of the car in sequence 4

4 Experiments and Analysis

4.1 Speed Estimator

Because of lacking the ground truth information of vehicle speed, previous work only benchmark their model on a few cars or even bicycles [1,5,9]. To tackle this problem, we propose a novel way to get the ground truth labels for each car. Concretely, we take the road ground signs, for example road arrows, as the reference object whose length is usually known. Thus, the transformation from the pixel space to world space is derived by the road arrow prior. We find that this approach also works when the drone moves along with the vehicle flow.

To fully evaluate the performance of our speed estimator, we fly the UAV in diverse combinations of drone height, camera tile angle, drone speed and traffic flow direction. The *Avg.accuracy* is adopted as the main metric for a test video and it is the mean of speed accuracy of all cars in the video. Note that, the codes of previous work are unavailable, we thereby can not compare directly with them under the same dataset.

We now discuss the results in Table 2. Overall, our speed estimator achieves the accuracy ranging from 97.0% to 98.8% over the 7 sequences and 698 test vehicles, showing the effectiveness and consistency of our model.

We can further summarise three observations from the table. First, the contrast of sequences 1–3 and sequences 5–8 indicates that measuring the speed from a hovering drone is slightly more stable than measuring from a moving UAV. This is in line with our anticipation because estimating the parameters of the drone state suffers from extra noise when the drone moves. Second, the change of drone height and camera tile angle does not affect the performance of our model. This conclusion can be drawn from the consistent accuracy results of sequences 5–8 where drone height varies from 41 m to 55 m and the camera angle fluctuates between 15° and 35°. Third, the centroid shift of the bounding box of the car lead to non-negligible error in video 4, where the traffic flow moves vertically and the camera angle is non-zero. It is clear from Fig. 4 that the speed is getting more and more underestimated while the car is moving away from the camera.

Table 2. Experiment design and results of speed estimator

	Drone height	Camera angle	Drone speed	Flow direction	Car amount	Avg. gt speed	Avg. error (km/h)	Avg. accuracy
Seq 1	49 m	0°	0	Left and Right	119	62.4	0.76	98.8%
Seq 2	49 m	0°	0	Up and Down	152	59.6	0.79	98.7%
Seq 3	49 m	25°	0	Left and Right	109	60.8	1.38	97.7%
Seq 4	49 m	25°	0	Up and Down	–	–	–	–
Seq 5	55 m	15°	20	Left and Right	95	64.6	1.7	97.3%
Seq 6	45 m	15°	20	Left and Right	77	72.1	1.54	97.9%
Seq 7	45 m	35°	20	Left and Right	74	60.6	1.65	97.3%
Seq 8	41 m	35°	20	Left and Right	72	66.2	1.64	97.5%
Mean	–	–	–	–	698(total)	–	–	**98.0%**

The reason for the speed error is visualised in Fig. 5 which reveals the movement of the centre point of the bounding box (car) in the image plane. Since we compute the displacement of the car by tracking its centre point, the measurement is therefore always smaller than the ground truth in the scenario of video 4. However, this flaw of our algorithm can be solved either by traditional image registration [38] or deep learning image correction [30]. We leave this improvement to our future work.

4.2 Flow Analysor

The main functions of Flow Analysor are main flow identification, outlier classification and traffic volume estimation. We test the performance of Flow Analysor on another two video sequences where the traffic scene is noisy and there exist many cars out of the main road (red boxes in Fig. 1). The results in Table 3 present the high accuracy 99.1% of our models with only one inlier being classified as an outlier. We investigate the wrongly classified sample and find that it is actually caused by a jitter of the UAV.

Table 3. Experiment results of flow analysor

Model	Outliers	Inliers	Predicted outliers	Precision	Recall	Accuracy
Outlier classification	44	187	45	97.7%	100%	99.1%
Traffic volume estimation				100%	99.5%	99.1%

5 Conclusion

In this paper, we address the problem of heavy constraints in visual speed estimation from UAVs and fill in the gap of the end-to-end ITMS that integrates speed estimation and flow analysis. First, we have shown that a speed estimator modelled by simple geometric relationships can perform accurately in diverse drone settings. Combining the flow analysis module, we constructed a unified traffic monitoring system SeeFar to provide informative perception data of vehicles. Finally, a low-cost way of estimating the true vehicle speed is presented and is promising to be a labelling method for large-scale datasets of vehicle speed estimation. However, our visual speed estimator suffers from the centroid shift error of object detection in one specific scenario and the detector and tracker of SeeFar are the bottlenecks regarding the computational cost. In future work, we will investigate eliminating the centroid shift error by image registration and achieve real-time SeeFar by optimising the object detector and tracker.

References

1. Balamuralidhar, N., Tilon, S., Nex, F.: MultEye: monitoring system for real-time vehicle detection, tracking and speed estimation from UAV imagery on edge-computing platforms. Remote Sens. **13**(4), 573 (2021)
2. Bergmann, P., Meinhardt, T., Leal-Taixe, L.: Tracking without bells and whistles. In: Proceedings of the IEEE/CVF International Conference on Computer Vision, pp. 941–951 (2019)
3. Bewley, A., Ge, Z., Ott, L., Ramos, F., Upcroft, B.: Simple online and realtime tracking. In: 2016 IEEE International Conference on Image Processing (ICIP), pp. 3464–3468. IEEE (2016)
4. Biswas, D., Su, H., Wang, C., Blankenship, J., Stevanovic, A.: An automatic car counting system using overfeat framework. Sensors **17**(7), 1535 (2017)
5. Biswas, D., Su, H., Wang, C., Stevanovic, A.: Speed estimation of multiple moving objects from a moving UAV platform. ISPRS Int. J. Geo Inf. **8**(6), 259 (2019)
6. Biswas, D., Su, H., Wang, C., Stevanovic, A., Wang, W.: An automatic traffic density estimation using single shot detection (SSD) and mobilenet-SSD. Phys. Chem. Earth Parts A/B/C **110**, 176–184 (2019)
7. Bochkovskiy, A., Wang, C.Y., Liao, H.Y.M.: YOLOv4: optimal speed and accuracy of object detection. arXiv:2004.10934 (2020)
8. Chen, L., Ai, H., Zhuang, Z., Shang, C.: Real-time multiple people tracking with deeply learned candidate selection and person re-identification. In: 2018 IEEE International Conference on Multimedia and Expo (ICME), pp. 1–6. IEEE (2018)

9. Chen, Y., Zhao, D., Er, M.J., Zhuang, Y., Hu, H.: A novel vehicle tracking and speed estimation with varying UAV altitude and video resolution. Int. J. Remote Sens. **42**(12), 4441–4466 (2021)
10. Ciaparrone, G., Sánchez, F.L., Tabik, S., Troiano, L., Tagliaferri, R., Herrera, F.: Deep learning in video multi-object tracking: a survey. Neurocomputing **381**, 61–88 (2020)
11. Du, D., et al.: The unmanned aerial vehicle benchmark: object detection and tracking. In: Proceedings of the European Conference on Computer Vision (ECCV), pp. 370–386 (2018)
12. Duan, K., Bai, S., Xie, L., Qi, H., Huang, Q., Tian, Q.: CenterNet: keypoint triplets for object detection. In: Proceedings of the IEEE/CVF International Conference on Computer Vision, pp. 6569–6578 (2019)
13. Elloumi, M., Dhaou, R., Escrig, B., Idoudi, H., Saidane, L.A.: Monitoring road traffic with a UAV-based system. In: 2018 IEEE Wireless Communications and Networking Conference (WCNC), pp. 1–6. IEEE (2018)
14. Franklin, R.J., et al.: Traffic signal violation detection using artificial intelligence and deep learning. In: 2020 5th International Conference on Communication and Electronics Systems (ICCES), pp. 839–844. IEEE (2020)
15. Ge, Z., Liu, S., Wang, F., Li, Z., Sun, J.: YOLOx: exceeding YOLO series in 2021. arXiv:2107.08430 (2021)
16. Girshick, R.: Fast R-CNN. In: Proceedings of the IEEE International Conference on Computer Vision, pp. 1440–1448 (2015)
17. Girshick, R., Donahue, J., Darrell, T., Malik, J.: Rich feature hierarchies for accurate object detection and semantic segmentation. In: Proceedings of the IEEE Conference on Computer Vision and Pattern Recognition, pp. 580–587 (2014)
18. Guo, M., Wang, P., Chan, C.Y., Askary, S.: A reinforcement learning approach for intelligent traffic signal control at urban intersections. In: 2019 IEEE Intelligent Transportation Systems Conference (ITSC), pp. 4242–4247. IEEE (2019)
19. He, K., Gkioxari, G., Dollár, P., Girshick, R.: Mask R-CNN. In: Proceedings of the IEEE International Conference on Computer Vision, pp. 2961–2969 (2017)
20. glenn jocher: yolov5. github (2021)
21. Kalman, R.E.: A new approach to linear filtering and prediction problems (1960)
22. Khan, N.A., Jhanjhi, N., Brohi, S.N., Usmani, R.S.A., Nayyar, A.: Smart traffic monitoring system using unmanned aerial vehicles (UAVS). Comput. Commun. **157**, 434–443 (2020)
23. Kuhn, H.W.: The hungarian method for the assignment problem. Naval Res. Logist. Quar. **2**(1–2), 83–97 (1955)
24. Li, J., Chen, S., Zhang, F., Li, E., Yang, T., Lu, Z.: An adaptive framework for multi-vehicle ground speed estimation in airborne videos. Remote Sens. **11**(10), 1241 (2019)
25. Liu, S., Qi, L., Qin, H., Shi, J., Jia, J.: Path aggregation network for instance segmentation. In: Proceedings of the IEEE Conference on Computer Vision and Pattern Recognition, pp. 8759–8768 (2018)
26. Liu, X., Liu, W., Mei, T., Ma, H.: A deep learning-based approach to progressive vehicle re-identification for urban surveillance. In: Leibe, B., Matas, J., Sebe, N., Welling, M. (eds.) ECCV 2016. LNCS, vol. 9906, pp. 869–884. Springer, Cham (2016). https://doi.org/10.1007/978-3-319-46475-6_53
27. McLachlan, G.J.: Mahalanobis distance. Resonance **4**(6), 20–26 (1999)
28. Meinhardt, T., Kirillov, A., Leal-Taixe, L., Feichtenhofer, C.: TrackFormer: multi-object tracking with transformers. arXiv:2101.02702 (2021)

29. Milan, A., Rezatofighi, S.H., Dick, A., Reid, I., Schindler, K.: Online multi-target tracking using recurrent neural networks. In: Thirty-First AAAI Conference on Artificial Intelligence (2017)
30. Palazzi, A., Borghi, G., Abati, D., Calderara, S., Cucchiara, R.: Learning to map vehicles into bird's eye view. In: Battiato, S., Gallo, G., Schettini, R., Stanco, F. (eds.) ICIAP 2017. LNCS, vol. 10484, pp. 233–243. Springer, Cham (2017). https://doi.org/10.1007/978-3-319-68560-1_21
31. Ran, N., Kong, L., Wang, Y., Liu, Q.: A robust multi-athlete tracking algorithm by exploiting discriminant features and long-term dependencies. In: Kompatsiaris, I., Huet, B., Mezaris, V., Gurrin, C., Cheng, W.-H., Vrochidis, S. (eds.) MMM 2019. LNCS, vol. 11295, pp. 411–423. Springer, Cham (2019). https://doi.org/10.1007/978-3-030-05710-7_34
32. Redmon, J., Divvala, S., Girshick, R., Farhadi, A.: You only look once: unified, real-time object detection. In: Proceedings of the IEEE Conference on Computer Vision and Pattern Recognition, pp. 779–788 (2016)
33. Redmon, J., Farhadi, A.: YOLO9000: better, faster, stronger. In: Proceedings of the IEEE Conference on Computer Vision and Pattern Recognition, pp. 7263–7271 (2017)
34. Redmon, J., Farhadi, A.: YOLOv3: an incremental improvement. arXiv:1804.02767 (2018)
35. Rego, A., Garcia, L., Sendra, S., Lloret, J.: Software defined network-based control system for an efficient traffic management for emergency situations in smart cities. Futur. Gener. Comput. Syst. **88**, 243–253 (2018)
36. Ren, S., He, K., Girshick, R., Sun, J.: Faster R-CNN: towards real-time object detection with region proposal networks. Adv. Neural Inf. Process. Syst. **28** (2015)
37. Sun, P., et al.: Transtrack: Multiple object tracking with transformer. arXiv:2012.15460 (2020)
38. Szeliski, R., et al.: Image alignment and stitching: a tutorial. Found. Trends® Comput. Graph. Vis. **2**(1), 1–104 (2007)
39. Tian, Z., Shen, C., Chen, H., He, T.: Fcos: fully convolutional one-stage object detection. In: Proceedings of the IEEE/CVF International Conference on Computer Vision, pp. 9627–9636 (2019)
40. Vaswani, A., et al.: Attention is all you need. Adv. Neural Inf. Process. Syst. **30** (2017)
41. Wang, C.Y., et al.: Cspnet: A new backbone that can enhance learning capability of CNN. In: Proceedings of the IEEE/CVF Conference on Computer Vision and Pattern Recognition Workshops, pp. 390–391 (2020)
42. Wang, Z., Zheng, L., Liu, Y., Li, Y., Wang, S.: Towards real-time multi-object tracking. In: Vedaldi, A., Bischof, H., Brox, T., Frahm, J.-Mil. (eds.) ECCV 2020. LNCS, vol. 12356, pp. 107–122. Springer, Cham (2020). https://doi.org/10.1007/978-3-030-58621-8_7
43. Wojke, N., Bewley, A., Paulus, D.: Simple online and realtime tracking with a deep association metric. In: 2017 IEEE International Conference on Image Processing (ICIP), pp. 3645–3649. IEEE (2017)
44. Zhiwei, H., Yuanyuan, L., Xueyi, Y.: Models of vehicle speeds measurement with a single camera. In: 2007 International Conference on Computational Intelligence and Security Workshops (CISW 2007), pp. 283–286. IEEE (2007)

Spatial-Temporal Autoencoder with Attention Network for Video Compression

Neetu Sigger[1(✉)], Naseer Al-Jawed[1], and Tuan Nguyen[2]

[1] School of Computing, The University of Buckingham, Buckingham, UK
{neetu.sigger,naseeraljawed}@buckingham.ac.uk

[2] School of Computer and Mathematical Sciences, University of Greenwich, London, UK
tuan.nguyen@greenwich.ac.uk

Abstract. Deep learning-based approaches are now state of the art in numerous tasks, including video compression, and are having a revolutionary influence in video processing. Recently, learned video compression methods exhibit a fast development trend with promising results. In this paper, taking advantage of the powerful non-linear representation ability of neural networks, we replace each standard component of video compression with a neural network. We propose a spatial-temporal video compression network (STVC) using the spatial-temporal priors with an attention module (STPA). On the one hand, joint spatial-temporal priors are used for generating latent representations and reconstructing compressed outputs because efficient temporal and spatial information representation plays a crucial role in video coding. On the other hand, we also added an efficient and effective Attention module such that the model pays more effort on restoring the artifact-rich areas. Moreover, we formalize the rate-distortion optimization into a single loss function, in which the network learns to leverage the Spatial-temporal redundancy presented in the frames and decreases the bit rate while maintaining visual quality in the decoded frames. The experiment results show that our approach delivers the state-of-the-art learning video compression performance in terms of MS-SSIM and PSNR.

Keywords: Video compression · Deep learning · Auto-encoder · Rate-distortion optimization · Attention mechanism

1 Introduction

There have been high demands on video compression over recent years, with the efficient transmission of top resolution and high-quality video data over the bandwidth-limited Internet. Especially during the COVID pandemic, the increasing data traffic was used for online classes, virtual meetings, Netflix, YouTube, online gaming, etc. As a result, there's a growing demand for more practical video compression schemes to speed up the method of exchanging visual media over the bandwidth-limited Internet. By utilizing redundancies within the data to produce a more miniature representation, compression methods are employed to handle the expanding sizes of stored media.

S. Sclaroff et al. (Eds.): ICIAP 2022, LNCS 13233, pp. 290–300, 2022.
https://doi.org/10.1007/978-3-031-06433-3_25

Deep Learning (DL) is revolutionizing image and video processing, and DL-based techniques are state-of-the-art in several related problems such as classification, detection, or compression [1, 2].

In the past few decades, several video compression algorithms have been standardized, e.g., MPEG [3], H.264 [4], and H.265 [5]. These standards are hand-made, and the components in the compression framework cannot be optimized together. Motivated by the success of the deep neural networks (DNN) in improving image compression rate-distortion performance [6, 7] various DNN based video compression architectures [8, 9] have been developed. In these learned video compression methods, the whole network is optimized in an end-to-end manner. The DL-based image compression relies on the ability of DNN to extract meaningful representations of two-dimensional data because the latent space of an image represented by the network must contain information about the most important features and structures in the image. The convolutional auto-encoder is particularly suitable for image processing because it can take advantage of the spatial redundancy in the image.

It is still a difficult task to figure out how to generate and compress motion information that is optimized for video compression. To decrease the spatial-temporal redundancy in video sequences, video compression algorithms mainly rely on optical flow information. Furthermore, developing a DL-based video compression system that minimizes the rate-distortion-based objective for both residual and motion information is also another challenge. Recently, image compression algorithms [10, 11] based on machine learning methods have shown great superiority in coding efficiency for spatial redundancy removal compared with conventional codecs. These models get benefit from non-linear transforms, DNN based conditional entropy model, and a joint rate-distortion optimization (RDO) under an end-to-end learning strategy. Learned video compression can be extended from the image compression approach by further exploiting the temporal redundancy or correlation. Besides, the artifacts in frames also harm the performance of video-oriented tasks (e.g., video summarization [12], action recognition, and localization [13]). Accordingly, artifact reduction in frames, which aims to reduce the artifacts, recover missed details from the frame, and pay more attention to complex regions to improve coding performance, becomes a hot topic in the multimedia field [14].

Therefore, this paper proposes a Spatial-Temporal Video Compression network (STVC) using the Spatial-Temporal Priors with an Attention module (STPA). As shown in Fig. 1, the proposed STVC approach uses convolutional networks for representing inputs, reconstructing compressed outputs. Specifically, the proposed STPA network contains an attention network in both the encoder and decoder. Given a sequence of inputs $\{x_1, x_2, \ldots, x_t\}$, the encoder of STPA generates the latent representations $\{y_1, y_2, \ldots, y_t\}$, and the decoder also reconstructs the compressed outputs $\{\hat{x}_1, \hat{x}_2, \ldots, x_t\}$ from $\{y_1, y_2, \ldots, y_t\}$. Besides, we also add an attention module into STPA network architecture. Attention modules can make learned models pay more attention to complex regions to improve our coding performance. The attention mechanism is also embedded to generate a more compact representation for both latent features and hyperpriors. Our attention model is applied at different layers (not only for quantized features at the bottleneck), to adapt intelligently through the end-to-end learning framework.

Our main aim is to improve the flaws of the traditional video compression methods by replacing each traditional aspect with its minimalistic neural network equivalent. The following is a summary of the contributions:

1. We propose a Spatial-Temporal Video Compression network (STVC) that jointly learns motion compression, motion estimation, and residual compression. It optimizes all the components simultaneously under the scrutiny of a single loss feature.
2. We propose the Spatial-Temporal Priors with an Attention module (STPA). To compress the corresponding motion and residual to consider existing spatial and temporal data and compress them in a quantized latent representation using autoencoders, multi-scale connections, and convolutions. We also apply the attention module together with the autoencoder to reduce the artifacts and recover missing details from the frame and pay more attention to complex regions.

Our experiments validate that compressing the motion information using our approach can significantly improve the compression performance. Our framework outperforms the DVC [8], H.264 [4], and H.265 [5] when measured by MS-SSIM and PSNR. In the following, Sect. 2 presents the related works. The proposed STVC and STPA are introduced in Sect. 3. Then, the experiments in Sect. 4 validate the performance of the proposed STVC approach to the existing learned video compression approaches. Finally, concluding remarks and future works are described in Sect. 5.

2 Related Work

In this section, we discuss several deep learning methods related to the image and video compressions that are highly relevant to our work.

2.1 Learned Image Compression

DNN-based image compression methods are generally based on automatic encoders. For the first time [15], it is proposed to use a recurrent encoder to progressively encode image compression. In recent years, convolutional autoencoders have been studied extensively, including non-linear transformations (e.g., generalized division normalization) [7], differentiable quantization (e.g., soft-to-hard quantization, and uniform noise approximation), hyper-prior [10] probability models to estimate entropy in end-to-end DNN image compression frameworks. Recently, the context-adaptive [16] and coarse to fine hyper-prior [10] entropy models have been designed to further improve the distortion rate performance and exceed the traditional image codec. Rate-Distortion Optimization is applied to minimize the Lagrangian cost $R + \lambda D$ in end-to-end training. Here, R is the entropy rate, and D is the distortion measured by Mean Square Error (MSE) or Multi-Scale Structural SIMilarity (MS-SSIM).

2.2 Learned Video Compression

Deep learning is also getting more and more attention in video compression. To improve the coding efficiency of manual standards (e.g., H.264 and H.265), many methods have been proposed [6, 17] to replace components in H.264 with DNN. Among them, [18] used DNN in motion compensation score interpolation, [19] proposed a DNN model for frame predictions. In addition, [1, 20] use DNN to enhance the H.265 loop filter. However, these methods can only improve the performance of a specific component of the video compression framework, but they fail to optimize the video frames together. Inspired by the success of learning image compressions, some learning-based video compression methods are proposed in [2, 23]. However, [19] and [1] still use some manual strategies, such as block matching for motion estimation and compensation, so the entire compressed frame cannot be optimized in an overall manner. Recently, several end-to-end DNN frames have been proposed for video compression [22, 23]. Specifically, [8] proposed a deep video compression (DVC) method, which uses optical flow for the motion estimation and uses two auto-encoders to compress motion and residual separately. DVC outperformed on H.264 mainly at high bit rates, but the coding efficiency dropped unexpectedly at low bit rates as reported. Our model proposes a Spatial-temporal compression method for videos, and it outperforms from low to high bitrates as compared to H.264.

2.3 Attention Mechanism

In general, attention can be thought of as a guideline for allocating available processing resources to the most informative aspects. To rescale the feature maps, it's frequently paired with a gating function (e.g., sigmoid). With a trunk-and-mask attention mechanism, [24] presented a residual attention network for image classification. The attention module is also used by [25] in video compression to improve performance. Overall, the goal of these efforts is to direct the network's attention to the regions of interest. However, there has been an investigation into using the attention mechanism to learn the spatial correlation of different sampling densities to increase video compression efficiency.

3 The Proposed STVC Approach

3.1 Framework

Figure 1 shows the main components of the framework and the relationships among them. The framework is inspired by traditional video codecs H.264 and H.265. A video can be thought of as a collection of frames $f_1, f_2, \ldots, f_{t-1}, f_t$. So, we define the current video sequences and compressed frames as $\{f_t\}_{t=1}^{T}$ and $\left\{\hat{f}_t\right\}_{t=1}^{T}$, respectively. To reduce the redundancy in frames, motion estimation is required. We apply the optical flow network [26] to estimate the temporal motion between the current frame and the previously compressed frame. In our framework, we use the same motion compensation method as [8]. In the following, the residual x_t^r, between f_t and the motion compensated frame $\overline{f}_t$ can be obtained and compressed by another STPA. Using the compressed residual as $\hat{x}_t^r$, we can reconstruct the compressed frame $\hat{f}_t = \overline{f}_t + \hat{x}_t^r$.

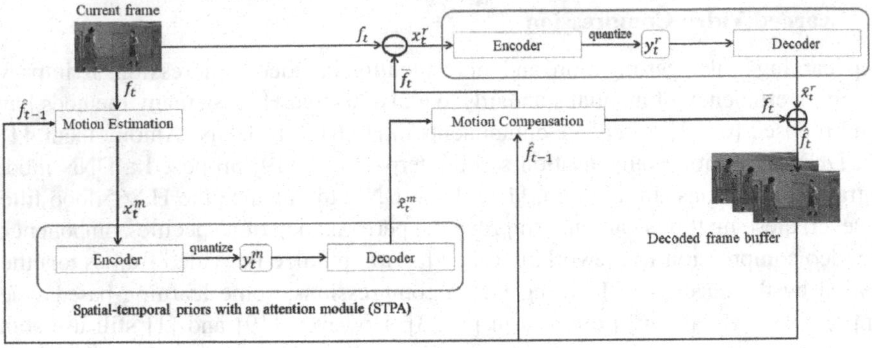

Fig. 1. The framework of our STVC approach. The modules with red colour box are proposed STPA network shown in Fig. 2. We use hyperprior, which is illustrated in Fig. 2, is applied on the latent representations y_t^m and y_t^r to reduce the spatial redundancy.

3.2 Spatial-Temporal Priors with an Attention Module (STPA)

In STVC, we apply two STPAs to compress x_t^m and x_t^r for motion and residual compression, respectively. Since the two STPAs share the same architecture, we denote both x_t^m and x_t^r by xt in this section for simplicity. The compression process in STPA can be formulated by

$$y_t = E(x_t; \phi_E)$$

$$\hat{y}_t = Q(y_t)$$

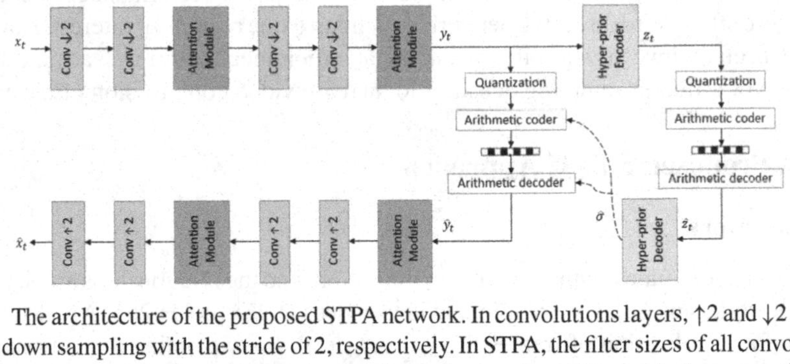

Fig. 2. The architecture of the proposed STPA network. In convolutions layers, ↑2 and ↓2 indicate up and down sampling with the stride of 2, respectively. In STPA, the filter sizes of all convolutional layers are set as 3 × 3 when compressing motion and set as 5 × 5 for residual compression. The filter number of each layer is set as 128. Attention modules are the same as in [27].

$$\hat{x}_t = D(\hat{y}_t; \theta_D) \tag{1}$$

Here $x_t, \hat{x}_t, y_t, \hat{y}_t$ denote the current input frame, reconstructed frame, the latent variables, and the quantized latent variables, respectively. Notation E and D denote the encoder and decoder, respectively, and ϕ and θ correspond to their parameters. Notation Q denotes real round-based quantization in the inference stage.

In the training process, considering that non-differentiable quantization will result in the inability to backpropagate the gradient, the work uses a uniform noise to replace the quantization here. When compressing the t-th frame, the auto-encoders map the input x_t to a latent representation.

$$\tilde{\mathrm{y}}_{\mathrm{t}} = \mathrm{U}(\mathrm{y}_{\mathrm{t}})$$

$$\tilde{x}_t = D(\tilde{y}_t; \theta_D) \tag{2}$$

where $\hat{y}_t$ and $\hat{x}_t$ represent the latent variables with uniform noise added and its decoding reconstruction. Notation U denotes adding uniform noise in the training stage.

Taking the inputs of only the current x_t and y_t to the encoder and decoder, they fail to take advantage of the spatial redundancy and are not able to recover missed details in the latent variables y.

On the contrary, the proposed STPA includes hyperprior and attention in both the encoder and decoder. The architecture of the STPA network is illustrated in Fig. 2. We follow [10] to use four 2 × down-sampling convolutional layers with the activation function of GDN [7] in the encoder of STPA. In the middle of the four convolutional layers and bottleneck of quantized features, we insert an attention module [27] to reduce the artifacts and recover missed details. Therefore, the model can pay more attention to complex regions to improve coding performance. Therefore, the proposed STPA generates latent representation based on the current frame as well the as previous frame. Similarly, the STPA decoder also has an attention module with the four 2 × up-sampling convolutional layers with IGDN, and thus also reconstructs x_t from both the current and previous latent representations.

To reduce the spatial redundancy in the latent variables y_t, an auxiliary hyperprior network [10] encodes its structural information z_t. Formulated by

$$z_t = E_h(y_t; \phi_h)$$

$$\hat{z}_t = \mathrm{Q}(\mathrm{z}_{\mathrm{t}})$$

$$p_{\hat{y}_t|\hat{z}_t}(\hat{y}_t|\hat{z}_t) \leftarrow D_h(\hat{z}_t; \theta_h) \tag{3}$$

where E_h and D_h denote the encoder and decoder of this hyperprior network, and ϕh and θh correspond to their trainable parameters. $p_{\hat{y}_t|\hat{z}_t}(\hat{y}t|\hat{z}_t)$ are estimated distributions conditioned on z_t. There is no prior for z_t, so a factorized density model ψ is used to encode z_t as

$$p_{\hat{z}_{i\,t}|\psi}(\hat{z}_{it}|\psi) = \Pi_i\left(p_{z_{it}|\psi}(\psi) * U\left(-\frac{1}{2}, \frac{1}{2}\right)\right)(\hat{z}_{it}) \tag{4}$$

where zit denotes the i-th element of z at time t, and i specifies the position of each element or each signal.

3.3 Loss Function

The purpose of our video compression framework is to reduce distortion between the original input frame, f_t and the compressed frame $\hat{f}_t$ while reducing the number of bits utilized for encoding the video. As a result, we propose the rate-distortion optimization problem below.

$$L = \lambda D + R = \lambda d(f_t, \hat{f}_t) + \hat{y}_t^m + \hat{y}_t^r \tag{5}$$

We employ Mean Square Error (MSE) in our approach, and $d(f_t, \hat{f}_t)$ specifies the distortion between f_t and $\hat{f}_t$. The amount of bits utilized to encode the representations is represented by I(). Both the residual representation $\hat{y}_t^r$ and the motion representation $\hat{y}_t^m$ should be encoded into the bitstreams in our technique. λ is the Lagrange multiplier that determines the trade-off between the number of bits and distortion.

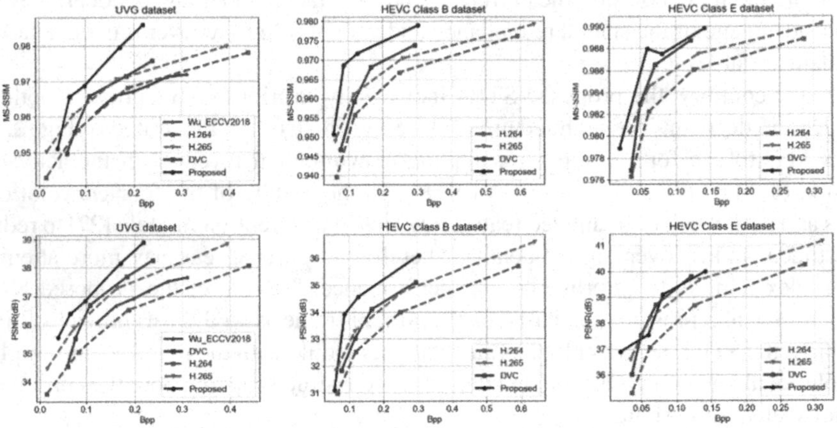

Fig. 3. Rate-distortion performance of the proposed approach compared with the learning-based video codec in [28], DVC [8], and traditional video compression method H.264 [4] and H.265 [5] approaches on the UVG and JCT-VC datasets.

4 Experiments

4.1 Training the Proposed Network

Our proposed approach is trained using the Vimeo-90k dataset [29]. The dataset is built for video denoising, deblocking, and super-resolution. Each training sample video has seven frames. The first frame is compressed as an I-frame, while the remaining six frames are compressed as P-frames. To evaluate compression quality, we use the Multiscale Structural Similarity (MS-SSIM) [30] index and the Peak Signal-to-Noise Ratio (PSNR), and then train the model with settings that are optimized for MS-SSIM and PSNR, respectively. To train the PSNR model, it uses the Mean Square Error (MSE). We set $\lambda = 1024$ for MS-SSIM and PSNR. The Adam optimizer [33] is utilized for training. The initial learning rate is set as 10^{-4} for loss function. We use Bpp (bits per pixel) to express the required bits for each pixel in the current frame to measure the number of bits for encoding the representations.

4.2 Evaluating the Performance

The evaluations are being carried out to ensure that our proposed model is effective. The performance is evaluated using the JCT-VC [4] (Classes B and E) and UVG [32] datasets. JCT-VC Classes B and UVG have 1920 × 1080 resolutions. The UVG dataset has a GOP (group of pictures) size of 12, and the HEVC dataset has a GOP size of 10, respectively. We compare our method to that of DVC [8, 28], H.264 [4], and H.265 [5]. We use FFmpeg in very fast mode and the settings in [28] to make compressed frames using H.264 and H.265.

Table 1. BDBR performances with Proposed, DVC, and H.265 model when compared with H.264.

Dataset	MS-SSIM			PSNR		
	Proposed	DVC	H.265	Proposed	DVC	H.265
UVG (Average)	**−38.62**	−16.46	−26.19	**−39.24**	−37.34	−36.19
Class B (Average)	**−29.59**	−29.09	−28.31	**−33.17**	−27.92	−31.73
Class E (Average)	**−35.51**	−33.16	−29.52	**−36.01**	−22.23	−35.54
Average	**−34.57**	−26.24	−28.17	**−36.14**	−29.26	−34.30

Rate-Distortion Curve. The rate-distortion curves for the JCT-VC and UVG datasets are shown in Fig. 3. The bit rate is determined using Bpp, while the quality is measured using MS-SSIM and PSNR. Overall, the proposed MS-SSIM model outperforms DVC [8, 28], H.264, and H.265, as illustrated in Fig. 3. DVC is comparable with H.265 at low bit rates on the JCT-VC dataset but our approach's rate-distortion curves clearly outperform DVC [8], from low to high bitrates on the JCT-VC dataset. Also, it can be demonstrated that our PSNR model outperforms DVC [8, 28], H.264, and H.265 on all videos (average).

Bit-Rate Difference. In addition, we evaluated Bjøntegaard Delta Bitrate (BDBR) [33] using H.264 anchors. BDBR calculates the average bit rate difference compared to the anchor point, and a lower BDBR value indicates better performance. Table 1 shows the BDBR calculated using MS-SSIM and PSNR, where a negative number indicates that the bit rate is lower than that of the anchor, which means that it is better than H.264, and the bold number is the best result among all learning methods.

In Table 1, in order to fairly compare MS-SSIM with optimized methods DVC [8], H.264, and H.265, we first report the BDBR of our model based on MS-SSIM. In UVG and JCT-VC types B and E, our model is even significantly better than the DVC method in terms of MS-SSIM. As shown in Table 1, our model is better than H.264 in MS-SSIM, with an average BDBR of 34.57%, which is also better than DVC (BDBR = 26.24%). In terms of PSNR, Table 1 shows that our PSNR model outperforms H.264 (very fast LDP) from low to high bit rates on the UVG and JCT-VC test sets. The BDBR results calculated by PSNR in Table 1 also show that the bitrate achieved by our method is

36.14% lower than H.264 (very fast LDP). Please note that as far we know, there are no learned video compression methods that have exceeded the default setting of H.265 in PSNR. Our proposed method can be further developed to improve the performance of the next generation learning video compression and to help gradually catch up with manual standards.

5 Conclusion and Future Work

This paper has proposed a spatial-temporal video compression network (STVC). Specifically, we proposed an auto-encoders style network with an attention module to compress motion and residual, fully exploring the spatial and temporal correlation in video frames. We proposed a rate-distortion optimization framework to train a single Spatial-temporal autoencoder for reconstruction loss. Key novelty laid on the accurate motion representation for exploiting temporal correlation and hyper-priors is leveraged to improve the spatial correlation and entropy coding efficiency.

We evaluated our methods and reported the performances among the traditional H.264/AVC, H.265/HEVC, and learning-based DVC and [28]. Our approach offered consistent gains over existing methods across a variety of contents and bit rates, but the PSNR model on the default setting of H.265 did not beat the performance, so it needs further research and development. Moreover, the proposed approach achieved significant performance at the cost of higher encoding complexity. Another possible focus for future work is to study reducing complexity and the trade-off between complexity and rate-distortion performance. For example, the proposed network may be sped up by reducing the number of layers and channels in the autoencoders and the motion compensation network or by utilizing a more time-efficient optical flow network for motion prediction [1].

References

1. Chen, T., Liu, H., Shen, Q., Yue, T., Cao, X., Ma, Z.: DeepCoder: A Deep Neural Network Based Video Compression (2017)
2. Chen, Z., He, T., Jin, X., Wu, F.: Learning for Video Compression, April 2018. https://doi.org/10.1109/TCSVT.2019.2892608
3. Aramvith, S., Sun, M.-T.: MPEG-1 AND MPEG-2 Video Standards (1999)
4. Wiegand, T., Sullivan, G.J., Bjøntegaard, G., Luthra, A.: Overview of the H.264/AVC video coding standard. IEEE Trans. Circuits Syst. Video Technol. **13**(7), 560–576 (2003). https://doi.org/10.1109/TCSVT.2003.815165
5. Sullivan, G.J., Ohm, J.R., Han, W.J., Wiegand, T.: Overview of the high efficiency video coding (HEVC) standard. IEEE Trans. Circuits Syst. Video Technol. **22**(12), 1649–1668 (2012). https://doi.org/10.1109/TCSVT.2012.2221191
6. Xu, M., Li, T., Wang, Z., Deng, X., Yang, R., Guan, Z.: Reducing Complexity of HEVC: A Deep Learning Approach, September 2017. https://doi.org/10.1109/TIP.2018.2847035
7. Ballé, J., Laparra, V., Simoncelli, E.P.: End-to-end Optimized Image Compression, November 2016. http://arxiv.org/abs/1611.01704
8. Lu, G., Ouyang, W., Xu, D., Zhang, X., Cai, C., Gao, Z.: DVC: An End-toend Deep Video Compression Framework, November 2018. http://arxiv.org/abs/1812.00101

9. Pessoa, J., Aidos, H., Tomas, P., Figueiredo, M.A.T.: End-to-end learning of video compression using spatio-temporal autoencoders. In: IEEE Workshop on Signal Processing Systems, SiPS: Design and Implementation, October 2020. https://doi.org/10.1109/SiPS50750.2020.9195249
10. Ballé, J., Minnen, D., Singh, S., Hwang, S.J., Johnston, N.: Variational image compression with a scale hyperprior, January 2018. http://arxiv.org/abs/1802.01436
11. Hu, Y., Yang, W., Liu, J.: Coarse-to-Fine Hyper-Prior Modeling for Learned Image Compression. https://huzi96.github.io/coarse-to-fine-compression.html
12. Apostolidis, E., Adamantidou, E., Metsai, A.I., Mezaris, V., Patras, I.: Video Summarization Using Deep Neural Networks: A Survey, January 2021. http://arxiv.org/abs/2101.06072
13. Xu, Y., et al.: GIF Thumbnails: Attract More Clicks to Your Videos (2021). www.aaai.org
14. Zou, N., et al.: End-to-End Learning for Video Frame Compression with SelfAttention, April 2020. http://arxiv.org/abs/2004.09226
15. Toderici, G., et al.: Full Resolution Image Compression with Recurrent Neural Networks, August 2016. http://arxiv.org/abs/1608.05148
16. Lee, J., Cho, S., Beack, S.-K.: Context-adaptive entropy model for end-to-end optimized image compression. https://github.com/JooyoungLeeETRI/CA_Entropy_Model
17. Li, T., Xu, M., Zhu, C., Yang, R., Wang, Z., Guan, Z.: A deep learning approach for multi-frame in-loop filter of HEVC. IEEE Trans. Image Process. **28**(11), 5663–5678 (2019). https://doi.org/10.1109/TIP.2019.2921877
18. Liu, J., Xia, S., Yang, W., Li, M., Liu, D.: One-for-all: grouped variation network-based fractional interpolation in video coding. IEEE Trans. Image Process. **28**(5), 2140–2151 (2019). https://doi.org/10.1109/TIP.2018.2882923
19. Choi, H., Bajic, I.V.: Deep Frame Prediction for Video Coding, December 2018. http://arxiv.org/abs/1901.00062
20. Dai, Y., Liu, D., Wu, F.: A Convolutional Neural Network Approach for Post-Processing in HEVC Intra Coding, August 2016. https://doi.org/10.1007/978-3-31951811-4_3
21. Li, T., Xu, M., Yang, R., Tao, X.: A DenseNet based approach for multiframe in-loop filter in HEVC. In: Data Compression Conference Proceedings, vol. 2019-March, pp. 270–279, May 2019. https://doi.org/10.1109/DCC.2019.00035
22. Cheng, Z., Sun, H., Takeuchi, M., Katto, J.: Learning Image and Video Compression through Spatial-Temporal Energy Compaction (2019)
23. Yang, R., Mentzer, F., van Gool, L., Timofte, R.: Learning for Video Compression with Hierarchical Quality and Recurrent Enhancement, March 2020. http://arxiv.org/abs/2003.01966
24. Wang, F., et al.: Residual Attention Network for Image Classification, April 2017. http://arxiv.org/abs/1704.06904
25. Zhao, M., Xu, Y., Zhou, S.: Recursive fusion and deformable spatiotemporal attention for video compression artifact reduction. In: MM 2021 - Proceedings of the 29th ACM International Conference on Multimedia, pp. 5646–5654, October 2021. https://doi.org/10.1145/3474085.3475710
26. Ranjan, A., Black, M.J.: Optical Flow Estimation using a Spatial Pyramid Network, November 2016. http://arxiv.org/abs/1611.00850
27. Cheng, Z., Sun, H., Takeuchi, M., Katto, J.: Learned Image Compression with Discretized Gaussian Mixture Likelihoods and Attention Modules, January 2020. http://arxiv.org/abs/2001.01568
28. Wu, C.-Y., Singhal, N., Krähenbühl, P.: Video Compression through Image Interpolation, April 2018. http://arxiv.org/abs/1804.06919
29. Xue, T., Chen, B., Wu, J., Wei, D., Freeman, W.T.: Video enhancement with task-oriented flow. Int. J. Comput. Vis. **127**(8), 1106–1125 (2019). https://doi.org/10.1007/s11263-018-01144-2

30. Wang, Z., Simoncelli, E.P., Bovik, A.C.: Multi-scale structural similarity for image quality assessment (2003)
31. Kingma, D.P., Ba, J.: Adam: A Method for Stochastic Optimization, December 2014. http://arxiv.org/abs/1412.6980
32. Mercat, A., Viitanen, M., Vanne, J.: UVG dataset: 50/120fps 4K sequences for video codec analysis and development. In: MMSys 2020 - Proceedings of the 2020 Multimedia Systems Conference, pp. 297–302, May 2020. https://doi.org/10.1145/3339825.3394937
33. Hanhart, P., Ebrahimi, T.: Calculation of average coding efficiency based on subjective quality scores. http://mmspg.epfl.ch/scenic

On the Evaluation of Video-Based Crowd Counting Models

Emanuele Ledda[1(✉)], Lorenzo Putzu[2], Rita Delussu[2], Giorgio Fumera[2], and Fabio Roli[1,2]

[1] Department of Informatics, Bioengineering, Robotics, and Systems Engineering, University of Genova, Genoa, Italy
emanuele.ledda@uniroma1.it, fabio.roli@unige.it

[2] Department of Electrical and Electronic Engineering, University of Cagliari, Cagliari, Italy
{lorenzo.putzu,rita.delussu,fumera}@unica.it

Abstract. Crowd counting is a challenging and relevant computer vision task. Most of the existing methods are image-based, i.e., they only exploit the spatial information of a single image to estimate the corresponding people count. Recently, video-based methods have been proposed to improve counting accuracy by also exploiting temporal information coming from the correlation between adjacent frames. In this work, we point out the need to properly evaluate the temporal information's specific contribution over the spatial one. This issue has not been discussed by existing work, and in some cases such evaluation has been carried out in a way that may lead to overestimating the contribution of the temporal information. To address this issue we propose a categorisation of existing video-based models, discuss how the contribution of the temporal information has been evaluated by existing work, and propose an evaluation approach aimed at providing a more complete evaluation for two different categories of video-based methods. We finally illustrate our approach, for a specific category, through experiments on several benchmark video data sets.

Keywords: Video-based crowd counting and density estimation · Spatial-temporal information

1 Introduction

In recent years, in the computer vision field there has been a growing interest in crowd counting and density estimation for its potential security-related applications, in particular to video surveillance [15,19], e.g., to monitor the number and distribution of people in public areas during mass gatherings events. This is still a challenging task due to issues such as perspective distortions, scale and illumination variations, the complexity of the scene background, and occlusions. State-of-the-art methods are based on convolutional neural networks

S. Sclaroff et al. (Eds.): ICIAP 2022, LNCS 13233, pp. 301–311, 2022.
https://doi.org/10.1007/978-3-031-06433-3_26

(CNNs), and most of them are image-based, i.e., they estimate the crowd count of individual images or frames [6]. However, in application scenarios where crowd counting has to be carried out on a video sequence (e.g., for real-time crowd monitoring), image-based methods disregard the temporal correlation between adjacent frames, which contains potentially useful information to improve counting accuracy [18,23]. Accordingly, some video-based methods have recently been proposed, aimed at exploiting both the spatial information of a single frame and the temporal information over a sequence of adjacent frames [6,13,17,18,20,23,30].

In this work, we focus on how to properly evaluate the specific contribution provided by the temporal information to the accuracy of video-based methods. Our interest in this issue is motivated by the fact that it has not been explicitly addressed by existing work; moreover, some of the existing works did not provide a specific evaluation, whereas the evaluation carried out in some works may incur the risk of *overestimating* the temporal contribution, since part of the reported performance gain may be actually due to an improved capability of exploiting the *spatial* information by the CNN module devoted to processing the temporal one. To address this issue, after providing the necessary background on crowd counting (Sect. 2), we first propose a categorisation of existing video-based CNN models, based on how they process spatial and temporal information (Sect. 3). Based on our categorisation, we discuss how the contribution of the spatial information has been evaluated in the respective works. We then point out issues arising for two categories of models (multi-thread sequential and multi-thread parallel), and propose a more complete evaluation method for these categories, aimed at providing a more precise quantification of the actual gain achieved by the temporal information (Sect. 4). This can help to better understand the effectiveness of video-based crowd counting models. We illustrate the above issue and the proposed evaluation method through experiments on five benchmark video data sets for crowd counting, focusing on multi-thread sequential models (Sect. 5). Conclusions are finally drawn in Sect. 6.

2 Background

In this section, we briefly survey existing crowd counting methods. Most of the existing methods, including all of the early ones, are **image-based** [6]. Early methods can be categorised into detection-based and regression-based. The former is based on detecting pedestrians or their body parts (typically, heads) and is in principle more accurate, but is effective only for small and sparse crowds, without occlusions [6,15,19]. Regression-based methods estimate people count using low-level local or global image features. State-of-the art methods are based on CNNs and are mostly regression-based; they usually estimate the crowd density map as an intermediate step, which enables them to capture the spatial distribution of a crowd. The ground truth density map required for training is usually obtained by superimposing 2D Gaussian kernels centred on the manually annotated head positions of each pedestrian [27]. Given that each kernel is normalised to sum to one, the estimated people count can be easily obtained during inference [6,19]. We describe here two representative CNN architectures, which will be used in the experiments of Sect. 5. The Multi-Column CNN (MCNN) architecture [27] consists of three branches aimed at addressing large scale variations typical of crowd images. Their outputs are merged and mapped to the density map. Given the complexity of training a whole network, recent models exploit a pre-trained backbone network, typically VGG-16 [12] or VGG-19 [16]. In particular, Congested Scene Recognition Network (CSRNet) [12] is a simple architecture consisting of a VGG-16 network (except for the fully-connected layers) aimed at extracting low-level features followed by a back-end module composed of dilated convolutional layers aimed at extracting deeper information. Other CNN models exploit a backbone network and a back-end module, but they include intermediate additional modules aimed at addressing specific issues, such as variations in crowd distribution [25], scale [11,24,26,28,29] and background [14].

Recently, some **video-based** methods have been proposed [6]. They use different approaches ranging from Recurrent Neural Networks (RNNs) [23], 3D convolutions [13,17,18,20,30], spatial transformers [7,8], optical flow [1,2] and other approaches [21,22]. In the next section we propose a categorisation of the existing video-based models, which we shall exploit later to discuss how the temporal contribution is evaluated.

3 A Categorisation of Video-Based Crowd Counting Models

Since all of the existing video-based crowd counting models use the density map as an intermediate step for predicting the crowd count, we adopt the formulation of [27] to describe them. During inference, **image-based** models take as input a single image or frame $x \in \mathbb{R}^{C \times W \times H}$, where C, W and H denote the number of channels (typically, $C = 3$ for a RGB image), width and height, respectively, and output its estimated density map $y = f(x; \theta) \in \mathbb{R}^{1 \times W \times H}$, where θ denotes a vector of model parameters. To this aim, the training set consists of a set of pairs $(x_i, y_i) \in \mathbb{R}^{C \times W \times H} \times \mathbb{R}^{1 \times W \times H}$, where the ground truth y_i is computed as described in Sect. 2. Note that training images can belong to different, unrelated scenes, as happens for some benchmark data sets with the aim of promoting generalisation capability to different target scenes [27]. **Video-based** models, instead, estimate during inference the density map $y^{(t)}$ of the t-th input frame of a video sequence, $x^{(t)}$, in several ways. A typical approach is to consider a sequence of K consecutive frames, for a given value of K: $y^{(t)} = f(x^{(t-K+1)}, x^{(t-K+2)}, \ldots, x^{(t)}; \theta)$ [6]. In this case the training set is made up of pairs $(X_i, y_i) \in \mathbb{R}^{K \times C \times W \times H} \times \mathbb{R}^{1 \times 1 \times W \times H}$, where X_i denotes a sequence of K frames and y_i the corresponding ground truth density maps. Other choices are also possible (see e.g. [23]).

We now propose a categorisation of existing video-based CNN models for crowd counting, based on how they process spatial and temporal information (see Fig. 1). The main distinction we make is among **single-** and **multi-thread**

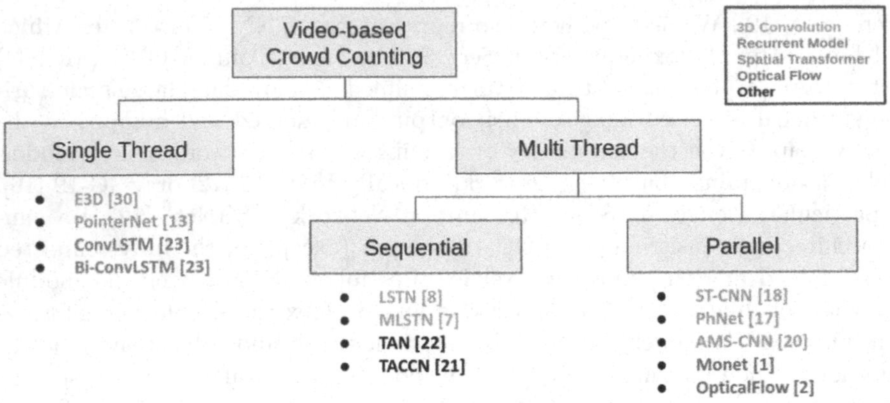

Fig. 1. Proposed taxonomy of existing video-based methods (see text for the details). For each specific method, the colour map indicates the kind of solution used to exploit temporal information.

models. Single-thread models use a "monolithic" CNN architecture for processing both kinds of information; [13,23,30] multi-thread models use distinct modules, instead [1,2,7,8,17,18,20–22]. In particular, they contain a *spatial* module that independently processes one or more frames to produce either density maps or low-level spatial features to be used in subsequent processing steps, and a *temporal* module that processes K adjacent frames; the final output is typically the estimated density map of the K-th frame. Multi-thread models can in turn be categorised into **sequential** models, if the spatial module is placed *before* the temporal one [7,8,21,22], and **parallel** models, if the two modules work in parallel [1,2,17,18,20]. Note that parallel models need a further module to fuse the outputs of the spatial and temporal ones.

Regardless of the category, different solutions have been proposed to exploit temporal information, including recurrent network architectures [23], 3D convolutions [13,17,18,20,30], spatial transformers [7,8], optical flow [1,2], and other ad hoc methods [21,22].

Since the rationale of video-based models is to integrate temporal information for improving counting accuracy with respect to using spatial information alone, their performance has been compared in the respective works with several image-based models. However, no work has explicitly addressed the issue of how to properly assess the specific contribution provided by the temporal information over the spatial one. In particular, all works compare the proposed video-based model with state-of-the-art image-based ones, whose architectures are however unrelated to each other: this does not allow a direct evaluation of the gain achieved by exploiting the temporal information. A more focused evaluation is given in some works, where the performance of the proposed video-based model is evaluated for different numbers K of consecutive frames [7,17,22,30]. Interestingly, among the single-thread models some work considered also the limit case of $K = 1$, which is obtained by a "degenerate" version of the proposed video-based architecture to an image-based one [7,23,30]. A similar approach has been used also for some multi-thread models: in this case, the "degenerate" model consisted in the corresponding spatial module alone [18,22].

4 Evaluation of the Temporal Contribution

In this section, we discuss the different approaches used in the literature to evaluate video-based models, subdividing them according to our taxonomy.

Multi-thread Sequential Models. Some works [8,21,22] compared the performance of the proposed video-based model V, made up of a sequence of a spatial (S) and a temporal module (T), with the one attained by an image-based counterpart made up of the spatial module alone (S'), which is re-trained on the image-based version of the same training set used for V. The corresponding performance improvement was considered as the contribution provided by the temporal information. However, we argue that such an improvement may overestimate the *actual* temporal contribution. To see why this could be the case, consider another possible image-based counterpart, denoted by V', defined as a "degenerate" version of a given video-based model where the spatial module is reduced to processing a *single* input frame ($K = 1$ in the notation of Sect. 3), and the input connections of the temporal module are modified accordingly. For instance, spatial modules consisting of K identical branches with shared weights [8,21,22] would be simply replaced by a single branch. Assume now that the original video-based model V outperforms the spatial module alone, S', and that the performance of the "degenerate" model V' (re-trained on the image-based version of the same training set as S'), is close to that of V. Would it be still correct to state that the performance gain of the video-based model over S' is entirely due to the temporal information exploited by T? We believe instead that, in this case, the temporal model T, although originally conceived for processing temporal information, also provided a *spatial* contribution; in other words, it mainly improved the capability of the spatial module S of exploiting the spatial information.

Accordingly, we believe that a more complete account of the behaviour of multi-thread sequential video-based models can be attained by comparing them with the specific image-based counterpart V' defined above, i.e., with a "degenerate", image-based version of the original architecture V that also includes the temporal module.

Multi-thread Parallel Models. In this category, the modules S and T work in parallel, and also S processes the input video frames; accordingly, in [18] the evaluation of the contribution of temporal information was carried out by comparing the performance of the original model V with the ones of two distinct counterparts: the first one is image-based and consists of the spatial module alone (S'), whose input was reduced to a single frame similarly to sequential models; the second one is video-based, and consists of the temporal module alone (T'), whose input was kept to K adjacent frames as in V. Another work [17] uses V in two different setups, at first feeding it with a sequence of consecutive frames and then using the same frame multiple times to force the network to provide a prediction using only spatial information. We point out that also for this kind of models a more complete evaluation could be obtained by considering a "degenerate" version V' made up of the same modules S and T working in parallel, reduced to an image-based model which processes a single input frame ($K = 1$), and trained accordingly. Our argument is similar to the one presented above: if the performance of V' is close to that of V, one should conclude that the main contribution of the module T is to improve the capability of S of exploiting the *spatial* information.

Single-Thread Models. In the case of "monolithic" architectures, without a separation between spatial and temporal modules, some works evaluated the contribution of the temporal information by devising a "degenerate" image-based version of the whole model, with a single frame ($K = 1$) as the input. For the recurrent video-based architecture ConvLSTM-nt [23], which is capable of processing an input frame sequence of *variable* size K, a degenerate version V' was obtained by simply using during inference a single input frame (i.e., $K = 1$). An analogous approach was used for the E3D model, which is based on 3D convolutions [30]: it consists of using a degenerate version (named E2D), obtained by replacing 3D convolutional layers with 2D ones, thus using a single frame ($K = 1$) as input both in training and in the inference phase. We believe that the above approach is correct, provided that in the model of [23] the degenerate version V' is re-trained in an image-based fashion, instead of using as V' the same model V and just testing it using single frames as the input (it is not clear which of the two solutions has been used in [23]). In general, we point out that each specific single-thread model may require an ad hoc definition of the corresponding degenerate version.

In the next section, we give an example of the proposed evaluation approach, focusing on multi-thread sequential models, through experiments on several benchmark video data sets.

5 Experiments

The goal of the experiments reported in this section is to show an example of the evaluation approach of the temporal contribution of video-based crowd counting models we proposed in Sect. 4. We only focus on the multi-thread sequential architecture, given its lower implementation complexity. To this aim, we implemented two instances of this kind of architecture, using as the spatial module two state-of-the-art image-based models whose code was made available by the authors, MCNN [27] and CSRNet [12] (see Sect. 2 for their description). We also defined a simple temporal module, suitable to the purpose of our experiments. We point out that the considered video-based models were not conceived to attain a competitive performance against existing ones on benchmark data

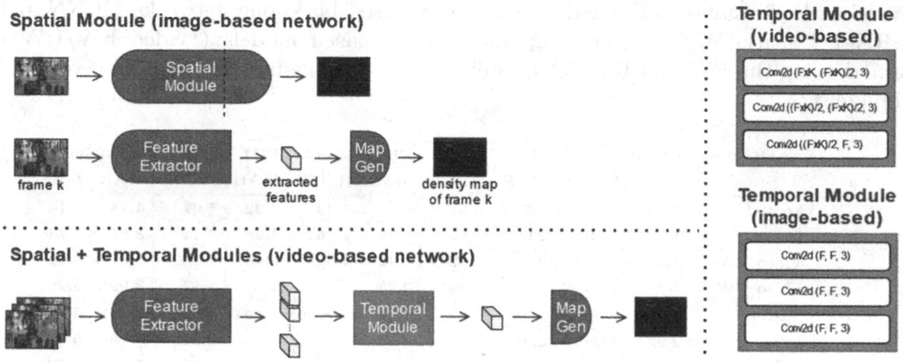

Fig. 2. Top-left: architecture of the image-based models (MCNN and CSRNet) used to implement the spatial module, shown both as a single black box and as the feature extractor and map generator sub-modules. Bottom-left: the video-based architecture obtained by K copies of the spatial module (with shared weights) and the temporal module. Right: details of the temporal module architecture: for each layer, number of input channels, output channels and filter size (F is the number of channels of the feature extractor).

sets, but only to allow us to give a simple example of the proposed evaluation approach.

5.1 Experimental Setting

Network Architecture. Figure 2 (top) shows the details of the considered spatial module when used as a stand-alone image-based crowd counting model, highlighting the inner modules for feature extraction and density map generation. The video-based model was obtained by using K copies of the feature extractor with shared weights, and by inserting the temporal module between the feature extractor and the map generator (Fig. 2, bottom). This way, the input of the temporal module consists of the concatenation of the features extracted by K adjacent frames. We then implemented a simple temporal module using three convolutional layers made up of 3×3 kernels and a number of input and output channels proportional to K, followed by a ReLU activation function, as shown in Fig. 2.

Data Sets. Only some of the benchmark data sets for crowd counting made up of frame sequences acquired by the same camera contain at least one annotated frame per second, which can be considered the minimal requirement for effective exploitation of temporal information. For instance, the well known Shanghai World Expo 2010 [19] contains only one annotation every 30 s. Accordingly, we used the following data sets that fulfil the above requirement: Mall [4], UCSD [3] and PETS [9]. **Mall** was collected from a single surveillance camera in a shopping mall. It contains 2,000 frames that, as in previous works [15,19] we split

Table 1. Performances obtained by the image-based backbone networks MCNN and CSRNet ("spatial"), by the corresponding video-based models ("video FW/UW") trained using the FW and UW set-up, and by their image-based, degenerate versions ("image").

	Architecture	UCSD		Mall		PETSview1		PETSview2		PETSview3	
		MAE	RMSE	MAE	RMSE	MAE	RMSE	MAE	RMSE	MAE	RMSE
MCNN	Spatial	3.95	4.65	5.33	6.17	6.26	7.93	4.22	5.07	4.18	5.14
	Image	1.83	2.30	2.65	3.36	3.42	4.16	4.28	5.14	3.87	4.74
	Video FW	1.79	2.29	2.63	3.32	3.92	4.86	3.46	4.20	3.37	4.21
	Video UW	1.90	2.43	2.33	2.96	4.23	5.56	4.01	5.17	3.99	5.29
CSRNet	Spatial	4.38	5.66	2.58	3.22	7.27	8.91	10.94	12.61	3.66	4.47
	Image	2.97	4.08	2.45	3.03	4.03	5.08	5.76	7.54	4.79	5.62
	Video FW	2.36	3.17	2.20	2.75	4.23	4.92	5.59	7.08	5.11	6.02
	Video UW	2.54	3.35	2.47	3.07	7.01	8.21	7.19	7.94	5.21	6.00

into 600, 200 and 1,200 frames for training, validation and testing, respectively. **UCSD** was collected with a low-resolution camera in a pedestrian walkway at a university campus. As in previous works [15,19], we used a subset of 2,000 frames and split them into 600, 200 and 1,200 training, validation and testing frames, respectively. **PETS2009** is a benchmark for several visual surveillance tasks [9]. It is composed of different video sequences acquired with various cameras at different times. We grouped the video sequences from the same cameras to create three data sets named PETSview1, PETSview2 and PETSview3 [5]. Each of them contains 1,229 frames that we split into 361, 128 and 740 frames for training, validation and testing, respectively.

Training Procedure and Performance Metrics. For each benchmark data set, we trained the two video-based models described above under two different set-ups as in previous work [8,22]. We trained first the spatial module, using the classic image-based set-up, with the hyper-parameters suggested in the respective papers [12,27]. We then trained the whole video-based model by either updating only the weights of the temporal module ("frozen spatial weights", FW) or updating the weights of the spatial module ("updated spatial weights", UW). The latter set-up allows, in principle, to obtain a "temporal-aware" spatial feature extractor, whereas the former keeps the capability of processing temporal information confined to the temporal module. To train the video-based model, we set $K = 10$, i.e., we considered a sequence of 10 adjacent frames to predict the density map of the last one supervised the ground truth density map of the same last frame during training. To adopt the proposed evaluation approach, we separately trained also a degenerate, image-based version using a single feature extraction sub-module ($K = 1$). During training, we used the GAME metric [10], to simulate a parametrized patch-level loss. Finally, we evaluated the performance on the corresponding testing frames for each benchmark data set, using two common metrics: Mean Absolute Error (MAE) and Root Mean Square Error (RMSE), computed between the true and estimated people count over all testing images.

5.2 Results

Table 1 reports the results obtained by the image-based spatial network ("spatial", either MCNN or CSRNet), the video-based model under the two different training set-up ("video FW" and "video UW"), and its degenerate, image-based version ("image").

Consider first the performance attained by the video-based models under the FW and UW training set-up. Since in all data sets the best performance was attained under the FW set-up, in the subsequent comparison we focus on this version. Table 1 shows that the FW video-based models outperformed the image-based spatial module alone (which is the counterpart considered in other work) on all data sets, except for PETSview3 when CSRNet was used as the spatial module. In some cases, the performance gap is remarkable, e.g., on Mall, when MCNN was used as the spatial module. According to the criterion used in previous work, this would show that the video-based model could effectively exploit the temporal information. However, by comparing the performance of the video-based models with the corresponding degenerate versions, it can be seen that the performance gap is considerably smaller, or even negligible, although the former also outperform the latter besides the image-based spatial module (with the same exception mentioned above); notably, this is the case of Mall, when CNN was used as the spatial module. This is a clear example of the phenomenon we hypothesised in Sect. 4, i.e., in this case, one should say that the additional module, originally devised to exploit temporal information, actually provides also a *spatial* contribution, which in some cases appears even predominant over the temporal one. That is to say; the temporal module improved the capability of the spatial module of exploiting the spatial information, such that the video-based model benefited from the temporal information only to a lower extent than it seemed by comparing its performance with the image-based spatial module. These results provide clear support to our suggestion to extend the evaluation of multi-thread sequential video-based crowd counting models, including their specific degenerate version described in Sect. 4.

6 Conclusions and Future Work

In this work, we considered recently proposed video-based crowd counting models and pointed out an issue that we believe has not received enough attention so far, i.e., how to provide a complete account of the contribution of temporal information over the spatial one to their overall accuracy. To this aim, we provided a categorisation of existing video-based models and discussed how they had been evaluated in the respective works, pointing out a possible problem in the evaluation of the temporal contribution for models in the multi-thread sequential and parallel categories; we then proposed a more complete evaluation method for such categories. Focusing on multi-thread sequential models, we provided empirical evidence that the evaluation method used in previous work can overestimate the actual contribution of temporal information, whereas our method highlights that the overall performance of video-based models can be mainly due to an improved capability of exploiting the *spatial* information by the module originally devoted to process the temporal one. A necessary follow-up of this work is to extend the empirical evaluation of the proposed approach to multi-thread parallel models.

Acknowledgement. This work was supported by the projects "Law Enforcement agencies human factor methods and Toolkit for the Security and protection of CROWDs in mass gatherings" (LETSCROWD), EU Horizon 2020 programme, grant agreement No. 740466, and "IMaging MAnagement Guidelines and Informatics Network for law enforcement Agencies" (IMMAGINA), European Space Agency, ARTES Integrated Applications Promotion Programme, contract No. 4000133110/20/NL/AF.

Emanuele Ledda is affiliated with the Italian National PhD in Artificial Intelligence, Sapienza University of Rome. He also acknowledges the cooperation with and the support from the Pattern Recognition and Applications Lab of the University of Cagliari.

References

1. Bai, H., Chan, S.G.: Motion-guided non-local spatial-temporal network for video crowd counting. CoRR abs/2104.13946 (2021)
2. Bandyopadhyay, S.: Optical flow based crowd counting in video frames. In: 10th International Conference on Computing, Communication and Networking Technologies, ICCCNT, pp. 1–6 (2019)
3. Chan, A.B., Liang, Z.S.J., Vasconcelos, N.: Privacy preserving crowd monitoring: Counting people without people models or tracking. In: CVPR, pp. 1–7 (2008)
4. Chen, K., Loy, C.C., Gong, S., Xiang, T.: Feature mining for localised crowd counting. In: BMVC, pp. 1–11 (2012)
5. Delussu, R., Putzu, L., Fumera, G.: Investigating synthetic data sets for crowd counting in cross-scene scenarios. In: VISIGRAPP, pp. 365–372 (2020)
6. Fan, Z., Zhang, H., Zhang, Z., Lu, G., Zhang, Y., Wang, Y.: A survey of crowd counting and density estimation based on convolutional neural network. Neurocomputing **472**, 224–251 (2022)
7. Fang, Y., Gao, S., Li, J., Luo, W., He, L., Hu, B.: Multi-level feature fusion based locality-constrained spatial transformer network for video crowd counting. Neurocomputing **392**, 98–107 (2020)
8. Fang, Y., Zhan, B., Cai, W., Gao, S., Hu, B.: Locality-constrained spatial transformer network for video crowd counting. In: International Conference on Multimedia and Expo, ICME, pp. 814–819 (2019)
9. Ferryman, J., Shahrokni, A.: Pets 2009: dataset and challenge. In: IEEE International Workshop on PETS, pp. 1–6 (2009)
10. Guerrero-Gómez-Olmedo, R., Torre-Jiménez, B., López-Sastre, R., Maldonado-Bascón, S., Oñoro-Rubio, D.: Extremely overlapping vehicle counting. In: Paredes, R., Cardoso, J.S., Pardo, X.M. (eds.) IbPRIA 2015. LNCS, vol. 9117, pp. 423–431. Springer, Cham (2015). https://doi.org/10.1007/978-3-319-19390-8_48
11. Guo, Q., Zeng, X., Hu, S., Phoummixay, S., Ye, Y.: Learning a deep network with cross-hierarchy aggregation for crowd counting. Knowl. Based Syst. **213**, 106691 (2021)
12. Li, Y., Zhang, X., Chen, D.: CSRNet : dilated convolutional neural networks for understanding the highly congested scenes. In: IEEE Conference on Computer Vision and Pattern Recognition, CVPR, pp. 1091–1100. Computer Vision Foundation/IEEE Computer Society (2018)

13. Liu, C., Huang, Y., Mu, Y., Yu, X.: Pointernet: spatiotemporal modeling for crowd counting in videos. In: ICDLT: 5th International Conference on Deep Learning Technologies, pp. 26–31 (2021)
14. Liu, Y.-B., Jia, R.-S., Liu, Q.-M., Zhang, X.-L., Sun, H.-M.: Crowd counting method based on the self-attention residual network. Appl. Intell. **51**(1), 427–440 (2020). https://doi.org/10.1007/s10489-020-01842-w
15. Loy, C.C., Chen, K., Gong, S., Xiang, T.: Crowd counting and profiling: methodology and evaluation. In: Ali, S., Nishino, K., Manocha, D., Shah, M. (eds.) Modeling, Simulation and Visual Analysis of Crowds. TISVC, vol. 11, pp. 347–382. Springer, New York (2013). https://doi.org/10.1007/978-1-4614-8483-7_14
16. Ma, Z., Wei, X., Hong, X., Gong, Y.: Bayesian loss for crowd count estimation with point supervision. In: ICCV, pp. 6141–6150 (2019)
17. Meng, S., Li, J., Guo, W., Ye, L., Jiang, J.: PHNet: parasite-host network for video crowd counting. In: 25th International Conference on Pattern Recognition, ICPR, pp. 1956–1963 (2020)
18. Miao, Y., Han, J., Gao, Y., Zhang, B.: ST-CNN: spatial-temporal convolutional neural network for crowd counting in videos. Pattern Recognit. Lett. **125**, 113–118 (2019)
19. Sindagi, V., Patel, V.M.: A survey of recent advances in CNN-based single image crowd counting and density estimation. Pattern Recognit. Lett. **107**, 3–16 (2017)
20. Tripathy, S.K., Srivastava, R.: AMS-CNN: attentive multi-stream CNN for video-based crowd counting. Int. J. Multimedia. Inf. Retr. **10**(4), 239–254 (2021). https://doi.org/10.1007/s13735-021-00220-7
21. Wu, Q., Zhang, C., Kong, X., Zhao, M., Chen, Y.: Triple attention for robust video crowd counting. In: IEEE ICIP, pp. 1966–1970 (2020)
22. Wu, X., Xu, B., Zheng, Y., Ye, H., Yang, J., He, L.: Fast video crowd counting with a temporal aware network. Neurocomputing **403**, 13–20 (2020)
23. Xiong, F., Shi, X., Yeung, D.: Spatiotemporal modeling for crowd counting in videos. In: International Conference on Computer Vision, ICCV, pp. 5161–5169 (2017)
24. Xu, C., Liang, D., Xu, Y., et al.: AutoScale: learning to scale for crowd counting. Int. J. Comput. Vis. **130**, 405–434 (2022). https://doi.org/10.1007/s11263-021-01542-z
25. Yu, Y., Zhu, H., Wang, L., Pedrycz, W.: Dense crowd counting based on adaptive scene division. Int. J. Mach. Learn. Cybern. **12**(4), 931–942 (2020). https://doi.org/10.1007/s13042-020-01212-5
26. Zhang, S., Zhang, X., Li, H., He, H., Song, D., Wang, L.: Hierarchical pyramid attentive network with spatial separable convolution for crowd counting. Eng. Appl. Artif. Intell. **108**, 104563 (2022)
27. Zhang, Y., Zhou, D., Chen, S., Gao, S., Ma, Y.: Single-image crowd counting via multi-column convolutional neural network. In: CVPR, pp. 589–597 (2016)
28. Zhu, F., Yan, H., Chen, X., Li, T.: Real-time crowd counting via lightweight scale-aware network. Neurocomputing **472**, 54–67 (2022)
29. Zhu, F., Yan, H., Chen, X., Li, T., Zhang, Z.: A multi-scale and multi-level feature aggregation network for crowd counting. Neurocomputing **423**, 46–56 (2021)
30. Zou, Z., Shao, H., Qu, X., Wei, W., Zhou, P.: Enhanced 3D convolutional networks for crowd counting. In: 30th British Machine Vision Conference, BMVC, p. 250 (2019)

Frame-Wise Action Recognition Training Framework for Skeleton-Based Anomaly Behavior Detection

Hiroaki Tani(✉) and Tomoyuki Shibata

Corporate Research and Development Center, Toshiba Corporation, Kawasaki, Japan
{hiroaki4.tani,tomoyuki1.shibata}@toshiba.co.jp

Abstract. We propose a novel training framework for frame-wise action recognition from a single video frame. Most existing action recognition methods employ sequence-wise features extraction, but these approaches constrain the setting of the camera's frame rate and are not always effective in the variety of existing network environments. To this end, our framework employs temporal attention obtained from a pretrained action recognition model, allowing training action recognition from a single video frame selected based on its "action-ness," even from existing training data without the need for additional data. In this paper, we demonstrate the effectiveness of our framework to address the challenging task of anomaly behavior detection from a single-frame skeleton. Our method is realized by using frame-wise features extracted from a skeleton-based action recognition model trained by our framework. By coupling our framework and our anomaly behavior detection method, we develop a powerful detector of anomaly behavior that humans can recognize from a single video frame. We evaluate our method with the "ShanghaiTech Campus" anomaly behavior detection benchmark dataset, and confirm its effectiveness when the input consists of single-frame skeletons.

Keywords: Action recognition · Anomaly detection · Surveillance camera · Graph convolutional network

1 Introduction

Action recognition has attracted considerable attention and has been widely investigated [2,7,8,25,29]. It continues to improve in performance by recognizing even small human motion. These technologies have a wide range of applications.

Automatically detecting anomaly behaviors such as "falling" and "fighting" in surveillance videos is called "anomaly behavior detection," and it is expected to be useful for early detection and resolution of accidents using surveillance camera installations. Therefore, anomaly behavior detection, which can achieve safety and security for many people, has attracted more attention in recent years [19,26]. However, most existing methods have the weakness of requiring temporal information. In some situations, it is not possible to use stable high frame rate

S. Sclaroff et al. (Eds.): ICIAP 2022, LNCS 13233, pp. 312–323, 2022.
https://doi.org/10.1007/978-3-031-06433-3_27

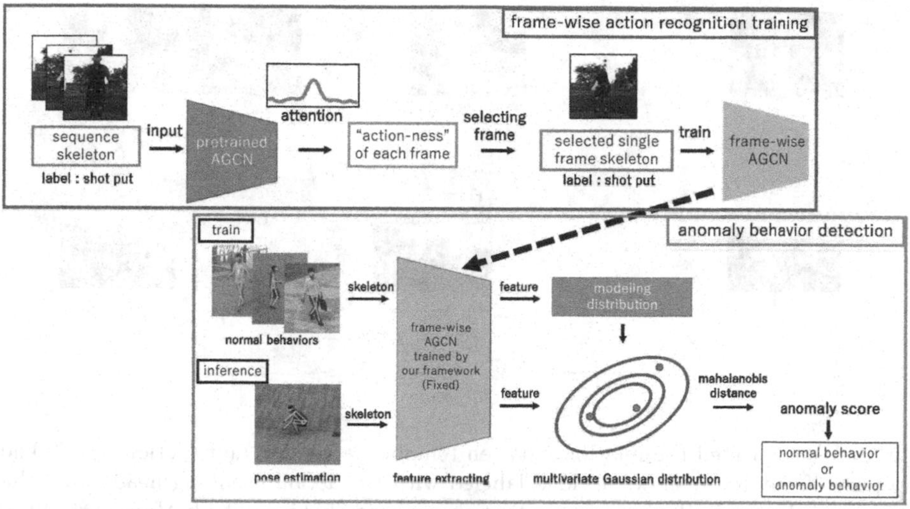

Fig. 1. Our framework employs the "action-ness" represented by the temporal attention obtained from a pretrained model, and we use the frames selected based on "action-ness" to train frame-wise models. Our method for anomaly behavior detection uses features extracted from a model trained by our framework. We use the multivariate Gaussian distribution and Mahalanobis distance for anomaly behavior detection.

videos because general surveillance cameras operate at low frame rates due to network limitations and camera settings. Existing methods that require fixed-width video sequences are thus difficult to operate.

Anomaly behavior detection can be roughly categorized into two approaches: extracting features directly from videos [1,4,5,10,16,20,21] and extracting features from action skeletons estimated by the pose estimator such as OpenPose [3,17,18,24]. Videos contain more information than do skeletons, but large amounts of information make the video approach computationally intensive and strongly susceptible to disturbance such as background features. Furthermore, privacy is a concern when collecting unprocessed surveillance videos. On the other hand, skeletons can be processed with less computation and protect privacy because they contain limited amounts of the information necessary for anomaly behavior detection, because subjects are represented as human joint coordinates. This study considers using the software as a service (SaaS) targeting method for skeleton-based anomaly behavior detection in industrial applications to enable operations in environments with different frame rates. This is an important issue that businesses need to solve to provide SaaS functionality in a generic way.

We propose a novel training framework for frame-wise action recognition from a single video frame, as shown at the top of Fig. 1. We leverage this challenge by using the key frames necessary for training. We thus employ temporal attention as obtained from a pretrained action recognition model, enabling us to train for action recognition from just a single video frame selected based on its "action-ness." Selecting single video frame with our framework realizes efficient

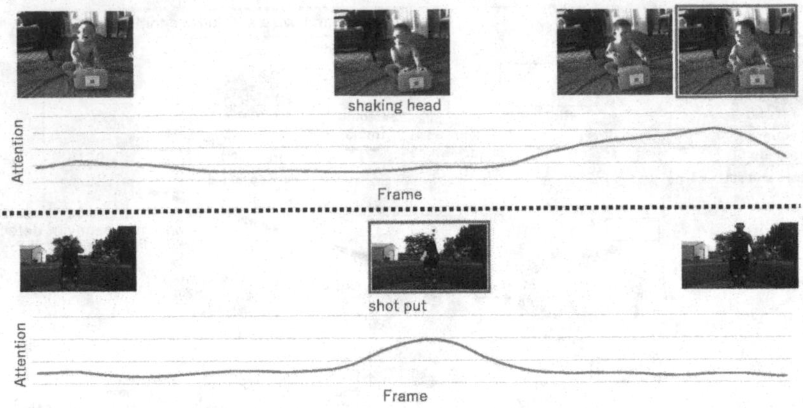

Fig. 2. We confirmed the relation between temporal attention and "action-ness." The top part of the figure shows a video labeled with the action "shaking head," and the attention is higher in the later frame that shows head-shaking. The bottom part of the figure shows a video labeled with the action "shot put," and the attention is higher in the middle frame where "shot put". The frame with the highest attention can be used for efficient learning in models for frame-wise action recognition.

training for frame-wise action recognition models without additional training data. Although the models trained by our framework cannot recognize the small human motions, we believe it can recognize coarse human motions.

We also propose a method for the challenging task of frame-wise anomaly behavior detection in combination with our framework, as shown at the bottom of Fig. 1. This method is inspired by the image-based method for detecting anomalies industrial inspection [23], and uses features extracted from a model trained by our framework. This method can detect anomaly behavior with high performance even when the input consists of single-frame skeletons.

2 Related Work

In recent years, there have been many studies in the field of action recognition models. The following provides an overview of skeleton-based action recognition models, which is related to the scope of this study. We also provide an overview of skeleton-based anomaly behavior detection.

2.1 Skeleton-Based Action Recognition Models

Categories of skeleton-based action recognition models include convolutional neural network models [6,13], long short-term memory (LSTM) models [15,28], and graph convolutional network (GCN) models [25,29]. Among these, GCN models have attracted significant attention in recent years because they outperform conventional methods [22]. The first skeleton-based action recognition model using a GCN was the spatial-temporal GCN (ST-GCN) proposed by [29]. Most proposed GCN models are based on ST-GCN, which uses an adjacency matrix to represent predefined human joint information. In contrast, adaptive graph convolutional networks (AGCN) [25] add an adjacency matrix to learn the relation between distant joints, and this model has high performance.

There are many datasets for action recognition [9,11,12,27,30], but few datasets for skeleton-based action recognition [14,29]. These skeleton-based datasets do not have segment annotation, which indicates a time period of actions. For these reasons, it is difficult to train a frame-wise skeleton-based model.

2.2 Skeleton-Based Anomaly Behavior Detection

Skeleton-based anomaly detection methods have been proposed only in recent years, so there are relatively few related studies on these methods compared with video-based methods [1,10,20,21]. [18] was the first to propose a skeleton-based anomaly behavior detection using LSTM. [24] proposed a method to identify short and long term anomaly behaviors by changing the range of the skeleton sequences. [17] proposed a method for detecting anomaly behavior using latent features in an autoencoder composed of GCNs.

These existing skeleton-based methods for detecting anomaly behavior are limited to semi-supervised approaches trained from scratch using only normal behaviors. However, datasets for anomaly behavior detection contain only normal behaviors such as "walking", but training without diverse behaviors is insufficient for learning action features. Furthermore, these methods assume the input of fixed-width skeleton sequences, so it is impossible to operate them from just a single-frame skeleton.

3 Method

Existing methods cannot achieve action recognition from a single video frame because they require temporal information to recognize human motions. We develop a novel training framework that makes frame-wise action recognition possible using temporal attention obtained by the pretrained models, as shown at the top of Fig. 1. As one use of our framework, we demonstrate that high performance skeleton-based anomaly behavior detection under the condition of single frame skeleton inputs.

Also, existing skeleton-based anomaly behavior detection methods cannot sufficiently learn various behavior features because these models learn from only simple behaviors such as "walking" and "standing". To this end, we also develop a method using features extracted from a model trained by our framework, as shown at the bottom of Fig. 1. The model trained by our framework have high performance in discriminating actions and detecting anomaly behaviors because they can learn the features of various actions in a large dataset. Therefore, it is expected that the features extracted from this trained model can be used to detect anomaly behavior with high performance.

3.1 Selecting Frames for Frame-Wise Action Recognition Training

Existing large datasets for action recognition, such as Kinetics [12], have only one action label annotation per short video, and it is unclear in which frames people are moving. Considering the case of training with a single video frame, training with frames that are not relevant to target actions would negatively affect performance. Clearly, training an action recognition model with randomly selected single frames cannot be effective. To this end, we use an effective frame selected based on "action-ness" for frame-wise action recognition training.

We propose a training framework to enable training of frame-wise action recognition from a single video frame selected based on temporal attention of a pretrained action recognition model. Figure 2 shows the relation between the temporal attention and videos when skeleton sequences are inputs of the pretrained AGCN [25]. The upper figure shows a video labeled with the "shaking head" action, and we can confirm that attention is larger in a later frame that shows head-shaking. The lower figure shows a video labeled with the "shot put" action, and we can also confirm that attention is larger in the middle frame showing that action. In other words, frames with large temporal attention are important frames for action recognition and can be considered as where the target is acting, that is, they signify "action-ness." Moreover, the large temporal attention frame can also be said to be important for frame-wise training, as it indicates that it is necessary for classification. Based on this idea, we train the frame-wise model by selecting the frame with the largest temporal attention to realize efficient training. Our framework is simple, yet effective for training frame-wise action recognition models.

In this paper, we use the AGCN [25], to which we apply our framework as a skeleton-based action recognition model. This model is pretrained on a large skeleton-based action recognition dataset "Kinetics-skeleton" from scratch. The AGCN has the convolution processes, global average pooling (GAP), and a fully connected layer for classification. We use up to convolution process of this model to obtain temporal attention, and after convolution processing $f(x)$ outputs $C \times T \times J$ feature maps. Here, J denotes the number of skeleton joints, T denotes the temporal length and C denotes the number of channels. We obtain the largest temporal attention from the following equation:

$$SelectedFrameIndex = Argmax(GAP(f(x))) \tag{1}$$

where GAP represents the GAP layer, which works with channel dimensions and skeleton joint dimensions, therefore $GAP(f(x))$ outputs a $1 \times T \times 1$ feature map. $Argmax$ selects the largest attention frame index from the temporal feature map. Thus, the single video frame corresponding to this index and is used for frame-wise action recognition training. We use the action labels as they were annotated on the original video for training.

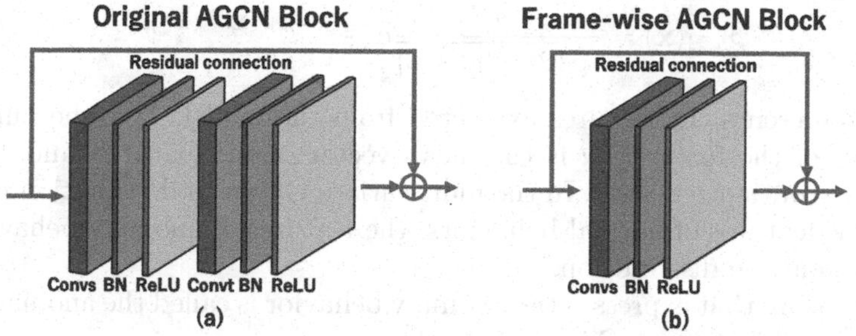

Fig. 3. (a) is an original AGCN Block; in (b) we remove the structure of the temporal processing from (a). Frame-wise AGCN consists of (b) blocks.

3.2 Frame-Wise AGCN

This section describes "frame-wise AGCN," which is a skeleton-based action recognition model without temporal dimension processing. Frame-wise AGCN is based on AGCN [25], which is a stack of nine basic processing blocks containing a spatial graph convolution (Convs) for skeleton joints, a temporal graph convolution (Convt) for the temporal dimension, a batch normalization (BN), and a ReLU, as shown a in Fig. 3(a). We remove a temporal graph convolution, BN, and ReLU from each block of AGCN to construct the "frame-wise AGCN," as shown in Fig. 3(b). This model has only spatial convolution, so that it can process single-frame skeleton. We perform experiments using only this model, but it is expected that our framework can be applied to other models as well.

To realize frame-wise action recognition training, we use selected frames by our framework to train frame-wise AGCN from scratch. The training method is almost the same as the existing skeleton-based action recognition models (e.g. AGCN), using single-frame skeleton inputs and corresponding action labels. The only difference from the existing models is whether inputs have a temporal dimension.

3.3 Anomaly Behavior Detection Method

We propose a method for anomaly behavior detection using features extracted from frame-wise AGCN trained by our framework. To calculate anomaly behavior scores, we use the multivariate Gaussian distribution of the feature extracted by frame-wise AGCN by inputting skeletons, as shown at the bottom of Fig. 1. We use 256-dimension features extracted from the output before the classification layer. When extracting features from the model, the parameters are fixed.

Inspired by the image-based anomaly detection method [23], we use the multivariate Gaussian distribution for anomaly behavior detection. Other probability models may be applicable, but they are future research. We directly model the multivariate Gaussian distribution of normal behavior data with features extracted from a model trained by our framework. The multivariate Gaussian distribution is defined as

$$\varphi_{\mu,\mathbf{\Sigma}}(\mathbf{x}) = \frac{1}{\sqrt{(2\pi)^D |\det\mathbf{\Sigma}|}} e^{-\frac{1}{2}(\mathbf{x}-\boldsymbol{\mu})^\top \mathbf{\Sigma}^{-1}(\mathbf{x}-\boldsymbol{\mu})} \tag{2}$$

where $\mathbf{x}$ represents the features extracted from the model, D is the number of dimensions of the features, μ is the mean vector of the features, and $\mathbf{\Sigma}$ is the symmetric covariance matrix. In the multivariate Gaussian distribution modeled by only the features of normal behaviors, the features of anomaly behaviors are distances from the distribution.

The measure that expresses the anomaly behavior is called the anomaly score. The Mahalanobis distance $M(\mathbf{x})$, defined as

$$M(\mathbf{x}) = \sqrt{(\mathbf{x}-\boldsymbol{\mu})^\top \mathbf{\Sigma}^{-1}(\mathbf{x}-\boldsymbol{\mu})} \tag{3}$$

is used to calculate the anomaly score under the multivariate Gaussian distribution with mean vector μ and symmetric covariance matrix $\mathbf{\Sigma}$. Here, small distances indicate normal behavior, and large distances indicate anomaly behavior. In this study, we do not set a threshold value, because we use the area under the curve (AUC) evaluation, which does not require a threshold value to evaluate anomaly behavior detection. We make this evaluation method the same as the other methods [17,18,24].

4 Experiments

4.1 Datasets and Evaluation

To apply our framework to a skeleton-based action recognition model for pre-training and frame-wise training, we use Kinetics-skeleton [29],which is a large skeleton dataset obtained from Kinetics-400 [12] using OpenPose [3]. Kinetics-400 consists of about 300,000 YouTube videos with 400 class labels, including daily activities and sports scenes. Kinetics-skeleton is the largest collection of skeleton actions and sufficient for training action features.

For anomaly behavior detection evaluation, we use the ShanghaiTech Campus dataset which is one of the largest and most diverse available surveillance camera datasets containing anomaly behaviors such as "falling," "fighting." It contains 130 videos captured in 13 scenes with complex lighting conditions and camera angles. This dataset has training data containing only normal behaviors (e.g. walking, and sitting), and evaluation data containing normal behaviors and anomaly behaviors. We use training data to model the multivariate Gaussian distribution and test data for evaluation. This evaluation is judged by whether an anomaly is occurring in each frame. We employ the AUC as one measure for evaluating anomaly detection performance. The closer the AUC is to 1, the higher the performance.

In our experiments of frame-wise anomaly behavior detection, we focus on the anomaly behaviors that humans can recognize from a single frame. Specifically, we evaluate the behaviors "throw," "theft," "swing," "jump," "fight," and "fall," as shown in Table 1. These behaviors are not annotated in the dataset, so we defined them manually.

Table 1. *Frame-wise Anomaly Behavior Detection Results:* We evaluated using ShanghaiTech Campus, with scores representing AUC. We evaluate on anomaly behaviors that humans can recognize from a single video frame. This table shows that our method can detect such anomaly behaviors and is superior to the comparison method. *indicates the results of our experiments using the public code.

	Throw	Theft	Swing	Jump	Fight	Fall	All
AE [17]*	0.788	0.673	0.350	0.616	0.619	0.535	0.596
Skeleton	0.664	0.647	0.819	0.629	0.774	0.742	0.723
PCA	0.657	0.646	0.679	0.587	0.788	0.746	0.701
Ours	**0.803**	**0.752**	**0.934**	**0.772**	**0.827**	**0.903**	**0.814**

Table 2. *Effectiveness of our framework:* We compare the accuracy of different selecting frame methods for our framework. The comparison of the selection method is "Random," "Center", "Low Att" and "Ours." Our method using "action-ness" as represented by temporal attention achieves high performance in the evaluation for most anomaly behaviors.

	Throw	Theft	Swing	Jump	Fight	Fall	All
Random	0.794	0.737	0.826	0.754	0.811	0.847	0.786
Center	0.800	0.736	0.932	0.747	**0.830**	0.850	0.802
Low Att	0.739	0.695	0.900	0.710	0.773	0.784	0.759
Ours	**0.803**	**0.752**	**0.934**	**0.772**	0.827	**0.903**	**0.814**

4.2 Experimental Setting

We employ the AGCN [25] which achieves high accuracy in Kinetics-skeleton action classification, and we use this model pretrained by Kinetics-skeleton. In selecting frames for frame-wise action recognition training, we extract temporal attention from the fixed parameters of the AGCN. We train the the frame-wise AGCN as in a typical classification using frames selected by our framework. Training details of frame-wise AGCN are the same as those of the AGCN [25]. In anomaly behavior detection, we extract 256-dimension features from the trained frame-wise AGCN at the outputs before the final classification layer. For skeleton preprocessing, we scale vertical and horizontal coordinate values by image size in a 0 to 1 value range, and normalize the data by subtracting the average skeleton coordinate. In the case of frame-wise anomaly behaviors detection, each frame is input. To prevent high anomaly scores due to failure of pose detection, we apply a Gaussian filter in the temporal direction to smooth calculated anomaly scores. Based on the method of [23], we use negative principal component analysis (NPCA) to extracted features. We cannnot measure the exact computation of our method at the inference, but we confirmed that it works sufficiently faster than a pose estimator.

4.3 Results

We conduct experiments to confirm the effectiveness of our framework and method. Other existing anomaly behavior detection methods cannot be compared in a fair environment, because their methods are structurally incapable of inputting a single-frame skeleton or their code is not publicly available. Therefore, we compare frame-wise anomaly behavior detection with only the method of [17] modified to frame-wise input and classic methods. All compared methods are limited to single-frame skeleton input. A description of the compared methods is given below.

GCN Autoencoder [17] **(AE).** [17] proposed a method for detecting latent features of anomaly behavior in an AE composed of GCNs. Originally this method is used to input sequence skeletons, but removing temporal processing, it makes inputting single-frame skeletons possible.

Skeleton Joints (Skeleton). We fit the multivariate Gaussian distribution using skeleton joints. This method uses skeleton joints as is instead of using features extracted from the trained model.

Principal Component Analysis (PCA) Compression. This method uses the compression in PCA for skeleton joints used above. PCA indicates Principal Component Analysis, which removes principal components that account for a total of 1% of smaller variance.

Table 1 shows the results of comparison experiments. Our frame-wise method outperforms the comparison methods for detecting anomaly behavior that humans can recognize from a single frame. Our method shows the highest performance for simple anomaly behaviors such as "swing" and "fall." These results show that we can detect major accidents by our method.

4.4 Ablation Study

Effectiveness of Our Framework. We confirmed the effectiveness of our framework, in which frame selection based on "action-ness" as represented by the largest attention obtained from the pretrained model. This evaluation is based on the frame-wise anomaly behavior detection using the ShanghaiTech Campus dataset in the same environments in Table 1. The comparisons are "Random," which randomly selected frames from the video sequence; "Center," which selected the middle of the video sequence, and "Low Att," which selected frames with the lowest attention. Table 2 shows the results. The frame-wise model trained by our framework shows high performance for many anomaly behaviors. "Center" shows the next-highest performance after our framework, likely because most videos show people acting in the middle of the video sequence. Poor results for "Low Att" indicate that "action-ness" in our framework is effective.

Sequence-Wise Anomaly Behavior Detection. To confirm our method using the features extracted from a trained model and the multivariate Gaussian

Table 3. *Sequence-wise anomaly behavior detection Results:* We evaluated using ShanghaiTech Campus and HR-ShanghaiTech Campus with scores representing AUC. Our method outperforms the current state of the art for all benchmarks. We can also confirm that our anomaly behavior detection method is effective. *indicates the results of our experiments using the public code.

	ShanghaiTech	HR-ShanghaiTech
Morais et al. [18]	0.734	0.754
Markovitz et al. [17]	0.752	0.722*
Rodrigues et al. [24]	0.760	0.770
Ours (Sequence-wise)	**0.775**	**0.782**

distribution as described in Sect. 3.3, we compared our sequence-wise method with those by [17,18,24]. Our sequence-wise method is a simple changing of the frame-wise AGCN to the pretrained AGCN, which originally accepts a sequential input. Because other existing methods are evaluated using "ShanghaiTech Campus" and "HR-ShanghaiTech Campus" [18] which is limited to some action in "ShanghaiTech Campus", the experiments are conducted under their same experiment conditions. This experiment input the sequence frame skeleton, and we also target anomaly behaviors that humans cannot recognize from a single frame such as "loitering" and "cycling" because we make this evaluation the same as the other methods. Therefore, Table 3 is a slightly different result than Table 1. We employ the pretrained AGCN as a sequence-wise model, and input 12-frame sequences size is as done by [17]. We employ the AUC as the evaluation protocol. Our sequence-wise method outperforms the state-of-the-art methods for all benchmarks in Table 3. This experiment shows that our anomaly behavior detection method, using features extracted from a pretrained model, is better than existing methods trained from scratch with a semi-supervised approach. We confirm that our anomaly behavior detection method is simple but more effective than the existing methods.

5 Conclusion

We proposed a training framework for frame-wise action recognition and a method for anomaly behavior detection. We evaluated whether the combination of our framework, and our method makes the challenging task of frame-wise anomaly behavior detection possible. Because our framework is not limited to anomaly behavior detection, we believe it can be applied to action recognition tasks which we will address in future research.

References

1. Abati, D., Porrello, A., Calderara, S., Cucchiara, R.: Latent space autoregression for novelty detection. In: Proceedings of the IEEE/CVF Conference on Computer Vision and Pattern Recognition, pp. 481–490 (2019)

2. Bertasius, G., Wang, H., Torresani, L.: Is space-time attention all you need for video understanding? arXiv preprint arXiv:2102.05095 (2021)
3. Cao, Z., Hidalgo, G., Simon, T., Wei, S.E., Sheikh, Y.: OpenPose: realtime multi-person 2D pose estimation using part affinity fields. IEEE Trans. Pattern Anal. Mach. Intell. **43**(1), 172–186 (2019)
4. Cohen, N., Hoshen, Y.: Sub-image anomaly detection with deep pyramid correspondences. arXiv preprint arXiv:2005.02357 (2020)
5. Defard, T., Setkov, A., Loesch, A., Audigier, R.: PaDiM: a patch distribution modeling framework for anomaly detection and localization. arXiv preprint arXiv:2011.08785 (2020)
6. Du, Y., Fu, Y., Wang, L.: Skeleton based action recognition with convolutional neural network. In: 2015 3rd IAPR Asian Conference on Pattern Recognition (ACPR), pp. 579–583. IEEE (2015)
7. Feichtenhofer, C.: X3D: expanding architectures for efficient video recognition. In: Proceedings of the IEEE/CVF Conference on Computer Vision and Pattern Recognition, pp. 203–213 (2020)
8. Feichtenhofer, C., Fan, H., Malik, J., He, K.: Slowfast networks for video recognition. In: Proceedings of the IEEE/CVF International Conference on Computer Vision, pp. 6202–6211 (2019)
9. Goyal, R., et al.: The "something something" video database for learning and evaluating visual common sense. In: Proceedings of the IEEE International Conference on Computer Vision, pp. 5842–5850 (2017)
10. Hasan, M., Choi, J., Neumann, J., Roy-Chowdhury, A.K., Davis, L.S.: Learning temporal regularity in video sequences. In: Proceedings of the IEEE Conference on Computer Vision and Pattern Recognition, pp. 733–742 (2016)
11. Karpathy, A., Toderici, G., Shetty, S., Leung, T., Sukthankar, R., Fei-Fei, L.: Large-scale video classification with convolutional neural networks. In: Proceedings of the IEEE conference on Computer Vision and Pattern Recognition, pp. 1725–1732 (2014)
12. Kay, W., et al.: The kinetics human action video dataset. arXiv preprint arXiv:1705.06950 (2017)
13. Kim, T.S., Reiter, A.: Interpretable 3D human action analysis with temporal convolutional networks. In: 2017 IEEE Conference on Computer Vision and Pattern Recognition Workshops (CVPRW), pp. 1623–1631. IEEE (2017)
14. Liu, J., Shahroudy, A., Perez, M., Wang, G., Duan, L.Y., Kot, A.C.: NTU RGB+D 120: a large-scale benchmark for 3D human activity understanding. IEEE Trans. Pattern Anal. Mach. Intell. (2019). https://doi.org/10.1109/TPAMI.2019.2916873
15. Liu, J., Shahroudy, A., Xu, D., Wang, G.: Spatio-temporal LSTM with trust gates for 3D human action recognition. In: Leibe, B., Matas, J., Sebe, N., Welling, M. (eds.) ECCV 2016, Part III. LNCS, vol. 9907, pp. 816–833. Springer, Cham (2016). https://doi.org/10.1007/978-3-319-46487-9_50
16. Liu, W., Luo, W., Lian, D., Gao, S.: Future frame prediction for anomaly detection-a new baseline. In: Proceedings of the IEEE Conference on Computer Vision and Pattern Recognition, pp. 6536–6545 (2018)
17. Markovitz, A., Sharir, G., Friedman, I., Zelnik-Manor, L., Avidan, S.: Graph embedded pose clustering for anomaly detection. In: Proceedings of the IEEE/CVF Conference on Computer Vision and Pattern Recognition, pp. 10539–10547 (2020)
18. Morais, R., Le, V., Tran, T., Saha, B., Mansour, M., Venkatesh, S.: Learning regularity in skeleton trajectories for anomaly detection in videos. In: Proceedings of the IEEE/CVF Conference on Computer Vision and Pattern Recognition, pp. 11996–12004 (2019)

19. Ramachandra, B., Jones, M.J., Vatsavai, R.R.: A survey of single-scene video anomaly detection (2020)
20. Ravanbakhsh, M., Nabi, M., Sangineto, E., Marcenaro, L., Regazzoni, C., Sebe, N.: Abnormal event detection in videos using generative adversarial nets. In: 2017 IEEE International Conference on Image Processing (ICIP), pp. 1577–1581. IEEE (2017)
21. Ravanbakhsh, M., Sangineto, E., Nabi, M., Sebe, N.: Training adversarial discriminators for cross-channel abnormal event detection in crowds. In: 2019 IEEE Winter Conference on Applications of Computer Vision (WACV), pp. 1896–1904. IEEE (2019)
22. Ren, B., Liu, M., Ding, R., Liu, H.: A survey on 3D skeleton-based action recognition using learning method. arXiv preprint arXiv:2002.05907 (2020)
23. Rippel, O., Mertens, P., Merhof, D.: Modeling the distribution of normal data in pre-trained deep features for anomaly detection. In: 2020 25th International Conference on Pattern Recognition (ICPR), pp. 6726–6733. IEEE (2021)
24. Rodrigues, R., Bhargava, N., Velmurugan, R., Chaudhuri, S.: Multi-timescale trajectory prediction for abnormal human activity detection. In: Proceedings of the IEEE/CVF Winter Conference on Applications of Computer Vision, pp. 2626–2634 (2020)
25. Shi, L., Zhang, Y., Cheng, J., Lu, H.: Two-stream adaptive graph convolutional networks for skeleton-based action recognition. In: Proceedings of the IEEE/CVF Conference on Computer Vision and Pattern Recognition, pp. 12026–12035 (2019)
26. Sodemann, A.A., Ross, M.P., Borghetti, B.J.: A review of anomaly detection in automated surveillance. IEEE Trans. Syst. Man Cybern Part C (Appl. Rev.) **42**(6), 1257–1272 (2012)
27. Soomro, K., Zamir, A.R., Shah, M.: UCF101: A dataset of 101 human actions classes from videos in the wild. arXiv preprint arXiv:1212.0402 (2012)
28. Veeriah, V., Zhuang, N., Qi, G.J.: Differential recurrent neural networks for action recognition. In: Proceedings of the IEEE International Conference on Computer Vision, pp. 4041–4049 (2015)
29. Yan, S., Xiong, Y., Lin, D.: Spatial temporal graph convolutional networks for skeleton-based action recognition. In: Thirty-Second AAAI Conference on Artificial Intelligence (2018)
30. Zhao, H., Torralba, A., Torresani, L., Yan, Z.: Hacs: Human action clips and segments dataset for recognition and temporal localization. In: Proceedings of the IEEE/CVF International Conference on Computer Vision, pp. 8668–8678 (2019)

The Automated Temporal Analysis of Gaze Following in a Visual Tracking Task

Vidushani Dhanawansa[1(✉)], Pradeepa Samarasinghe[1], Bryan Gardiner[2], Pratheepan Yogarajah[2], and Anuradha Karunasena[1]

[1] Faculty of Computing, Sri Lanka Institute of Information Technology, Malabe, Sri Lanka
vidushani.d@gmail.com, {pradeepa.s,anuradha.s}@sliit.lk
[2] School of Computing, Engineering and Intelligent Systems, Ulster University, Londonderry, UK
{b.gardiner,p.yogarajah}@ulster.ac.uk

Abstract. The attention assessment of an individual in following the motion of a target object provides valuable insights into understanding one's behavioural patterns in cognitive disorders including Autism Spectrum Disorder (ASD). Existing frameworks often require dedicated devices for gaze capture, focus on stationary target objects, or fails to conduct a temporal analysis of the participant's response. Thus, in order to address the persisting research gap in the analysis of video capture of a visual tracking task, this paper proposes a novel framework to analyse the temporal relationship between the 3D head pose angles and object displacement, and demonstrates its validity via application on the EYE-DIAP video dataset. The conducted multivariate time-series analysis is two-fold; the statistical correlation computes the similarity between the time series as an overall measure of attention; and the Dynamic Time Warping (DTW) algorithm aligns the two sequences, and computes relevant temporal metrics. The temporal features of latency and maximum time of focus retention enabled an intragroup comparison between the performance of the participants. Further analysis disclosed valuable insights into the behavioural response of participants, including the superior response to horizontal motion of the target and the improvement in retention of focus on the vertical motion over time, implying that following a vertical target initially proved a challenging task.

Keywords: Automated gaze analysis · Multivariate time-series analysis · Head pose

1 Introduction

The analysis of human gaze with respect to a dynamic target is vital to the perception of one's attention and engagement as per clinical documentation [6]. In contrast to manual qualitative evaluation, merits of automated analysis include

S. Sclaroff et al. (Eds.): ICIAP 2022, LNCS 13233, pp. 324–336, 2022.
https://doi.org/10.1007/978-3-031-06433-3_28

an unbiased judgement of the resultant gaze, enhanced by metrics to allow further prediction based on the level of attention. Frameworks for analysis of this calibre have been applied in a variety of domains ranging from social behaviour of adults [7,19,20] and infants [12] to classroom environments [3,4] and human-robot interaction [16].

Whilst gaze analysis on static images have been attempted with the aid of the line of sight [4] and heat maps [3,16] between the participant and target objects, only limited challenges in analysing a sequence of frames have been previously investigated. Lemaignan et al. [11] studied one's focus time on a target, determined by a broad field of attention estimated by the head pose, limiting the accurate alignment of the gaze to a small dynamic target. A study of joint attention [19] estimated the latency between the instruction and the resulting look, based on the presence of the gaze dot within the expected field of attention, whilst the latency, longest and shortest look for each object of interest has been previously assessed [20]. However, both these works rely on specialised sensors to capture the gaze and object position, as opposed to a single video.

To date, computer vision-based eye gaze estimation remains challenging due to variations in head pose, illumination, occlusions and the requirement for the capture of high-resolution images of the eye [9]. Thus, the input modality of the majority of work in this domain constitutes of dedicated devices to facilitate eye tracking, which deviates from a natural interaction and contributes towards one's discomfort and expense of data collection. The aforementioned impracticality in eye tracking within a resource-restricted environment has recently expedited the approach of gaze estimation via head pose. Prior work has extensively proven the feasibility of robust head pose estimation in a wild environment from 2D images [15]. In addition, the reliability of this gaze estimation has been adequately demonstrated by experiments [17] which concluded that the head pose contributes to an average of 88.7% to one's focus of attention. Therefore, the model proposed in this paper focuses on exploiting the relationship between the 3D head pose angles and position of the object within a visual tracking task.

Furthermore, the analysis of attention over a time frame in response to a dynamic physical target was a persisting research gap in the domain of attention analysis, given the invalidity of prior approaches primarily focusing on stationary targets [11,19,20]. As per the Diagnostic and Statistical Manual of Mental Disorders [2], children displaying prodromal symptoms of Autism Spectrum Disorder (ASD) show deficits in following another's pointing, eye gaze or movement of a

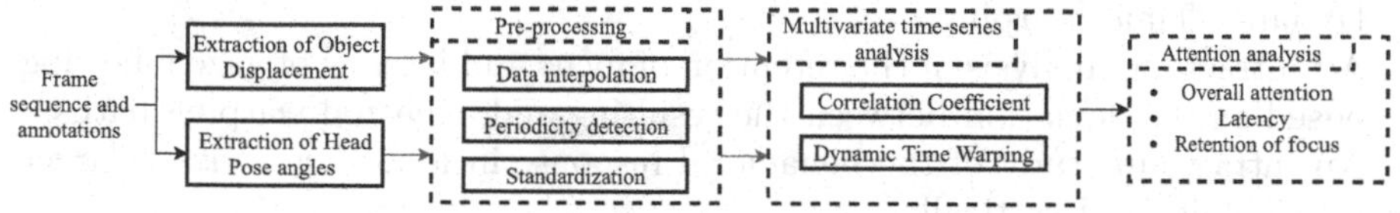

Fig. 1. System overview diagram of the proposed framework for the analysis of an object following task.

target object. For instance, the Autism Observation Scale for Infants (AOSI) [5], a standard set of semi-structured activities designed to detect and monitor early signs of autism, incorporates a visual tracking task which assesses the ability to track the trajectory of a rattle. The AOSI scoring system considers the smoothness of tracking the target object, the resultant delay in response, and ability to track the vertical movement of an object; factors which would be quantitatively assessed in the proposed approach.

Thus, an automated behavioural analysis of response to a dynamic target enhances the identification of symptoms of attention based disorders, and subsequent intervention to improve cognitive behaviour in children. To the best of the authors' knowledge, only a qualitative evaluation of a visual tracking task has been previously attempted [10], which visualises the variation of the estimated yaw to evaluate the smoothness of the participant's response. Furthermore, it fails to analyse the similarity between the head pose and the trajectory of the target object to assess how closely the target is followed, does not model the response to the vertical motion of the target, and is void of a temporal analysis of the response.

To address the aforementioned research gaps in the evaluation of an individual's engagement within a captured video of an object following task, the framework presented in Fig. 1 is proposed. The time-series data pertaining to the gaze and position of the target is subject to a multivariate time-series analysis, which employs: 1) statistical correlation to compute the similarity between the series, and 2) the Dynamic Time Warping (DTW) algorithm [13] to further analyse the temporal dynamics. In contrary to the traditional application of DTW as an exclusive similarity measure between time series in domains including gesture recognition [1] and gait analysis [18], the DTW algorithm was utilised in the proposed framework to compute the time lag, a parameter vital in drawing a conclusion on the attention extended by an individual. The attention in response to the motion of the stimulus along the horizontal and vertical planes was evaluated based upon the relationship between the yaw and x displacement; and pitch and y displacement.

The potential for application of the framework in a clinical behavioural study of a group, within a low-resource environment was demonstrated via the study on the EYEDIAP dataset [8]. The proposed framework is novel in that it is the first simple computational framework of this calibre, with the following contributions:

- A multivariate time-series analysis between the head pose angles and object displacement for periodic motion, based on statistical correlation and Dynamic Time Warping.
- An intragroup analysis of the retention of focus and latency of gaze following, based on the deviation between the resulting and expected warping paths.
- An intragroup study of the behavioral response in following a visual target, horizontally and vertically.

The rest of the paper is organized as follows: Sects. 2 and 3 detail the methodology followed in the proposed framework. The analysed results and discussion

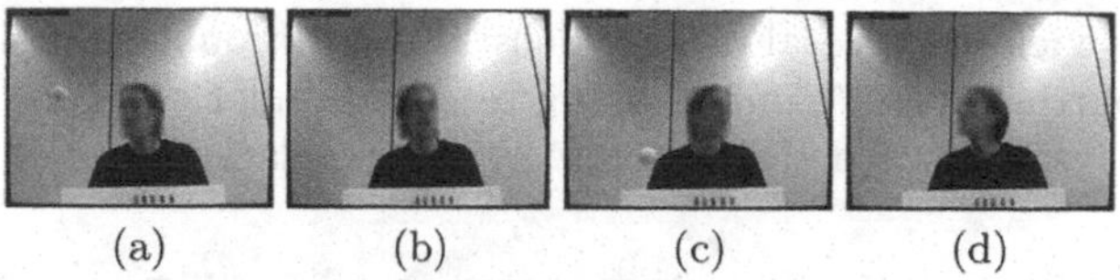

Fig. 2. Sample frames of video 10, with images modified to maintain privacy of participants: (a), (b) Following the horizontal motion of the target; (c), (d) Following the vertical motion of the target.

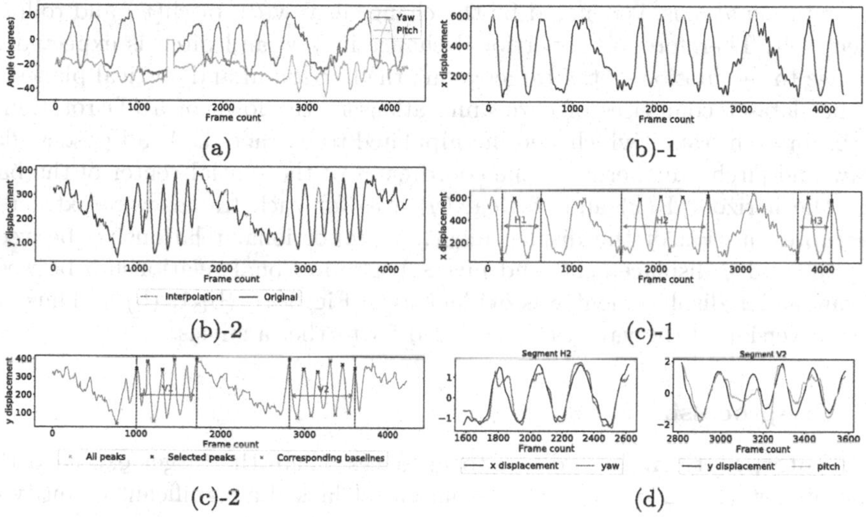

Fig. 3. Extracted data and pre-processing for video 9: (a) Extracted yaw and pitch angles; (b) Original and interpolated displacement; (c) Automatic periodicity detection and segmentation of displacement. Segments of horizontal and vertical motion are denoted as H1-H3, and V1-V2, respectively; (d) Relationship between standardized data for selected segments H2 and V2 from (c).

is presented in Sect. 4. Finally, the paper is concluded in Sect. 5 with directions for future work.

2 Extraction of Data and Pre-processing

2.1 EYEDIAP Video Dataset

The EYEDIAP dataset, to the best of the authors' knowledge, is the only available annotated video dataset capturing participants gazing at a dynamic visual target (Fig. 2). Those sessions of participants following a 3-D floating target (FT), while performing head movements (denoted as Mobile case, M) were extracted from the dataset. The selected video input consisted of a standard RGB stream at resolution 640×480 and frame rate of 30 fps. The dataset comprises of

16 participants yielding a total of 18 videos; two additional videos are included as a result of participants 15 and 16 being recorded under two conditions. Video of participant 4 was discarded since 60.9% of frames lacked annotations.

2.2 Extraction of Gaze and Object Displacement Data

The fine-grained head pose of an individual is interpreted as a continuous angular measurement across multiple Degrees of Freedom (DOF) [14]. The interpretation of 3 DOF is adopted in this work, where, motion about a vertical, horizontal and longitudinal axis, is denoted by the change in yaw(θ), pitch(ψ) and roll(ϕ), respectively. Therefore, a proportional change in yaw and pitch is expected in response to the motion of the target along the horizontal and vertical planes.

The dataset comprises of gaze annotations in the form of a 3D rotational matrix for each frame, which were manipulated to extract the head pose angles of yaw and pitch. Furthermore, the coordinates of the spatial center of the ball along the horizontal (x) and vertical (y) axes for each frame, were extracted as a representation of the displacement. A proportional relationship between the pitch and y displacement, and inversely proportional relationship between the yaw and x displacement was evident as in Fig. 3(a), (b)-1, (b)-2. Thus, an inverted version of the yaw was considered for further analysis.

2.3 Pre-processing

Data Interpolation. In addition to instances where the target exceeded the frame of view (Fig. 2(b), (d)), the displacement lacked a significant quantity of annotations. On average, 20% of frames per video lacks annotations as a result. Therefore, available data was interpolated as seen in Fig. 3(b)-1, (b)-2 to generate the x and y coordinates of the missing segments taking into account the gradients of the adjoining segments which represent the velocity of the target.

For a segment between $[t_1, t_2]$, midpoint $\bar{t}$, and gradients of adjoining segments m_1 and m_2, the interpolated displacement value $d_{int}[t]$ at a time instant t was generated by (1). For additional lacking annotations within the linear regions of displacement, linear interpolation was applied via (2).

$$d_{int}[t] = \begin{cases} m_1 - \frac{m_1.(t-t_1)}{\bar{t}-t_1} + d[t-1], & \text{if } t_1 < t < \bar{t}. \\ \frac{m_2.(t-\bar{t})}{t_2-\bar{t}} + d[t-1], & \text{if } \bar{t} \leq t < t_2. \end{cases} \tag{1}$$

$$d_{int}[t] = d[t_1](1-\mu) + d[t_2]\mu \tag{2}$$

where $\mu = \frac{t-t_1}{t_2-t_1}$. Segments exceeding 90 frames were not interpolated since the approximations made over a larger number of frames may be compromised.

Periodicity Detection and Segmentation. In order to analyse the attention of an individual over periodic motion of the target, the segments of strictly periodic motion along the horizontal or vertical axis were identified (Fig. 3(c)-1, (c)-2). All local maxima were identified by simple comparison with the neighboring values. The minimal horizontal distance between samples was restricted to 75, and the prominence was limited to 200 to eliminate minor peaks. A threshold of 360 frames was applied in identifying the consecutive peaks, since it was evident that the period of cycles of motion in the video dataset did not exceed this value.

Standardization. Given that the gaze angles and displacement lie within dissimilar ranges, all data segments were standardized by (3) where μ and σ denote the mean and standard deviation of the segment, respectively. This results in a series of a common mean 0, and standard deviation 1 as depicted in Fig. 3(d), facilitating comparison between the multivariate time series.

$$x_{std} = \frac{x - \mu}{\sigma} \tag{3}$$

3 Time-Series Analysis Between the Gaze and Object Displacement

3.1 Statistical Correlation

The Pearson's correlation coefficient r, computed by (4) evaluates the linear relationship between two continuous variables, x and y:

$$r = \frac{\sum_{i=1}^{n}(x_i - \bar{x})(y_i - \bar{y})}{\sqrt{\sum_{i=1}^{n}(x_i - \bar{x})^2 \sum_{i=1}^{n}(y_i - \bar{y})^2)}} \tag{4}$$

The coefficients between the yaw and x displacement, and pitch and y displacement, evaluated for each identified segment, served as an overall measure of the attention extended by the participant over the considered time frame. The coefficients for segments H2 and V2 in Fig. 3(d) are 0.941 and 0.816, respectively.

3.2 Dynamic Time Warping (DTW)

The Dynamic Time Warping (DTW) algorithm yields the optimal alignment between two non-linearly warped time-series by computing the corresponding points bearing the least cost between the two series. Thus, it facilitates further analysis of the time lag between the gaze and trajectory of the target.

Implementation. Let the two sequences be defined as $X = (x_1, ..., x_n)$ and $Y = (y_1, ..., y_m)$. The generated cost matrix $d_{n \times m}$ is computed as $d_{i,j} = |x_i - y_j|$ where i and j refer to the indices on the sequences X and Y, respectively. The warping path, W of length k, is defined as the sequence of points of minimum overall cost and is represented as $W = (w_1, ..w_t, .., w_k)$, where $w_t = (i_t, j_t)$ refers to a point lying on the warping path, W at time, t. The following restrictions apply in generating the warping path as in Fig. 4(a):

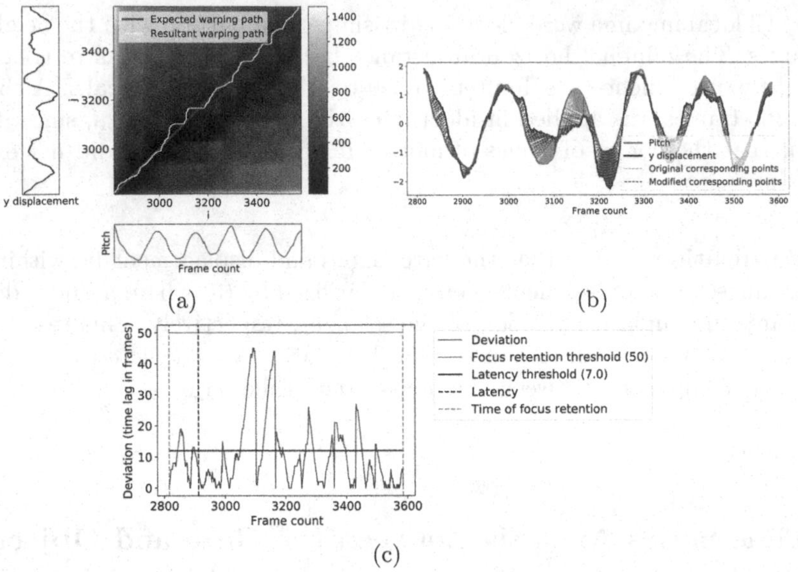

Fig. 4. Results of the DTW algorithm for segment V2 of video 9: (a) Generated cost matrix, expected warping path for perfect mapping, and the resultant warping path of minimum overall cost; (b) Original and modified alignment between the gaze and displacement sequences; (c) Time lag analysis of focus retention time and latency.

- Boundary condition: The path begins and ends with the starting and ending points of both signals, respectively, such that $w_1 = (1, 1)$, $w_k = (n, m)$.
- Monotonicity condition: The time order is preserved such that $i_{t-1} \leq i_t$ and $j_{t-1} \leq j_t$.
- Continuity condition: The translation of the path is restricted to adjacent points in the matrix. Therefore, $i_t - i_{t-1} \leq 1$ and $j_t - j_{t-1} \leq 1$.
- Warping window constraint: The window length (w), the maximum deviation of the warping path from the diagonal is restricted such that $|i-j| \leq w$. Based on empirical evidence, a window of 150 was applied.

Analysis of DTW Metrics. The accumulated cost of the warping path is a measure of the association between the two sequences aligned by the DTW approach. Therefore, a metric, Normalised Accumulated Cost (NAC) was defined as the derived cost normalised by the length of the warping path to account for sequences of varied lengths. Based on the continuity condition, the warping path includes all indices of each sequence, implying the repetition of indices and a path of variable length which hinders an accurate temporal analysis. Therefore, considering the displacement as the baseline, a modified warping path included only the corresponding index of minimum cost for each index of the displacement, as visualised in Fig. 4(b). Thereafter, the difference between the corresponding indices, normalised by the length of the warping path was calculated, and interpreted as the time lag. The retention of focus and latency is evaluated (Fig. 4(c)) based on this metric termed as Normalised Time Deviation (NTD).

The **time of focus retention** refers to the maximum duration an individual maintains a reasonable level of focus on the target object. Therefore, this was obtained by identifying the maximum number of consecutive frames of time lag less than a threshold of 45 frames, or 1.5 s within this framework. In addition, the percentage of focus retention over the number of frames of the segment was computed. This is similar in concept to the normalized ratio of time that the human interactant focuses its attention on the target, denoted as 'with-me-ness' in Lemaignan et al. [11]. The **latency** is the time elapsed prior to the gaze of the individual aligning closely with the path of the target object with certain accuracy, during which the time lag, or deviation should remain 0 over a time period of at least 1.5 s. However, a threshold equal to the 1st quartile of the time lag was considered to account for minor fluctuations. Interpreting Fig. 4(c), the individual retained the focus for the complete duration of 770 frames, i.e. 25.6 s. Furthermore, a latency of 3.23 s was recorded given the that 97 frames elapsed prior to the target being accurately followed. While the latency was computed for each segment, the overall latency of a participant was considered as that of the initial segment of the task.

4 Results and Discussion

4.1 Multivariate Time-Series Analysis Metrics

The metrics relevant to correlation and the DTW algorithm were compared to ensure the validity of the chosen approaches for the multivariate time-series analysis. It was hypothesised that an inverse relationship would result between the correlation coefficient, and the DTW metrics of NAC and NTD since a well-matched head pose series to the displacement yields a high correlation, and a minimum NAC and NTD. The scatter plot in Fig. 5 which includes metrics of all extracted segments in the dataset depicts an inversely proportional relationship between the relevant metrics, validating our hypothesis.

While it is understood that the two selected approaches of time-series analysis within this application corroborate each other, the significance of the approaches lie in their interpretation. The correlation coefficient, a single metric in the range [−1, +1] is an overall measure of the similarity between the head pose and the position of the target. The NAC is an enhanced similarity measure yielding the average cost between the sequences following the alignment by the DTW approach, and denotes how closely the trajectory of the head pose aligns with that of the target, removing dependancy on time. In contrast, the NTD is the average time lag between the aligned sequences, serving as a temporal measure between the time-series.

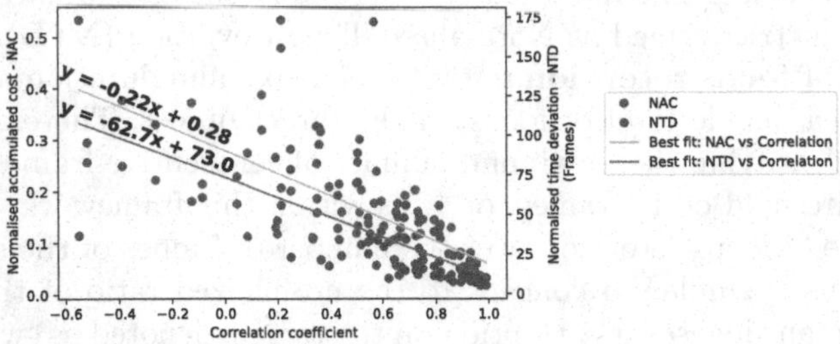

Fig. 5. Inverse relationship between Correlation Coefficient and DTW metrics (i.e. Normalized Accumulated Cost and Normalised Time Deviation) of all identified segments in the dataset.

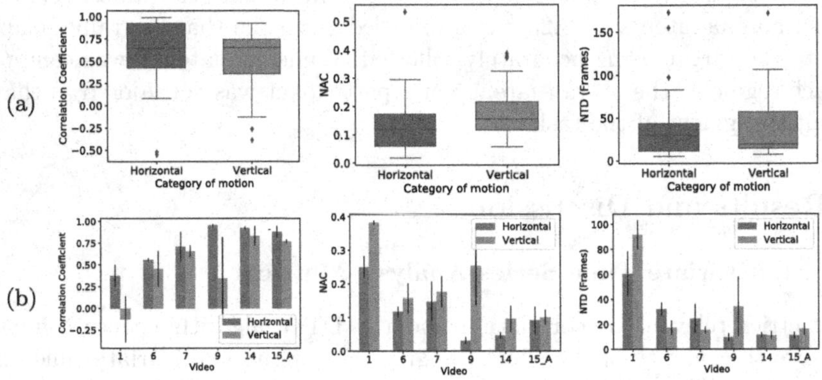

Fig. 6. Variation of Correlation Coefficient and DTW metrics of Normalised Accumulated Cost (NAC) and Normalised Time Deviation (NTD) in response to horizontal and vertical motion. (a) Box plots of all segments in the dataset; (b) Mean and standard deviation (represented by error bars) of all segments in sample videos. Selection of videos exhibit the possible variations in participant response within the dataset.

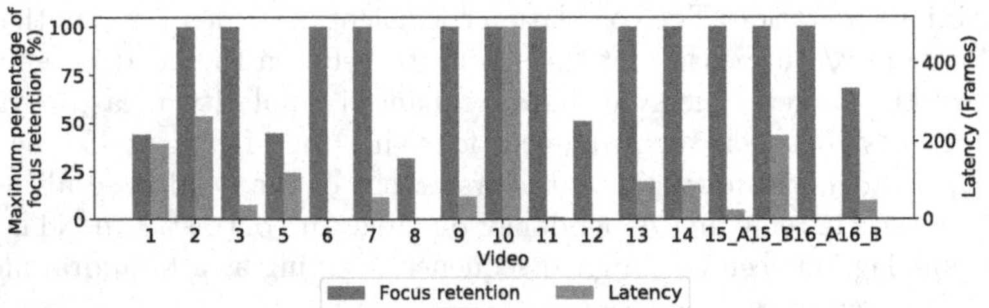

Fig. 7. Comparison of the maximum recorded focus retention percentage and the latency between participants.

4.2 Intragroup Comparison of Temporal Characteristics

A comparison between the performance of the participants in terms of the temporal features was conducted to further analyse the participants' level of attention. This considered the maximum percentage of focus retention amongst all segments within the video, and the latency of the initial segment as in Fig. 7. The focus retention measures the maximum time a participant is capable of maintaining focus on the target, with a minimal lag. It was evident that participants 1, 5, 8, 12 and 16_B were unable to maintain focus for more than 50% of a segment. In contrast, the latency is a strict measure of the time taken for the gaze of the participant to align with the expected trajectory of the target. The latency of the initial segment was utilised to extract the behaviour prior to any training the participant may accumulate during the task. Here, participants 1, 2, 5, 10, 15_B recorded comparably high latency values denoting that a significant time elapses prior to the gaze strictly aligning with the target position.

Taking into account both these factors, participants 1 and 5 were less successful in gaze following, bearing a low retention of focus and relatively high latency. Participants 6, 11 and 16_A demonstrated 100% of focus retention and negligible latency depicting excellent performance. Participant 10 demonstrated an interesting outcome with a focus retention of 100% and the maximum latency of the group, implying the maintenance of moderate focus throughout while failing to strictly follow the target.

An analysis of the latency and focus retention time of all segments revealed that in 58.8% of segments, the latency occurred before the focus retention time, signifying that individuals were more likely to retain focus continuously after having reached the expected trajectory even for a short period of time. In contrary, the latency occurred within the focus retention time in 29.4% of segments. This implies that certain individuals needed to focus on the target for a certain period of time prior to reaching the expected trajectory. The latency was not recorded in 11.8% segments indicating that participants failed to closely follow the target.

A comparison of the percentage of focus retention between the initial and final segments along the horizontal and vertical planes saw an increase in focus retention for vertical motion in 56.1% of participants. This notable improvement may imply that participants were involuntarily trained to focus on the target object over the time frame of the task. In contrast, 42.9% of participants demonstrated a declining focus retention percentage for horizontal motion.

4.3 Intragroup Comparison of Response to Horizontal and Vertical Motion

A study on the implications on the plane of motion of the target (Fig. 6) indicated that in general, participants were more responsive towards motion along a horizontal plane. Horizontal motion has resulted in a notably higher correlation coefficient and lower NAC throughout the dataset observing the 1st and 3rd quartiles in Fig. 6(a). The results of sample videos in Fig. 6(b) confirm this observation. This measure, along with the improvement in focus retention by the completion of the task may indicate that the vertical motion of the stimulus proves challenging, leading to reduced attention. In contrast, the declining or relatively constant focus could be justified by the monotonous nature of the horizontal motion, given that the object lies upon a constant level of sight.

It is also noteworthy that the NTD metric shows a less significant bias towards horizontal motion as evident through the median value of the boxplot figures. This is further highlighted by the sample results which depict that horizontal motion in videos 6 and 7 has yielded a higher NTD. This occurs since this metric emphasizes completely on the resulting time gap between the aligned sequences, whereas the correlation and NAC measures the similarity between the sequences. This suggests that the gaze of participants 6 and 7 follow the horizontal trajectory of the target, albeit with a significant lag.

5 Conclusion and Future Works

This paper proposes a novel framework for the automated evaluation of parameters pertaining to one's gaze response to the motion of a physical target. The framework addresses the persisting research gaps within gaze analysis focusing on a dynamic physical target. The time-series analysis between the head pose and the object displacement resulted in the correlation coefficient as an overall similarity measure. The analysis was enhanced by the similarity measure following the alignment of the time-series via the DTW algorithm, and the resultant time gap, upon which the time of focus retention and latency was computed. The application of the algorithm to the EYEDIAP video dataset demonstrated that the correlation and the metrics of DTW corroborated each other. Based on the temporal features, the gaze responses of participants were compared. While an increasing trend in focus retention between the initial and final segments of vertical motion was observed amongst participants, the response towards horizontal motion of the target outperformed that of vertical motion, as evident through all metrics.

Given the necessity of only a single video capture, the proposed framework shows prospects for implementation in low-resource environments such as ASD screening centers for children in developing countries. Thus, future work includes the application of the model on collected clinical data to evaluate the gaze response of children in the identification of deficits pertaining to ASD. A deep learning model for head pose estimation and a combination of object detection and tracking algorithms would replace the phase of extracting dataset annotations of gaze and displacement as conducted in this study. Furthermore, the proposed algorithm would incorporate an alternative framework based on the analysis of eye gaze angles, facilitating a comparison between the two modes for gaze analysis. Finally, the development of a model for the automated categorisation of a child's gaze response according to the AOSI standard, based on the computed correlation and temporal metrics, is worth investigating.

Acknowledgement. This research was supported by the Accelerating Higher Education Expansion and Development (AHEAD) Operation of the Ministry of Higher Education of Sri Lanka funded by the World Bank (https://ahead.lk/result-area-3/).

References

1. Alon, J., Athitsos, V., Yuan, Q., Sclaroff, S.: A unified framework for gesture recognition and spatiotemporal gesture segmentation. IEEE Trans. Pattern Anal. Mach. Intell. **31**(9), 1685–1699 (2008)
2. Association, A.P., Association, A.P., et al.: Diagnostic and Statistical Manual of Mental Disorders: DSM-5. Arlington, VA (2013)
3. Aung, A.M., Ramakrishnan, A., Whitehill, J.R.: Who are they looking at? Automatic eye gaze following for classroom observation video analysis. International Educational Data Mining Society (2018)
4. Bidwell, J., Fuchs, H.: Classroom analytics: measuring student engagement with automated gaze tracking. Behav. Res. Methods **49**, 113 (2011)
5. Bryson, S.E., Zwaigenbaum, L., McDermott, C., Rombough, V., Brian, J.: The autism observation scale for infants: scale development and reliability data. J. Autism Dev. Disord. **38**(4), 731–738 (2008)
6. Emery, N.J.: The eyes have it: the neuroethology, function and evolution of social gaze. Neurosci. Biobehav. Rev. **24**(6), 581–604 (2000)
7. Fathi, A., Hodgins, J.K., Rehg, J.M.: Social interactions: a first-person perspective. In: 2012 IEEE Conference on Computer Vision and Pattern Recognition, pp. 1226–1233. IEEE (2012)
8. Funes Mora, K.A., Monay, F., Odobez, J.M.: EYEDIAP: a database for the development and evaluation of gaze estimation algorithms from RGB and RGB-D cameras. In: Proceedings of the Symposium on Eye Tracking Research and Applications, pp. 255–258 (2014)
9. Ghosh, S., Dhall, A., Hayat, M., Knibbe, J., Ji, Q.: Automatic gaze analysis: a survey of deep learning based approaches. arXiv preprint arXiv:2108.05479 (2021)
10. Hashemi, J., et al.: A computer vision approach for the assessment of autism-related behavioral markers. In: 2012 IEEE International Conference on Development and Learning and Epigenetic Robotics (ICDL), pp. 1–7. IEEE (2012)
11. Lemaignan, S., Garcia, F., Jacq, A., Dillenbourg, P.: From real-time attention assessment to "with-me-ness" in human-robot interaction. In: 2016 11th ACM/IEEE International Conference on Human-Robot Interaction (HRI), pp. 157–164. IEEE (2016)
12. Michel, C., Kayhan, E., Pauen, S., Hoehl, S.: Effects of reinforcement learning on gaze following of gaze and head direction in early infancy: an interactive eye-tracking study. Child Development (2021)
13. Müller, M.: Information Retrieval for Music and Motion, vol. 2. Springer, Heidelberg (2007). https://doi.org/10.1007/978-3-540-74048-3
14. Neto, E.N.A., et al.: Real-time head pose estimation for mobile devices. In: Yin, H., Costa, J.A.F., Barreto, G. (eds.) IDEAL 2012. LNCS, vol. 7435, pp. 467–474. Springer, Heidelberg (2012). https://doi.org/10.1007/978-3-642-32639-4_57
15. Patacchiola, M., Cangelosi, A.: Head pose estimation in the wild using convolutional neural networks and adaptive gradient methods. Pattern Recogn. **71**, 132–143 (2017)

16. Saran, A., Majumdar, S., Short, E.S., Thomaz, A., Niekum, S.: Human gaze following for human-robot interaction. In: 2018 IEEE/RSJ International Conference on Intelligent Robots and Systems (IROS), pp. 8615–8621. IEEE (2018)
17. Stiefelhagen, R.: Tracking focus of attention in meetings. In: Proceedings. Fourth IEEE International Conference on Multimodal Interfaces, pp. 273–280. IEEE (2002)
18. Veeraraghavan, A., Roy-Chowdhury, A.K., Chellappa, R.: Matching shape sequences in video with applications in human movement analysis. IEEE Trans. Pattern Anal. Mach. Intell. **27**(12), 1896–1909 (2005)
19. Venuprasad, P., et al.: Characterizing joint attention behavior during real world interactions using automated object and gaze detection. In: Proceedings of the 11th ACM Symposium on Eye Tracking Research & Applications, pp. 1–8 (2019)
20. Venuprasad, P., Xu, L., Huang, E., Gilman, A., Chukoskie, L., Cosman, P.: Analyzing gaze behavior using object detection and unsupervised clustering. In: ACM Symposium on Eye Tracking Research and Applications, pp. 1–9 (2020)

Untrimmed Action Anticipation

Ivan Rodin[1,2], Antonino Furnari[1,3(✉)], Dimitrios Mavroeidis[2], and Giovanni Maria Farinella[1,3]

[1] University of Catania, Viale Andrea Doria 6, 95128 Catania, Italy
{ivan.rodin,antonino.furnari,giovanni.farinella}@unict.it
[2] Philips Research, High Tech Campus 34, 5656 AE Eindhoven, The Netherlands
dimitrios.mavroeidis@philips.com
[3] Next Vision s.r.l - Spinoff of the University of Catania, Catania, Italy
info@nextvisionlab.it
https://www.nextvisionlab.it/

Abstract. Egocentric action anticipation consists in predicting a future action the camera wearer will perform from egocentric video. While the task has recently attracted the attention of the research community, current approaches assume that the input videos are "trimmed", meaning that a short video sequence is sampled a fixed time before the beginning of the action. We argue that, despite the recent advances in the field, trimmed action anticipation has a limited applicability in real-world scenarios where it is important to deal with "untrimmed" video inputs and it cannot be assumed that the exact moment in which the action will begin is known at test time. To overcome such limitations, we propose an untrimmed action anticipation task, which, similarly to temporal action detection, assumes that the input video is untrimmed at test time, while still requiring predictions to be made before the actions actually take place. We propose an evaluation procedure for methods designed to address this novel task, and compare several baselines on the EPIC-KITCHENS-100 dataset. Experiments show that the performance of current models designed for trimmed action anticipation is very limited and more research on this task is required.

Keywords: Action anticipation · Untrimmed video processing · Egocentric vision

1 Introduction

Egocentric or first-person vision concerns the understanding of videos captured by wearable devices such as action cameras or smart glasses [1]. Egocentric videos provide visual observations from the unique point of view of the camera wearer, capturing hands, manipulated objects, head movements etc., which are useful to infer the behaviour of the users. The first-person point of view allows not only to detect and recognise the activities performed by a camera wearer, but also to model future intentions and activities [16]. The egocentric action anticipation task, in particular, consists in predicting an action performed by the camera

S. Sclaroff et al. (Eds.): ICIAP 2022, LNCS 13233, pp. 337–348, 2022.
https://doi.org/10.1007/978-3-031-06433-3_29

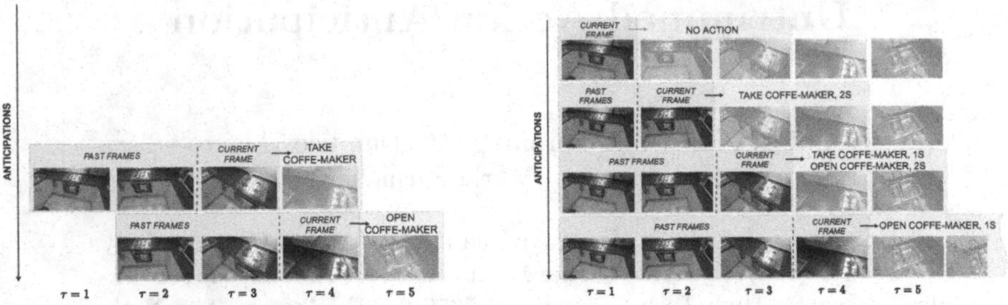

(a) Trimmed (b) Untrimmed

Fig. 1. Trimmed action anticipation assumes that a video sequence is sampled prior to the beginning of the action with a fixed anticipation time (usually set to 1 s). Untrimmed action anticipation processes video frames consequently and predicts all the actions beginning within a reasonable temporal horizon along with time-to-action estimates, or determine that no activities will be performed.

wearer before it occurs [4]. Following the notation proposed in [6], the task is to predict the class y of an action starting at time τ_s, by observing a *trimmed* video segment preceding the action, i.e., a video segment starting at time $\tau_s - (\tau_o + \tau_a)$ and ending at time $\tau_s - \tau_a$. Here, τ_o is the length of the observed trimmed video and τ_a is the anticipation time, denoting how many seconds in advance the anticipation should be performed. The above defined task, later referred to as "trimmed action anticipation" (Fig. 1(a)), has important limitations that restrict its applicability to real-life scenarios.

The first limitation arises from the observation that real action anticipation systems should be able to process video in an on-line fashion. More specifically, algorithms should be required to anticipate the occurrence of future actions at any time-step without knowing when the action will actually begin. We refer to this last task as "untrimmed action anticipation". For instance, *untrimmed* processing capabilities would be beneficial in such applications as guiding blind people by modelling their intentions and trajectories [15], or for human-robot interaction understanding [10,17], in which case a continuous video stream is expected as input to the activity anticipation module. In this paper, we show that methods targeted to trimmed action anticipation (e.g., [6]) achieve limited performance in the untrimmed anticipation scenario.

The second issue limiting the applicability of trimmed action anticipation in real scenarios is the fixed anticipation time τ_a, usually set to one second, which forces the action anticipation module to predict an action that will happen exactly τ_a seconds after the observed video segment. Indeed, setting a fixed anticipation time parameter τ_a does not follow the natural ability of humans to predict the future [2]. Rather than answering the question "what will happen next and when?", trimmed action anticipation answers the question "what will happen after τ_a seconds?". Taking in mind these considerations, it would be reasonable for an anticipation system to attempt to predict the pair $\widehat{y_a} = (\widehat{c_a}, \widehat{\tau_a})$

consisting of the next action $\widehat{c_a} = (verb_a, noun_a)$ to answer the "what" question, as well as the time-to-action $\widehat{\tau_a}$ to answer the "when" question.

A third issue is related to the constraint of the current definition of action anticipation to anticipate just one future action. Indeed, we believe that it is natural to predict, when possible, not only the next action $\widehat{c_{next}}$, which is the closest in the future to the current timestamp, but a sequence of future actions $\widehat{c_i}$ along with their respective time-to-action estimations $\widehat{\tau_i}$. This modified prediction scheme increases the amount of information returned by the system and can be important for modelling short, densely distributed actions. For instance, if the model misses the next very short action, it could still be able to handle the predictions of the remaining future actions, which is not evaluated in the current trimmed anticipation task.

In an attempt to overcome all of the above-mentioned problems, we propose an "untrimmed action anticipation" task (Fig. 1(b)), which assumes an untrimmed video as input and aims to predict at each time-step all actions beginning within a given temporal horizon (e.g., all actions beginning within 5 s). We propose an evaluation scheme which accounts for the different properties of the considered task and experimentally compare several baselines which are based on current trimmed action anticipation approaches. Results highlight that the proposed task is challenging for current methods designed to tackle trimmed action anticipation, suggesting that more research is required.

2 Related Works

Our work is related to previous investigations in the field. In [4], the task of trimmed egocentric action anticipation is presented. Despite previous studies on action anticipation [6,9,12,18] previous works have not explicitly considered the untrimmed version of the task. Nevertheless, some authors have investigated aspects closely related to this work.

The authors of [13] considered the problem of forecasting the time-to-collision between a suitcase-robot and a nearby pedestrian in a streaming video. The task aims to detect only collision events, thus the authors treat it as a regression task—predicting time-to-action at every timestamp. In [14], the task of single future event prediction is considered: if and when a future event will occur. Three metrics are used for evaluation: 1) Event prediction accuracy, assessing whether the event will occur within a prediction window; 2) Time-to-event error, which is the average absolute prediction error between the point estimate of the time-to-event distribution and the ground true time; 3) Model Surprise, indicating the quality and the confidence of the probabilistic output of the model.

Even if not concerned with future prediction, the task of online action detection [11] is relevant to our work. The task aims to predict an action class for the current frame of the video, how much time has passed since the beginning of the action and when the action will end. Three evaluation metrics are used to assess performance: 1) F1-score, for which correct detections are calculated based on the intersection over union (IoU) between prediction and ground truth

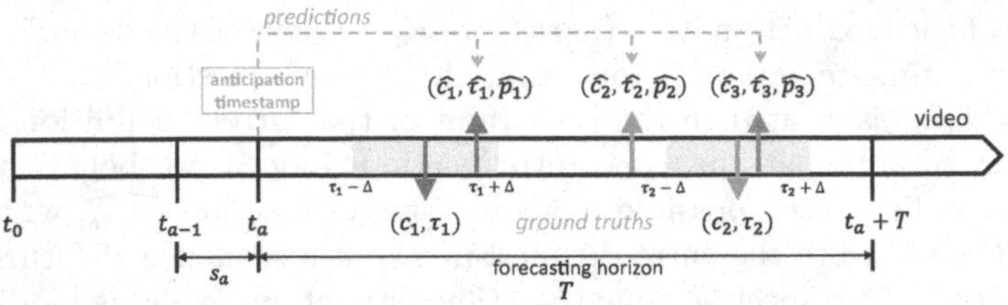

Fig. 2. Untrimmed Action Anticipation. Anticipations are evaluated every α seconds. For the anticipation timestamp t_a, the figure illustrates two ground truth future actions $y_a = \{(c_i, \tau_i)\}_{i=1}^{2}$ (downwards arrows) and three predictions $\widehat{y_a} = \{(\widehat{c_i}, \widehat{\tau_i}, \widehat{s_i})\}_{i=1}^{3}$ (upwards arrows) within the anticipation horizon T. Grey segments denote the Δ temporal offset around the ground-truth actions (see Sect. 4). Only $(\widehat{c_1}, \widehat{\tau_1})$ is a true positive prediction, since $(\widehat{c_2}, \widehat{\tau_2})$ does not lie within the temporal offset sampled around the ground truth action (c_2, τ_2), and $(\widehat{c_3}, \widehat{\tau_3})$ does not match the ground truth class label.

action intervals; 2) Start Localization Score (SL-Score) based on the relative distance between predicted and ground truth start times; 3) End Localization Score (EL-score), similar to SL-score but referred to the end of the action. Another relevant approach is StartNet, introduced in [8], which was developed to address the problem of Online Detection of Action Start (ODAS), where action classes and their starting times are detected in untrimmed streaming video input. Since the start time is predicted after the moment of action start is observed, also this work is not focused on future predictions. However, the untrimmed setting of the addressed problem, makes this paper relevant to our work. In particular, we will adopt the evaluation protocol based on point-level average precision (p-AP), used in [8] and proposed in [19].

3 Problem Formulation

We define untrimmed action anticipation as the task of predicting the actions beginning within a given anticipation horizon, together with their starting times relative to the current timestamp (time-to-action). Since evaluating the task at every frame in a video can be redundant and computationally expensive, for each video, we choose a set of anticipation timestamps $t_1, t_2, ..., t_n$ sampled every α seconds used for evaluation. For each anticipation timestamp t_a, we hence define a set of ground truth future actions y_a to be anticipated, represented as follows: $y_a = \{(c_i, \tau_i)\}_{i=1}^{N_a}$, where N_a is the number of future actions beginning within a fixed anticipation horizon T, c_i is the class label of the i-th future action and $\tau_i \in [0, T]$ is the time-to-action. The predictions at timestamp t_a are presented in the form $\widehat{y_a} = \{(\widehat{c_i}, \widehat{\tau_i}, \widehat{s_i})\}_{i=1}^{N_C}$, where N_C is the number of future actions predicted by the model, $\widehat{c_i}$ is the predicted class, $\widehat{\tau_i} \in \Re$ is the predicted time-to-action and $\widehat{s_i} \in [0, 1]$ is the model's confidence about the predicted "action, time-to-action" $(\widehat{c_i}, \widehat{\tau_i})$ pair. The proposed untrimmed action anticipation task is illustrated in Fig. 2.

It is worth noting that we require models to predict a set of future actions, rather than just the next one as in trimmed action anticipation. This is a natural choice considering that, in our scenario, models predict both the future action class and its time-to-action. For instance, a model systematically skipping the next action but correctly predicting labels and time-to-action of the subsequent ones would be penalised in the classic evaluation scheme, whereas it is evaluated more fairly with the proposed protocol.

4 Evaluation Protocol

Since our task formulation involves two subtasks i.e., predicting the class labels of future actions (classification) and predicting the time-to-action values (regression), the evaluation metric should take into account both problems. More than that, the metric should account for the situations in which no future action is going to take place within the temporal horizon, and, in fact, these situations are a big part of datasets of natural activities as discussed later. Considering all of the above, we propose to treat the untrimmed action anticipation task similarly to a detection problem and define an evaluation protocol which has commonalities with the tasks of action localisation and object detection. Indeed, there are many similarities between untrimmed action anticipation and object detection or action localisation. Both tasks have to deal with many "empty zones" either in the timeline or in the image where no action or object is present. Similarly to untrimmed action anticipation, tasks combine classification (i.e., predicting the action or object label) and regression (i.e., predicting the action start or the coordinates of the bounding boxes). We hence base our evaluation measure on the popular mean Average Precision (mAP) defined in the Pascal VOC Challenge [5] and adopted in many other works focused on object detection.

As in object detection, we first match each prediction to the most likely ground-truth future action annotation. Each prediction is then marked as either a true-positive or a false positive prediction. While in the Pascal VOC challenge the Intersection-Over-Union (IoU) was used to determine whether a detection is a true or a false positive by measuring the overlap between predicted and ground truth boxes, in untrimmed action anticipation, we propose to match a prediction to a ground truth action by measuring the temporal offset between the actual and predicted action start. Having the ground truth action y_a, its time-to-action τ_a and the predicted time-to-action $\hat{\tau_a}$, the temporal offset is calculated simply as $TO(\tau_a, \hat{\tau_a}) = |\tau_a - \hat{\tau_a}|$. The prediction $(\hat{c_i}, \hat{\tau_i}, \hat{s_i})$ is considered as a true positive if the class $\hat{c_i}$ matches the ground truth class c_i, and if the temporal offset is smaller than a specified threshold: $TO(\tau_i, \hat{\tau_i}) \leq \Delta$. An example of determining true or false positive predictions is illustrated in Fig. 2.

Once predictions have been matched to the ground truth, we compute precision-recall (PR) curve for each action class from the list of the ranked predictions. Chosen a rank, the fraction of positive examples matched with predictions which are above the given rank is defined as the rank-based recall, whereas the proportion of the predictions above that rank which have been matched to a ground truth annotation is defined as the rank-based precision. Rank-based precision and recall values are computed for each possible rank in order to form a PR curve. The average precision per class is calculated as the mean interpolated precision with respect to a set of uniformly sampled recall levels with a step of 0.1: $[0, 0.1, ..., 1]$. Thus, $AP = \frac{1}{11}\sum_{r\in[0,0.1,\ldots,1]} p_{interp}(r)$. Here, $p_{interp}(r)$ is the interpolated precision for the recall level r, computed by taking the maximum precision for which the corresponding recall exceeds r: $p_{interp}(r) = \max_{\widehat{r}:\widehat{r}\geq r} p(\widehat{r})$. The mean average precision (mAP) is obtained by averaging the AP values obtained for each of the considered action classes.

For our benchmark, we compute mAP considering multiple temporal thresholds $\Delta \in [0.25, 0.5, 0.75, 1.0]$ s. We also explore a relaxed variation of the proposed metric for untrimmed action anticipation which computes mean average precision scores assuming an infinite temporal offset thresholding ($\mathrm{mAP_{inf}}$).

5 Methods

In this section, we report the experiments performed on the EPIC-KITCHENS-100 dataset (EK-100) [3]. We perform two types of experiments. First, we test the sensibility of the evaluation protocol sensibility with respect to the task. To do so, we conduct a set of controlled experiments in which we generate baseline results by randomly perturbing ground truth annotations with different amounts of randomness. Second, we evaluate different baselines for untrimmed action anticipation based on models designed for the trimmed task.

5.1 Controlled Experiments

We propose a set of synthetic results obtained by randomly perturbing ground truth annotations using different perturbation parameters. These baselines are meant to provide a form of controlled experiments with predictors ranging from accurate to inaccurate based on the values chosen for the perturbation parameters. Our goal is to show that the proposed evaluation protocol does allow to estimate the model's ability to make good predictions. In these experiments, the random predictions $(\widehat{c_i}, \widehat{\tau_i}, \widehat{s_i})$ are generated from the ground-truth annotations (c_i, τ_i) as follows:

- With a "swapping" probability p_s, the ground-truth class label c_i is replaced with a label $\widehat{c_i}$ sampled from the class distribution in the training set:

$$\widehat{c_i} = \begin{cases} c_s \sim EDF(c), & \text{if } x \leq p_s \\ c_i, & \text{otherwise} \end{cases}$$

  where $x \sim U(0, 1)$, and $EDF(c)$ is the Empirical Distribution Function of action classes in the training set.

- The time-to-action $\widehat{\tau_i}$ is sampled from the normal distribution centred at the ground truth time-to-action with a pre-specified standard deviation σ: $\widehat{\tau_i} \sim N(\tau_i, \sigma)$
- The confidence score $\widehat{s_i}$ is set to $\widehat{s_i} = 1$

As a result, all action annotations $y_a = \{(c_i, \tau_i)\}_{i=1}^{N_a}$ are transformed to generated random predictions $\widehat{y} = (\widehat{c_i}, \widehat{\tau_i}, \widehat{s_i})_{i=1}^{N_a}$. For our experiments, we have explored different class swapping probabilities $x_s = [0.25, 0.5, 0.75]$ and different standard deviation values $\sigma = [0.33, 0.5, 1]$ for sampling time-to-action values.

5.2 Baselines Based on Trimmed Action Anticipation

We propose baselines for untrimmed action anticipation based on the Rolling Unrolling LSTM (RULSTM) model, which was originally designed for trimmed action anticipation [6]. The aim of these experiments is to explore the performance of current trimmed models when adapted to tackle the untrimmed action anticipation task. For simplicity, all these baselines rely on the RGB-inputs only.

RU: For the first baseline, we evaluate the original RULSTM model in an untrimmed manner, evaluating predictions every 0.25 s. The original RULSTM model was trained to predict a future action one second before it occurs. Thus, we fix the time-to-action for all the predictions to $\widehat{\tau_i} = 1$ s. For action label predictions we select all the classes with confidence larger than the threshold $\widehat{s_i} > 0.1$, providing the opportunity to predict multiple future actions. If the highest action prediction confidence is smaller than or equal to the threshold $\widehat{s_i} \leq 0.1$ we assume that the model has predicted no action. We consider this baseline to assess the suitability of current models optimised for trimmed action detection in the considered untrimmed scenario.

RU-no-act: For this baseline, we investigate whether re-training the RULSTM model on all timestamps (rather than on timestamps sampled exactly 1s before each action) and adding the opportunity to predict a "no action" as an additional label can provide better performance.

RU-sigmoid: In this baseline, we replace the final SoftMax layer in the RULSTM architecture with a sigmoid layer, and train the model using binary cross-entropy, thus adjusting the original model for a multi-class multi-label classification task.

RU-reg: This baseline is aimed to predict the time-to-action $\widehat{\tau_i}$ by exploiting an additional fully-connected layer attached to the RULSTM model and trained to solve the regression task with mean squared error as a loss function.

RU-5-clf: In order to infer the time-to-action predictions, this baseline makes use of five separate RULSTM models trained to predict future actions at various time scales of 1, 2, 3, 4 and 5 s.

6 Experimental Settings and Results

We perform our experiments on the EPIC-KITCHENS-100 dataset [3]. To make the dataset suitable for untrimmed action anticipation, we sampled timestamps every $\alpha = 0.25$ s, leading to 1.057.238 anticipation timestamps in the training set and 185.532 anticipation timestamps in the validation set.

The statistics of the retrieved anticipation timestamps indicate two important features of the task. First, 38% of anticipation timestamps annotations are not associated to future action labels within an anticipation horizon of five seconds, which highlights the necessity of modelling the *"no action"* predictions in real scenarios. Second, 36% of the anticipation timestamps contain at least two future action labels within the anticipation horizon, indicating the importance of being able to predict multiple future actions.

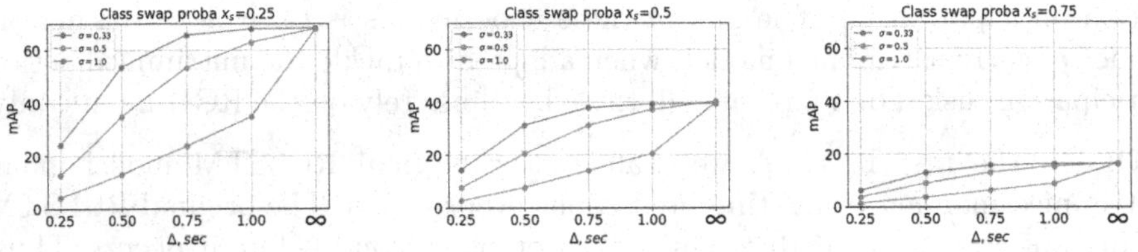

Fig. 3. Mean Average Precision Scores for the Random Baselines.

Table 1. Untrimmed anticipation results.

Action					
Model	$mAP_{0.25}$	$mAP_{0.5}$	$mAP_{0.75}$	$mAP_{1.0}$	mAP_{inf}
RU	**0.12**	**0.31**	**0.68**	0.84	2.05
RU-no-act	0.07	0.24	0.46	0.79	1.25
RU-sigmoid	0.01	0.02	0.03	0.06	0.11
RU-reg	0.01	0.01	0.02	0.04	0.07
RU-5-clf	0.11	**0.31**	0.66	**0.87**	**2.55**
Verb					
Model	$mAP_{0.25}$	$mAP_{0.5}$	$mAP_{0.75}$	$mAP_{1.0}$	mAP_{inf}
RU	0.25	0.67	1.28	2.01	4.82
RU-no-act	0.13	0.42	0.84	1.43	2.39
RU-sigmoid	0.02	0.03	0.05	0.10	0.19
RU-reg	0.01	0.02	0.03	0.05	0.11
RU-5-clf	**0.27**	**0.69**	**1.34**	**2.15**	**5.30**
Noun					
Model	$mAP_{0.25}$	$mAP_{0.5}$	$mAP_{0.75}$	$mAP_{1.0}$	mAP_{inf}
RU	0.23	0.60	1.14	1.79	4.25
RU-no-act	0.10	0.39	0.77	1.32	2.05
RU-sigmoid	0.02	0.03	0.04	0.09	0.18
RU-reg	0.01	0.01	0.02	0.04	0.10
RU-5-clf	**0.23**	**0.61**	**1.18**	**1.95**	**5.02**

Table 2. Results with respect to different sets of anticipation timestamps. "untrimmed ts": whole set of timestamps for the untrimmed task; "untrimmed ts, excl. no-act": all timestamps, excluding "no future action"; "trimmed ts": timestamps and action instances from the trimmed task.

$mAP_{1.0}$, action			
Model	untrimmed ts	untrimmed ts, excl. no-act	trimmed ts
RU	0.84	1.07	**4.72**
RU-no-act	0.79	1.02	2.98
RU-sigmoid	0.06	0.09	0.29
RU-reg	0.04	0.07	0.17
RU-5-clf	**0.87**	**1.12**	3.17

6.1 Results of the Controlled Experiments

Figure 3 reports the results of the controlled experiments for different choices of the perturbation parameters. The results show that the proposed evaluation protocol is sensitive to the perturbations of the parameters. In particular, the higher the class swapping probability is, the lower the metric scores are (compare the different plots in Fig. 3). Similar considerations apply to the perturbations of the time-to-action parameter: the larger the deviation from the ground truth time-to-action values of an action, the smaller the mAP score (compare the different series of the plots in Fig. 3). Naturally, the mAP_{inf} score (shown for $\Delta = \infty$) reports the same results for different values of σ, since the relaxed version of the metric does not penalises the predictions in the presence of wrong time-to-action estimates. Overall, the results suggest that the proposed evaluation protocol is suitable to evaluate models on this task. Indeed, a near-perfect method (blue line in the leftmost plot of Fig. 3) would achieve reasonable performance (mAP 30%) even for small values of the Δ parameter. Nevertheless, the task is challenging, as shown by the results in the rightmost plot of Fig. 3.

6.2 Trimmed Action Anticipation baseline Results

Table 1 reports the results of the models based on trimmed anticipation. Best results are highlighted in bold, whereas second-best scores are underlined. The results highlight that the considered untrimmed action anticipation task is challenging. Indeed, the standard RU model obtains modest results for action, verb and noun even for the relaxed mAP_{inf} metric (2.05, 4.82 and 4.25). Results are worse if more realistic measures are considered (e.g., see the $mAP_{0.5}$ column. Among the different baselines considered for comparisons, the combination of five RU models (RU-5-clf) consistently achieves best or second-best results. Simply adding a "no action" class (RU-no-act), replacing a softmax with a sigmoid to enable multiple detections (RU-sigmoid) or adding a regression layer (RU-reg) lead to decreased performance, which further highlight the challenging nature of the task and the limits of current models optimised on trimmed action detection. We observe that the trimmed models tested in the untrimmed task show a very large number of false positive predictions for the "no action" anticipation timestamps. In particular, exploratory analysis of the results shows that the best performing model (RU-5-clf) generates false positive predictions for 98% of the timestamps with no ground-truth action class, indicating the need of better ways to model the "no action" scenario appropriately. More attention should be paid to the time-to-action prediction. For all of the considered baselines the mAP_{inf} score is significantly higher than the $\text{mAP}_{1.0}$ score, indicating that in some cases the models could predict future actions, but they could not estimate the correct time-to-action even when a big temporal offset of one second is used.

Table 2 compares performances for different sets of anticipation timestamps with respect to $mAP_{1.0}, action$. The different sets include the untrimmed timestamps (same as the ones of Table 1), all timestamps excluding "no action" cases, and timestamps sampled 1 s. before the beginning of labelled actions (trimmed case). The results indicate that the models adapted from the trimmed task tend to obtain worst mAP scores when processing anticipation timestamps with multiple future actions or without defined future actions. Indeed, the task is much easier when trimmed timestamps are considered (last column).

The qualitative results of the experiments are presented in Fig. 4. The best performing model (RU-5-clf) is able to tackle the multiple actions prediction, as well as to predict the respected time-to-action estimates (Fig. 4a, frames 1, 3, 4, 5). Several predictions match with ground truth classes of future actions but have big error in time-to-action estimate (yellow), while sometimes the model is predicting the ongoing or past actions (*turn-off tap* on frames 2 and 3). It is also hard for model to predict the actions which are not aligned with common intuition of the human activities, i.e., predicting the *take juicer* action after filling the kettle with water, especially at the beginning when the juicer is out of the view (frames 2, 3, 4).

Much poorer performance is shown on the long sequences of the monotonic actions (Fig. 4b). The model tends to predict the beginning of the new action(s) within the anticipation horizon, however, the ongoing action does not terminates for a long time, and all the inferred predictions become false positives.

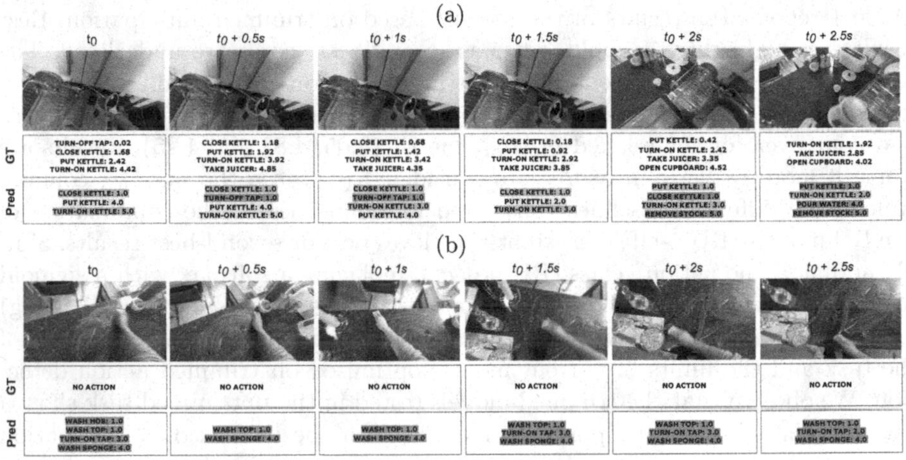

Fig. 4. Example sequences from untrimmed videos along with GT annotations of future actions and the predictions of the RU-5-clf baseline. True positive predictions are highlighted in green(considering $\Delta = 1$ s). Predictions with correct class, but inaccurate time-to-action estimates are reported in yellow. Wrong predictions are reported in red. (Color figure online)

7 Conclusion

We have proposed a new task of untrimmed action anticipation. Our findings have indicated that models optimised for trimmed anticipation have important limitations in this scenario, leading to many false-positive predictions and poor performance in the task of identifying the time-to-action. Retraining the same trimmed models on the video segments sampled with a sliding window, with the opportunities to predict the "no action" or multiple future actions also does not solve the problem if only the final layers of the model are changed. We believe more research is needed to build new solutions for the proposed task. We would also like to emphasise the attention on terminology: we refer to the proposed task as *untrimmed* action anticipation rather than *streaming* action anticipation [7]. Although we do not aim at improving the speed of the algorithms and do not consider real-time evaluation, we believe that streaming scenario is an important direction for future research in untrimmed action anticipation.

Acknowledgements. This research has been supported by Marie Skłodowska-Curie Innovative Training Networks - European Industrial Doctorates - PhilHumans Project, European Union - Grant agreement 812882 (http://www.philhumans.eu), project MEGABIT - PIAno di inCEntivi per la RIcerca di Ateneo 2020/2022 (PIACERI) - linea di intervento 2, DMI - University of Catania, and by the MISE - PON I&C 2014-2020 - Progetto ENIGMA - Prog n. F/190050/02/X44 - CUP: B61B19000520008.

References

1. Betancourt, A., Morerio, P., Regazzoni, C.S., Rauterberg, M.: The evolution of first person vision methods: a survey. IEEE Trans. Circ. Syst. Video Technol. **25**(5), 744–760 (2015)
2. Bubic, A., Von Cramon, D.Y., Schubotz, R.I.: Prediction, cognition and the brain. Front. Hum. Neurosci. **4**, 25 (2010)
3. Damen, D., et al.: Rescaling egocentric vision. arXiv preprint arXiv:2006.13256 (2020)
4. Damen, D., et al.: Scaling egocentric vision: the epic-kitchens dataset. In: Proceedings of the European Conference on Computer Vision (ECCV), pp. 720–736 (2018)
5. Everingham, M., Van Gool, L., Williams, C.K., Winn, J., Zisserman, A.: The pascal visual object classes (VOC) challenge. Inf. J. Comput. Vis. **88**(2), 303–338 (2010)
6. Furnari, A., Farinella, G.M.: What would you expect? Anticipating egocentric actions with rolling-unrolling LSTMS and modality attention. In: Proceedings of the IEEE International Conference on Computer Vision, pp. 6252–6261 (2019)
7. Furnari, A., Farinella, G.M.: Towards streaming egocentric action anticipation. arXiv preprint arXiv:2110.05386 (2021)
8. Gao, M., Xu, M., Davis, L.S., Socher, R., Xiong, C.: StartNet: online detection of action start in untrimmed videos. In: Proceedings of the IEEE/CVF International Conference on Computer Vision, pp. 5542–5551 (2019)
9. Ke, Q., Fritz, M., Schiele, B.: Time-conditioned action anticipation in one shot. In: Proceedings of the IEEE Conference on Computer Vision and Pattern Recognition, pp. 9925–9934 (2019)
10. Koppula, H.S., Saxena, A.: Anticipating human activities using object affordances for reactive robotic response. IEEE Trans. Pattern Anal. Mach. Intell. **38**(1), 14–29 (2015)

11. Li, Y., Lan, C., Xing, J., Zeng, W., Yuan, C., Liu, J.: Online human action detection using joint classification-regression recurrent neural networks. In: Leibe, B., Matas, J., Sebe, N., Welling, M. (eds.) ECCV 2016. LNCS, vol. 9911, pp. 203–220. Springer, Cham (2016). https://doi.org/10.1007/978-3-319-46478-7_13
12. Liu, M., Tang, S., Li, Y., Rehg, J.: Forecasting human object interaction: Joint prediction of motor attention and egocentric activity. arXiv:1911.10967 (2019)
13. Manglik, A., Weng, X., Ohn-Bar, E., Kitani, K.M.: Forecasting time-to-collision from monocular video: feasibility, dataset, and challenges. arXiv preprint arXiv:1903.09102 (2019)
14. Neumann, L., Zisserman, A., Vedaldi, A.: Future event prediction: if and when. In: Proceedings of the IEEE Conference on Computer Vision and Pattern Recognition Workshops (2019)
15. Ohn-Bar, E., Kitani, K., Asakawa, C.: Personalized dynamics models for adaptive assistive navigation systems. arXiv preprint arXiv:1804.04118 (2018)
16. Rodin, I., Furnari, A., Mavroeidis, D., Farinella, G.M.: Predicting the future from first person (egocentric) vision: a survey. Comput. Vis. Image Underst. **211**(5), 103252 (2021)
17. Ryoo, M., Fuchs, T.J., Xia, L., Aggarwal, J.K., Matthies, L.: Robot-centric activity prediction from first-person videos: what will they do to me? In: 2015 10th ACM/IEEE International Conference on Human-Robot Interaction (HRI), pp. 295–302. IEEE (2015)
18. Sener, F., Singhania, D., Yao, A.: Temporal aggregate representations for long term video understanding. arXiv:2006.00830 (2020)
19. Shou, Z., Pan, J., Chan, J., Miyazawa, K., Mansour, H., Vetro, A., Nieto, X.G., Chang, S.F.: Online action detection in untrimmed, streaming videos-modeling and evaluation. In: European Conference on Computer Vision (2018)

Forecasting Future Instance Segmentation with Learned Optical Flow and Warping

Andrea Ciamarra, Federico Becattini(✉), Lorenzo Seidenari, and Alberto Del Bimbo

University of Florence, Florence, Italy
{andrea.ciamarra,federico.becattini,lorenzo.seidenari, alberto.bimbo}@unifi.it

Abstract. For an autonomous vehicle it is essential to observe the ongoing dynamics of a scene and consequently predict imminent future scenarios to ensure safety to itself and others. This can be done using different sensors and modalities. In this paper we investigate the usage of optical flow for predicting future semantic segmentations. To do so we propose a model that forecasts flow fields autoregressively. Such predictions are then used to guide the inference of a learned warping function that moves instance segmentations on to future frames. Results on the Cityscapes dataset demonstrate the effectiveness of optical-flow methods.

Keywords: Video prediction · Instance segmentation · Deep learning

1 Introduction

Understanding urban environments from an ego-vehicle perspective is crucial for safe navigation. This problem can be declined under several points of view, such as detecting entities [9,25], understanding the road layout [2] or predicting the future [10,19,21]. At the same time, different modalities can be exploited to understand the environment ranging from RGB, which can produce useful representations such as semantic segmentations [18], to depth or even data produced by more complex sensors such as LiDARs [1], event cameras [25] or thermal cameras [13]. In this paper we focus on predicting future instance semantic segmentations of moving objects, since we believe it is the most informative representation for a machine planning motion. Most approaches are multimodal [8], integrating multiple modalities to achieve the highest accuracy. In general semantic segmentation [20], optical flow [30] and deep features [19] are considered as a source for future prediction. In this paper we study how can we rely on optical flow as unique source of information to forecast instance segmentations. We break down the problem in two steps: optical flow forecasting and learned instance warping. Our contribution is twofold:

- We first design an optical flow predictor to generate future flows based on convolutional LSTM and a UNet architecture.
- We train a warping neural network, with a specialized loss, that propagates current instance segmentations onto future frames, guided by predicted optical flows.

S. Sclaroff et al. (Eds.): ICIAP 2022, LNCS 13233, pp. 349–361, 2022.
https://doi.org/10.1007/978-3-031-06433-3_30

We show that, by choosing the appropriate loss and by learning a warp operator over noisy autoregressive inputs, we can obtain state-of-the art results on par with more complex multi-modal multi-scale approaches.

2 Related Work

Anticipating the future is a desirable asset for decision-making systems in variegate applications, like autonomous driving and robot navigation. In these scenarios, an artificial agent should have an intelligent component that forecasts motion of surrounding agents so to navigate safely. Huge progress has been done on anticipating the future by addressing different tasks, also exploiting multiple modalities. Early work addressed the video prediction task, where RGB past frames are used to synthesize future frames [24]. A large variety of methods was introduced, including autoregressive models [12], adversarial training [22], bidirectional flow [15], 3D convolutions [32] and recurrent networks [31] to improve prediction accuracy. Instead of predicting raw RGB values, in this work we are interested in directly forecasting more meaningful semantic-level contexts for future frames. Several works have been recently proposed addressing semantic segmentation and instance segmentation forecasting.

Future Semantic Segmentation. Scene understanding through future semantic segmentation can help an autonomous agent to take more intelligent decisions, e.g. forecasting specific category classes like pedestrians [3] or vehicles [4]. Luc et al. [20] proposed to learn semantic-level scene dynamics by mapping semantic segmentations, which are extracted from past frames, to the future through a deep CNN. In this work the authors also shown that directly predicting semantic segmentation is more effective than predicting RGB frames and then fed them into a semantic segmentation model. Terwilliger et al. [30] designed a flow-guided model, which first aggregates past flow features using a recurrent network to predict the future optical flow; then, the predicted flow is used to learn a warp layer to map the last semantic segmentation to the future next one. Chiu et al. [5] proposed a teacher-student learning network, which encodes past RGB frames to directly generate future segmentations. Other recent approaches extract features at different resolutions from past frames in order to map to the future ones, taking inspiration from the F2F architecture designed in [19]. Šarić et al. [27] used deformable convolutions to F2F network at a single-level in order to encode varied motion pattern. In [28], the same authors enriched the features extracted from past images with correlation coefficients between neighbouring frames, and performed the forecasting by warping observed features according to regressed feature flow.

We propose to learn motion dynamics through an optical flow predictive model, using an encoder-decoder model with skip connections followed by a ConvLSTM [34]. Differently from [30] and [28], we generate future semantic segmentations of moving objects, by aggregating together warped object masks, using the predicted flow and the semantic segmentation as inputs. Specifically, our model forecasts single instance segmentations, which allows a finer grained understanding of the scene for tasks such as path planning. In addition, we exploit a fully learnable warping module instead of relying on a differentiable, yet not-learnable, warping operator as in [30].

Future Instance Segmentation. The goal of the future instance segmentation task is to detect and segment individual objects of interest appearing in a video sequence. Semantic segmentation does not account for single objects, since they are fused together by assigning the same category label. Moreover instances can vary in number between frames and are not consistent across a video sequence. Most recent methods generate future instance and semantic segmentations, by forecasting features for unobserved frames given the past ones. For instance, Luc et al. [19] designed a multi-scale approach, named F2F, based on a convolutional network with Feature Pyramid Networks (FPN) [16]. Then, future instance segmentations are detected by running Mask R-CNN [9] from the predicted features, while future semantic segmentations, are computed by converting instance segmentation predictions according to the confidence score, thus proving better qualitative results than [20]. Sun et al. [29], following the same pipeline in [19], employs a set of connected ConvLSTMs [34], so to capture rich spatio-temporal contexts across different pyramid levels. Hu et al. [10] proposed a multi-stage optimization framework that exploits auto-path connections so to adaptively and selectively aggregate contexts from previous steps.

Lin et al. [18] designed a three-stage framework, able to forecast segmentation features from a decoder through predictive features outputted by an encoder, instead of directly predict pyramid segmentation features from input frames. Different than F2F [19], future predictions are finally generated by jointly training the whole system using Mask R-CNN [9] and Semantic FPN [14], for semantic segmentations and instance segmentations respectively. Graber et al. [8], proposed a more complete framework to forecast the near future, by decomposing a dynamic scene into *things* and *stuff*, i.e. individual objects and background, with multiple training stages and also considering odometry anticipation due to camera motion. In summary, recent works mainly address the future prediction task, both in terms of instances and semantics, following the typical pipeline based on F2F, using recurrent networks and complex connections. The predicted features are then fed in input to different segmentation models, so to provide instances and semantic predictions. Although taking several training stages, such frameworks are able to get better results.

Our method, instead, proposes to combine the predicted dense optical flow with an instance warping network, without taking several optimization stages. We also prove that the warping model can learn to forecast object motion, from the predicted scene flow and the current semantic map.

3 Our Approach

We introduce a novel framework for future instance segmentation that learns to predict future flow motion patterns. Such predictions are used along with the last observed semantic segmentation to warp detected objects to future unobserved

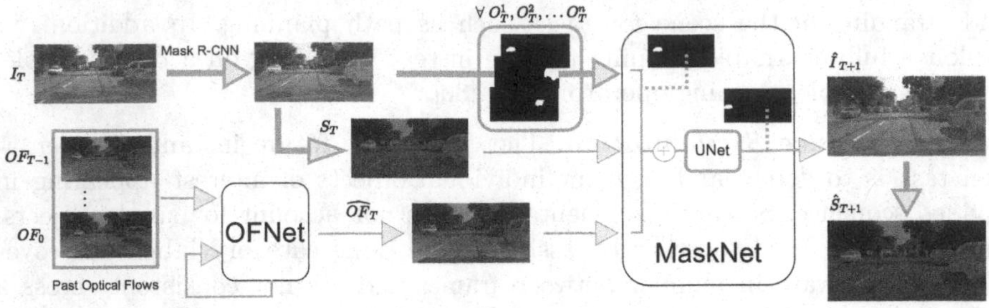

Fig. 1. OFNet forecasts future flows from past ones autoregressively. MaskNet warps instance segmentations to the future with the predicted flow. Future semantic segmentations are obtained by aggregating instances together.

frames (Fig. 1). This architecture consists of two main components: (i) OFNet, an optical flow predictive model, which forecasts motion dynamics through past optical flows, and (ii) MaskNet, a novel mask prediction model, trained to warp segmentation instances of moving objects to future frames.

Warping is performed by leveraging a region-based loss function instead of the typical cross-entropy loss, so to deal with small parts of interest since the background occupies most of the scene. To identify instances in a video sequence, we run Mask R-CNN [9] for each frame. Then, we exploit a bounding box association-based tracker based on [7] to generate object tracks and link instances for the sequence lifespan. inspired by [19], the proposed approach is also able to generate semantic segmentations by fusing instance mask predictions according to a rescoring function balancing detector score and mask size.

3.1 Optical Flow Forecasting

The first key part of the framework is OFNet, the optical flow prediction model. It is a supervised network that learns to generate future flows based on past ones. OFNet takes T dense optical flows from consecutive past frames and predicts the flow for the following time step.

Each optical flow is first estimated from consecutive pairs of frames, by running FlowNet2 [11], that produces a dense 2D flow field, with (u, v) horizontal and vertical displacements for each pixel. Then, we learn motion features from each of these optical flows through a time distributed UNet [26], a fully convolutional encoder-decoder network with a contracting path and an expanding path, connected with skip connections. From the original UNet we consider the entire structure up to the final prediction layer, so to output features with 64 channels (Fig. 2(a)). The sequence of extracted features generated by UNet are fed to a ConvLSTM [34] with 3×3 kernel followed by a 1×1 convolution, so to reduce the output channels to 2 as in a flow field. We train OFNet to forecast an output sequence shifted by one step ahead with respect to input, i.e. by generating the next flow for each corresponding input. This provides supervision for each timestep and allows us to use it in an autoregressive fashion.

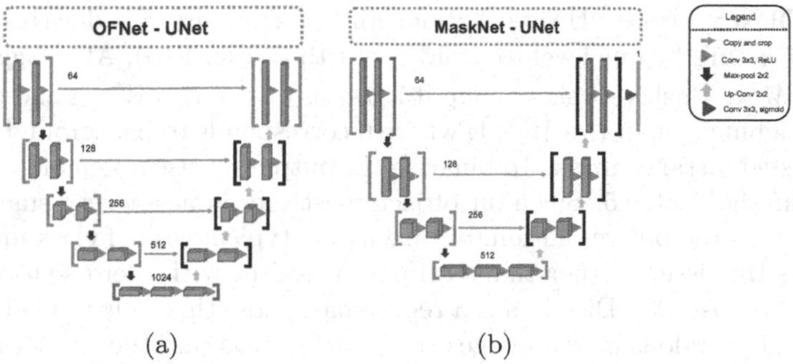

Fig. 2. UNet architectures used as feature extractor (a) to generate 64-channel flow features in OFNet, and (b) to produce future binary masks in MaskNet.

Formally, let OF_i be the optical flow produced by FlowNet2 given two consecutive frames X_i and X_{i+1}. We denote with F_i the motion features extracted from OF_i using our UNet-based backbone. We extend such notation referring to $F_{t:T}$ as the sequence of motion features spanning from t to T. Likewise, we use the same notation for optical flow sequences $OF_{t:T}$. At training time, given a sequence of T flow features $F_{t:t+T}$, OFNet learns to generate a sequence of T consecutive optical flows shifted by one timestep $\widehat{OF}_{t+1:t+T+1}$. Instead, at inference time we only retain the last prediction of the output sequence $\widehat{OF}_{t+T+1}$, that encodes the scene motion towards the first unseen future frame. OFNet is trained using an L2 loss function on the output sequence $\mathcal{L}_{\text{flow}}$:

$$\mathcal{L}_{\text{flow}} = \frac{1}{T} \sum_{k=1}^{T} \frac{\left(OF_k - \widehat{OF}_k\right)^2}{H \cdot W \cdot 2} \tag{1}$$

where T is the sequence length and H and W are the height and width of the feature map. We set $T = 6$ for all our experiments to provide sufficient information about past dynamics.

3.2 Instance Mask Forecasting

After generating future optical flows, we forecast future instance segmentations. To do that, we designed MaskNet, a novel model which leverages motions encoded in the predicted flows and the semantic segmentation of an observed frame, which provides context about the objects in the scene. Instead of globally warping semantic segmentations as done in [30], MaskNet is trained to understand future positions of individual objects. We aggregate information from the present frame X_t, i.e. the semantic segmentation of the scene S_t and a binary mask of the instance to be warped M_t^j, and from the estimated future, i.e. the predicted flow field $\widehat{OF}_t$. The data is concatenated channel-wise and fed to the network, which generates the future instance mask $\widehat{M}_{t+1}^j$ located in frame X_{t+1}.

As in OFNet, we use a UNet backbone for MaskNet but we reduce its encoder-decoder structure by one level to avoid overfitting (Fig. 2(b)). After aggregating $\{\widehat{OF}_t, S_t, M_t^j\}$, MaskNet takes as input a tensor of size $H \times W \times 4$ and produces as output a binary map $H \times W \times 1$, where 0 corresponds to background and 1 to the forecasted instance mask. To binarize the output we use a sigmoid activation and we threshold at 0.5. Since an object mostly occupies a very small region compared to the input resolution, we are in the typical case of class unbalance. To address this issue, rather than training MaskNet with a cross-entropy loss, we use a Dice loss [23]. Dice loss is a region-based loss that helps to better focus on overlapping regions in two images, which in our case are the predicted binary mask and the corresponding groundtruth instance. Thus, considering an N pixel image, we employ the mask loss $\mathcal{L}_{\text{mask}}$ defined through the Dice coefficient D:

$$\mathcal{L}_{\text{mask}} = 1 - D = 1 - \frac{2\sum_{i=1}^{N} \hat{M}_i \, {M^{GT}}_i}{\sum_{i=1}^{N} \hat{M}_i^2 + \sum_{i=1}^{N} {M^{GT}}_i^2} \tag{2}$$

where the Dice coefficient $D \in [0, 1]$ measures the similarity between the predicted mask $\hat{M}$ and the groundtruth one M^{GT}. Note that to compute the loss, we use the output of MaskNet after the sigmoid activation and before the final thresholding in order to work with real values in $[0, 1]$. Since our model works with single instances, we do not rely on a dedicated network to generate semantic segmentations of the image as most state of the art methods [8,18,30]. Instead, we simply group all object masks together maintaining their category label.

4 Experiments

We conduct experiments on the Cityscapes dataset [6], which contains recordings from a car driving in urban environments for a total of 5000 sequences, divided in 2975 for train, 1525 for testing and 500 for validation. Every sequence is long 1.8 s and consists of 30 frames at resolution 1024×2048. Groundtruth semantic and instance segmentations are available for the 20-th frame of each sequence. Our method is tested on 8 moving object categories: *person*, *rider*, *car*, *truck*, *bus*, *train*, *motorcycle*, *bicycle*. We evaluate the future predictions, both instance and semantic segmentations, 3 and 9 frames ahead, respectively denoted as short-term (about 0.17 s later) and mid-term (about 0.5 s later). Such evaluations are performed on annotated frames in the validation set, as done in literature. Predicted optical flows are obtained, by initially feeding $T = 6$ consecutive past optical flows $OF_{t-6:t-1}$ to OFNet and then, autoregressively forecasting each future flow OF_{t+k} with $k = 0, 1, \ldots, n$ ($n = 2$ for short-term and $n = 8$ for mid-term). Future semantic segmentation is evaluated by measuring the intersection-over-union (IoU) between predictions and the corresponding ground truth for each class and averaging among classes. Future instance segmentation instead is evaluated through two standard Cityscapes metrics: AP50 and AP. AP50 is average precision where an instance is correctly counted if it has at least 50% of IoU with the groundtruth. AP is average precision, averaged over ten equally

Table 1. Effect of different training regimes for MaskNet. Warping instances with predicted flows is included as a baseline.

Method	Train data		Finetuned	Short term $\Delta t = 3$			Mid term $\Delta t = 9$		
	FlowNet	OFNet	Layers	AP	AP50	IoU	AP	AP50	IoU
Warping	✗	✗	–	15.9	34.3	60.7	2.8	8.8	39.4
MaskNet	✓	✗	–	18.7	39.4	65.5	4.0	12.3	44.5
	✗	✓	All	18.5	39.3	65.2	4.5	14.9	41.9
	✓	✓	3	18.7	39.2	65.7	5.1	16.0	44.8
	✓	✓	2	**18.9**	**39.5**	**65.9**	**5.9**	**17.4**	**45.5**

spaced IoU thresholds from 50% to 95%. Since these metrics require a confidence score for each prediction, we use the score provided by the detector for each object. However, small objects are likely to be false positives generated by the detector. To mitigate this issue, we reduce the score by 0.5 for objects smaller than 64×64px and by 0.3 for objects smaller than 128×128px.

4.1 Implementation Details and Ablation Study

We compute the optical flow at every pair of consecutive frames for each video sequence, both for training and validation sets, through FlowNet-c with pre-trained weights from FlowNet2 [11]. Frames are resized to 128×256, as in [30]. We detect all objects from each video frame, by running the Mask R-CNN model provided in Detectron2 [33], pretrained on COCO [17] and finetuned on Cityscapes. We train OFNet with Adam and learning rate 0.0001. Our architecture is inspired by UNet, but in order to perform regression we change the number of channels to 2 in the last 1×1 convolutional layer and we use a linear activation function. We set the batch size to 3 and we employ the L2 loss as in Eq. 1 and we train the network for 100 epochs. MaskNet instead is trained for 4 epochs using Adam with learning rate 0.0001. Inputs are resized to 256×512. First, we pretrain the model using optical flows computed with FlowNet [11] instead of the predicted ones. Then, we finetune for 3 additional epochs the last two layers using the predicted flows autoregressively, for both short-term and mid-term models. This two-step optimization allows the model to learn how to warp instances using clean flows and then to adapt itself to handle noisy flow fields.

We observe that using a Cross-Entropy loss instead of a Dice loss, MaskNet converges to trivial solutions predicting background for every pixel of the frame. This underlines the importance of using a region based loss for this kind of task.

In Table 1 we conduct an ablation study by exploring different training strategies. As a first control experiment, we output future segmentations by directly warping masks with flows predicted by OFNet. In this case MaskNet is not used. Then we evaluate the effect of pretraining MaskNet using ground truth flows generated by FlowNet [11]. Different fine-tuning strategies are also analyzed.

Table 2. Flow prediction MSE up to mid-term (t+9) on Cityscapes val set. We also show a break down of MSE for the u and v optical flow components.

	t+1	t+2	t+3	t+4	t+5	t+6	t+7	t+8	t+9
MSE	0.36	0.54	0.60	0.78	0.84	0.99	1.08	1.28	1.19
MSE_u	0.45	0.67	0.70	0.94	1.04	1.24	1.31	1.64	1.55
MSE_v	0.26	0.41	0.50	0.63	0.63	0.75	0.84	0.91	0.82

As expected using a learned warping function, MaskNet, yields superior results with respect to simply warping masks. It also emerges that the two step training is fundamental. Training MaskNet only on precomputed flows yields good results at short term but exhibits a severe drop in performance at mid-term, since the model is not trained to deal with noisy predictions as inputs. Similarly, using only predicted flows, i.e. without pretraining, the model fails to warp masks appropriately up to mid-term. We also study the effect of using OFNet in an autoregressive way or by feeding FlowNet flows to it in a teacher-forcing way. This allows us to better understand the impact of flow quality on MaskNet's ability to warp masks. Interestingly, when OFNet is fed autoregressively with predicted flows, MaskNet exhibits a large improvement in average precision for mid-term predictions. In fact, using a teacher forcing approach, i.e. feeding only clean flows to OFNet, the produced flows are not sufficiently realistic. This is confirmed by the fact that this model performs similarly to the one trained only on real flows. Finally, we compare performances by finetuning the last 3 layers instead of 2. We found that 2 layers to train is the best choice.

5 Results

We first analyse the prediction capabilities of OFNet. Since we want to use it to generate future flows up to mid-term (9 frames in the future), we show in Table 2 how error accumulates while predicting flows autoregressively. As expected, the MSE values increase as the time horizon shifts forward, although the error accumulation appears to be bounded by a linear growth. Interestingly, the horizontal component u exhibits a higher error. This is caused by the fact that most objects move horizontally in front of the camera capturing an urban scene from an ego-vehicle perspective.

We then provide in Fig. 3 a qualitative analysis of the warping capabilities of MaskNet feeding clean future flows using FlowNet2 as an oracle. In the following we will refer to this model as *MaskNet-Oracle.*

This model can provide accurate results among different categories. Interestingly, it can correctly warp small non-rigid parts of an instance, like legs, as well as objects at different scales like approaching cars. These results prove that MaskNet can understand most motions of individual objects in a scene by simply exploiting optical flow, without further motion information. MaskNet-Oracle, being fed with ground truth optical flows, acts as an upper bound of

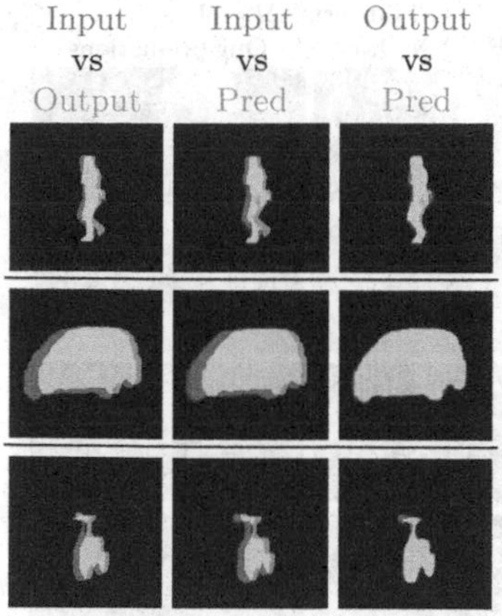

Fig. 3. Instance predictions: input vs gt output, input vs prediction, gt output vs prediction. Mask intersections in yellow. (Color figure online)

Table 3. Comparison results for future instance segmentation (AP and AP50) and future semantic segmentation (IoU) of moving objects on Cityscapes val set.

Method	Short term $\Delta t = 3$			Mid term $\Delta t = 9$		
	AP	AP50	IoU	AP	AP50	IoU
Mask RCNN oracle	34.6	57.4	73.8	34.6	57.4	73.8
MaskNet-oracle	24.8	47.2	69.6	16.5	35.2	61.4
Copy-last segm. [19]	10.1	24.1	45.7	1.8	6.6	29.1
Optical-flow shift [19]	16.0	37.0	56.7	2.9	9.7	36.7
Optical-flow warp [19]	16.5	36.8	58.8	4.1	11.1	41.4
Mask H2F [19]	11.8	25.5	46.2	5.1	14.2	30.5
S2S [20]	–	–	55.4	–	–	42.4
F2F [19]	19.4	39.9	61.2	7.7	19.4	41.2
LSTM M2M [30]	–	–	65.1	–	–	46.3
Deform F2F [27]	–	–	63.8	–	–	49.9
CPConvLSTM [29]	22.1	44.3	–	11.2	25.6	–
F2MF [28]	–	–	67.7	–	–	54.6
PSF [8]	17.8	38.4	60.8	10.0	22.3	52.1
APANet [10]	23.2	46.1	64.9	12.9	29.2	51.4
PFL [18]	24.9	48.7	69.2	14.8	30.5	56.7
MaskNet (Ours)	19.5	40.5	65.9	6.4	18.4	45.5

our system. We also evaluate a Mask R-CNN oracle, by directly running Mask R-CNN on future frames. We compare our approach, that uses the predicted optical flows from OFNet, with such upper bounds and with recent approaches from the state of the art (Table 3). We exceed some state of the methods, such as F2F and PSF [8] at short-term, while mid-term predictions are very close to the feature-to-feature approach introduced by Luc et al. [19]. This confirms our choice of using motion cues, instead of learning different Feature Pyramid Networks. However, recent frameworks, such as CPConvLSTM [29], APANet [10] and PFL [18], reported better results, where more complex information from intra-level and inter-level connections are learned across FPNs from past frames. This requires several training stages to improve both short and mid predictions.

We also generate future semantic segmentations for moving objects (Table 3). We obtain the future semantic segmentations by grouping by category the predicted instance masks. MaskNet obtains good results and overcomes most of those state-of-the-art approaches that provide forecasts in terms of both instance and semantic segmentations. In general our approach performs well at short-term, but gets worse at mid-term compared to methods that predict future FPNs after several training stages and learning specific connections, like the auto-path connections as in APANet [10], or that use segmentation models properly trained on predicted features, as done in PFL [18]. Some qualitative results are shown for instance and semantic segmentation predictions, both at short-term and mid-term, in Fig. 4.

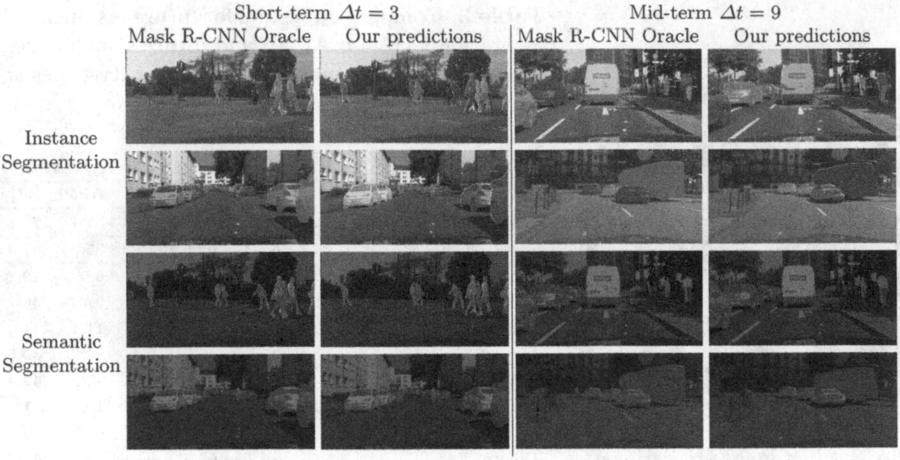

Fig. 4. Visualization results of future predictions on Cityscapes val set.

Overall, compared to the other methods, our approach is simple and without complex optimization stages can reach competing performances. Nonetheless, our approach underlines the efficacy and the limitations of relying solely on optical flow data, which is easy to obtain even with any typical on-board sensor whether it is as RGB camera, an event camera or a LiDAR. However, we believe that there is still margin for improvement using a flow-based predictor like ours. This stems from the results of our MaskNet-Oracle, as using clean flows estimated by FlowNet2 leads to overcome all state-of-the-art results at both short-term and mid-term. We believe that integrating information such as depth-maps could lead to significant improvements in the state of the art. However, this is out of the scope of our paper and we leave it as future work.

6 Conclusions

In this work we addressed the future instance segmentation task using predicted optical flows and a learned warping operator, which also allows to produce future semantic segmentations without additional training. We designed a framework made of a predictive optical flow model able to forecast future flows given past ones through a UNet+ConvLSTM architecture. A second component, MaskNet, warps observed objects to future frames, using the current semantic map and the predicted optical flow. We conducted experiments on Cityscapes, demonstrating the effectiveness of optical flow based methods.

Acknowledgement. This work was supported by the European Commission under European Horizon 2020 Programme, grant number 951911 - AI4Media

References

1. Beltrán, J., Guindel, C., Moreno, F.M., Cruzado, D., Garcia, F., De La Escalera, A.: BirdNet: a 3D object detection framework from LiDAR information. In: 2018 21st International Conference on Intelligent Transportation Systems (ITSC), pp. 3517–3523. IEEE (2018)
2. Berlincioni, L., Becattini, F., Galteri, L., Seidenari, L., Bimbo, A.D.: Road layout understanding by generative adversarial inpainting. In: Escalera, S., Ayache, S., Wan, J., Madadi, M., Güçlü, U., Baró, X. (eds.) Inpainting and Denoising Challenges. TSSCML, pp. 111–128. Springer, Cham (2019). https://doi.org/10.1007/978-3-030-25614-2_10
3. Brazil, G., Yin, X., Liu, X.: Illuminating pedestrians via simultaneous detection and segmentation. In: Proceedings of the IEEE International Conference on Computer Vision, pp. 4950–4959 (2017)
4. Chen, X., Kundu, K., Zhang, Z., Ma, H., Fidler, S., Urtasun, R.: Monocular 3D object detection for autonomous driving. In: Proceedings of the IEEE Conference on Computer Vision and Pattern Recognition, pp. 2147–2156 (2016)
5. Chiu, H.k., Adeli, E., Niebles, J.C.: Segmenting the future. IEEE Robot. Autom. Lett. **5**(3), 4202–4209 (2020)
6. Cordts, M., et al.: The cityscapes dataset for semantic urban scene understanding. In: Proceedings of the IEEE Conference on Computer Vision and Pattern Recognition, pp. 3213–3223 (2016)
7. Cuffaro, G., Becattini, F., Baecchi, C., Seidenari, L., Del Bimbo, A.: Segmentation free object discovery in video. In: Hua, G., Jégou, H. (eds.) ECCV 2016. LNCS, vol. 9915, pp. 25–31. Springer, Cham (2016). https://doi.org/10.1007/978-3-319-49409-8_4
8. Graber, C., Tsai, G., Firman, M., Brostow, G., Schwing, A.G.: Panoptic segmentation forecasting. In: Proceedings of the IEEE/CVF Conference on Computer Vision and Pattern Recognition, pp. 12517–12526 (2021)
9. He, K., Gkioxari, G., Dollár, P., Girshick, R.: Mask R-CNN. In: Proceedings of the IEEE International Conference on Computer Vision, pp. 2961–2969 (2017)
10. Hu, J.F., Sun, J., Lin, Z., Lai, J.H., Zeng, W., Zheng, W.S.: Apanet: auto-path aggregation for future instance segmentation prediction. In: IEEE Transactions on Pattern Analysis and Machine Intelligence (2021)
11. Ilg, E., Mayer, N., Saikia, T., Keuper, M., Dosovitskiy, A., Brox, T.: Flownet 2.0: evolution of optical flow estimation with deep networks. In: Proceedings of the IEEE conference on computer vision and pattern recognition, pp. 2462–2470 (2017)
12. Kalchbrenner, N., et al.: Video pixel networks. In: International Conference on Machine Learning, pp. 1771–1779. PMLR (2017)
13. Kieu, M., Bagdanov, AD., Bertini, M., del Bimbo, A.: Task-conditioned domain adaptation for pedestrian detection in thermal imagery. In: Vedaldi, A., Bischof, H., Brox, T., Frahm, J.-M. (eds.) ECCV 2020. LNCS, vol. 12367, pp. 546–562. Springer, Cham (2020). https://doi.org/10.1007/978-3-030-58542-6_33
14. Kirillov, A., Girshick, R., He, K., Dollár, P.: Panoptic feature pyramid networks. In: Proceedings of the IEEE/CVF Conference on Computer Vision and Pattern Recognition, pp. 6399–6408 (2019)
15. Kwon, Y.H., Park, M.G.: Predicting future frames using retrospective cycle gan. In: Proceedings of the IEEE/CVF Conference on Computer Vision and Pattern Recognition, pp. 1811–1820 (2019)

16. Lin, T.Y., Dollár, P., Girshick, R., He, K., Hariharan, B., Belongie, S.: Feature pyramid networks for object detection. In: Proceedings of the IEEE Conference on Computer Vision and Pattern Recognition, pp. 2117–2125 (2017)
17. Lin, T.-Y., et al.: Microsoft COCO: common objects in context. In: Fleet, D., Pajdla, T., Schiele, B., Tuytelaars, T. (eds.) ECCV 2014. LNCS, vol. 8693, pp. 740–755. Springer, Cham (2014). https://doi.org/10.1007/978-3-319-10602-1_48
18. Lin, Z., Sun, J., Hu, J.F., Yu, Q., Lai, J.H., Zheng, W.S.: Predictive feature learning for future segmentation prediction. In: Proceedings of the IEEE/CVF International Conference on Computer Vision, pp. 7365–7374 (2021)
19. Luc, P., Couprie, C., Lecun, Y., Verbeek, J.: Predicting future instance segmentation by forecasting convolutional features. In: Proceedings of the European Conference on Computer Vision (ECCV), pp. 584–599 (2018)
20. Luc, P., Neverova, N., Couprie, C., Verbeek, J., LeCun, Y.: Predicting deeper into the future of semantic segmentation. In: Proceedings of the IEEE International Conference on Computer Vision, pp. 648–657 (2017)
21. Marchetti, F., Becattini, F., Seidenari, L., Del Bimbo, A.: Multiple trajectory prediction of moving agents with memory augmented networks. In: IEEE Transactions on Pattern Analysis and Machine Intelligence (2020)
22. Mathieu, M., Couprie, C., LeCun, Y.: Deep multi-scale video prediction beyond mean square error. arXiv:1511.05440 (2015)
23. Milletari, F., Navab, N., Ahmadi, S.A.: V-net: fully convolutional neural networks for volumetric medical image segmentation. In: 2016 Fourth International Conference on 3D Vision (3DV), pp. 565–571. IEEE (2016)
24. Oprea, S., et al.: A review on deep learning techniques for video prediction. In: IEEE Transactions on Pattern Analysis and Machine Intelligence (2020)
25. Perot, E., de Tournemire, P., Nitti, D., Masci, J., Sironi, A.: Learning to detect objects with a 1 megapixel event camera. Adv. Neural. Inf. Process. Syst. **33**, 16639–16652 (2020)
26. Ronneberger, O., Fischer, P., Brox, T.: U-Net: convolutional networks for biomedical image segmentation. In: Navab, N., Hornegger, J., Wells, W.M., Frangi, Alejandro F. (eds.) MICCAI 2015. LNCS, vol. 9351, pp. 234–241. Springer, Cham (2015). https://doi.org/10.1007/978-3-319-24574-4_28
27. Šarić, J., Oršić, M., Antunović, T., Vražić, S., Šegvić, S.: Single level feature-to-feature forecasting with deformable convolutions. In: Fink, G.A., Frintrop, S., Jiang, X. (eds.) DAGM GCPR 2019. LNCS, vol. 11824, pp. 189–202. Springer, Cham (2019). https://doi.org/10.1007/978-3-030-33676-9_13
28. Saric, J., Orsic, M., Antunovic, T., Vrazic, S., Segvic, S.: Warp to the future: joint forecasting of features and feature motion. In: Proceedings of the IEEE/CVF Conference on Computer Vision and Pattern Recognition, pp. 10648–10657 (2020)
29. Sun, J., et al.: Predicting future instance segmentation with contextual pyramid convlstms. In: Proceedings of the 27th ACM International Conference on Multimedia, pp. 2043–2051 (2019)
30. Terwilliger, A., Brazil, G., Liu, X.: Recurrent flow-guided semantic forecasting. In: 2019 IEEE Winter Conference on Applications of Computer Vision (WACV), pp. 1703–1712. IEEE (2019)
31. Villegas, R., Yang, J., Hong, S., Lin, X., Lee, H.: Decomposing motion and content for natural video sequence prediction. arXiv:1706.08033 (2017)
32. Wang, Y., Jiang, L., Yang, M.H., Li, L.J., Long, M., Fei-Fei, L.: Eidetic 3D LSTM: a model for video prediction and beyond. In: International Conference on Learning Representations (2018)

33. Wu, Y., Kirillov, A., Massa, F., Lo, W.Y., Girshick, R.: Detectron2. https://github.com/facebookresearch/detectron2 (2019)
34. Xingjian, S., Chen, Z., Wang, H., Yeung, D.Y., Wong, W.K., Woo, W.C.: Convolutional LSTM network: a machine learning approach for precipitation nowcasting. In: Advances in Neural Information Processing Systems, pp. 802–810 (2015)

Depth-Aware Multi-object Tracking in Spherical Videos

Liliana Lo Presti[1(✉)], Giuseppe Mazzola[2], Guido Averna[1], Edoardo Ardizzone[1], and Marco La Cascia[1]

[1] Engineering Department, University of Palermo, Palermo, Italy
liliana.lopresti@unipa.it
[2] Department of Humanities, University of Palermo, Palermo, Italy
giuseppe.mazzola@unipa.it

Abstract. This paper deals with the multi-object tracking (MOT) problem in videos acquired by 360-degree cameras. Targets are tracked by a frame-by-frame association strategy. At each frame, candidate targets are detected by a pre-trained state-of-the-art deep model. Associations to the targets known till the previous frame are found by solving a data association problem considering the locations of the targets in the scene. In case of a missing detection, a Kalman filter is used to track the target. Differently than works at the state-of-the-art, the proposed tracker considers the depth of the targets in the scene. The distance of the targets from the camera can be estimated by geometrical facts peculiar to the adopted 360-degree camera and by assuming targets move on the ground-plane. Distance estimates are used to model the location of the targets in the scene, solve the data association problem, and handle missing detection. Experimental results on publicly available data demonstrate the effectiveness of the adopted approach.

Keywords: Multi-object tracking · Equirectangular image · 360° videos · Depth

1 Introduction

Multi-object tracking (MOT) is the problem of analyzing a video, namely a sequence of images, and detecting the locations on the image plane of targets moving in the environment over time. The problem can be extended to 3D when a world coordinate system is known or is estimated, and is especially challenging in a crowd scenario where objects with similar appearances are difficult to discriminate from each other. In these cases, not all targets are detected due to partial/total occlusions, and false positives have to be handled. Handling occlusions can take advantage of knowing the distance of the targets from the camera. Objects close to the camera can still be detected while the farthest ones require algorithms to predict their location on the image plane.

Recently, the work in [16] showed how to estimate the distance from the camera of objects of interest given only the height of the 360° camera and the

S. Sclaroff et al. (Eds.): ICIAP 2022, LNCS 13233, pp. 362–374, 2022.
https://doi.org/10.1007/978-3-031-06433-3_31

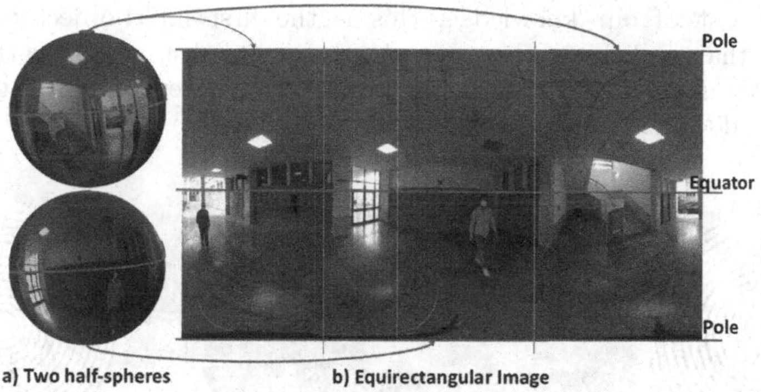

Fig. 1. The figure shows on the left two half-spheres, each one acquired by a lens in a dual lens 360° camera. On the right, the equirectangular image on which the spherical one is projected. The middle line represents the equator of the spherical image, while the uppest and lowest rows are the sphere poles.

coordinates, on the spherical image, of the contact point of the target with the ground plane. 360° camera devices acquire panoramic images with a view spanning 360° horizontally and 180° vertically. Recent devices are typically formed by a system of two wide-angle lenses, each of which can shoot (more than) half of the scene. In the acquired spherical images, pixels are mapped onto a sphere centered into the camera. Spherical images can be stored by applying equirectangular projections [6], after correcting the distortion introduced by lenses, if their shape is known. As shown in Fig. 1, the central row of the equirectangular is the equator, and the upper and lower rows are the poles. This projection introduces a distortion, which is more visible approaching the poles.

Tracking in equirectangular images is challenging due to the image's circularity and high resolution. The former is a consequence of the sphericity of the image and implies that a target walking around the camera and depicted close to the left/right border of the image may re-appear at the opposite right/left side of the image itself. High resolution of the image is required to represent the entire scene without losing too many details.

We propose to exploit geometrical facts about the equirectangular projection to estimate and track the targets' locations onto the ground plane by taking advantage of the estimated distances of the targets from the camera. In Fig. 2, four targets are moving around the scene. Green bounding boxes represent ground-truth information, while the colored ones are the tracker's predictions (in this case only two). The radial plot on the left shows the estimated targets' locations on the ground plane. The center of the plot is the camera location. Concentric circles are the loci of points equidistant from the camera. Circles are exactly one meter apart from each other. Numbers are identifiers (color-coded) associated with the targets and are close to their location on the ground.

To the best of our knowledge, this is the first multi-object tracker for 360° videos that explicitly estimates and uses targets' distances from the camera to track the targets. In this respect, we propose a novel depth-aware multi-object tracker that does not need any calibration procedure.

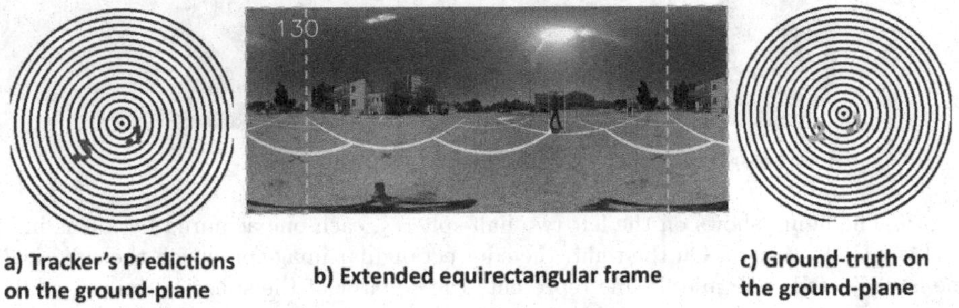

Fig. 2. At the center, an extended equirectangular image (dotted areas are included to enhance target detection). Green bounding-boxes are ground-truth, colored ones are tracker's predictions. The image shows two pedestrians far more than 15 m from the camera. They are annotated but not detected. On the left, the plot shows the locations on the ground-plane of the two tracked objects. On the right, the plot shows the locations of the annotated objects. Pedestrians distant more than 15 m from the camera are not shown. (Color figure online)

The only assumption of our approach is planar motion of the targets on the ground-plane. We stress that if the camera height is known, exact targets' locations on the ground plane can be derived. This characteristic is very appealing in several applications, especially in surveillance but also in behavior understanding applications to better understand how people interact. We tested our tracker on the publicly available dataset [16] by adapting the MOT evaluation protocol to spherical images. We also implemented a baseline technique to compare with. The plan of the paper is as follow. In Sect. 2 we summarize state-of-the-art work in multi-object tracking; In Sects. 3 and 4 we present our novel approach and implementation details respectively. Finally, in Sect. 5 we discuss experimental results and in Sect. 6 we present conclusions and future work.

2 Related Work

Solution of MOT problem relies on the matching, at time t, between observed objects and targets detected at time $t-1$. Information used to establish the matches can be about object appearance, 2D/3D location, or trajectory and is incrementally refined over time, including the possibility of dynamically changing the number of targets to account for objects entering/exiting in/from the scene.

Matching can be found by two main approaches [1]: frame-by-frame associations of observations to targets and deferred-logic tracking. With the former sequential tracking strategy, at each frame, the most likely associations between object detections and targets are estimated and not modified anymore. The problem is modeled as a data association (DA) one, and often linear programming techniques involving constrained optimization problems are adopted [2,11,12].

Global Nearest Neighbor Standard Filter [1,11] considers all assignments within a region of interest and solves a maximal bipartite matching problem to find the best assignments between detections and targets, generally by the Munkres algorithm.

In deferred-logic tracking, observation-to-target associations are delayed until evidence of their correctness is accumulated. Multiple Hypotheses Tracking (MHT) [20] builds a hypothesis tree whose root-leaf paths represent all possible combinations of detection-target associations through time. The exponential growth of the tree is avoided by pruning heuristics, and the path with the highest likelihood is selected as the correct target track.

In this paper, we adopt a frame-by-frame association strategy. At each frame, we detect candidate targets and associate them with the identities known until the previous frame. Considering that we are processing spherical images, and we can estimate the distance from the camera of the detected targets, we solve the DA problem through the Munkres algorithm by considering the targets' depth while accounting for the image circularity.

A recent survey of MOT works using deep learning is in [5]. In general, these works [4,23,24] use a deep model for solving the detection problem, the Munkres algorithm to solve the DA problem, and prediction methods such as Kalman filter to track the targets through the occlusions. Deep models can additionally be used to estimate similarities [23] or model motion [4].

Works about tracking in 360-degree videos [17] focus mostly on single object tracking by applying deep learning strategies [9,15] and Kalman or particle filters [3,18,19]. The work most similar to ours is the one in [14], which is built upon DeepSORT [22]. DeepSORT solves the DA problem by considering deep features extracted by the detector (YOLO v2). In contrast to our approach, in [14], detections are taken on image slices to account for the high image resolution. Similar to our approach, the method computes the detections on the extended frame in order to account for the image circularity. With respect to this work, we model tracking by considering the distances from the camera when tracking pedestrians, and use them to handle occlusions.

3 Modeling Targets' Locations on Spherical Images

Despite targets are detected on the equirectangular image, in our approach tracking is performed on the ground-plane. On the image plane, the target's ground-touching point is approximated by the middle point of the lower side of the bounding box detected on the image. By geometrical facts, the distance of the target from the camera is estimated and, together with the azimuth angle, provides the polar coordinates of the target on the ground in a coordinate system centered onto the camera ground location. In this section we provide details on the adopted reference systems.

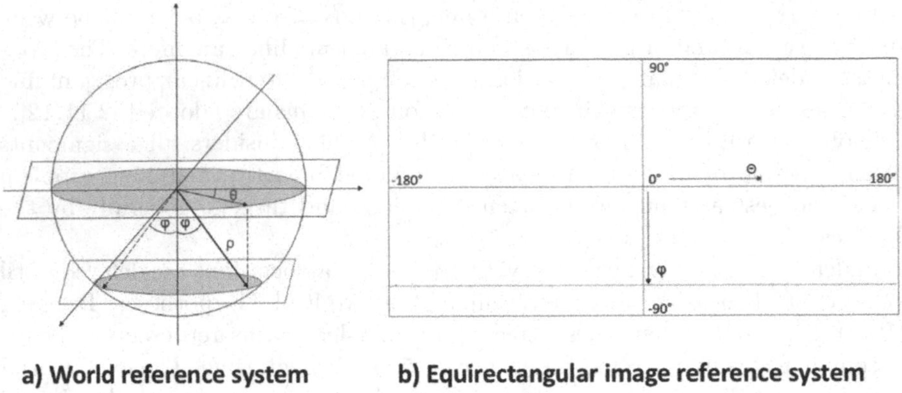

Fig. 3. The image shows the world coordinate system on the left, and the equirectangular coordinate system on the right.

3.1 360° Videos and Equirectangular Images

As shown in Fig. 1, a 360° camera device can acquire panoramic images with a view spanning 360° horizontally and 180° vertically, and can represent the surrounding environment at each shot. This is of interest in several fields such as video surveillance, robotics, and cultural heritage applications.

Regardless of the type of 360° device, pixels of the images are mapped onto a sphere centered into the 360° camera. Equirectangular and cubic projections are often adopted to allow displaying the image on monitors or viewers [6]. While cubic projections map the spherical points onto the plane tangent to the sphere, equirectangular projection maps the whole sphere to a single image. In particular, the central row of the equirectangular image represents the sphere equator; the uppermost and lowermost rows correspond to the sphere poles. In general, each row of the equirectangular image corresponds to the intersection between the sphere and a plane parallel to the horizontal plane of the camera [16].

Pixel coordinates (x_r, y_r) on the equirectangular image represent normalized values of polar and azimuth angles of the corresponding point on the sphere surface. The angles can be recovered from the pixel coordinates by a simple rescaling and shifting such that the polar angle ϕ ranges in $[-90°, 90°]$, while the azimuth angle θ ranges in $[-180°, 180°]$ (see Fig. 3). Of course, by this projection, the radial coordinate of the spherical coordinate system cannot be preserved.

3.2 Estimating Targets' Distances and Locations

The algorithm to estimate the targets' distances from the camera [16] is based on pure geometrical facts and is uncalibrated, as there is no need to estimate the camera parameters. The algorithm takes advantage from a simple fact: all points of a plane parallel to the horizontal camera plane and equally distant from the camera are projected onto the same row of the equirectangular image.

The only hypothesis that must hold to apply the method is that target move on the ground plane, and that the ground plane is parallel to the horizontal camera plane. Furthermore, it must be possible to measure the touching point of the target with the ground on the spherical image (the target must be visible). We define the touching point $P = (x_L, y_L)$ as the middle one in the lowest side of the bounding box enclosing the target on the equirectangular image.

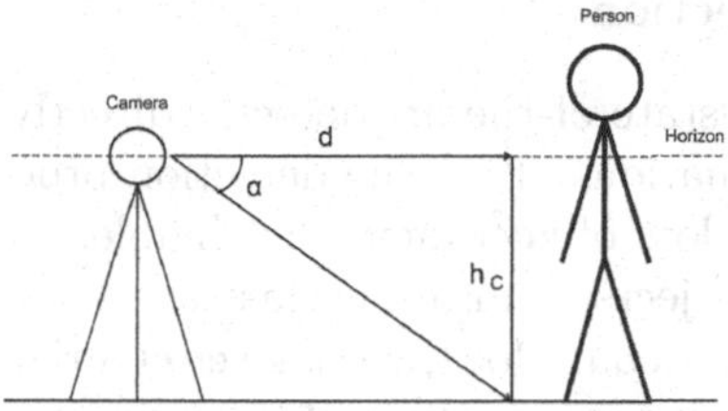

Fig. 4. The figure shows how the distance of the target from the camera can be estimated. h_c is the camera height (in meters). α is the angle between the camera horizontal plane and the line passing through the target's ground-touching point and the camera center. d is the distance of the target from the camera and can be estimated by trigonometrical equations.

As shown in Fig. 4, the distance d of the object of interest from the camera can be estimated as [16]:

$$d = h_c \cot \alpha \tag{1}$$

where h_C is the camera height and α is the angle between the camera plane passing through the sphere equator and the line through the camera center and the target's touching point (see Fig. 4). With this formulation, the distance d is the length of one of the two catheti of the resulting right triangle, and is related to the other one by the trigonometric formula 1.

The angle α is estimated from the point P on the equirectangular image:

$$\alpha = \frac{\frac{h}{2} - y_L}{\frac{h}{2}} \cdot 90^\circ \tag{2}$$

where h is the equirectangular image height (in pixels).

For each target, we can model its location considering the polar coordinates (d, θ) of its touching point. We then model the location of the target in Cartesian coordinates as $l = (d \cos \theta, d \sin \theta)$ on the ground-plane.

4 Proposed MOT Algorithm

Our tracking strategy relies on a simple and very common MOT strategy that aims at updating the targets' locations by associating the targets' predicted locations with the ones provided by a pre-trained pedestrian detector. Hence, the main steps in our algorithm are: pedestrian detection, prediction of the targets' location, and data association. Differently than other works, in our algorithm locations are modeled directly on the ground and expressed in a coordinate system centered in the camera's ground-touching point. Another advantage of our formulation is that it naturally accounts for the circularity of the equirectangular image due to the adopted reference coordinate system.

4.1 Pedestrian Detection

Similar to [14], we run a state-of-the-art detector directly on the equirectangular image. Of course, deformations affect the detection process but we have experimentally found that the loss of accuracy is negligible. Since we are interested in pedestrians, we ignore objects from other classes.

A simple approach to account for the image circularity is expanding the image as shown in Fig. 2 and removing duplicated bounding boxes. Let us assume that the image is originally $W \times H$ while the extended image is $W' \times H'$. Duplicated bounding boxes are removed by shifting all of them to the left by W. Then, those with positive coordinates on the image plane are retained and matched with the non-shifted ones by intersection-over-union. Finally, the most internal ones are selected. In our experiments, we have adopted the Faster-RCNN [8] detector.

4.2 Modeling and Predicting Targets' Locations

We model targets' trajectories as discrete-time linear dynamical systems. The state of the system represents the location of the target on the ground-plane and its velocity $s_t = (x_t, y_t, v_t^x, v_t^y)$, modeled as continuous variables. We use Kalman filter (KF) to make the state evolve over time in two steps: prediction and update steps. The prediction step allows to make the state evolve over time based on the knowledge of the past state. At each discrete time, the state s_{t-1} is linearly combined to generate the new state s_t by also accounting for some Gaussian noise $w_{t-1} \sim N(0, \Sigma_s)$. In our model, no external control signal is needed and we assumed targets move based on a uniform linear motion model.

The update step uses the difference between predictions and observations to refine the state estimate. KF is a recursive filter where prediction and update steps alternate to progressively refine the current state estimation. In particular, as detailed in [7], the filter can be viewed as a weighted average estimator where noise covariance matrices and Kalman-Gain matrix are iteratively computed to refine the state estimations. Whenever an observation is unavailable, no updating step is performed and KF is used to predict the target location. In the long run, this can of course yield to drifting of the tracker.

4.3 Data Association

Since we use a frame-by-frame association approach, we need to associate detections with known targets at each frame. We also need to decide when a new target has to be included in the pool of tracked objects, and when a target is exiting the scene. Associations are estimated by using ground-plane coordinates found through the estimates of targets' distances from the camera. As already explained, this approach naturally accounts for the image circularity.

As usually done, we model MOT as a minimum weight matching problem in bipartite graphs. In a bipartite graph, vertices are grouped into two disjoint sets such that no two vertices within the same set are adjacent. A matching is a subset of edges of the graph that do not share common vertices. The minimum weight matching has minimal sum of the edge weights.

In our framework, the vertices of one set of the bipartite graph represent the targets' locations predicted at time t by Kalman filter, while the vertices in the other set represent the locations of the pedestrian found by the detector on the image plane and converted into ground coordinates as detailed in Sect. 3.2. Each vertex of the first set can potentially match any other vertex in the second set. Edge weights measure the dissimilarity between targets and detections considering both their locations on the ground plane and the appearance features. Appearance features are extracted by using the first flattened layer of a pre-trained CNN (ResNet-50 [10] trained on Imagenet in our experiments), but any other appearance descriptor may be used as well.

We fuse appearance features and locations by considering that the two data vectors have different lengths and values varying in different ranges. Given a target with a predicted location $P_t^T \in R^2$ and a detected bounding box with estimated touching point $P_t^D \in R^2$ at time t, the dissimilarity $l(T, D)$ between the two locations is defined as

$$l(T, D) = \frac{1}{2}||P_t^T - P_t^D||_1. \tag{3}$$

Assuming the appearance features of the target and the detection are $F_{t-1}^T \in R^L$ and $F_t^D \in R^L$ respectively, the dissimilarity $f(T, D)$ between the two feature vectors is defined as

$$f(T, D) = \frac{1}{L}||F_{t-1}^T - F_t^D||_1. \tag{4}$$

In the above dissimilarity scores, L1-norm is used since it is more meaningful and efficient than L2-norm in case of high-dimensional data. We compute $l(T, D)$ and $f(T, D)$ for all pairs of n targets and m detected bounding boxes. To make the L1-norm values comparable, we estimate the z-scores of the two sets of $n \times m$ norm values, $z^l(T, D)$ and $z^f(T, D)$ respectively. Hence, the final dissimilarity score for a pair is obtained as the minimum between $z^l(T, D)$ and $z^f(T, D)$. With this score, the tracker associates a target to a detection or because the locations are close or because the appearance features are similar.

4.4 Trackers' Birth, Death and Updating

At each frame, two cases can arise: new targets enter the scene (or false positives are detected) or some targets exit the scene (or are not detected, for instance due to occlusions). To account for such cases, we adopt the strategy in [11], and augment the two sets of vertices in the bipartite graph with "fake" vertices. The vertices added to the target set represent potentially new targets entering the scene. The vertices added to the detection set account for missing. By including these fake vertices, each set will count $n + m$ nodes.

With these additional vertices, it is necessary to set a default weight value for the fake edges. This value is especially important for the success of the tracking strategy since it represents an error tolerance on the dissimilarity score of the matches and, hence, defines the search area on the ground-plane and in the appearance feature space. We have experimentally found that a value of 2 works pretty well. This value is the superior limit of the 95% of the confidence interval of the estimated z-scores.

Once the data association problem is solved, we use the associations to update the targets' locations and the corresponding Kalman filters. When no association is found, we maintain the Kalman filter prediction as target's location. We keep the target alive for T frames ($T = 90$ in our experiments) such that the target can be tracked through potential occlusions. However, this strategy increases the risks of tracking false positives. To limit this issue, when a new target identity is discovered, we wait for K frame ($K = 3$ in our experiments) before adding this new target to the pool of tracked objects.

5 Experimental Results

We performed tests on the publicly available CVIP360 dataset [16]. The dataset includes two sets of videos. The first set includes 11 videos acquired indoor, while the second one includes 6 videos collected outdoor. Overall, the dataset includes about 18K frames with more than 50K annotated bounding boxes.

We use the CLEAR metrics [13], including MOTA, FP, FN, ID-switches to assess the tracking quality. We also report the IDF1 score [21] that evaluates the identity preservation ability and focuses more on the association performance.

Performance have been evaluated with a modified version of the py-motmetrics library. Since images are circular, we assessed tracking on the ground-plane.

Our baseline method, *Baseline (image)*, implements a standard MOT technique by using the same detector and data association technique as our tracker but models trajectories on the image plane. It is not able to account for the image circularity. In case of missing, the target's location is not updated.

We present ablation studies by constructing our tracker step-by-step. The tracker *Ours (image, circularity)* tracks targets on the image plane and accounts for the image circularity by an ad-hoc matching process. Duplicated bounding boxes on the extended frame are not filtered out but are used to solve data association by retaining the minimum dissimilarity score. This approach requires more comparisons but accounts for the image circularity. This method is relatively close to [14], except for the detection strategy, as explained in Sect. 2. In our ablation study, we compare trackers on equal terms of pedestrian detector. We detect pedestrians on the whole image, focusing on the tracking strategy.

The tracker *Ours (ground)* tracks on the ground-plane and can naturally account for the image circularity. Duplicated bounding boxes are not used for data association. The tracker *Ours (ground + KF)* uses Kalman filter to track targets in case of missing detection. Results of the above techniques are presented with and without using appearance features.

5.1 Results

Tables 1 and 2 show our experimental results in indoor and outdoor videos respectively. The tables report average values of the metrics over the test videos. FP, FN and ISDW are average raw values (not percentage). We note that the number of FP and FN (missing) in the various experiments are similar because they are run on equal terms of detector. In indoor videos, best results are achieved by modeling tracking on the ground.

Despite the appearance model is weak, it contributes to a small improvement of the performance. In outdoor videos, all methods are comparable. Without appearance features, KF improves the IDF1 score but not the MOTA due to a small increase of the ID-switches. In outdoor videos, appearance does not help the tracker and, while MOTA scores are comparable, IDF1 decreases. This might be due to the fact that generally appearance features are sensitive to illumination variations. The decreases of the performance when using ground-plane coordinates in outdoor videos might be due to the uncertainty in the touching-point estimate on the image plane, especially for the farthest pedestrians. Indeed, in outdoor, distance from the camera can be more than 10 m. One of such cases is shown in Fig. 2 (pedestrian with ID 3). The detector may be unable to detect the pedestrian and, if it does, the bounding box is not much reliable.

Despite the limitations of the detection step, all experiments show that tracking on the ground-plane decreases the number of ID-switches. In practice, especially in case of occlusions, is much easier for the tracker to solve ambiguities.

Table 1. Results on the indoor video dataset

Tracker	IDF1↑	MOTA↑	FP↓	FN↓	IDSW↓
Baseline (image)	92.78	93.22	163.09	49.27	41.18
Ours (image, circularity)	92.86	94.95	74.36	49.27	35.09
Ours (ground)	93.66	**96.78**	61.73	49.27	7
Ours (ground + KF)	**94.21**	96.10	61.72	49.27	**5.73**
Baseline (image + appearance)	93.82	95.88	60.45	49.63	18.54
Ours (image + appearance, circularity)	93.82	95.89	60.45	49.64	17.36
Ours (ground + appearance)	94.00	96.05	61.82	49.82	**8**
Ours (ground + appearance + KF)	**94.61**	**96.19**	53.63	49.81	10.18

Table 2. Results on the outdoor video dataset

Tracker	IDF1↑	MOTA↑	FP↓	FN↓	IDSW↓
Baseline (image)	86.5	88.8	81.16	317.67	67.33
Ours (image, circularity)	86.83	88.18	120.67	338.33	58.3
Ours (ground)	86.6	**89.21**	47.5	393.3	**12.3**
Ours (ground + KF)	**89.52**	89.69	33.5	393.33	12.83
Baseline (image + appearance)	**88.6**	**89.62**	49	337.5	34.3
Ours (image + appearance, circularity)	88.30	89.38	49	365.67	24.17
Ours (ground + appearance)	85.8	88.90	64	393.22	16
Ours (ground + appearance + KF)	87.28	89.43	49.33	393.33	**15.17**

6 Conclusions and Future Work

This paper proposes a MOT tracker that estimates targets' locations on the ground-plane by using the distance of the targets from the 360° camera. Such distance is estimated by the method in [16], which is uncalibrated and works under mild hypotheses. Targets' locations on the ground-plane are used to solve data association and model trajectories. We experimentally found that measuring the location on the ground is required to properly measure performance of any tracker in 360° videos. Our preliminary results show that, when the targets' touching point can be reliably estimated on the image, ground-plane coordinates improve tracking especially in case of occlusions.

In future work we will study how to improve the detection of farthest objects in order to have a better estimation of the touching point. We will also improve appearance and motion models to handle occlusions.

Acknowledgement. This research was partially funded by MIUR grant number PRIN I-MALL 2017BH297_004 and Italian PON IDEHA grant ARS01_00421.

References

1. Betke, M., Wu, Z.: Data association for multi-object visual tracking. Synth. Lect. Comput. Vis. **6**(2), 1–120 (2016)
2. Castanon, D.A.: Efficient algorithms for finding the k best paths through a trellis. IEEE Trans. Aerosp. Electron. Syst. **26**(2), 405–410 (1990)
3. Chen, G., St-Charles, P.L., Bouachir, W., Bilodeau, G.A., Bergevin, R.: Reproducible evaluation of pan-tilt-zoom tracking. In: 2015 IEEE International Conference on Image Processing (ICIP), pp. 2055–2059. IEEE (2015)
4. Chu, Q., Ouyang, W., Li, H., Wang, X., Liu, B., Yu, N.: Online multi-object tracking using CNN-based single object tracker with spatial-temporal attention mechanism. In: Proceedings of the International Conference on Computer Vision, pp. 4836–4845 (2017)

5. Ciaparrone, G., Sánchez, F.L., Tabik, S., Troiano, L., Tagliaferri, R., Herrera, F.: Deep learning in video multi-object tracking: a survey. Neurocomputing **381**, 61–88 (2019)
6. Corbillon, X., Simon, G., Devlic, A., Chakareski, J.: Viewport-adaptive navigable 360-degree video delivery. In: 2017 IEEE International Conference on Communications (ICC), pp. 1–7. IEEE (2017)
7. Funk, N.: A study of the kalman filter applied to visual tracking. University of Alberta, Project for CMPUT 652(6) (2003)
8. Girshick, R.: Fast R-CNN. In: Proceedings of the International Conference on Computer Vision, pp. 1440–1448 (2015)
9. He, K., Gkioxari, G., Dollár, P., Girshick, R.: Mask R-CNN. In: Proceedings of International Conference on Computer Vision, pp. 2961–2969 (2017)
10. He, K., Zhang, X., Ren, S., Sun, J.: Deep residual learning for image recognition. In: Proceedings of the IEEE Conference on Computer Vision and Pattern Recognition, pp. 770–778 (2016)
11. Huang, T., Russell, S.: Object identification in a Bayesian context. In: IJCAI, vol. 97, pp. 1276–1282. Citeseer (1997)
12. Jiang, H., Fels, S., Little, J.J.: A linear programming approach for multiple object tracking. In: Conference on Computer Vision and Pattern Recognition, pp. 1–8. IEEE (2007)
13. Leal-Taixé, L., Milan, A., Reid, I., Roth, S., Schindler, K.: Motchallenge 2015: towards a benchmark for multi-target tracking. arXiv preprint arXiv:1504.01942 (2015)
14. Liu, K.C., Shen, Y.T., Chen, L.G.: Simple online and realtime tracking with spherical panoramic camera. In: 2018 IEEE International Conference on Consumer Electronics (ICCE), pp. 1–6. IEEE (2018)
15. Lo Presti, L., La Cascia, M.: Deep motion model for pedestrian tracking in 360 degrees videos. In: Ricci, E., Rota Bulò, S., Snoek, C., Lanz, O., Messelodi, S., Sebe, N. (eds.) ICIAP 2019. LNCS, vol. 11751, pp. 36–47. Springer, Cham (2019). https://doi.org/10.1007/978-3-030-30642-7_4
16. Mazzola, G., Lo Presti, L., Ardizzone, E., La Cascia, M.: A dataset of annotated omnidirectional videos for distancing applications. J. Imaging **7**(8), 158 (2021)
17. Mi, T.W., Yang, M.T.: Comparison of tracking techniques on 360-degree videos. Appl. Sci. **9**(16), 3336 (2019)
18. Monteleone, V., Lo Presti, L., La Cascia, M.: Pedestrian tracking in 360 video by virtual PTZ cameras. In: 2018 IEEE 4th International Forum on Research and Technology for Society and Industry (RTSI), pp. 1–6. IEEE (2018)
19. Monteleone, V., Lo Presti, L., La Cascia, M.: Particle filtering for tracking in 360 degrees videos using virtual PTZ cameras. In: Ricci, E., Rota Bulò, S., Snoek, C., Lanz, O., Messelodi, S., Sebe, N. (eds.) ICIAP 2019. LNCS, vol. 11751, pp. 71–81. Springer, Cham (2019). https://doi.org/10.1007/978-3-030-30642-7_7
20. Reid, D.: An algorithm for tracking multiple targets. IEEE Trans. Autom. Control **24**(6), 843–854 (1979)
21. Ristani, E., Solera, F., Zou, R., Cucchiara, R., Tomasi, C.: Performance measures and a data set for multi-target, multi-camera tracking. In: Hua, G., Jégou, H. (eds.) ECCV 2016. LNCS, vol. 9914, pp. 17–35. Springer, Cham (2016). https://doi.org/10.1007/978-3-319-48881-3_2
22. Wojke, N., Bewley, A., Paulus, D.: Simple online and realtime tracking with a deep association metric. In: 2017 IEEE International Conference on Image Processing (ICIP), pp. 3645–3649. IEEE (2017)

23. Xu, J., Cao, Y., Zhang, Z., Hu, H.: Spatial-temporal relation networks for multi-object tracking. In: International Conference on Computer Vision, October 2019
24. Yu, F., Li, W., Li, Q., Liu, Yu., Shi, X., Yan, J.: POI: multiple object tracking with high performance detection and appearance feature. In: Hua, G., Jégou, H. (eds.) ECCV 2016. LNCS, vol. 9914, pp. 36–42. Springer, Cham (2016). https://doi.org/10.1007/978-3-319-48881-3_3

FasterVideo: Efficient Online Joint Object Detection and Tracking

Issa Mouawad[1,2(✉)] and Francesca Odone[1,2]

[1] MaLGa Machine Learning Genoa Center, Genoa, Italy
[2] DIBRIS - Universitá degli Studi di Genova, Genoa, Italy
issa.mouawad@dibris.unige.it, francesca.odone@unige.it

Abstract. Object detection and tracking in videos represent essential and computationally demanding building blocks for current and future visual perception systems. In order to reduce the efficiency gap between available methods and computational requirements of real-world applications, we propose to re-think one of the most successful methods for image object detection, Faster R-CNN, and extend it to the video domain. Specifically, we extend the detection framework to learn instance-level embeddings which prove beneficial for data association and re-identification purposes. Focusing on the computational aspects of detection and tracking, our proposed method reaches a very high computational efficiency necessary for relevant applications, while still managing to compete with recent and state-of-the-art methods as shown in the experiments we conduct on standard object tracking benchmarks (Code available at https://github.com/Malga-Vision/fastervideo).

Keywords: Multiple-object tracking · Joint detection and tracking

1 Introduction

Detecting and tracking multiple objects in video sequences is a core building block for several applications. Recently, deep learning based methods achieved unprecedented success in detecting objects in both general-purpose and application-oriented settings [18,19,36]. Utilizing such methods in the video domain remain challenging due to the inefficiency of per-frame processing and the lack of a temporally consistent understanding of objects trajectories.

Multiple-object tracking deals with the task of tracking several targets locations across video frames and is able to derive trajectories across time by associating tracks and detected objects. Multiple-object tracking, however, is usually handled in the tracking-by-detection framework, which assumes that objects in each frame are detected using a separate algorithm and only addresses association [28,33]. This separation causes an additional computational cost, and prohibits the sharing of representations and information between the two tasks.

S. Sclaroff et al. (Eds.): ICIAP 2022, LNCS 13233, pp. 375–387, 2022.
https://doi.org/10.1007/978-3-031-06433-3_32

Recently, it has been noted that joint architectures reduce the computational overhead by deriving the outputs of multiple tasks simultaneously, proving in the same time the performance benefits of this holistic processing where computation and hidden representations are shared among tasks [6,25,34].

Building on this line of research, we address detection and tracking jointly, relying on a state of the art image object detector, Faster R-CNN [19], which we extend to the video domain.

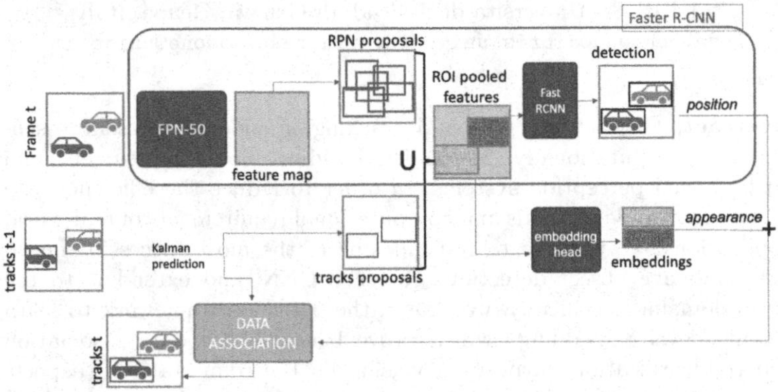

Fig. 1. Overview of our proposed method for joint detection and tracking

The modular structure of two-stages detectors, allows us to control the computational cost of the detector, while exploiting additional information provided by the video input: we control the complexity of Faster R-CNN Region Proposal Network (RPN) by reducing of the number of image-based proposals, while adding proposals originated by the previous video frames. Also, the two-stage design can be extended to learn appropriate embeddings which are effective to improve the tracking association step—see Fig. 1.

In summary, the contributions of our paper can be summarized as follows:

- We provide a modular extension of a 2-stage image-based object detector to address jointly video object detection and tracking.
- We re-use the detector learned representations to simplify the tracking task by learning embeddings which are then employed to boost the data association accuracy across frames.
- We use temporal prior of objects location to both guide the detector and reduce its computational cost
- Our pipeline achieves accuracy results comparable to state-of-the-art methods, while consistently delivering several fold efficiency improvement , highlighting under-explored accuracy-speed trade-off points.

The rest of this paper is organized as follows: first, we review related works tackling detection and tracking tasks either separately or jointly. Next, we describe in details our proposed pipeline targeted at joint detection and tracking. Finally, we conduct experiments using KITTI [7] and MOT [5,16] datasets and we compare with other similar methods in terms of both MOT metrics and inference time to highlight speed performance trade-offs.

2 Related Work

2.1 Video Object Detection

Unlike image-based methods which can be applied in video settings on separate frames independently, several works try to rely on additional video prior to guide video object detection. This direction is gaining interest due to the challenges posed by videos to image-based detectors such as motion blur and focus loss.

In the work of [37], optical flow is used as to warp features from consecutive frames, while a recurrent neural network is used in [31] to aggregate frames and learn to detect objects across time. These methods, however, do not construct trajectories of detected objects.

2.2 Multi-object Tracking

Due to the high accuracy achieved by recent object detectors, many works in the literature follow the tracking-by-detection approach. Such approach assumes that detections are obtained for each frame separately, and focuses on association. Association between detections and tracks is often formulated as a bipartite graph and an association cost is identified for each possible match. The problem is usually solved using the Hungarian algorithm [12]. In addition to intersection over union (IoU) metric, several approaches propose the use of appearance-based similarity for matching, obtained using optical flow, low-level feature descriptors, or motion-based features [3,9,11]. While other methods try to learn similarity metrics from data using contrastive learning [17] or triplet losses [1,4] Different data representation is additionally explored in recent works [28,29,33] allowing end-to-end learning of MOT methods.

2.3 Joint Object Detection and Tracking

This research direction focuses, instead, jointly on both detection and tracking tasks. These methods, in particular, promise to simplify perception tasks and yield more efficient pipelines. In [15], the authors propose a scheduler network which is used to alternate between running a full detector or simply locating already-seen objects. Another direction aims to transform typical object detectors to perform jointly detection and tracking. Some of these methods extend a two-stage detector (mainly Faster R-CNN) [1,26], while others focus on single-stage detectors [14,35]. Most of these methods, however, introduce additional modules which reduce efficiency of the joint pipeline, even below that of the baseline detector.

3 Proposed Method

In this section we present our architecture based on the well established Faster R-CNN object detector, which we extend to address detection and tracking jointly and efficiently. By reducing the number of proposals and relying on tracked objects as an additional source of proposals, we improve the speed-accuracy trade-off. Additionally, we introduce an embedding network branch that allows us to learn an appropriate and descriptive appearance representation for tracked instances boosting data association accuracy. Figure 1 shows the building blocks of the proposed method.

3.1 Video Object Detection

We adopt as an image-based object detector Faster R-CNN with a FPN-50 backbone and pyramid layers [13]. To optimize it for the video-based object detection, we propose two modifications.

Sparser RPN Proposals. Sparser proposals allow us to control the computational cost of the detector. Figures 2 and 3 provide an experimental evidence: Fig. 2, reports the accuracy (measured in both mAP and AP) on COCO2017 validation set using different number of proposals compared with the original 1k proposals [19]. Comparable accuracies are obtained with 1/10 of the proposals. Figure 3 shows a slow yet consistent decrease in the inference time incurred by decreasing the number of proposals. To maintain a high accuracy, we exploit space-time continuity of objects instances in subsequent frames, as we detail in the following.

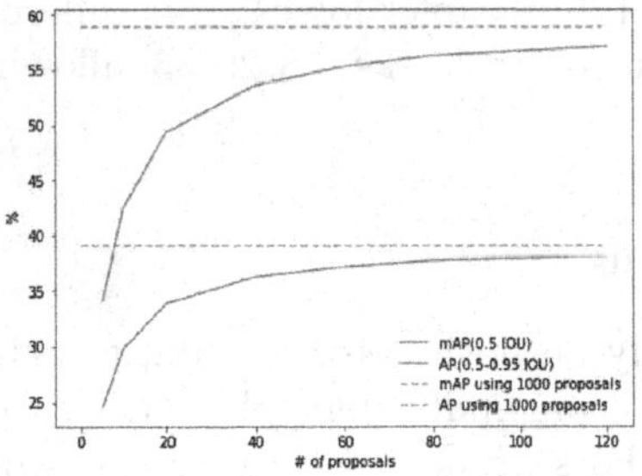

Fig. 2. Accuracy obtained with different number of proposals (COCO2017 validation set).

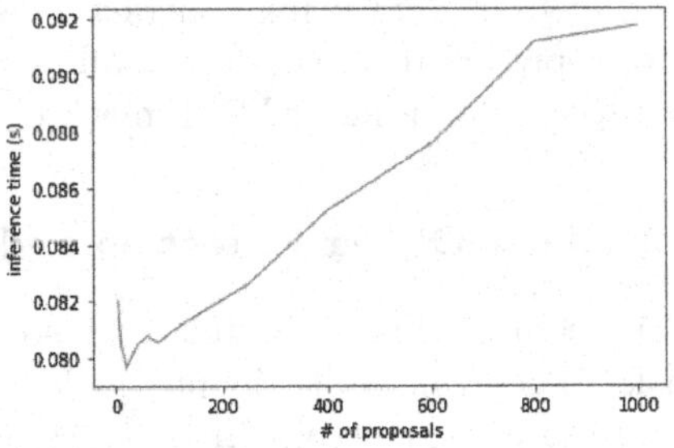

Fig. 3. Inference speed with different number of proposals reoprted on Quadro P-5000

Tracks Proposals. In order to compensate for the reduced number of proposals, and address occlusions and illumination changes, we provide a reverse feedback from the tracker to the detector. We perform a Kalman prediction step on tracks from the frame $t-1$ (with two opposing bounding box corners in the Kalman state) and add these predicted bounding boxes as proposals for the frame at time t. This provides an attention mechanism for the detector to focus on important parts of the image where previously seen objects are expected to be found.

3.2 Embeddings Learning

Deep features associated with the entire image or a specific instance provide a robust descriptor which withstands a certain viewpoint or illumination change [20]. For this, we add an embedding learning branch to the classical Faster R-CNN RoI Heads, extracting a representation for any object instance which we will employ for data association and re-identification. Re-identification plays an important role in object tracking systems [10] but it is typically addressed as a separate task [1]. We, instead, integrate the embedding module within our joint detection and tracking framework. This allows the network to robustly handle occlusions and fuzzy associations relying on objects encoding learned specifically to be viewpoint and illumination invariant.

The new embedding learning branch is placed on top of the RoI-pooled features using two fully connected layers separated by a ReLU non-linearity and batch normalization.

The resulting network is trained jointly for both detection and embedding learning tasks and final loss is composed of three losses: the original RPN and Fast R-CNN losses [19], plus the embedding loss for which we use the triplet loss [21]. While RPN and detection losses are calculated as usual for all the ground truth objects present in the images, in the case of the triplet loss we need to select meaningful triplets to keep convergence under control [21].

An instance x_i^a of a specific object i (an anchor), should be closer to the positive example x_i^p (another instance of the same object) than to a negative one x_i^n (an instance of another object), by a margin α:

$$\|E(x_i^a) - E(x_i^p)\|_2^2 + \alpha < \|E(x_i^a) - E(x_i^n)\|_2^2. \tag{1}$$

Then the triplet loss seeks to maximize the distance between the encoding of the anchor example x_i^a and the negative example x_i^n, while, at the same time, minimizing the distance between the anchor example and the positive example x_i^p, as depicted in Fig. 4.

The choice of the triplets is crucial as it is infeasible to consider all combinations of the whole training set. We apply a batch-wise hard example mining. First, we choose a batch of $B = 8$ images at random from a consecutive $D = 16$ frames, considering only batches where at least $P = 8$ objects are present at least $K = 4$ times. This allows for a robust calculation of the triplet loss; the intuition behind sampling the batch at random from neighborhood of frames is to mimic missed detections and increase the robustness of the learned embeddings against

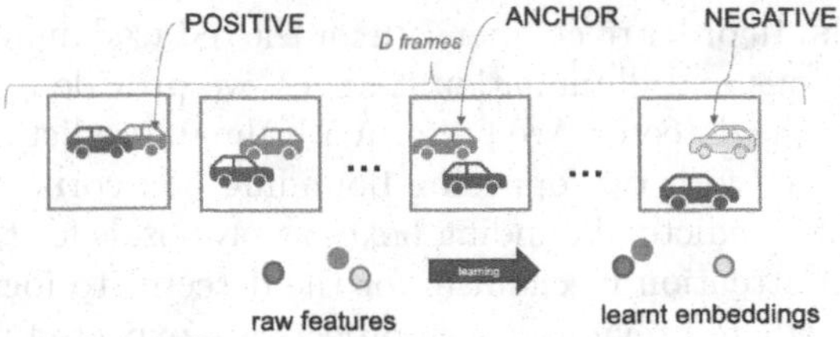

Fig. 4. The distance of the learned embeddings is smaller for instances of the same object (anchor and positive) and larger for different objects (anchor and negative)

severe appearance changes. Then, for each anchor example, we pick the hardest positive and negative examples within the batch B such that:

$$x^p = \arg\max_{x_i^p \in B} \|E(x_i^a) - E(x_i^p)\|_2^2 \quad \text{and} \quad x^n = \arg\min_{x_i^n \in B} \|E(x_i^a) - E(x_i^n)\|_2^2 \tag{2}$$

Finally, the loss is formulated using triplets (x_i^a, x_i^p, x_i^n) for all the P objects:

$$L_{embedding} = \frac{1}{n} \sum_{i=1}^{P} [\|E(x_i^a) - E(x_i^p)\|_2^2 - \|E(x_i^a) - E(x_i^n)\|_2^2 + \alpha]_+ \tag{3}$$

3.3 Data Association and Tracking

During inference, RoI-pooling is performed as the original Faster R-CNN using the proposals generated by RPN to extract instance-level features which are fed to the two heads (the Fast R-CNN one and the proposed embedding head depicted in Fig. 1). Usual class and box predictions are calculated as in [8], while the final boxes are used to pool features which are fed to the embedding head. In order to assign identifiers to detected object, we formulate a data association step which matches the detections at time t with the tracklets from $t-1$. We define, inspired by [23], a matching cost between detections and tracklets as the linear combination of two distances: the Jaccard or IoU Distance to capture *position or spatial proximity* between bounding boxes and the cosine distance (defined as 1−cosine similarity) between the two objects embeddings which captures *appearance similarity*:

$$Cost = \alpha * dist_{position} + \beta * dist_{appearance} \tag{4}$$

(α and β are weighting factors, we both set to 0.5). Next, we rely on the Hungarian algorithm [12] to perform a minimal-cost matching between the detector output and the tracks leading to matches and possibly mismatches. In order to avoid forced one-to-one weak matches, we set a maximum cost for any match which we discard if violated.

In Fig. 5, we show some examples of re-identification events accomplished utilizing objects embeddings. We report on the figure the embedding distance

Fig. 5. Examples of occluded object re-identification using object embeddings and cosine distance (best seen in color), # indicate frame number

between pair of objects across different frames showing the ability of the embedding head to generate similar embeddings for the same object even across a wide time frame and for challenging appearance shifts and occlusions.

4 Experimental Analysis

4.1 KITTI Benchmark

Object Detector Initial Training: We focus in our experiments on car objects for both detection and tracking. We use Faster R-CNN official implementation Detectron2 [30] with weights of FPN trained on COCOtrain2017. Next, we finetune the model on KITTI's object detection benchmark focusing only on the car class. After discarding images common to the tracking benchmark, we are left with 4k images we split to 3k images for training the object detector for 5 epochs and 1k images for validation. We keep the original number of proposals during training, but reduce it during evaluation to 20 proposals on KITTI.
Joint Detection and Tracking: We attach the embedding head and we finetune the pipeline using a multi-task loss (formed by the original detection loss [8], and the embedding loss from Equation (3)) on a split of the tracking training set (as defined in [27]) using a batch size of 8 images for 5k iterations.
Oracle Detections: We conduct a series of experiments to highlight the performance of the tracking branch and the embedding head. To this end, we use ground truth boxes as detections source (oracle detections) and we compare the resulting tracking results using different combinations of matching costs. Table 1 shows that the association based on position provides superior performance with respect to the learnt appearance. Combining the two, however, boosts this performance further which indicates that the learned embeddings help simplify the association even with accurate boxes.

Table 1. Data association cost assessment - KITTI Benchmark on a held-out set using Oracle detections (reporting Multiple Object Tracking Accuracy - MOTA; MOT Precision - MOTP; False Positives - FP; False Negatives - FN; IDentity switches - ID)

Matching method	MOTA↑	MOTP↑	FP↓	FN↓	IDs↓
Position (IoU distance)	96.0	90.5	169	71	26
Appearance (embedding distance)	94.9	90.8	214	88	38
Both	**97.2**	90.8	**151**	**31**	**2**

Ablation Study: We use the validation set to tune hyper-parameters and study the pipeline ablations measured by the tracking accuracy and inference time[1] and report the results in Table 2. The table highlights the effect of each design choice on both FPS (frames-per-second FPS)and IDs (identity switches), where using specifically-learned features and temporal prior help to recover many identity switches with the best trade-off achieved using the proposed method.

Table 2. KITTI Benchmark ablation study: the proposed method (first row); without tracks proposals (second row); alternative data association: position-based association (third row), position plus appearance based on raw Faster R-CNN features (fourth row)

Ablations	MOTA↑	MOTP↑	P %↑	R %↑	IDs↓	FPS↑
Proposed method	**81.2**	80.0	93.0	91.1	16	13.5
Proposals						
No track proposals	80.0	79.9	92.8	90.5	35	13.4
Data association						
Position only	79.4	80.0	91.9	91.9	46	**13.9**
Raw ROI-pooled feat.	80.1	80.3	93.3	89.6	**12**	10.5

Comparative Analysis: To provide a fair comparison with the state of the art, we mainly focus on published methods which have access to comparable data and annotation. Thus, we omit monocular 3D tracking methods and methods which use LiDAR point clouds.

For evaluation, we use KITTI test set and submit results to the evaluation server. Table 3 provides a general overview on how our proposed method compares to other vision based methods published in the literature. For all methods not incorporating detection time in their performance evaluation we added the

[1] For Detection, an NVIDIA Quadro P5000 GPU has been used to obtain the time measurements.

Table 3. KITTI Tracking benchmark from KITTI evaluation server after submitting the results on the test set for the proposed method

Method	MOTA%↑	MOTP%↑	MT%↑	IDs↓	3D GT	Online	FPS↑
Our method	81.6	80.1	68.3	401		✓	**15**
SMAT [9]	83.6	85.9	62.8	198		✓	5
TuSimple [3]	86.3	84.1	71.1	292			5
QD [17]	84.9	84.9	69.5	313		✓	5.8
MASS [11]	84.6	85.4	74.0	353		✓	10
MOTBP [22]	82.7	85.5	72.6	934	✓	✓	2.5

cost of Faster R-CNN. Results show that our proposed method is able to compete with other well performing methods, while, at the same time, achieving near real-time inference for both detection and tracking, highlighting the advantage of addressing the tasks jointly. Our method, additionally, is fully online, and requires no additional labels.

4.2 MOT Benchmark

Joint Detection and Tracking. We use MOT challenge [16] benchmarks which are the de facto standard in object tracking literature. We focus on MOT17 and MOT20 which offer different levels of difficulty and crowded scenes. We use the base detector trained on COCOtrain2017 and fine-tune it on MOT17Det, for MOT17 and fine-tune it again on MOT20 training set for MOT20. Similarly to the above experiment, we attach the embedding head and fine-tune on tracking ground-truth in both cases. In order to account for the crowded scenes, we set the number of proposals to 50 in MOT17 and 100 for MOT20 experiments.
Comparative Analysis: in Table 4, to provide a fair comparison, we report our results on the test set alongside results achieved by Tracktor++ and SORT [2] after accounting for the detection time. Results of MOT17 benchmark suggest that our proposed method is able to achieve a responsive performance for both tasks jointly while maintaining a speed-accuracy trade-off. While results on MOT20, albeit consistent in terms of comparative analysis, demonstrate a clear degradation in the inference time for all methods, caused by the high resolution video frames and the large number of objects (31 in MOT17 vs 170 in MOT20). Tackling such dense scenes is the aim of our future work efforts.

Table 4. MOT17 and MOT20 Tracking benchmarks obtained from MOT CHALLENGE server after submitting the results on the test set for the proposed method

Method	MOTA↑	IDF1↑	MOTP↑	P %↑	R %↑	IDs ↓	FPS ↑
MOT17							
Ours	49.4	45.1	77	88.3	58.1	5589	5.37
T.++ [1]	53.3	52.3	78	96.3	56	2072	1.25
SORT [2]	43.1	39.8	77.8	90.7	49	4852	7.1
MOT20							
Ours	44.7	39.1	76.2	92.5	49.5	4171	2.3
T++ [1]	50.8	52.1	76.8	84.7	62.7	2751	0.19
SORT [2]	42.7	45.1	78.5	90.2	48.8	4470	6.6

4.3 Inference Time Analysis

In order to better understand the computational cost breakdown across different datasets, we provide a finer-level analysis of the inference time of the joint framework to highlight detection time and tracking time and other parameters which have an influence on the overall time (see Table 5). While detection time *d_time* relies solely on the image dimensions and the number of proposals used, tracking time *t_time* is also affected by the average number of objects. Tracking time is dominated by building the distance matrix, and calculating the cosine distance between high-dimensional vectors (the embeddings), while solving the linear assignment using the Hungarian algorithm adds only a marginal cost.

Table 5. Average inference time breakdown and other average indicators measured on the test sets of each of the experimented dataset

Dataset	Resolution (average)	# tracks	# proposals	d_time (ms)	t_time (ms)	FPS
KITTI	1242 × 375	7.4	27.4	60.2	5.2	15
MOT17	1737 × 994	22.2	72.2	94	50	5.3
MOT20	1394 × 907	64.9	164	84	315	2.3

5 Conclusion and Future Work

In this work, we have proposed a novel and efficient joint object detection and tracking algorithm and discussed the importance of multi-task learning in solving similar visual tasks. Such joint processing is being increasingly adopted in the recent literature with additional inspirations coming from learning using privileged information framework [32]. New large scale datasets [24] provide multi-task annotations to fuel these methods with unprecedented amount of data in the autonomous driving domain.

Our obtained results demonstrate the benefit of using a simple method with a modular internal structure such as Faster R-CNN in striking a reasonable speed-accuracy trade off, and thus, achieving efficient inference (consistently several-fold faster than other methods) while in the same time delivering competitive accuracy. Future work will tackle specifically real-world scenarios, in the autonomous navigation field. This will allow us to fully appreciate the computational benefits of our approach, compared with competing methods. Additionally, wider object categories will introduce additional clutter in the scene, for which efficiency aspects need to be pushed further to account for such real-world scenarios.

References

1. Bergmann, P., Meinhardt, T., Leal-Taixe, L.: Tracking without bells and whistles. In: Proceedings of IEEE ICCV (2019). https://doi.org/10.1109/ICCV.2019.00103
2. Bewley, A., Ge, Z., Ott, L., Ramos, F., Upcroft, B.: Simple online and realtime tracking. In: 2016 IEEE ICIP. IEEE (2016). https://doi.org/10.1109/ICIP.2016.7533003
3. Choi, W.: Near-online multi-target tracking with aggregated local flow descriptor. In: Proceedings of IEEE ICCV (2015). https://doi.org/10.1109/ICCV.2015.347
4. Dai, P., Weng, R., Choi, W., Zhang, C., He, Z., Ding, W.: Learning a proposal classifier for multiple object tracking. In: Proceedings of IEEE CVPR (2021). https://doi.org/10.1109/CVPR46437.2021.00247
5. Dendorfer, P., et al.: MOT20: a benchmark for multi object tracking in crowded scenes. arXiv preprint arXiv:2003.09003 (2020)
6. Dvornik, N., Shmelkov, K., Mairal, J., Schmid, C.: Blitznet: a real-time deep network for scene understanding. In: Proceedings of IEEE ICCV (2017). https://doi.org/10.1109/ICCV.2017.447
7. Geiger, A., Lenz, P., Urtasun, R.: Are we ready for autonomous driving? The KITTI vision benchmark suite. In: IEEE Conference CVPR (2012). https://doi.org/10.1109/CVPR.2012.6248074
8. Girshick, R.: Fast R-CNN. In: Proceedings of IEEE ICCV, pp. 1440–1448 (2015). https://doi.org/10.1109/ICCV.2015.169
9. Gonzalez, N.F., Ospina, A., Calvez, P.: SMAT: smart multiple affinity metrics for multiple object tracking. In: Campilho, A., Karray, F., Wang, Z. (eds.) ICIAR 2020. LNCS, vol. 12132, pp. 48–62. Springer, Cham (2020). https://doi.org/10.1007/978-3-030-50516-5_5
10. Hermans, A., Beyer, L., Leibe, B.: In defense of the triplet loss for person re-identification. arXiv preprint arXiv:1703.07737 (2017)
11. Karunasekera, H., Wang, H., Zhang, H.: Multiple object tracking with attention to appearance, structure, motion and size. IEEE Access. **7**, 104423–104434 (2019). https://doi.org/10.1109/ACCESS.2019.2932301
12. Kuhn, H.W.: The Hungarian method for the assignment problem. Naval Res. Logist. Q. **2**(1–2), 83–97 (1955)
13. Lin, T.Y., Dollár, P., Girshick, R., He, K., Hariharan, B., Belongie, S.: Feature pyramid networks for object detection. In: Proceedings of IEEE CVPR (2017). https://doi.org/10.1109/CVPR.2017.106

14. Lu, Z., Rathod, V., Votel, R., Huang, J.: Retinatrack: online single stage joint detection and tracking. In: Proceedings of IEEE CVPR (2020). https://doi.org/10.1109/CVPR42600.2020.01468
15. Luo, H., Xie, W., Wang, X., Zeng, W.: Detect or track: towards cost-effective video object detection/tracking. In: Proceedings of AAAI, vol. 33 (2019). https://doi.org/10.1609/aaai.v33i01.33018803
16. Milan, A., Leal-Taixé, L., Reid, I., Roth, S., Schindler, K.: MOT16: a benchmark for multi-object tracking. arXiv:1603.00831 (2016)
17. Pang, J., Qiu, L., Li, X., Chen, H., Li, Q., Darrell, T., Yu, F.: Quasi-dense similarity learning for multiple object tracking. In: Proceedings of IEEE CVPR (2021). https://doi.org/10.1109/CVPR46437.2021.00023
18. Redmon, J., Divvala, S., Girshick, R., Farhadi, A.: You only look once: unified, real-time object detection. In: Proceedings of IEEE CVPR (2016). https://doi.org/10.1109/CVPR.2016.91
19. Ren, S., He, K., Girshick, R., Sun, J.: Faster R-CNN: towards real-time object detection with region proposal networks. Adv. Neural Inf. Process. Syst. **28** (2015). https://doi.org/10.1109/TPAMI.2016.2577031
20. Salvador, A., Giró-i Nieto, X., Marqués, F., Satoh, S.: Faster R-CNN features for instance search. In: Proceedings of CVPR (2016). https://doi.org/10.1109/CVPRW.2016.56
21. Schroff, F., Kalenichenko, D., Philbin, J.: Facenet: A unified embedding for face recognition and clustering. In: Proceedings of IEEE CVPR, pp. 815–823 (2015)
22. Sharma, S., Ansari, J.A., Murthy, J.K., Krishna, K.M.: Beyond pixels: leveraging geometry and shape cues for online multi-object tracking. In: IEEE ICRA. IEEE (2018). https://doi.org/10.1109/ICRA.2018.8461018
23. Sorial, M., Mouawad, I., Simetti, E., Odone, F., Casalino, G.: Towards a real time obstacle detection system for unmanned surface vehicles. In: Oceans 2019 MTS/IEEE Seattle (2019). https://doi.org/10.23919/OCEANS40490.2019.8962685
24. Sun, P., et al.: Scalability in perception for autonomous driving: Waymo open dataset. In: Proceedings of IEEE CVPR (2020). https://doi.org/10.1109/CVPR42600.2020.00252
25. Teichmann, M., Weber, M., Zoellner, M., Cipolla, R., Urtasun, R.: Multinet: real-time joint semantic reasoning for autonomous driving. In: 2018 IEEE Intelligent Vehicles Symposium (IV). IEEE (2018). https://doi.org/10.1109/IVS.2018.8500504
26. Voigtlaender, P., et al.: MOTS: multi-object tracking and segmentation. In: Proceedings of IEEE CVPR (2019). https://doi.org/10.1109/CVPR.2019.00813
27. Weng, X., Kitani, K.: A baseline for 3D multi-object tracking. arXiv preprint arXiv:1907.03961 (2019)
28. Weng, X., Wang, Y., Man, Y., Kitani, K.M.: GNN3DMOT: graph neural network for 3d multi-object tracking with 2D–3D multi-feature learning. In: Proceedings of IEEE CVPR (2020). https://doi.org/10.1109/CVPR42600.2020.00653
29. Weng, X., Yuan, Y., Kitani, K.: Joint 3d tracking and forecasting with graph neural network and diversity sampling. arXiv preprint arXiv:2003.07847 (2020)
30. Wu, Y., Kirillov, A., Massa, F., Lo, W.Y., Girshick, R.: Detectron2 (2019). https://github.com/facebookresearch/detectron2
31. Xiao, F., Lee, Y.J.: Video object detection with an aligned spatial-temporal memory. In: Proceedings of ECCV (2018). https://doi.org/10.1007/978-3-030-01237-3_30

32. Xu, H., Gao, Y., Yu, F., Darrell, T.: End-to-end learning of driving models from large-scale video datasets. In: Proceedings of IEEE CVPR (2017). https://doi.org/10.1109/CVPR.2017.376
33. Xu, Y., Osep, A., Ban, Y., Horaud, R., Leal-Taixé, L., Alameda-Pineda, X.: How to train your deep multi-object tracker. In: Proceedings of CVPR (2020). https://doi.org/10.1109/CVPR42600.2020.00682
34. Zamir, A.R., Sax, A., Shen, W., Guibas, L.J., Malik, J., Savarese, S.: Taskonomy: Disentangling task transfer learning. In: Proceedings of IEEE CVPR (2018). https://doi.org/10.1109/CVPR.2018.00391
35. Zhou, X., Koltun, V., Krähenbühl, P.: Tracking objects as points. In: Vedaldi, A., Bischof, H., Brox, T., Frahm, J.-M. (eds.) ECCV 2020. LNCS, vol. 12349, pp. 474–490. Springer, Cham (2020). https://doi.org/10.1007/978-3-030-58548-8_28
36. Zhou, X., Wang, D., Krähenbühl, P.: Objects as points. arXiv preprint arXiv:1904.07850 (2019)
37. Zhu, X., Dai, J., Yuan, L., Wei, Y.: Towards high performance video object detection. In: IEEE Conference on CVPR, June 2018. https://doi.org/10.1109/CVPR.2018.00753

A Large-scale TV Dataset for Partial Video Copy Detection

Van-Hao Le(✉), Mathieu Delalandre, and Donatello Conte

LIFAT Laboratory, Tours, France
{VanHao.Le,Mathieu.Delalandre,Donatello.Conte}@univ-tours.fr

Abstract. This paper is interested with the performance evaluation of the partial video copy detection. Several public datasets exist designed from web videos. The detection problem is inherent to the continuous video broadcasting. The alternative is then to process with TV datasets offering a deeper scalability and a control of degradations for a fine performance evaluation. We propose in this paper a TV dataset called STVD. It is designed with a protocol ensuring a scalable capture and robust groundtruthing. STVD is the largest public dataset on the task with a near 83k videos having a total duration of 10,660 h. Performance evaluation results of representative methods on the dataset are reported in the paper for a baseline comparison.

Keywords: Partial video copy · Detection · TV · Dataset · Evaluation

1 Introduction

This paper is interested with the Partial Video Copy Detection (PVCD). It aims to find one or more segments of a reference video which have transformed copies. It is a well-known topic in the computer vision field [12]. The recent works investigate the detection methods robust to the spatial & temporal deformations [5,6,15] or real-time [4,13,19]. A key aspect for any computer vision task is to design public datasets for performance evaluation. A few has been proposed in the literature for the PVCD [8,9,21]. They have been mainly designed from Web videos ensuring realistic degradations. However, this approach raises different problems such as **(i)** a huge user interaction **(ii)** errors in the groundtruth **(iii)** a low-level of scalability **(iv)** an unbalance distribution of test sets **(v)** a difficulty to challenge the methods on a particular detection problem.

The PVCD is inherent to the continuous video broadcasting. An alternative is to process with TV datasets offering meaningful data and having a low-level of noise. This ensures a deeper scalability, a robust groundtruthing with the help of TV metadata and a fine control of video degradations. We propose in this paper a large-Scale TV Dataset for the PVCD, called STVD. It is made public available for the needs of the research community[1]. Section 2 describes the

[1] http://mathieu.delalandre.free.fr/projects/stvd/index.html.

S. Sclaroff et al. (Eds.): ICIAP 2022, LNCS 13233, pp. 388–399, 2022.
https://doi.org/10.1007/978-3-031-06433-3_33

related work. Section 3 presents our protocol and pipeline for the video capture and groundtruthing. Experiments to design the dataset are reported in Sect. 4 with performance evaluation results of representative methods. Section 5 provides conclusions and perspectives. For convenience, Table 1 gives the meaning of the main symbols used in the paper.

Table 1. Main symbols and terms used in the paper

Symbols	Meaning
$\mathbf{t}, \hat{\mathbf{t}}$	the scheduled and detected timestamp for a TV program
$\mathbf{L} \in [\mathbf{L}_{\min}, \mathbf{L}_{\max}]$	$\mathbf{L} = \hat{\mathbf{t}} - \mathbf{t}$ is the latency, $\mathbf{L}_{\min} < \mathbf{0}$, $\mathbf{L}_{\max} > \mathbf{0}$ the min and max
$\mathbf{L}^- < 0, \mathbf{L}^+ > 0$	a negative and positive latency, respectively
$\mathbf{W} = \mathbf{W}^- + \mathbf{W}^+$	the capture window
$\mathbf{D} \in [\mathbf{D}_{\min}, \mathbf{D}_{\max}]$	$\mathbf{D} > 0$ is a program duration, $\mathbf{D}_{\min}$, $\mathbf{D}_{\max}$ the min and max
$\mathbf{T_0}, \ldots, \mathbf{T_6}$	the video degradations and transformations
$\mathcal{S}$	a $\mathbf{T_0}$ sequence starting at $\mathbf{s} = \mathbf{t} + \lvert\mathbf{L}^-\rvert$ and ending at $\mathbf{e} = \mathbf{t} + \mathbf{D} + \mathbf{L}^+$
α, β	the parameters to control the degradation level

2 Related Work

Several datasets have been proposed in the literature for the performance evaluation of the PVCD. They are listed in Table 2. These datasets provide video files with a groundtruth. The groundtruth labels the partial video copies. The datasets can be used to characterize the tasks of video detection or retrieval. They are constituted by two main sets of (i) query and (ii) testing videos.

Table 2. Comparison of datasets for the PVCD performance evaluation

Datasets	TV_2007	CC_WEB	TRECVID	TV_2014	VCDB	SVD	STVD
Reference	[10]	[21]	[16]	[2]	[9]	[8]	Ours
Year	2007	2009	2010	2014	2016	2019	2021
Query videos	100	24	1,608	N/A	28	1,206	243
Positive videos	500	3,481	134	20,000,000	528	10,211	19,280
Negative videos	N/A	9,309	7,866	N/A	100,000	26,927	64,040
Duration (h)	60,000	537	200	380,000	2,030	197	10,660
Annotation cost (m-h)	N/A	N/A	N/A	N/A	700	800	105
Source of capture	TV	Web	Web	TV	Web	Web	TV
Degradation methods	Synthetic	Real	Synthetic	Synthetic	Real	Synthetic	Synthetic
Public available	No	Yes	No	No	Yes	Yes	Yes

The (h), (m-h) and N/A stand for (in hours), (in man-hours) and (not available), respectively.

The testing set groups negative and positive videos. The negative videos are not appearing in the query set. The positive videos contain copies of the queries. Some datasets have a small size [8,16,21]. Another limitation is the unbalanced distribution of positive/negative videos [9,16]. This is explained by the groundtruthing approaches requiring a huge user interaction [8,9]. Several datasets are not public available due to the intellectual property [2,10,16].

The positive videos are queries with degradations. Depending the datasets, the degradations could result from a real noise [9,21] or produced with synthetic methods [2,8,10,16]. The real noise results from the video processing pipeline (i.e., capture, networking, editing). The datasets could be obtained from a TV [2, 10] or a Web [8,9,16,21] capture. As a general trend, the TV capture guaranties a lowest level of noise. Synthetic methods could be applied next for degradation. This allows a fine control for performance evaluation.

The main public dataset in the literature is VCDB [9]. It presents several limitations such as **(i)** an average scalability challenge **(ii)** a weak balance for the positive/negative videos **(iii)** a huge level of noise at the capture making unable to drive the performance evaluation on a particular detection problem.

We propose in this paper a new dataset and protocol for the TV video capture and groundtruthing. This dataset is public available for the needs of the research community[1]. It is the biggest public dataset in the literature with a near 83k videos and having a total duration of 10,660 hours Table 2. Our capture is obtained with a low-level of degradation for a fine performance evaluation. Our protocol and dataset are presented in next Sect. 3.

3 STVD: A Large-scale TV Dataset

We present in this section our protocol to design our large-Scale TV Dataset (STVD). Figure 1 details our pipeline where 3 main components are used. We drive first a TV video capture (**C1**) that extracts positive/negative video candidates. This component processes with TV metadata. This requires a user interaction to constitute the query set and a video detection for verification driven in the component (**C2**). A final component (**C3**) is used for degradation.

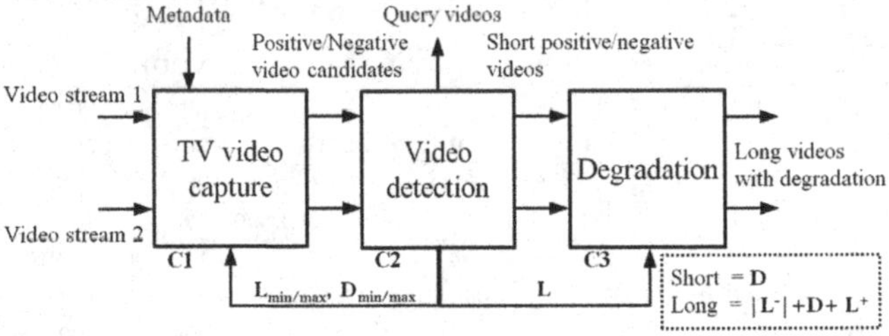

Fig. 1. Pipeline for constructing the STVD dataset

Our component (**C1**) is described into the publication [14]. It is mainly related to the hardware architecture and web crawling topics out of the scope of this paper. Section 3.1 reminds it for short. The components (**C2**) and (**C3**), for the video detection and degradation, constitute the new and key contributions of this paper. They are presented in details in Sects. 3.2 and 3.3, respectively.

3.1 TV Video Capture (C1)

Our component (**C1**) [14] captures the TV programs with a workstation [3,13,14]. This workstation can record daily video files on 8 TV channels simultaneously. We have captured 24 public channels during a period of 3 months. We have obtained a root database composed of 14, 400 h of TV programs at a resolution 240×320 and having a total size of 3.46 TB. The resolution 240×320 constitutes a best tradeoff between the memory cost and video degradation.

We have processed next the TV metadata to capture positive/negative video candidates. These metadata are gathered by a Web crawler. This crawler targets only the daily and weekly programs having the maximum occurrence for the needs of the PVCD. A robust hashing method and user interaction are employed to guaranty a unique hash code for every TV program.

Every program in the metadata is delivered with a timestamp $\mathbf{t}$ to notify when it starts. However, no information is given about the exact location and duration of the repeated content. In addition, the TV broadcasting suffers from latency. To solve these problems, we have triggered the capture to get the jingles only appearing at the kickoff of programs Fig. 2. We have used a window having a size $\mathbf{W} = \mathbf{W}^- + \mathbf{W}^+$. The parameter $\mathbf{W}^-$ guaranties the minimum latency with the TV broadcasting $\mathbf{W}^- \geq |\mathbf{L}_{\min}|$. $\mathbf{W}^+$ is set with the maximum latency and jingle duration $\mathbf{W}^+ \geq \mathbf{D}_{\max} + \mathbf{L}_{\max}$. The capture is then done on the interval $[\mathbf{t} - \mathbf{W}^-, \mathbf{t} + \mathbf{W}^+]$. The $\mathbf{D}_{\max}$, $\mathbf{L}_{\min\,/\,\max}$ parameters are set with a loop-based methodology from the video detection (**C2**) as shown in Fig. 1.

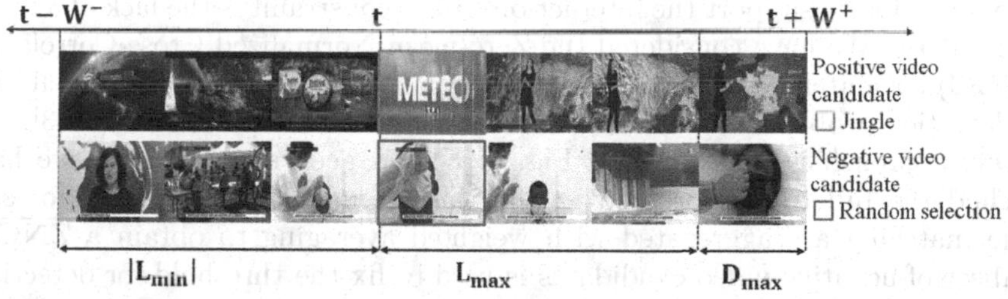

Fig. 2. TV video capture

Our component (**C1**) captures too the negative video candidates. These videos are not supposed to appear in the query and positive sets. For reliability, similar to the strategy deployed in [10] we have used two separate streams for the capture Fig. 1. For a better robustness, we have selected program contents apart of jingles. For every TV program, we have made idle for selection all the sequences where a jingle could appear in the range $[\mathbf{t} - \mathbf{W}^-, \mathbf{t} + \mathbf{W}^+]$. Any valid/not idle video sequence has been split into successive intervals having a duration $\mathbf{W}$. Within any interval, a selection is obtained at $\mathbf{t} = \mathbf{W}^-$ with a random duration $\mathbf{D} \in [\mathbf{D}_{\min}, \mathbf{D}_{\max}]$. For more details, please refer to [14].

3.2 Video Detection (C2)

The component (**C1**) captures positive/negative video candidates. The negative video candidates are made consistent with the strategy deployed at the capture. For the positive video candidates, they have to be used to constitute the query set and validated. This is processed by our component (**C2**) Fig. 1.

A domain knowledge is required to detect the jingles. Indeed, repeated content not related to the programs could appear as the advertising. We have driven the detection with a user interaction and a GUI. Different error-prone cases could occur during the interaction: a jingle could be absent due to errors in the metadata, a jingle could present a different visual content Fig. 3(a) the jingle could have a near-duplicate jingle appearing in a different channel Fig. 3(b) or for a different program within a same channel Fig. 3(c).

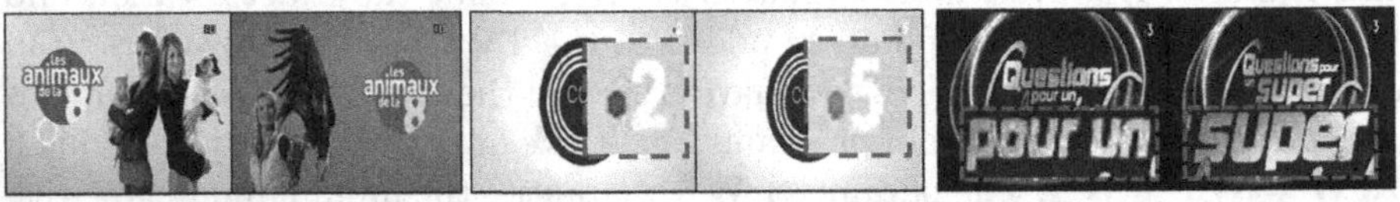

(a) (b) (c)

Fig. 3. Jingles (a) with a different visual content (b) (c) that are near-duplicate

This involves a large amount of visual inspection requiring an automatic video detection to support the interaction. A key constraint is the lack of a training database. We have considered the Zero-mean Normalized Cross-Correlation (**ZNCC**) for matching as a learning-free method [20]. The **ZNCC** fits well with the detection problem as it is robust to noise and contrast-invariant [4,13].

Our approach is illustrated in Fig. 4. For an accurate detection, we have matched the full frames ordered in the time domain. The **ZNCC** scores of frame matching are aggregated with weighted averaging to obtain a $\overline{\mathbf{ZNCC}}$. A subset of negative video candidates is used to fix the threshold for detection. The maximum score gives the timestamp for detection $\hat{\mathbf{t}}$. The difference with the scheduled timestamp is the latency $\mathbf{L} = \hat{\mathbf{t}} - \mathbf{t}$. The overall detection is supported with a GPU and time-efficient implementations suitable for the user interaction. As shown in Sect. 4.1, we have obtained a separability with this approach. The $\mathbf{D}_{\mathrm{min}\,/\,\mathrm{max}}$, $\mathbf{L}_{\mathrm{min}\,/\,\mathrm{max}}$ parameters and the latency $\mathbf{L}$ are used for setting in the components (**C1**) and (**C3**) as shown in Fig. 1.

The user interaction could be time consuming. We have adopted a strategy for bounding. All the hashcodes of TV programs are marked first as unlabelled. The hashcode with the maximum number of occurrence is still selected for inspection. It is labelled when a jingle is detected and validated by the user. The detection cases of Fig. 3 serve to correct the hashcodes. The case (a) splits the hashcode whereas the cases (b) (c) merge two hashcodes. This strategy guaranties a low-level of interaction compared to the other approaches Table 2.

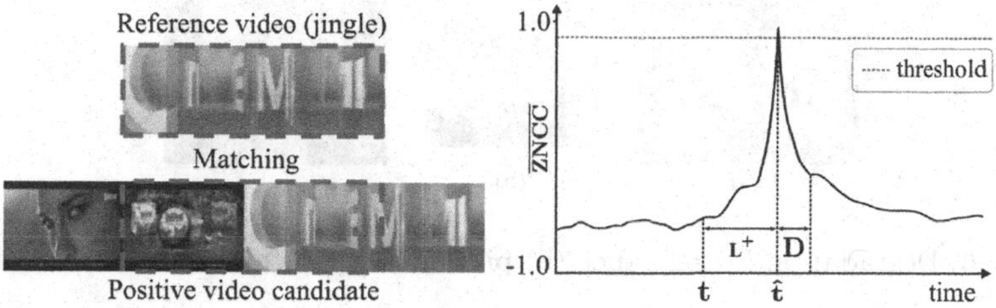

Fig. 4. Video detection

3.3 Video Degradation (C3)

The positive/negative videos obtained with the components (**C1**), (**C2**) correspond to true-life captures with real noisy conditions. For the needs of performance evaluation, a common issue is to apply additional synthetic methods to degrade the videos [2,8,10,16]. By performing transformations, a fine performance evaluation can be handled and more challenging datasets can be designed in order to stress the methods for detection.

Similar to the works [2,8,10,16], we have selected a set of representative methods detailed in Table 3 labelled $\mathbf{T_0}$ to $\mathbf{T_6}$. These are applied both to the positive and negative videos. For the performance evaluation of PVCD methods, we use first a transformation $\mathbf{T_0}$ to get long videos embedding the positive videos. Then, the methods enter in two categories Fig. 5 for pixel attack $\mathbf{T_{1-2}}$ (b) and global transformations $\mathbf{T_{3-5}}$ (c). A final transformation $\mathbf{T_6}$ is used for video speeding.

Table 3. Degradation methods for video transformation

Label	Method	Parameters
$\mathbf{T_0}$	Video cut	Uses the latency distribution to cut segments before/after the video and having a duration $\|\mathbf{L}^-\|, \mathbf{L}^+$, respectively
$\mathbf{T_1}$	Down-scaling	Applies a random down-scaling $\alpha \in [0.1, 0.9]$ to get frames from 24 × 32 up to 216 × 288 for a robust matching with time optimization [18]
$\mathbf{T_2}$	Compression	Processes with a parameter $\frac{1}{\beta}$ with $\beta \in [1, 80]$ applied to the recommended kbps $\in \{140, 280, 420\}$ for capture [1] such as $\frac{1}{\beta} \times$ kbps
$\mathbf{T_3}$	Flipping	Applies randomly (yes/no) a flipping transformation to the video
$\mathbf{T_4}$	Rotating	Applies a random vertical/horizontal rotation $\in \{0, \frac{\Pi}{2}, \Pi, \frac{3}{2}\Pi\}$
$\mathbf{T_5}$	Black border & stretching	Selects an aspect ratio $\frac{w}{h} \in \{0.46, 0.56, 0.63, 0.75, 1.33, 1.6, 1.78, 2.17\}$ to introduce left/right borders ($\frac{w}{h} < 1$) or to stretch the image ($\frac{w}{h} > 1$)
$\mathbf{T_6}$	Video speeding	Speeds down the videos at a FPS $\in [15, 25]$

Fig. 5. Degradation (a) reference (b) pixel attack (c) global transformations

For the needs of the PVCD, short negative/positive videos must be embedded into longest sequences $\mathcal{S}$. We use a specific transformation $\mathbf{T_0}$ in our approach designed with our latency measure $\mathbf{L}$ Fig. 4. $\mathbf{T_0}$ extracts additional left/rigth video segments within the window of size $\mathbf{W}$ Fig. 6(a). The duration of $\mathcal{S}$ must be fixed, we have set $\mathbf{T_0}$ with the latency distribution obtained with the component (**C2**) as illustrated in Fig 1. Considering a short negative video timestamped at $\mathbf{t} = \mathbf{W}^-$ in (**C1**), or a query/jingle detected at $\hat{\mathbf{t}}$ in (**C2**), $\mathcal{S}_\mathbf{i}$ is obtained by cutting a long video at $\mathbf{s_i} = \mathbf{t_i} - |\mathbf{L}^-|$ and $\mathbf{e_i} = \mathbf{t_i} + \mathbf{D_i} + \mathbf{L}^+$ (and with $\hat{\mathbf{t}}_\mathbf{i}$ respectively) with $\mathbf{L}^-, \mathbf{L}^+$ random negative/positive latency values.

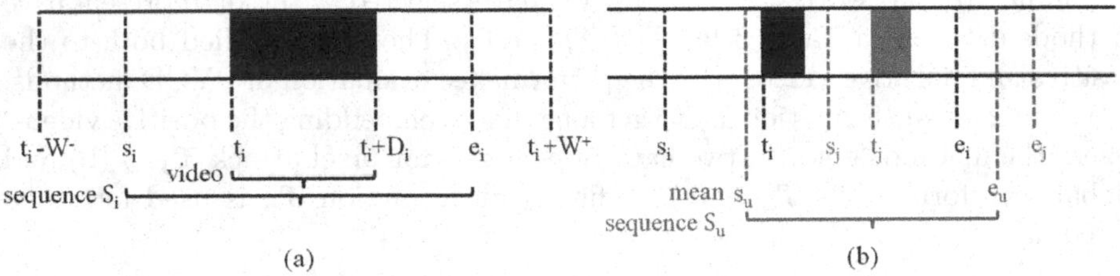

Fig. 6. (a) a sequence $\mathcal{S}$ (b) covering case

A sequence $\mathcal{S}$ could be extracted for any short negative video. Indeed, a selection at $\mathbf{t} = \mathbf{W}^-$ within a window of size $\mathbf{W}$ by (**C1**) cannot result in a covering case while using the latency in $\mathbf{T_0}$. However, such a case could appear with the short positive videos Fig. 6(b). Considering two videos timestamped at $\mathbf{t_i}, \mathbf{t_j}$ with $\mathbf{t_i} < \mathbf{t_j}$, we could met a case where $\mathbf{t_j} - \mathbf{t_i} \ll \mathbf{W}^+$. A mean sequence $\mathcal{S}_\mathbf{u}$ must be computed and preserved if $\forall \mathcal{S}_\mathbf{i} \in \mathcal{S}_\mathbf{u}$, $\mathbf{s_u} < \mathbf{t_i} + \mathbf{D_i} < \mathbf{e_u}$. That is, a long positive video for testing could embed several queries.

We apply next a set of baseline video processing $\mathbf{T_{1-6}}$ for degradation. $\mathbf{T_1}, \mathbf{T_2}$ are set with recommended parameters for robust low-resolution video processing [18] and capture [1]. Two parameters α, β control the level of degradation. $\mathbf{T_3}$ and $\mathbf{T_4}$ apply realistic geometric transformations for video rendering as the flipping and the horizontal/vertical rotations. The aspect ratio parameters in $\mathbf{T_5}$ have been fixed using the standard screen resolutions[2]. $\mathbf{T_6}$ speeds down the videos with predefined FPS similar to [8,21].

[2] For desktop, tablet and phone https://gs.statcounter.com/.

We have combined the degradations $\mathbf{T_0}$ to $\mathbf{T_6}$ to generate the test sets A to F as detailed in Table 4. The test set A gives a root capture while applying only $\mathbf{T_0}$. It is given for the needs of tuning a performance evaluation task. The test sets B and C apply a pixel attack with $\mathbf{T_1}$ and $\mathbf{T_2}$ at two levels of degradation with the control of parameters α, β. The test set B has a low-level of distortion and scalability and constitutes a "hello world" benchmark. The test set C presents a hard pixel attack. The test set D is related to the global transformations with $\mathbf{T_3}$ to $\mathbf{T_5}$ whereas the test set E applies $\mathbf{T_6}$ for video speeding. For storage optimization, $\mathbf{T_1}$ and $\mathbf{T_2}$ are used with predefined parameters α, β ensuring a negligible degradation. At last, the test set F combines the sets C, D and E.

Table 4. Test sets

Test set	$\mathbf{T_0}$	$\mathbf{T_{1-2}}$	$\alpha \in$	$\beta \in$	$\mathbf{T_{3-5}}$	$\mathbf{T_6}$	Description
Set A	√						Root capture for tuning
Set B	√	√	$[0.25, 0.9[$	$[1, 40[$			"Hello world" test set
Set C	√	√	$[0.1, 0.25]$	$[40, 80]$			Pixel attack with scalability
Set D	√	√	0.6	20	√		Global transformations with scalability
Set E	√	√	0.6	20		√	Video speeding with scalability
Set F	√	√	$[0.1, 0.25]$	$[40, 80]$	√	√	Combination of sets C, D and E

4 Experiments

4.1 Dataset and Groundtruthing

We report in this section experiments to generate the dataset with the groundtruth. Table 5 details the dataset organization where the components (**C1**), (**C2**) and (**C3**) have been used to generate the positive/negative videos with degradations.

Table 5. STVD dataset

	Root capture		C1, C2		C3		
	Channels	Duration	Videos	Duration	Test sets	Videos	Duration
Positive videos	8	4,800 h	3,780	6 h	6	19,280	2,515 h
Negative videos	16	9,600 h	12,165	21 h	6	64,040	8,145 h

We have split the root database obtained with (**C1**) into two subsets of 4,800 and 9,600 h for the positive/negative videos. We have captured then a near 3k and 12k positive/negative video candidates with the metadata[3]. The positive video candidates have been processed with the component (**C2**). We have

[3] A full analysis and experiments with the metadata are reported into [14].

extracted 243 distinct jingles with the GUI. They have been matched against the positive video candidates and a subset of negative videos as detailed in Sect. 3.2. We have observed a separability between the interclass/intraclass $\overline{\mathbf{ZNCC}}$ distributions $\in [0.79, 0.90]$ Fig. 7(a). This ensures none false positive case.

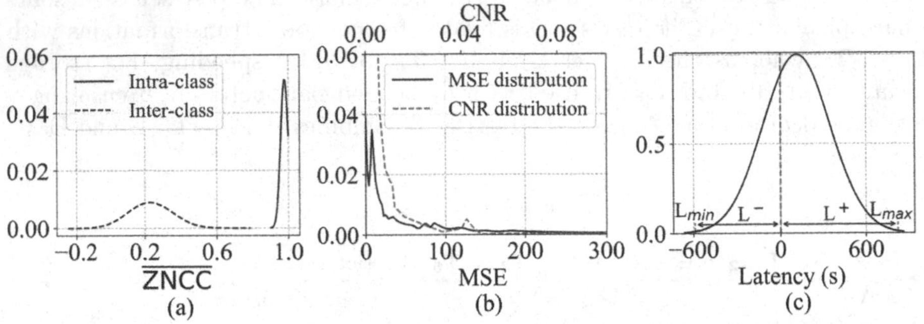

Fig. 7. Distributions of (a) $\overline{\mathbf{ZNCC}}$ (b) MSE/CNR (c) latency

For a further investigation, Fig. 7(b) gives a characterization of the interclass $\overline{\mathbf{ZNCC}}$ distribution in terms of compression noise and contrast deviation. We have employed the standard metrics MSE[4] [7] and CNR (see footnote 4) [11], respectively. The distributions are compact with a MSE < 20 and CNR < 0.01 for most of the matching cases. This reflects the $\overline{\mathbf{ZNCC}}$ distributions obtained in Fig. 7(a) and the sources of degradation for the TV capture.

We have analysed then the two time aspects illustrated in Fig. 4. We have obtained a total video duration of 6 h. Table 5 as the jingles have a short duration $\mathbf{D} \in [1, 25]$ s. For the latency, we have observed an almost-Gaussian distribution with $\mathbf{L} \in [-590, 820]$ seconds at $\pm \mathbf{3}\sigma$ Fig. 7(c). We have used the duration and latency distributions to set the $\mathbf{D}_{\text{min / max}}$, $\mathbf{L}_{\text{min / max}}$ parameters and the random model in $\mathbf{T_0}$ for (**C1**) and (**C3**) as shown in Fig. 1.

We have then applied the component (**C3**) to get the 6 test sets as discussed in Sect. 3.3. For the test sets A, C, D, E and F we have obtained $5 \times 3,780 = 18,900$ and $5 \times 12,165 = 60,825$ positive/negative videos, respectively. The test set B has been generated in balance for a low scalability with a total number of $2 \times 3,780 = 7,560$ videos. We have observed a $\simeq$15% of covering cases with the positive videos Fig. 6(b). We have then obtained a total amount of $\simeq$83k videos composed of 19,280 and 64,040 positive/negative videos Table 5.

Considering the latency distribution Fig. 7(c), the application of $\mathbf{T_0}$ has resulted in an average duration $|\mathbf{L}^-| + \mathbf{L}^+$ of 7.5 min. The total duration of the dataset is 10,660 h with 2,515 h and 8,145 h for the positive/negative videos, respectively Table 5. Each test set C to F contains $\simeq$1,960 h of testing video for scalability competitive with the VCDB dataset [9] Table 2. Considering the all test sets A to F, STVD is the largest dataset of the literature $\times 5$ larger than

[4] Mean Square Error, Contrast Noise Ratio.

VCDB. STVD is made public available (see footnote 1) for the needs of the research community.

4.2 Performance Evaluation

We present in this section performance evaluation results on the STVD dataset of representative PVCD methods [22–24]. These methods process in two steps for key-frame extraction and matching. The key-frame extraction selects candidate frames for matching based on sampling methods [22,23] or temporal features [24]. The matching processes with features (SIFT [24], BRIEF [23] and CNN [22]) and optimization components for the time processing requirement.

We have applied a protocol for a fair comparison. We have normalized the key-frame extraction step within all the methods. The SIFT and BRIEF features are not supporting the global transformations. We have bounded the evaluation to the test sets B and C only. We have characterized the methods in a learning-free/pre-trained mode. Only the SIFT and BRIEF features of query frames have been stored for comparison. The CNN features have been obtained from a pre-trained VGG16 network from the ILSVRC dataset [17]. We have also removed the optimization components for a strongest accuracy. The F_1 score has been used as it is common to characterize the PVCD methods [4–6,9,15,19].

We have evaluated first the method [23] on the test set B. We have obtained a score $F_1 = 0.98$ highlighting the "hello world" ability. Further experiments have been investigated on the test set C Fig. 8. We have constituted first a subset with in balance 3k +3k positive/negative videos. Figure 8(a) gives the F_1 scores against the normalized thresholds for all the methods. We have obtained optimum scores $F_1 \in [0.73, 0.83]$ with a top $F_1 = 0.83$ for the method [23]. A gap $\simeq$0.15 appears for [23] between the test sets B and C due to the pixel attack. Figure 8(b) reports the results of the top methods [22,23] on the full test set C while increasing the negative videos up to 12k. We have observed a gap $\simeq$0.25 for the F_1 score due to the scalability with a better robustness for the CNN features [22].

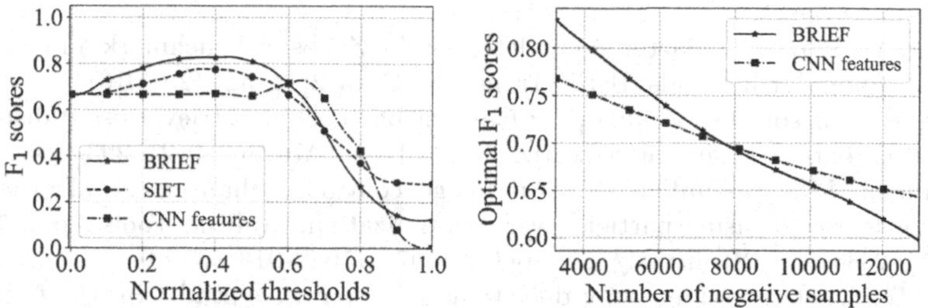

Fig. 8. F_1 scores on the test set C (a) comparison of methods [22–24] (b) performance with scalability for [22,23]

5 Conclusions and Perspectives

We propose in this paper a new dataset to evaluate the PVCD methods called STVD. This dataset is designed with a protocol ensuring a scalable capture and robust groundtruthing. STVD is today the largest public dataset on the task. It covers a near 83k videos for a total duration of 10,660 h. Performance evaluation results of representative methods on the dataset are reported in the paper for a baseline comparison. A key issue next will be to promote the dataset in the research community. Additional test sets should be included to address specific PVCD tasks such as the real-time or near-duplicate detection.

References

1. AVerMedia: AVerMedia capture card software developement kit. Tech. Rep. 4.2, AVerMedia Technologies, Inc. (2015). http://www.avermedia.com
2. Chenot, J., Daigneault, G.: A large-scale audio and video fingerprints-generated database of TV repeated contents. In: Workshop on Content-Based Multimedia Indexing (CBMI), pp. 1–6 (2014)
3. Delalandre, M.: A workstation for real-time processing of multi-channel TV. In: International Workshop on AI for Smart TV Content Production (AI4TV), pp. 53–54 (2019)
4. Guzman-Zavaleta, Z.J., Feregrino-Uribe, C.: Partial-copy detection of non-simulated videos using learning at decision level. Multim. Tools Appl. **78**(2), 2427–2446 (2018). https://doi.org/10.1007/s11042-018-6345-2
5. Han, Z., He, X., Tang, M., Lv, Y.: Video similarity and alignment learning on partial video copy detection. In: ACM International Conference on Multimedia (MM), pp. 4165–4173 (2021)
6. Hu, Y., Mu, Z., Ai, X.: STRNN: end-to-end deep learning framework for video partial copy detection. J. Phys. Conf. Ser. **1237**(2), 022112 (2019)
7. Ieremeiev, O., Lukin, V., Okarma, K., Egiazarian, K.: Full-reference quality metric based on neural network to assess the visual quality of remote sensing images. Remote Sens. **12**(15), 2349(2020)
8. Jiang, Q., et al.: SVD: a large-scale short video dataset for near-duplicate video retrieval. In: International Conference on Computer Vision (ICCV), pp. 5280–5288 (2019)
9. Jiang, Y., Wang, J.: Partial copy detection in videos: a benchmark and an evaluation of popular methods. IEEE Trans. Big Data **2**(1), 32–42 (2016)
10. Joly, A., Buisson, O., Frélicot, C.: Content-based copy retrieval using distortion-based probabilistic similarity search. IEEE Trans. Multim. **9**(2), 293–306 (2007)
11. Kanmani, M., Narsimhan, V.: An image contrast enhancement algorithm for grayscale images using particle swarm optimization. Multim. Tools Appl. **77**(18), 23371–23387 (2018). https://doi.org/10.1007/s11042-018-5650-0
12. Law-To, J., et al.: Video copy detection: a comparative study. In: Conference on Image and Video Retrieval (CIVR), pp. 371–378 (2007)
13. Le, V., Delalandre, M., Conte, D.: Real-time detection of partial video copy on TV workstation. In: International Conference on Content-Based Multimedia Indexing (CBMI), pp. 1–6 (2021)

14. Le, V., Delalandre, M., Conte, D.: Une large base de données pour la détection de segments de vidéos tv. In: Journées Francophones des Jeunes Chercheurs en Vision par Ordinateur (ORASIS) (2021)
15. Liu, X., Feng, X., Pan, P.: GANN: a graph alignment neural network for video partial copy detection. In: Conference on Big Data Security on Cloud (BigDataSecurity), Conference on High Performance and Smart Computing (HPSC), Conference on Intelligent Data and Security (IDS), pp. 191–196 (2021)
16. Over, P., et al.: TRECVID 2010 - an overview of the goals, tasks, data, evaluation mechanisms and metrics. NIST (2010). https://www.nist.gov/
17. Russakovsky, O., et al.: ImageNet large scale visual recognition challenge. Int. J. Comput. Vis. (IJCV) **115**(3), 211–252 (2015)
18. Su, J., Vargas, D., Sakurai, K.: One pixel attack for fooling deep neural networks. Trans. Evol. Comput. (TEVC) **23**(5), 828–841 (2019)
19. Tan, W., Guo, H., Liu, R.: A fast partial video copy detection using KNN and global feature database. In: Preprint arXiv. No. 2105.01713 (2021)
20. Wang, X., Wang, X., Han, L.: A novel parallel architecture for template matching based on zero-mean normalized cross-correlation. IEEE Access **7**, 186626–186636 (2019)
21. Wu, X., Ngo, C., Hauptmann, A., Tan, H.: Real-time near-duplicate elimination for web video search with content and context. IEEE Trans. Multim. **11**(2), 196–207 (2009)
22. Zhang, C., et al.: Large-scale video retrieval via deep local convolutional features. Adv. Multim. **7862894**, 1687–5680 (2020)
23. Zhang, Y., Zhang, X.: Effective real-scenario video copy detection. In: International Conference on Pattern Recognition (ICPR), pp. 3951–3956 (2016)
24. Zhu, Y., Huang, X., Huang, Q., Tian, Q.: Large-scale video copy retrieval with temporal-concentration sift. Neurocomputing **187**, 83–91 (2016)

Poker Bluff Detection Dataset Based on Facial Analysis

Jacob Feinland, Jacob Barkovitch(✉), Dokyu Lee, Alex Kaforey, and Umur Aybars Ciftci

Binghamton University, Binghamton, NY, USA
{jfeinla1,jbarkov1,akafore1,uciftci}@binghamton.edu

Abstract. Unstaged data with people acting naturally in real-world scenarios is essential for high-stakes deception detection (HSDD) research. Unfortunately, multiple HSDD studies involve staged scenarios in controlled settings with subjects who were instructed to lie. Using in-the-wild footage of subjects and analyzing facial expressions instead of invasive tracking of biological processes enables the collection of real-world data. Poker is a high-stakes game involving a deceptive strategy called bluffing and is an ideal research subject for improving HSDD techniques. Videos of professional poker tournaments online provide a convenient data source. Because proficiency in HSDD generalizes well for different high-stakes situations, findings from poker bluff detection research have the potential to transfer well to other more practical HSDD applications like interrogations and customs inspections. In the hopes of encouraging additional research on real-world HSDD, we present a novel in-the-wild dataset for poker bluff detection. To verify the quality of our dataset, we test its regression accuracy and achieve a Mean Square Error of 0.0288 with an InceptionV3 model.

Keywords: Deception detection · Poker bluff detection · Facial expression recognition

1 Introduction

Facial expression recognition is a crucial task to explore for understanding nonverbal communication and how it affects people. Its applications include entertainment [4], marketing [2], surveillance [29], and psychology [20]. As this task has grown in popularity due to its many potential applications, researchers have attempted using facial expression recognition to interpret nonverbal cues in human-computer interactions [16,17].

Research on non-contact high-stakes deception detection (HSDD) using facial expression recognition is one of these attempts to interpret nonverbal cues [26]. Facial expression recognition's potential for non-contact deception detection has propelled research in this field since the applications in police interrogations and airport security would benefit greatly from it [10]. To our knowledge, there is no publicly accessible database of real-world data in these applications. Much

S. Sclaroff et al. (Eds.): ICIAP 2022, LNCS 13233, pp. 400–410, 2022.
https://doi.org/10.1007/978-3-031-06433-3_34

of the existing deception detection research relies on data from staged scenarios and controlled settings [8,11,33].

To overcome the scarcity of real-world HSDD data, we present a new HSDD dataset based on online poker footage. Poker is a type of card game where each player is given a set number of cards, called a hand, and they bet (usually money) when they think their cards are strong [25]. After everyone bets, the player that can make the strongest combination out of their own cards and the shared cards (or community cards) wins the round and all the money in the current betting pool (also called the pot). In the process, it is common for players to bluff, or bet large amounts of money with a weak hand to feign having a strong hand and deceive others. This increases the likelihood that opponents will fold, or back out to avoid losing money because they believe their hands are unlikely to win. In this paper, we focus on the common Texas hold'em variant of poker, and further uses of the term "poker" in this paper refer to this variant. In professional Texas hold'em poker tournaments, the act of bluffing becomes an example of high-stakes deception because the amount of money in the pot tends to be very large [25].

Poker tournament footage provides certain inherent advantages compared to other sources of HSDD data and yields results that we expect to transfer well to other HSDD applications. We expect this because a person's skill in HSDD has been shown to extend across different fields of HSDD [9]. In other words, someone who is skilled at seeing through deception in one high-stakes scenario is typically also skilled at seeing through deception in a different high-stakes scenario. This implies the HSDD task is similar for different applications. As a result, we believe HSDD research based on poker tournament footage will support the development of broadly applicable HSDD methods.

One major advantage poker tournament footage has over other HSDD sources is the abundance of publicly accessible data online. This data does not involve sensitive information (as opposed to criminal interrogation footage), making it much easier to gain access to. The non-invasive nature of recording video footage, unlike monitoring heart rate and breathing rate, enables the inexpensive and efficient collection of much more data from many players. Because poker is a very popular game and is akin to a spectator sport, tournaments are held frequently around the world and many are recorded and uploaded to free video sharing sites. The game information displayed on-screen for spectators of tournament broadcasts (but hidden from players) provides convenient deception labels with each player's own apparent likelihood of winning and their bets. Because poker tournament footage is so accessible, our dataset could easily be expanded as needed.

Another major advantage of poker tournament footage is that it provides natural data. This in-the-wild footage has a variety of face angles, lighting conditions, face resolutions, and occlusion levels. Access to training data with varied conditions that match the input data is important for creating a robust model that is accurate in real-world situations [7,18,22]. Additionally, bluffs in poker tournaments are more natural than the staged deception scenarios used in studies where subjects are instructed to lie [18,33]. Non-invasive means of recording data also allow subjects to behave more naturally [18].

In this paper, by analyzing bluffs in poker footage, we provide a high-stakes deception detection dataset and a systematic method for expanding it on demand. In summary, the contributions of this paper are as follows:

- We present a new in-the-wild dataset for high-stakes deception detection using poker tournament footage.
- We propose a partially-automated labeling process to expand the dataset as needed.
- We prove the usefulness of the proposed dataset for the bluff detection task.

2 Related Works

A variety of deception detection approaches have been proposed by researchers over the years [1,12,18,33]. One commonly used deception detection approach is using the polygraph machine. However, this has been widely criticized in the academic community for being inaccurate up to 90% of the time [12]. Despite this, the polygraph is still considered relevant credible evidence in legal cases and can result in up to 351 false positives out of 950 people [12].

Other researchers have proposed non-invasive or minimally-invasive deception detection data sources including video [1,11], thermal imaging [33], audio [19], text transcripts [32], or a combination of these modalities [23,30]. In [18], a custom arm-band that measures heart rate and skin conductance was designed to reduce subjects' discomfort during monitoring as they play a deception-based card game.

In [11], the researchers created a video-based dataset for poker fold prediction, where pairs of amateur players played a simplified poker variant online with their faces and bets digitally recorded. This provided precise timestamps for game events based on mouse clicks. Predictions for multiple video frames over a certain duration are compared to make a final prediction for whether a player folds. Because the subjects did not play for money, this dataset cannot be used for HSDD.

Despite the similarity between deception in interrogations and bluffs in poker as examples of HSDD, limited prior research has focused on high-stakes poker for deception detection. There is a need for a large in-the-wild HSDD dataset, and our method of generating a poker tournament dataset aims to fill this niche.

3 Methods/Technical Details

Our process for bluff labeling and detection is as follows: first, we use optical character recognition (OCR) to obtain on-screen data from video input, such as players' hands, community cards, bets, and the pot amount. This data is then used in a Monte Carlo simulation [24] to infer whether a player is bluffing or not. Next, face recognition matches each detected face image with the corresponding bluff label. This labeling process can be seen in Fig. 1. We use the dataset to train a convolutional neural network (CNN) for detecting bluffs.

3.1 Parsing Game Information

To obtain sufficient detail for facial-expression-based deception detection, we use online videos with high-definition resolution. Additionally, the footage must include a very clear poker game state display containing all the game information, including each player's hand, who is betting, every bet amount, and the pot amount. The visual format must also stay consistent throughout the game.

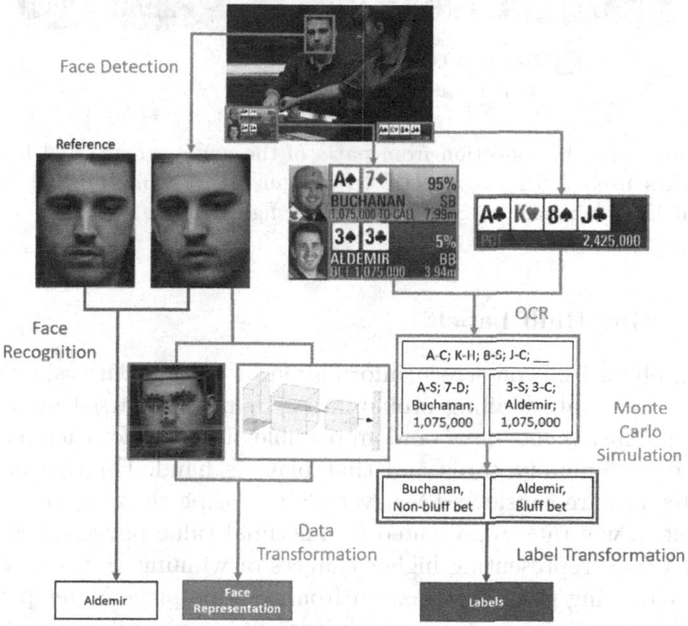

Fig. 1. Face images are extracted from a video frame and compared to reference images of the players to identify the face by name. Game display information is parsed from the screen using OCR. A Monte Carlo simulation generates win rates from game data. Bluff event labels are transformed into frame-by-frame labels.

The uniform formatting of the game state display is crucial for the labeling process since it allows for reliable automatic parsing of textual game data. The OCR algorithm Tesseract [13] is used for this task. To spot each bluff event, or change in the game state, relevant regions of video frames are cropped for cards and other game information based on the format of the video's game display. To achieve this, bounding boxes were manually set for each unique game state format, as seen in Fig. 2. OCR is applied to on-screen text while OpenCV template matching [3] is used to recognize card suits.

The parsing process generates a list of changes in on-screen data. Using this automatically generated list, we quickly scanned through each video and verified the accuracy of the labels in the list, correcting any rank or suit that was misread. Next, we removed any consecutive listings of identical events caused by OCR misreads. For games where the hands are not displayed on-screen until the end of the game, we manually inputted these hands. On-screen game information besides bets were also removed.

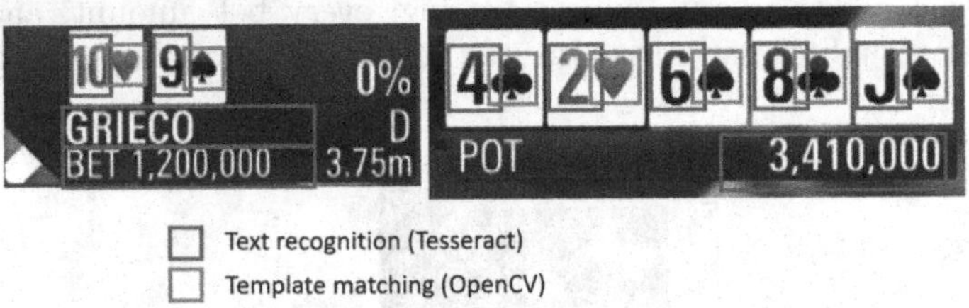

Fig. 2. Example of data collection from parts of the game screen. Red boxes signify boundary boxes used for Tesseract [13] to recognize text while green boxes are for template matching to recognize card type. (Color figure online)

3.2 Generating Bluff Labels

For labeling bluffs from on-screen information, a Monte Carlo simulation [24] is used to approximate bluffing probabilities. Using the parsed information as inputs, the simulation generates random possible outcomes for each player based on the shown community cards and that player's hand. For this paper, 4000 random outcomes are considered for every label. Using the outcomes, the simulation outputs a win rate, represented as a decimal value between zero and one, with higher values representing higher chances of winning. Since this output is only calculated using game information from a single player's perspective, it is a reasonable approximation to a professional poker player's estimate of their own chances of winning. Therefore, when a player makes a high bet with a low predicted win rate, this can be considered a bluff.

The labeled bluff data can be paired with corresponding cropped and aligned face images through face recognition. In this paper, InsightFace [31] is used to identify players' faces by applying the state-of-the-art face recognition model ArcFace [6]. We first used InsightFace to extract random faces from the video via the pre-trained ArcFace model. We then manually identified 10–20 images of each player's face with as much variation in head angle, lighting, and facial expressions as possible to fine-tune the model. After training, InsightFace can identify players' faces and output cropped and aligned face images, each labeled with the frame number and name. These images are matched to the bluff labels using the names displayed on-screen. Figure 3 is an example of face recognition in this process.

Fig. 3. Face recognition predictions against the player's faces in validation set. Act = actual win rate at frame, pred = model prediction of win rate at frame.

3.3 Dataset Details

For the proposed dataset, we identified bluff events from 30 min of footage (about 50,000 frames). Each event means a player in the video made a bet. The face images of a given player from 5 s before to 2 s after each of their bluff events are labeled with that event's Monte Carlo value. The resulting dataset includes a total of 15 unique faces with 2,320 face images. These images are labeled based on 65 events in total.

4 Bluff Detection Analysis

In order to test the effectiveness of our dataset, we used it to fine-tune an InceptionV3 [27] CNN model that was pre-trained on ImageNet [5]. For our research, we converted InceptionV3 to a regression model since the outputs of our Monte Carlo simulations are continuous values between 0 and 1, as described in Sect. 3.2. This leads to an increase of output precision since a regression model on our dataset outputs a percent certainty instead of a yes or no answer which a classification model would output. This required us to use mean squared error to compute the loss and to change the number of output features to one. We also used the Adam optimizer [14] with a learning rate of 0.001 and a weight decay of 5×10^{-6}.

Our results were obtained by training the model for 20 epochs with a batch size of 64. Figure 4 compares the distributions of actual bluff values and predicted values. We use 57% of the data for the training set and 43% of the data for the validation set. Our baseline model achieves a best Mean Square Error of 0.0288 on the validation set. The median error is 19.22% with an average relative error (RE) of 20.31%. Figure 5 shows a clear linear association between bluff predictions and true values, with the spread indicating the degree of error. Our results show that our data is reliable since the model achieves high accuracy for predicting bluff likelihood based on facial analysis. Equation 1 shows the formula we use to calculate RE. v_P is the set of predicted values, v_E is the set of expected (true) values, and δ is the set of relative errors. Each item i correlates to the

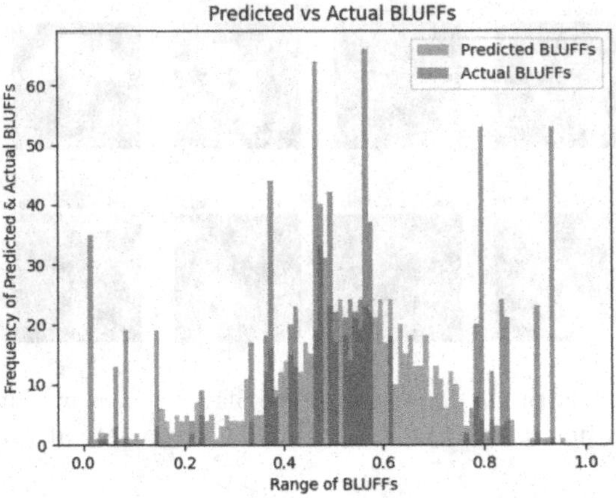

Fig. 4. Comparison of our baseline model's predictions plotted against the actual values in a histogram. The figure shows how the model can predict the full range of bluff probabilities.

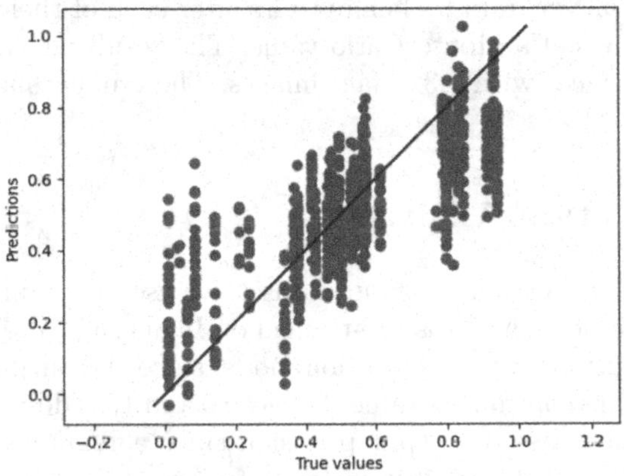

Fig. 5. True bluff values vs predicted bluff values. The y-axis are the outputs from the model on each image and the x-axis are the actual bluff values of each image. The blue line shows the best fit based on the points. The upward trend shows that the model can differentiate between a high and low probability of bluffing based on facial images.

value at that index in each set from 0 to $|v| - 1$, where $|v|$ is the total number of values. Equation 2 shows the formula for average RE where it is the sum of each RE divided by the length of the RE set since $|\delta| = |v|$.

$$\delta_i = \left| \frac{v_{P_i} - v_{E_i}}{v_{E_i}} \right| \quad (1)$$

$$\overline{\delta} = \frac{1}{|v|} \sum_{i=0}^{|v|-1} \delta_i \quad (2)$$

The heatmap in Fig. 6 shows the areas of the face that are most important for the model to produce its regression result, with red indicating the most important regions and purple indicating the least important. Many of the images indicate high importance in the forehead and moderate importance in the nose, which may result from brow wrinkling as a significant indicator of bluffing and nose wrinkling as a minor factor. The edges and corners of the image are purple showing that the model is mainly looking at the faces and not the surroundings. The heatmap is generated using the Score-CAM function for CNNs from [28] and [21].

Fig. 6. Heatmap from Inception V3 parameters. Red = area of high importance for predictions. Purple = area of low importance for predictions. (Color figure online)

In addition to InceptionV3 we test our dataset using the deep CNN AlexNet [15] to show that other state-of-the-art models can also have good accuracy and similar results on our data. This shows that we have high-quality data since multiple models can extract important features from our dataset. Figure 7 shows the comparison of the relative errors from the two models. The graphs show that both models achieve similar low error rates and perform about the same on our dataset. Both model's average REs are within 16% of each other.

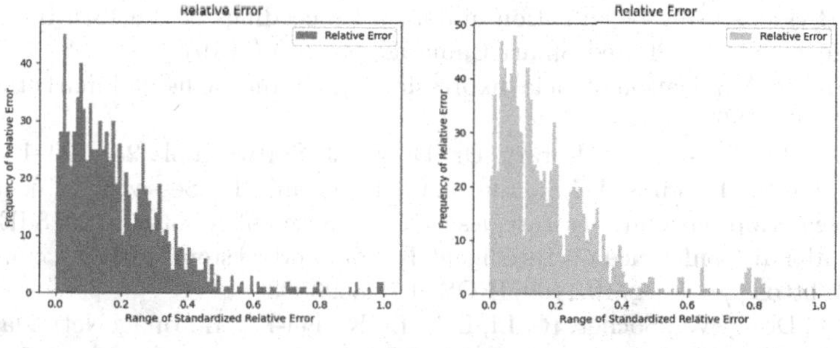

Fig. 7. Comparison of relative errors for InceptionV3 (left), and AlexNet (right). The similar low relative error of both models show that models besides InceptionV3 can learn from our dataset.

5 Conclusion and Future Work

This paper demonstrates the value of using professional poker tournament footage to study high-stakes deception detection and provides a bluff detection dataset with a unique labeling system. Since no large, diverse, and publicly available video-based bluff detection dataset yet exists, the creation of such a dataset is essential for research in this field. Using the partially-automated labeling system presented in this paper, expanding our dataset to achieve this goal would be relatively simple. An important consideration for the extension of our dataset is acquiring data with equal representation by gender and race. This is an inherent drawback of many professional poker tournaments: the players are predominantly white men so other genders and racial groups are underrepresented. By selecting a large number of videos, manually creating more game information formats to include more tournaments, and carefully selecting players from videos, this large and diverse bluff detection dataset is possible.

Binary classification could be an intuitive way to compare bluff bets and non-bluff bets. To categorize bets as bluffs or not, manually labeled classification data could be used for setting a bluff threshold. By analyzing win rates and bet amounts for different hands, the bluff threshold used for automatic labeling can be set dynamically. Bet to pot ratio or bet to player's chip ratio could also be incorporated to further increase the reliability of the bluff threshold.

One way to make our research more useful in real-world HSDD applications would be by developing a simple-to-use HSDD application with real-time functionality. Because HSDD is a transferrable skill for different scenarios, police interrogations and airport security could benefit from a strong model based on our dataset that rapidly determines the likelihood that a subject is being deceptive.

References

1. Avola, D., Cinque, L., Foresti, G.L., Pannone, D.: Automatic deception detection in RGB videos using facial action units. In: Proceedings of the 13th International Conference on Distributed Smart Cameras, pp. 1–6 (2019)
2. Barreto, A.: Application of facial expression studies on the field of marketing. Brain Face **9**, 163–189 (2017)
3. Bradski, G.: The openCV library. Dr. Dobb's J. Softw. Tools **25**, 120–123 (2000)
4. Cosentino, S., Randria, E.I.S., Lin, J.Y., Pellegrini, T., Sessa, S., Takanishi, A.: Group emotion recognition strategies for entertainment robots. In: 2018 IEEE/RSJ International Conference on Intelligent Robots and Systems (IROS), pp. 813–818 (2018). https://doi.org/10.1109/IROS.2018.8593503
5. Deng, J., Dong, W., Socher, R., Li, L.J., Li, K., Fei-Fei, L.: ImageNet: a large-scale hierarchical image database. In: 2009 IEEE Conference on Computer Vision and Pattern Recognition, pp. 248–255. IEEE (2009)
6. Deng, J., Guo, J., Xue, N., Zafeiriou, S.: ArcFace: additive angular margin loss for deep face recognition. In: Proceedings of the IEEE/CVF Conference on Computer Vision and Pattern Recognition (CVPR), June 2019

7. Dhall, A., Goecke, R., Joshi, J., Sikka, K., Gedeon, T.: Emotion recognition in the wild challenge 2014: baseline, data and protocol. In: Proceedings of the 16th International Conference on Multimodal Interaction, pp. 461–466. ICMI 2014, Association for Computing Machinery, New York, NY, USA (2014)
8. Dubey, N.S.S.: Deception detection using artificial neural network and support vector machine. In: 2018 Second International Conference on Electronics, Communication; Aerospace Technology (ICECA), pp. 1205–1208 (2018)
9. Frank, M.G., Ekman, P.: The ability to detect deceit generalizes across different types of high-stake lies. J. Person. Soc. Psychol. **72**(6), 1429–1439 (1997). https://doi.org/10.1037/0022-3514.72.6.1429
10. Geisheimer, J., Greneker, E.: A non-contact lie detector using radar vital signs monitor (RVSM) technology. IEEE Aerosp. Electron. Syst. Mag. **16**(8), 10–14 (2001). https://doi.org/10.1109/62.942213
11. Vinkemeier, D., Valstar, M., Gratch, J.: Predicting folds in poker using action unit detectors and decision trees. In:2018 13th IEEE International Conference on Automatic Face & Gesture Recognition (FG 2018), pp. 504–511 (2018). https://doi.org/10.1109/FG.2018.00081
12. Gregory, D.D.B.T.: Evolution flies. Science (1999)
13. Kay, A.: Tesseract: an open-source optical character recognition engine. Linux J. **2007**(159), 2 (2007)
14. Kingma, D.P., Ba, J.: Adam: a method for stochastic optimization. In: Bengio, Y., LeCun, Y. (eds.) 3rd International Conference on Learning Representations, ICLR 2015, San Diego, CA, USA, 7–9 May 2015, Conference Track Proceedings (2015). http://arxiv.org/abs/1412.6980
15. Krizhevsky, A., Sutskever, I., Hinton, G.E.: ImageNet classification with deep convolutional neural networks. In: Pereira, F., Burges, C.J.C., Bottou, L., Weinberger, K.Q. (eds.) Advances in Neural Information Processing Systems, vol. 25, Curran Associates, Inc. (2012)
16. Le, T.L., Dong, V.T.: Toward a vietnamese facial expression recognition system for human-robot interaction. In: The 2011 International Conference on Advanced Technologies for Communications (ATC 2011), pp. 252–255 (2011)
17. Liu, Z., et al.: A facial expression emotion recognition based human-robot interaction system. IEEE/CAA J. Autom. Sin. **4**(4), 668–676 (2017)
18. Liu, Z., Luo, P., Wang, X., Tang, X.: Deep learning face attributes in the wild. In: 2015 IEEE International Conference on Computer Vision (ICCV), pp. 3730–3738 (2015). https://doi.org/10.1109/ICCV.2015.425
19. Mathur, L., Matarić, M.J.: Unsupervised audio-visual subspace alignment for high-stakes deception detection. In: ICASSP 2021–2021 IEEE International Conference on Acoustics, Speech and Signal Processing (ICASSP), pp. 2255–2259 (2021)
20. Moetesum, M., Aslam, T., Saeed, H., Siddiqi, I., Masroor, U.: Sketch-based facial expression recognition for human figure drawing psychological test. In: 2017 International Conference on Frontiers of Information Technology (FIT), pp. 258–263 (2017). https://doi.org/10.1109/FIT.2017.00053
21. Ozbulak, U.: PyTorch CNN visualizations (2019). https://github.com/utkuozbulak/pytorch-cnn-visualizations
22. Payne, G.P.J.: The Hawthorne effect. In: Key Concepts in Social Research, pp. 108–111. SAGE Publications, Ltd. (2004). https://doi.org/10.4135/9781849209397.n22
23. Rill-Garcia, R., Jair Escalante, H., Villasenor-Pineda, L., Reyes-Meza, V.: High-level features for multimodal deception detection in videos. In: Proceedings of the IEEE/CVF Conference on Computer Vision and Pattern Recognition (CVPR) Workshops, June 2019

24. Samik, R.: Introduction to Monte Carlo simulation. In: 2008 Winter Simulation Conference, pp. 91–100 (2008). https://doi.org/10.1109/WSC.2008.4736059
25. State of California Department of Justice: Texas hold'em. https://oag.ca.gov/sites/all/files/agweb/pdfs/gambling/BGC_texas.pdf. Accessed 15 Sep 2021
26. Su, L., Levine, M.D.: High-stakes deception detection based on facial expressions. In: 2014 22nd International Conference on Pattern Recognition, pp. 2519–2524 (2014). https://doi.org/10.1109/ICPR.2014.435
27. Szegedy, C., Vanhoucke, V., Ioffe, S., Shlens, J., Wojna, Z.: Rethinking the inception architecture for computer vision. CoRR abs/1512.00567 (2015). http://arxiv.org/abs/1512.00567
28. Wang, H., Du, M., Yang, F., Zhang, Z.: Score-CAM: improved visual explanations via score-weighted class activation mapping. CoRR abs/1910.01279 (2019). http://arxiv.org/abs/1910.01279
29. Wang, Q., Jia, K., Liu, P.: Design and implementation of remote facial expression recognition surveillance system based on PCA and KNN algorithms. In: 2015 International Conference on Intelligent Information Hiding and Multimedia Signal Processing (IIH-MSP), pp. 314–317 (2015). https://doi.org/10.1109/IIH-MSP.2015.54
30. Wu, Z., Singh, B., Davis, L., Subrahmanian, V.: Deception detection in videos. In: Proceedings of the AAAI Conference on Artificial Intelligence, vol. 32(1), April 2018
31. Guo, J., Deng, J., An, X., Yu, J.: Insightface. https://github.com/deepinsight/insightface. Accessed 21 Aug 2020
32. Zhou, L., Twitchell, D.P., Qin, T., Burgoon, J.K., Nunamaker, J.F.: An exploratory study into deception detection in text-based computer-mediated communication. In: Proceedings of the 36th Annual Hawaii International Conference on System Sciences, 2003, p. 10. IEEE (2003)
33. Rajoub, B.A., Zwiggelaar, R.: Thermal facial analysis for deception detection. IEEE Trans. Inf. Forens. Secur. **9**(6), 1015–1023 (2014)

Engagement Detection with Multi-Task Training in E-Learning Environments

Onur Copur[1(✉)], Mert Nakıp[2], Simone Scardapane[1], and Jürgen Slowack[3]

[1] Sapienza University of Rome, 00185 Rome, Italy
onurcopur12@gmail.com, simone.scardapane@uniroma1.it
[2] Institute of Theoretical and Applied Informatics, Polish Academy of Sciences, 44-100 Gliwice, Poland
mnakip@iitis.pl
[3] Barco NV, 8500 Kortrijk, Belgium
jurgen.slowack@barco.com

Abstract. Recognition of user interaction, in particular engagement detection, became highly crucial for online working and learning environments, especially during the COVID-19 outbreak. Such recognition and detection systems significantly improve the user experience and efficiency by providing valuable feedback. In this paper, we propose a novel Engagement Detection with Multi-Task Training (ED-MTT) system which minimizes mean squared error and triplet loss together to determine the engagement level of students in an e-learning environment. The performance of this system is evaluated and compared against the state-of-the-art on a publicly available dataset as well as videos collected from real-life scenarios. The results show that ED-MTT achieves 6% lower MSE than the best state-of-the-art performance with highly acceptable training time and lightweight feature extraction.

Keywords: Engagement detection · Activity recognition · E-learning · Triplet loss · Multi-task training

1 Introduction

During the COVID-19 outbreak, nearly all of the learning activities, as other meeting activities, transferred to online environments [32]. Online learners participate in various educational activities including reading, writing, watching video tutorials, online exams, and online meetings. During the participation in these educational activities, participants show various engagement levels, e.g. boredom, confusion, and frustration [11]. To provide feedback to both instructors and students, online educators need to detect their online learners' engagement status precisely and efficiently. For example, the teacher can adapt and make lessons more interesting by increasing interaction, such as asking questions to involve non-interacting students. Since, in e-learning environments, students are not speaking most of the time, the engagement detection systems should extract

S. Sclaroff et al. (Eds.): ICIAP 2022, LNCS 13233, pp. 411–422, 2022.
https://doi.org/10.1007/978-3-031-06433-3_35

valuable information from only visual input [29]. This makes the problem non-trivial and subjective because annotators can perceive different engagement levels from the same input video. The reliability of the dataset labels is a big concern in this setting but often is ignored by the current methods [29,30,32]. Because of this, deep learning models overfit to the uncertain samples and perform poorly on validation and test sets.

In this paper, we propose a system called Engagement Detection with Multi-Task Training (ED-MTT)[1] to detect the engagement level of the participants in an e-learning environment. The proposed system first extracts features with *OpenFace* [2], then aggregates frames in a window for calculating feature statistics as additional features. Finally, it uses Bidirectional Long Short-Term memory (Bi-LSTM) [13] unit for generating vector embeddings from input sequences. In this system, we introduce a triplet loss as an auxiliary task and design the system as a multi-task training framework by taking inspiration from [22], where self-supervised contrastive learning of multi-view facial expressions was introduced. The reason for the triplet loss usage is based on the ability to utilize more elements for training via the combination of original samples. In this way, it avoids overfitting and makes the feature representation more discriminative [9]. To the best of our knowledge, this is a novel approach in the context of engagement detection. The key novelty of this work is the multi-task training framework using triplet loss together with Mean Squared Error (MSE). The main advantages of this approach are as follows:

- Multi-task training with triplet and MSE losses introduces an additional regularization and reduces possibly over-fitting due to very small sample size.
- Using triplet loss mitigates the label reliability problem since it measures relative similarity between samples.
- A system with lightweight feature extraction is efficient and highly suitable for real-life applications.

Furthermore, we evaluate the performance of ED-MTT on a publicly available "Engagement in The Wild" dataset [7], which is comprised of separated training and validation sets. In our experimental work, we first analyze the importance of feature sets to select the best set of features for the resulting trained ED-MTT system. Then, we compare the performance of ED-MTT with 9 different works [1,5,15,20,24,25,27,31,32] from the state-of-the-art which will be reviewed in the next section. Our results show that ED-MTT outperforms these state-of-the-art methods with at least 6% improvement on MSE.

The rest of this paper is organized as follows: Sect. 2 reviews the related works in the literature. Section 3 explains the architectural design of ED-MTT. Section 4 presents experimental results for the performance evaluation of ED-MTT and comparison with the state-of-the-art methods. Section 5 conclude our work and experimental results.

[1] Code and pretrained model are available at https://github.com/CopurOnur/ED-MTT.

2 Related Works

One of the first attempts to investigate the relationships between facial features, conversational cues, and emotional expressions with engagement detection is presented by D'Mello et al. in [8]. The authors in [10,28] used the Facial Action Coding System (FACS) which is a measure of discrete emotions with facial muscle movements, and point out the relation between specific engagement labels and facial actions. In Reference [28], Whitehill et al. showed that automated engagement detectors perform with comparable accuracy to humans. In [3], Booth et al. compared the performance of a Long-Short Term Memory (LSTM) based method with SVM and KNN methods with non-verbal features. In [6], Dewan et al. proposed a Local Directional Pattern (LDP) to extract person-independent edge features which are fed to a Deep Belief Network. Huang et al. [14] proposed a model, called Deep Engagement Recognition Network (DERN), which combines temporal convolution, bidirectional LSTM, and an attention mechanism to identify the degree of engagement based on the features captured by OpenFace [2]. Moreover in [18], Liao et al. proposed Deep Facial Spatiotemporal Network (DFSTN) which is developed based on extracting facial spatial features and global attention for sequence modeling with LSTM. Finally, in [19,21,23], authors used models which are based on Convolutional Neural Networks (CNN) and Residual Networks (ResNet) [12]. All the works above considered the engagement detection problem as a multi-class classification problem. In contrast, in this paper, we follow a more recent line of research that considers engagement detection as a regression problem, where MSE loss is used to measure a continuous distance between predicted and ground truth engagement levels.

Yang et al. [31] also used MSE loss and developed a method that ensembles four separate LSTMs using facial features extracted from four different sources. In [20], Niu et al. combined the outputs of three Gated Recurrent Units (GRU) based on a 117-dimensional feature vector composed of eye gaze action units and head pose features. In [24], Thomas et al. used Temporal Convolutional Network (TCN) on the same set of features as in [20]. In previous works [29,32], the most common ways to overcome over-fitting is data augmentation and cross-validation training. Some other works [1,27] consider imbalanced sampling [17] and using weighted/ranked loss functions. Moreover, some works also consider spatial dropout and batch normalization as a regularization technique [5,24]. All the previous studies focus on small sample sizes and imbalanced labels but none of them consider the reliability of the labels. On the other hand, in this paper, ED-MTT aims to handle both overfitting and label reliability at the same time via multi-task training with triplet loss.

3 Architectural Design for Engagement Detection with Multi-Task Training

We now present our architectural design as well as the multi-tasking with the combination of MSE and triplet loss for training, which are the main contributions of this work. To this end, Fig. 1 displays the training architecture of

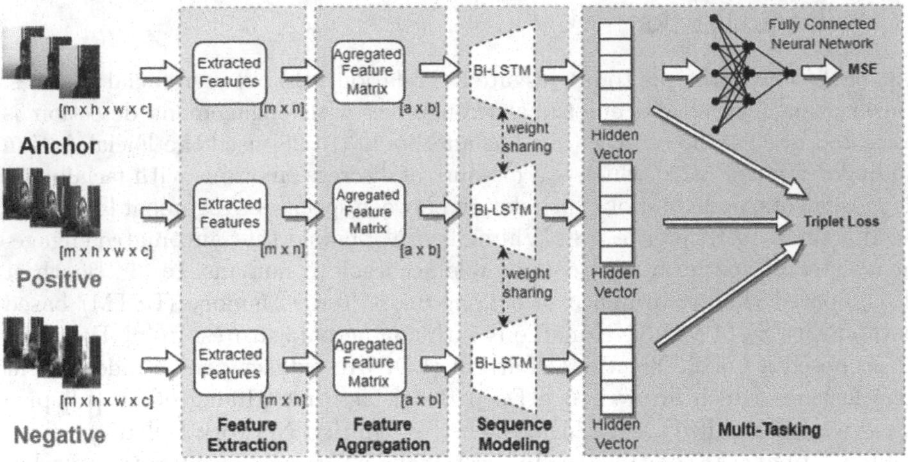

Fig. 1. The training architecture for ED-MTT.

ED-MTT that consists of four main parts: Feature Extraction, Frame Aggregation, Sequence Modeling, and Multi-Tasking. The inputs of this architecture are three batches of samples as Anchor, Positive and Negative. In each batch, each sample is the sequence of images which is obtained by segmenting a video into m frames each of size $h \times w \times c$, where h denotes the height in pixels, w denotes the width in pixels and c denotes the number of color channels of each frame, where RGB color space is used. During the training with this approach, each sample s in the anchor batch is assumed to have a labeled engagement level E^s between 0 and 1. For each s, E^s is assigned into either low engagement or high engagement classes. To this end, if $E^s < 0.5$, s is assigned into the low engagement class; otherwise, i.e. $E^s \geq 0.5$, s is assigned into the high engagement class. Then, for each sample s in the anchor batch, the positive batch contains a random sample from the same engagement class of s while the negative batch contains a random sample from the opposite engagement class of s.

Furthermore, the outputs of the architecture in Fig. 1 are the MSE and Triplet Loss which are combined to train the Bi-LSTM model. Note that during inference, the engagement level prediction is the output of the fully connected neural network. While creating a multi-task learning problem through triplet loss, which aims to prevent overfitting due to the very few samples available for engagement detection during e-learning, we are able to perform regression for continuous engagement levels using MSE. In the rest of this section, we explain each part of the training architecture.

3.1 Feature Extraction

In order to narrow down the feature space by extracting the important features from the sequence of video frames, we first determine the features that are related to the engagement level of a subject. Accordingly, as done in [14,20,24,29,31], we consider 29 features which are related to eye gaze, head pose, head rotation, and facial action units. We extract these features with OpenFace which provides many different facial features [2] and can be described as

$$\mathbf{Y}^s_{m\times n} = \text{OpenFace}(\mathbf{X}^s_{mhwc}), \tag{1}$$

where $\mathbf{X}^s_{mhwc}$ is the tensor of frame sequences at sample s, and $\mathbf{Y}^s_{m\times n}$ is the matrix of sequence of features at sample s, where the (i,j)-th element of $\mathbf{Y}^s_{m\times n}$ the feature i for frame j.

In the result of feature extraction, the eye gaze-related features are, *gaze_0_x, gaze_0_y, gaze_0_z* which are eye gaze direction vectors in world coordinates for the left eye and *gaze_1_x, gaze_1_y, gaze_1_z* for the right eye in the image. The head pose-related features are *pose_Tx, pose_Ty, pose_Tz* representing the location of the head with respect to the camera in millimeters (positive Z is away from the camera). *pose_Rx, pose_Ry, pose_Rz* indicates the rotation of the head in radians around x, y, z axes. This can be seen as pitch (Rx), yaw (Ry), and roll (Rz). The rotation is in world coordinates with the camera being the origin. Finally, the following 17 facial action unit intensities varying in the range 0−5 are used: *AU01_r, AU02_r, AU04_r, AU05_r, AU06_r, AU07_r, AU09_r, AU10_r, AU12_r, AU14_r, AU15_r, AU17_r, AU20_r, AU23_r, AU25_r, AU26_r, AU45_r.*

3.2 Feature Aggregation over Time Windows

We now explain the aggregation of feature statistics over time windows with multiple video frames. In this way, the number of features (which was equal to n at the end of the Feature Extraction phase) is increased to b in order to provide more information to the Sequence Model.

Let the operation of the "Feature Aggregation over Time Windows" be shown as

$$\mathbf{Z}^s_{a\times b} = \text{Aggregate}(\mathbf{Y}^s_{m\times n}), \tag{2}$$

where $\mathbf{Z}^s_{a\times b}$ is the matrix of the b feature statistics for a aggregated frames. Let z be the number of frames in each time window that are considered for feature aggregation, where $m = a \times z$. Then, in each of a windows, we compute the *mean*, *variance*, *standard deviation*, *minimum*, and *maximum* of each feature over the consecutive z frames resulting in b feature statistics, where $b = 5 \times n$.

3.3 Sequence Modeling Combined with Multi-Tasking

Multi-task learning aims to learn multiple different tasks simultaneously while maximizing performance on one or all of the tasks [4]. The suggested architecture contains two tasks: The first task is predicting the multi-level engagement label by optimizing the MSE loss between actual and predicted labels. The second task is learning hidden vector embeddings by optimizing the triplet loss.

As shown in Fig. 1, during sequence modeling, we use three parallel (siamese) Bi-LSTM models with weight sharing to compute the hidden vectors for Triplet

Loss and for MSE loss as cascaded to the Fully Connected Neural Network. However, note that training is performed for only one Bi-LSTM model since the Bi-LSTM models in Fig. 1 are used with weight sharing for triplet loss. We call the Bi-LSTM model for the aggregated feature matrix $\mathbf{Z}^s_{a\times b}$ as

$$T^s_v = \text{Bi-LSTM}(\mathbf{Z}^s_{a\times b}), \tag{3}$$

where T^s_v is the hidden vector, which is the hidden state of the last layer of Bi-LSTM model. Thus, the length of this vector, denoted by v, is equal to twice the number of hidden units of the last layer of the Bi-LSTM.

Triplet loss is a loss function where a baseline (anchor) sample is compared with a positive and negative sample. The distance between the anchor and the positive sample is minimized and the distance between the anchor and the negative is maximized. We use the triplet loss function which is presented in [26] and defined as

$$\begin{aligned} \ell(\text{Anchor}, \text{Positive}, \text{Negative}) &= L = \{l_1, \dots l_s \dots, l_S\}^\top, \\ l_s &= \max\{d(\text{Anchor}_s, \text{Positive}_s) - d(\text{Anchor}_s, \text{Negative}_s) + \text{margin}, 0\}, \end{aligned} \tag{4}$$

where S is the number of samples in a batch, d is the euclidean distance, and $margin$ is a non-negative margin representing the minimum difference between the positive and negative distances that are required for the loss to be 0. Moreover, Anchor_s, Positive_s and Negative_s denote the Anchor, Positive and Negative batches for sample s, respectively.

In addition to the triplet loss, we also minimize the MSE loss which measures the error for the engagement regression. To this end, we cascade the Bi-LSTM model to the Fully Connected Neural Network whose output is the engagement level. Recall that the engagement regression is the main task during the real-time application. Accordingly, during training, the minimization of MSE can be considered as the main task while the minimization of Triplet loss is the auxiliary task.

4 Experimental Results

4.1 Dataset

For the performance evaluation of the proposed technique, we use both training and validation datasets published at "Emotion Recognition in the Wild" (EmotiW 2020) challenge [7] where the engagement regression is a sub-task. The dataset is comprised of 78 subjects (25 females and 53 males) whose ages range from 19 to 27. Each subject is recorded while watching an approximately 5 min long stimulus video of a Korean Language lecture. This procedure results in a collection of 195 videos, where the environment varies over videos and the subjects are not disturbed during recording. The engagement level of each video recording is labeled by a team of five between 0 and 3 resulting in the distribution shown in Fig. 2.

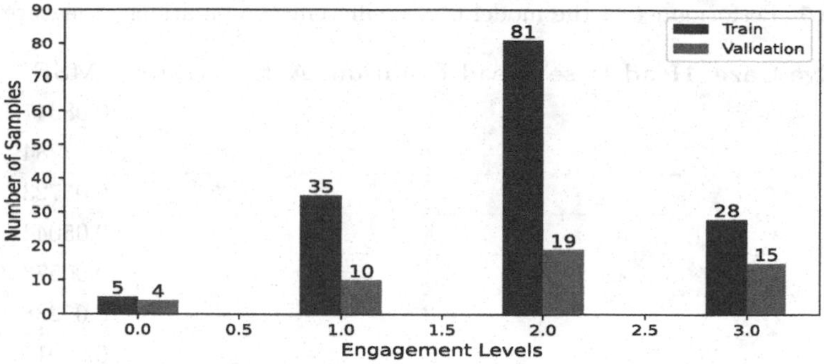

Fig. 2. He distribution of the engagement classes for each of the training and validation sets. In this figure, we see that the dataset is highly imbalanced, in particular there is a lack of low level engagement class samples.

4.2 Experimental Setup and Hyperparameter Settings

We implemented ED-MTT by using PyTorch on Python 3.7.12. The experiments are executed on the Google Colab platform where the operating system is Linux-5.4.144, and the GPU device is Tesla P100-PCIE-16 GB. The model is trained via the adam optimizer [16] for 500 epochs with 5×10^{-5} initial learning rate and batch size of 16.

Furthermore, during our experiments, we first fixed the number of aggregated frames $a = 100$. At the input of Bi-LSTM, we used a batch normalization with an imbalanced sampler from the "imbalanced-learn" library of Python [17]. Then in order to determine the architectural hyperparameters of the sequential model, we performed a random search for the number of Bi-LSTM layers, the size of the hidden state as well as the number of neurons at each of two fully connected neural network layers. The random search sets are as follows: $\{1, 2, 3\}$ for the number of Bi-LSTM layers, $\{128, 256, 512, 1024\}$ for the size of hidden state of each Bi-LSTM layer[2], $\{256, 128, 64\}$ for the first layer of the fully connected neural network, and $\{32, 16, 8\}$ for the second layer of fully connected neural network. At the end of this search, the resulting architecture is comprised of 2 Bi-LSTM layers each of whose hidden state size is 1024, and two sequential fully connected layers with 64 and 32 neurons respectively.

4.3 Performance Evaluation

We now evaluate the performance of ED-MTT for engagement detection on a publicly available "Engagement in The Wild" dataset. During performance evaluation, we first aim to select the subset of facial and head position features with respect to their effects on the performance of our system. To this end, Table 1 displays the performance of the model under different combinations of feature sets, where the combinations are selected empirically to achieve high

[2] Note that the size of the hidden state is constant across all Bi-LSTM layers.

Table 1. Performance of the model under different combinations of feature sets

Eye Gaze	Head Pose	Head Rotation	Action Units	MSE
✓	✗	✗	✗	0.08347
✗	✓	✗	✗	0.07784
✗	✗	✓	✗	0.05723
✗	✗	✗	✓	0.05044
✗	✗	✓	✓	0.06578
✗	✓	✓	✗	0.07238
✓	✗	✓	✗	0.06915
✓	✓	✗	✗	0.06036
✓	✓	✓	✗	0.06973
✓	✗	✓	✓	0.05681
✓	✓	✗	✓	**0.04271**
✓	✓	✓	✓	0.05431

performance. Recall that the number of features in each feature set is as follows: 6 features in *Eye Gaze*, 3 features in *Head Pose*, 3 features in *Head Rotation*, and 17 features in *Action Units*. According to our observations on the results presented in this table, we may draw the following conclusions:

- The best performance is achieved by using all features except *Head Rotation* features. Accordingly, in the rest of our results, we use the combination of *Eye Gaze*, *Head Pose*, and *Action Unit* features.
- The most effective individual feature set is *Action Units*.
- The MSE loss significantly decreases for the majority of the cases when *Action Unit* features are included in the selected features.

Furthermore, in Fig. 3, we present the comparison of ED-MTT against the state-of-the-art engagement regression methods that are evaluated on the Engagement in The Wild dataset. In this figure, the MSE scores of the state of the art methods are taken from their original papers. The results show that ED-MTT achieves the best performance with 0.0427 MSE loss on the validation set. Although the performances of all methods are highly competitive with each other, the ED-MTT improved the best performance (Chang et al. [5]) in the literature by 6%. In addition, the training time of ED-MTT is around 38 min for 149 samples for 500 epochs.

Figure 4 displays the box plot of the predicted engagement levels on the validation sets which are classified with respect to the ground truth engagement labels in the dataset. In this figure, from median and percentiles of predicted engagement levels, one may see that the continuous predictions of ED-MTT distinctly reflects the four level of engagement classes in the ground truth labels. Moreover, ED-MTT can easily distinguish between classes 0, 0.33, and 0.66 while the difference between 0.66 and 1.0 is more subtle.

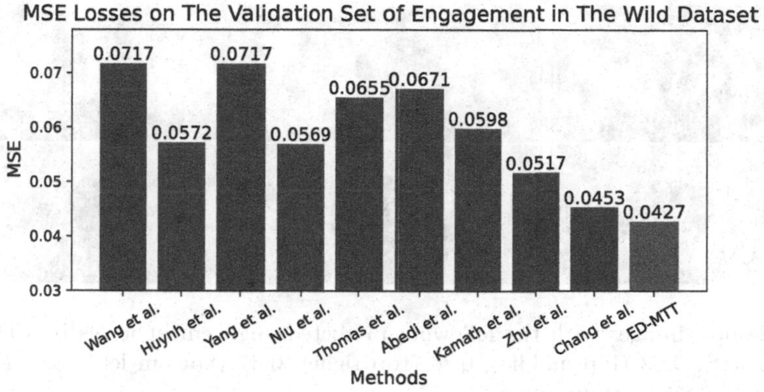

Fig. 3. The performance comparison of ED-MTT against the state-of-the-art methods, where the MSE scores are presented as in the original papers. *Wang et al.* [27] score: 0.0717, *Huynh et al.* [25] score: 0.0572, *Yang et al.* [31] score: 0.0717, *Niu et al.* [20] score: 0.0569, *Thomas et al.* [24] score: 0.0655, *Abedi et al.* [1] score: 0.0671, *Kamath et al.* [15] score: 0.0598, *Zhu et al.* [32] score: 0.0517, *Chang et al.* [5] score: 0.0453, ***ED-MTT* score: 0.0427**

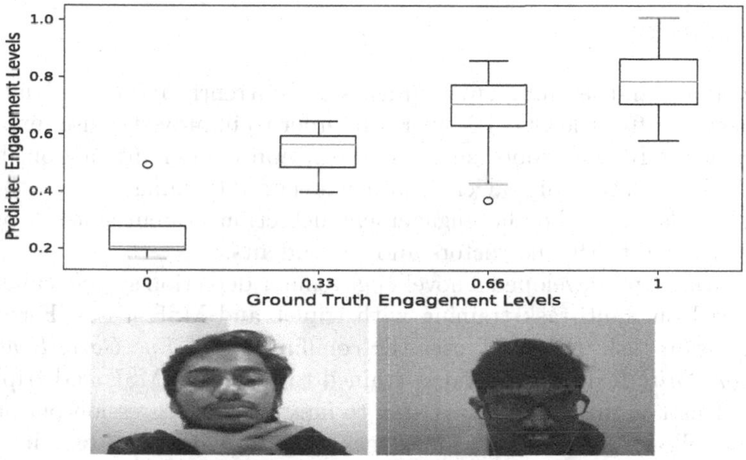

Fig. 4. The figure presents (top) the box plot of predicted engagement levels for each class in the ground truth engagement levels and (bottom) sample images from [7] correspond to outliers in the box plot, respectively.

4.4 Qualitative Results

Finally, ED-MTT is also tested on a preliminary real-life engagement detection tasks for which the prediction results are presented in Fig. 5. These results show that the proposed model, ED-MTT, trained on Engagement in The Wild dataset is able to provide highly successful predictions in real-life use-cases, which are totally different than the cases in the training set. According to our observations

Fig. 5. Sample images with the following predicted engagement levels by ED-MTT: 0.35 (top left), 0.53 (top middle), 0.86 (top right), 0.47 (bottom left), 0.61 (bottom middle), and 0.82 (bottom right).

on the prediction results for a total (approximately) 12 min long videos including 8 people, the model can successfully distinguish different levels of engagement (very low, low, high, and very high engagement levels). However, the predicted engagement levels lie between 0.2 and 0.92, which forces to determine smaller quantization intervals to classify engagement levels in real-life use-cases.

5 Conclusion

Online working and learning environments are currently more essential in our lives, especially after the COVID-19 era. In order to improve the user experience and efficiency, advanced tools, such as recognition of user interaction, became highly crucial in these digital environments. For e-learning, one of the most important tools might be the engagement detection system since it provides valuable feedback to the instructors and/or students.

In this paper, we developed a novel engagement detection system called "ED-MTT" based on multi-task training with triplet and MSE losses. For engagement regression task, ED-MTT uses the combination of *Eye Gaze*, *Head Pose*, and *Action Units* feature sets and is trained to minimize MSE and triplet loss together. This training approach is able to improve the regression performance due to the following reasons; 1) multi-task training with two losses introduces an additional regularization and reduces over-fitting due to very small sample size, 2) triplet loss measures relative similarity between samples to mitigate the label reliability problem. 3) minimization of MSE ensures that the main loss considered for the regression problem is minimized alongside the triplet loss.

The performance of ED-MTT is evaluated and compared against the performances of the state-of-the-art methods on the publicly available Engagement in The Wild dataset which is comprised of separated training and validation sets. Our results showed that the novel ED-MTT method achieves 6% lower MSE than the lowest MSE achieved by the state-of-the-art while the training of ED-MTT takes around 38 min for 149 samples for 500 epochs. We tested the performance of ED-MTT for real-life use cases with 8 different participants, and the prediction results for majority of these cases were shown to be highly successful.

References

1. Abedi, A., Khan, S.: Affect-driven engagement measurement from videos. arXiv preprint arXiv:2106.10882 (2021)
2. Baltrusaitis, T., Zadeh, A., Lim, Y.C., Morency, L.P.: OpenFace 2.0: facial behavior analysis toolkit. In: 2018 13th IEEE International Conference on Automatic Face Gesture Recognition (FG 2018), pp. 59–66 (2018)
3. Booth, B.M., Ali, A.M., Narayanan, S.S., Bennett, I., Farag, A.A.: Toward active and unobtrusive engagement assessment of distance learners. In: 2017 Seventh International Conference on Affective Computing and Intelligent Interaction (ACII), pp. 470–476 (2017)
4. Caruana, R.: Multitask learning. Mach. Learn. **28**(1), 41–75 (1997)
5. Chang, C., Zhang, C., Chen, L., Liu, Y.: An ensemble model using face and body tracking for engagement detection. In: Proceedings of the 20th ACM International Conference on Multimodal Interaction, pp. 616–622 (2018)
6. Dewan, M.A.A., Lin, F., Wen, D., Murshed, M., Uddin, Z.: A deep learning approach to detecting engagement of online learners. In: 2018 IEEE SmartWorld/SCALCOM/UIC/ATC/CBDCom/IOP/SCI, pp. 1895–1902 (2018)
7. Dhall, A., Sharma, G., Goecke, R., Gedeon, T.: EmotiW 2020: driver gaze, group emotion, student engagement and physiological signal based challenges. In: Proceedings of the 2020 International Conference on Multimodal Interaction, pp. 784–789 (2020)
8. D'Mello, S.K., Craig, S.D., Graesser, A.C.: Multimethod assessment of affective experience and expression during deep learning. Int. J. Learn. Technol. **4**(3–4), 165–187 (2009)
9. Dong, X., Shen, J.: Triplet loss in siamese network for object tracking. In: Proceedings of the European Conference on Computer Vision (ECCV), September 2018
10. Grafsgaard, J., Wiggins, J.B., Boyer, K.E., Wiebe, E.N., Lester, J.: Automatically recognizing facial expression: predicting engagement and frustration. In: Educational Data Mining 2013 (2013)
11. Gupta, A., D'Cunha, A., Awasthi, K., Balasubramanian, V.: DAiSEE: towards user engagement recognition in the wild. arXiv preprint arXiv:1609.01885 (2016)
12. He, K., Zhang, X., Ren, S., Sun, J.: Deep residual learning for image recognition. In: Proceedings of the IEEE Conference on Computer Vision and Pattern Recognition, pp. 770–778 (2016)
13. Hochreiter, S., Schmidhuber, J.: Long short-term memory. Neural Comput. **9**(8), 1735–1780 (1997)
14. Huang, T., Mei, Y., Zhang, H., Liu, S., Yang, H.: Fine-grained engagement recognition in online learning environment. In: 2019 IEEE 9th International Conference on Electronics Information and Emergency Communication (ICEIEC), pp. 338–341 (2019)
15. Kamath, S., Singhal, P., Jeevan, G., Annappa, B.: Engagement analysis of students in online learning environments. In: Misra, R., Shyamasundar, R.K., Chaturvedi, A., Omer, R. (eds.) ICMLBDA 2021. LNNS, vol. 256, pp. 34–47. Springer, Cham (2022). https://doi.org/10.1007/978-3-030-82469-3_4
16. Kingma, D.P., Ba, J.: Adam: a method for stochastic optimization. arXiv preprint arXiv:1412.6980 (2014)
17. Lemaître, G., Nogueira, F., Aridas, C.K.: Imbalanced-learn: a python toolbox to tackle the curse of imbalanced datasets in machine learning. J. Mach. Learn. Res. **18**(17), 1–5 (2017)

18. Liao, J., Liang, Y., Pan, J.: Deep facial spatiotemporal network for engagement prediction in online learning. Appl. Intell. **51**, 1–13 (2021)
19. Murshed, M., Dewan, M.A.A., Lin, F., Wen, D.: Engagement detection in e-learning environments using convolutional neural networks. In: 2019 IEEE (DASC/PiCom/CBDCom/CyberSciTech), pp. 80–86. IEEE (2019)
20. Niu, X., et al.: Automatic engagement prediction with gap feature. In: Proceedings of the 20th ACM International Conference on Multimodal Interaction, pp. 599–603 (2018)
21. Rao, K.P., Rao, M.C.S.: Recognition of learners' cognitive states using facial expressions in e-learning environments. J. Univ. Shanghai Sci. Technol. 93–103 (2020)
22. Roy, S., Etemad, A.: Self-supervised contrastive learning of multi-view facial expressions. In: Proceedings of the 2021 International Conference on Multimodal Interaction, pp. 253–257 (2021)
23. Thiruthuvanathan, M., Krishnan, B., Rangaswamy, M.A.D.: Engagement detection through facial emotional recognition using a shallow residual convolutional neural networks. Int. J. Intell. Eng. Syst. **14**, 236–247 (2021)
24. Thomas, C., Nair, N., Jayagopi, D.B.: Predicting engagement intensity in the wild using temporal convolutional network. In: Proceedings of the 20th ACM International Conference on Multimodal Interaction, pp. 604–610 (2018)
25. Thong Huynh, V., Kim, S.H., Lee, G.S., Yang, H.J.: Engagement intensity prediction withfacial behavior features. In: 2019 International Conference on Multimodal Interaction, pp. 567–571 (2019)
26. Balntas, V., Riba, E., Ponsa, D., Mikolajczyk, K.: Learning local feature descriptors with triplets and shallow convolutional neural networks. In: Proceedings of the British Machine Vision Conference (BMVC), pp. 119.1–119.11. BMVA Press (2016)
27. Wang, K., Yang, J., Guo, D., Zhang, K., Peng, X., Qiao, Y.: Bootstrap model ensemble and rank loss for engagement intensity regression. In: 2019 International Conference on Multimodal Interaction, pp. 551–556 (2019)
28. Whitehill, J., Serpell, Z., Lin, Y., Foster, A., Movellan, J.R.: The faces of engagement: automatic recognition of student engagementfrom facial expressions. IEEE Trans. Affect. Comput. **5**(1), 86–98 (2014)
29. Wu, J., Yang, B., Wang, Y., Hattori, G.: Advanced multi-instance learning method with multi-features engineering and conservative optimization for engagement intensity prediction. In: Proceedings of the 2020 International Conference on Multimodal Interaction, pp. 777–783 (2020)
30. Wu, J., Zhou, Z., Wang, Y., Li, Y., Xu, X., Uchida, Y.: Multi-feature and multi-instance learning with anti-overfitting strategy for engagement intensity prediction. In: 2019 International Conference on Multimodal Interaction, pp. 582–588 (2019)
31. Yang, J., Wang, K., Peng, X., Qiao, Y.: Deep recurrent multi-instance learning with spatio-temporal features for engagement intensity prediction. In: Proceedings of the 20th ACM International Conference on Multimodal Interaction, pp. 594–598 (2018)
32. Zhu, B., Lan, X., Guo, X., Barner, K.E., Boncelet, C.: Multi-rate attention based GRU model for engagement prediction. In: Proceedings of the 2020 International Conference on Multimodal Interaction, pp. 841–848 (2020)

Special Session

Ship Detection and Tracking Based on a Custom Aerial Dataset

Luigi Paiano[1(✉)], Francesca Calabrese[1(✉)], Marco Cataldo[1(✉)], Luca Sebastiani[2], and Nicola Leonardi[2]

[1] Apphia s.r.l., Edificio DHiTech, Via per Monteroni, s.n.c., 73100 Lecce, Italy
{luigi.paiano,francesca.calabrese,marco.cataldo}@apphia.it
[2] Fincantieri NexTech S.p.A., Head Office in Milano, Milan, Italy
{luca.sebastiani,nicola.leonardi}@fincantierinxt.it

Abstract. This paper presents an approach based on machine learning techniques for detection and tracking ship in marine environment monitoring, with focus on a custom large data set based on aerial images. The work is placed in the context of autonomous navigation by the use of an unmanned surface naval platform assisted by an aerial drone. The work is according to a data-centric Artificial Intelligence (AI) approach, which involves building AI systems with quality data with a focus on ensuring that the data clearly conveys what the AI must learn. The application of machine learning techniques is used for automatic target detection and tracking. Target information in the surrounding environment allows context-awareness and obstacle identification and it can support naval platform in the management of collision avoidance. The paper focuses on the need of large amounts of data for the training stage to perform robust detections and tracking even in critical glare and waves variations. The paper presents a custom data set which includes fine-tuned public ship aerial images and images acquired by UAV flights over different maritime scenarios. The network's training results are described and the detection and tracking performance is evaluated in different video sequences from UAV flights over such scenarios.

Keywords: Aerial ship dataset · Data-centric approach · YOLOv4 · Ship tracking

1 Introduction

The sea transport sector is becoming increasingly important in modern life. The constant effort of the scientific community in the marine sector is aimed at allowing, in whole or in part, autonomous navigation. The detection of nearby objects and boats is essential for the development of systems for safety in navigation. Over time, more and more sensors have been developed to support surveillance operations in general. Situational awareness is a key aspect of decision management in complex and dynamic areas, such as air travel, air traffic control,

S. Sclaroff et al. (Eds.): ICIAP 2022, LNCS 13233, pp. 425–436, 2022.
https://doi.org/10.1007/978-3-031-06433-3_36

navigation, power plant operations, military command and control, and emergency services. In the recent years, there has been a huge development in the use of Unmanned Aerial Vehicles (UAVs) both for recreational and business purposes. Our work focuses on maritime surveillance scenarios, more specifically on the automatic detection and tracking of vessels using video from UAV flights.

The application areas of AI methods in maritime navigation and vessel situational awareness are identified as object identification, localization, and trajectory analysis [1]. In the literature, there are several studies conducted on ship detection and tracking in maritime environments. The methods used in these studies range from classical algorithms to neural networks. Cruz et al. in [2] and [3], present a method to detect boat in a maritme surveillance scenario using Convolutional Neural Netorks (CNN) to perform robust detections even in the presence of distractors like wave crests, sun glare and scale variation. Huixuan Fh et al. in [4] proposed the modified YOLOv4 neural network to improve the accuracy and speed of detection of marine targets. Within the scope of the researches [5,6], optimized Faster-R-CNN-based sea ship recognition algorithms are proposed. Zou Y et al. in [7] proposed an improved SSD algorithm based on the MobilenetV2 CNN for target detection and recognition in vessel images. Huang H et al. proposed an improved target detection algorithm based on YOLOv3 [8], improving its accuracy and robustness with the premise of ensuring real-time performance. Chen X et al. proposed a YOLO-based framework for detecting ships from ocean images [9]. Huang Z et al. proposed an improved YOLOv3 network [10] for intelligent detection and classification of ship images/videos. The authors in [11] used an image dataset to train and test the two popular single-stage object detection models, YOLOv4 and Tiny YOLOv4. As a result of the study, the best performance of the YOLOv4 model was confirmed. In the researches [12–17] there is the design of systems for the automated detection of ships from optical satellite images, using different techniques, such as statistical methods, signal processing algorithms, mathematical morphology, unsupervised learning techniques, and convolutional neural networks. Yang-Lang Chang et al. in [18] proposed a method to detect ship from SAR images through the YOLOv2. Yuwen Chen et al. in [19] used and trained the YOLOv5 neural network, validating its use in the field applied to ship detection. The authors in [20] presented ship detection and tracking using the improved YOLOv3 and Deep Simple Real Tracking SORT algorithm. Deep Sort has shown that it performs better in complex scenes and is resistant to interference such as occlusion and camera movement, compared to cutting-edge algorithms such as KCF, MIL, MOVES, TLD, and median flux. Zhang Sr. et al. in [21] proposed an algorithm for ship detection and tracking using improved YOLO. Won-Jae Lee et al. in [22] proposed detection, location and tracking methods applied to videos taken from real boats. The results obtained were compared with AIS data and showed that the proposed algorithm could be used effectively for the awareness of the surrounding environment. An approach aimed at identifying and tracking objects in the maritime environment by association of probabilistic data is reported in the study proposed in Haghbayan et al. [23].

1.1 MARIN Project

The study presented in this paper is part of the research and development activities of the "MARIN - Monitoraggio Ambientale Remoto Integrato su piattaforma Navale" project (project Code: KATGSO3 - "Programma operativo FESR 2014–2020 Obiettivo Convergenza" - Regolamento Regionale n. 17/2014 - Titolo II Capo 1 - "Aiuti ai programmi di investimento delle grandi imprese"), co-funded by Regione Puglia within the framework of "Contratti di Programma". Project beneficiaries are Fincantieri NexTech S.p.A., RINA Consulting S.p.A. and Co.M.Media s.r.l.. The objective of MARIN project is to set-up and a technological demonstrator of the enabling technologies for autonomous navigation and test it in marine environment. The demonstrator platform is the TESEO I, an experimental vessel jointly developed by NAVTEC and Tringali Shipyard as part of a previous research project, with subsequent adaptation to the needs of the MARIN project.

In this context, the study focuses on elaboration of data coming from the optical system on an aerial drone for supporting an unmanned platform in obstacle identification. The aim is the identification of a set of targets within the operating scenario that represents the set of external objects (fixed or mobile), which can be assimilated to obstacles to avoid, as they are subject to possible collision with the unmanned platform.

The information extracted from elaboration on UAV data, together with the data coming from other systems, could contribute to the multisensor data fusion process aimed at avoiding collisions. In this project, the authors propose ship detection and tracking based on YOLOv4 and Deep Sort algorithms. The Bounding Boxes of the various identified objects in the video acquired by the drone and the position information of the drone are used to estimate the position of the targets.

1.2 Contributions

The main contributions of this paper are the introduction of a custom dataset of thousands of aerial images in maritime surveillance scenarios and the evaluation of the performance of the state-of-the-art algorithms on real maritime scenarios using data captured by UAV flights.

Collecting and labeling these datasets is a laborious task. Our dataset presents a considerable amount of samples representative of challenging situations, some of them acquired in real world scenarios, with a strong presence of glare, wave crests, wakes, and with objects of interest of different types, scales and shapes, isolated or multiple and close to others, in open sea and in quay.

The dataset does not include Synthetic aperture radar (SAR) imagery.

The main characteristics of the dataset are:

- more than 1700 aerial public fine-tuned images (note that some of these images include more ships in the same image);
- about 50 images acquired on the field;

- objects of interest were labeled by a human operator;
- different types of objects of interest were observed, namely cargo ships, smaller boats, sailing yachts, dinghies.

Data acquisition by UAV is performed with the camera facing downwards, with fixed zoom and non-changeable angle of view.

1.3 Outline of the Paper

This work is organized as follows. Section 2 introduces the background on the machine learning techniques used in the research. Section 3 describes dataset sources, with details on UAVs used to capture the dataset and the scenarios considered. It presents data preparation, focusing on the labeling and data augmentation processes. Section 4 contains information on the experimental environment used, presents and discusses the results of training, detection and tracking. Finally, in Sect. 5, we present some concluding remarks.

2 Background

2.1 You Only Look Once (YOLO)

YOLO is a very fast and accurate object detection architecture, which was created by Joseph Redmond [24]. YOLO consists of a convolutional neural network (CNN). Its speed was attributed to its one-shot detection mechanism, in which it simultaneously predicted bounding box coordinates and class probabilities from an image. It is ideal for use in real-time video detection. The main implementation of YOLO is based on Darknet, which is an open source neural network framework written in C and CUDA. As mentioned, the object detection algorithm used in this study is the fourth version of the YOLO architecture.

2.2 Deep Simple Online and Real Time Tracking with Convolutional Neural Networks

The Deep SORT (Simple Online Real time Tracking) algorithm [26] has proven to be one of the fastest and most robust approaches for tracking multiple objects. The Deep SORT algorithm solves the problems present in SORT in which there were limitations in the presence of occlusions and in the change of points of view. Deep SORT [27] uses convolutional neural networks for object detection, the Hungarian algorithm for data association and the Kalman filter for motion prediction. Deep SORT, in data association, not only relies on motion-based metrics, but integrates a deep aspect-based metric via a convolutional neural network. This modification leads to greater robustness in problems relating to occlusion, changing the point of view and the use of a non-fixed camera.

3 Dataset Construction

3.1 Dataset Sources

3.1.1 Public Dataset

The goal was to identify a set of aerial images containing several types of ships in order to ensure a diversification of the dataset. For the construction of the dataset, the search for aerial images was carried out directly on the web, identifying different datasets listed below:

- Seagull Dataset [28], an example is shown in Fig. 1(a),
- Aerial Maritime Drone Dataset - Roboflow, an example in Fig. 1(b),
- Ship Detection from Aerial Images - Kaggle, an example in Fig. 1(c),
- MAritime SATellite Imagery (MASATI) Dataset [29], an example in Fig. 1(d).

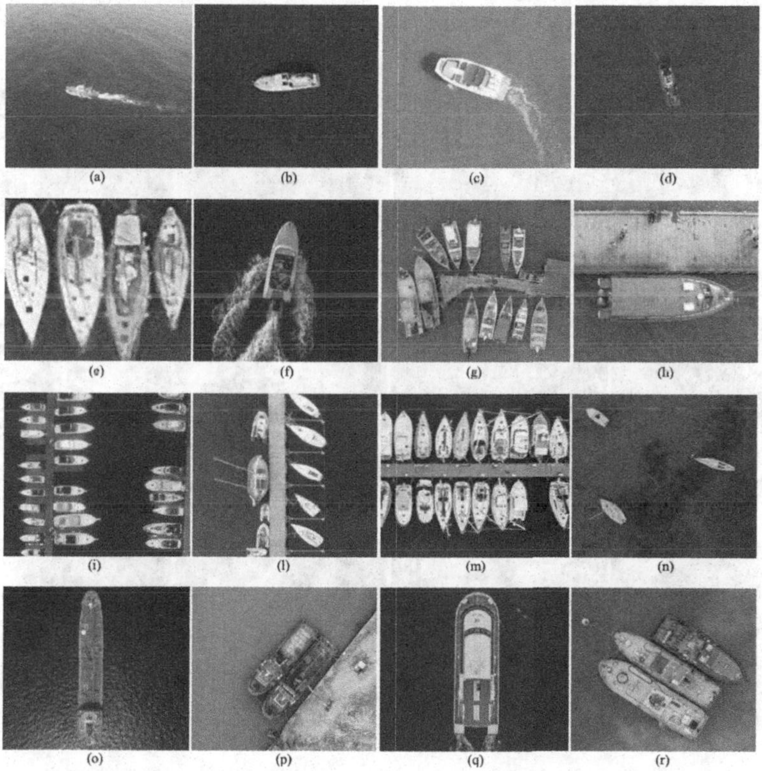

Fig. 1. Public ship dataset

In addition, images containing ships were extracted from web and selected from several free stocks images (with hight definition images):

- Google Earth, an example in Fig. 1(e),
- Pexels Free Stock Photos, example images are shown in Fig. 1(f–h),
- iStock, example images are shown in Fig. 1(i–l),
- Shutterstock, example images are shown in Fig. 1(m–n, o–r).

The result of the activities of research and selection of images has led to the definition of a dataset images related to pleasure and merchant ships. Considering the operational scenario (e.g. type and size of the ship), a selection of images from the various sources is made, for a total of over 1700 images containing one or more ships.

3.1.2 Field Data Collection

Within the MARIN project, field recordings were made using the con DJI Inspire 2 and DJI Matrice 300 RTK drone. Camera on Matrice 300 RTK is Zenmuse H20 Series camera, with an optical sensor 1/2.3″ CMOS-12 MP, a field of view (DFOV) of 82.9°, a focal length of 4.5 mm, a video resolution of 1920 × 1080 px and a frame per second (FPS) of 30.

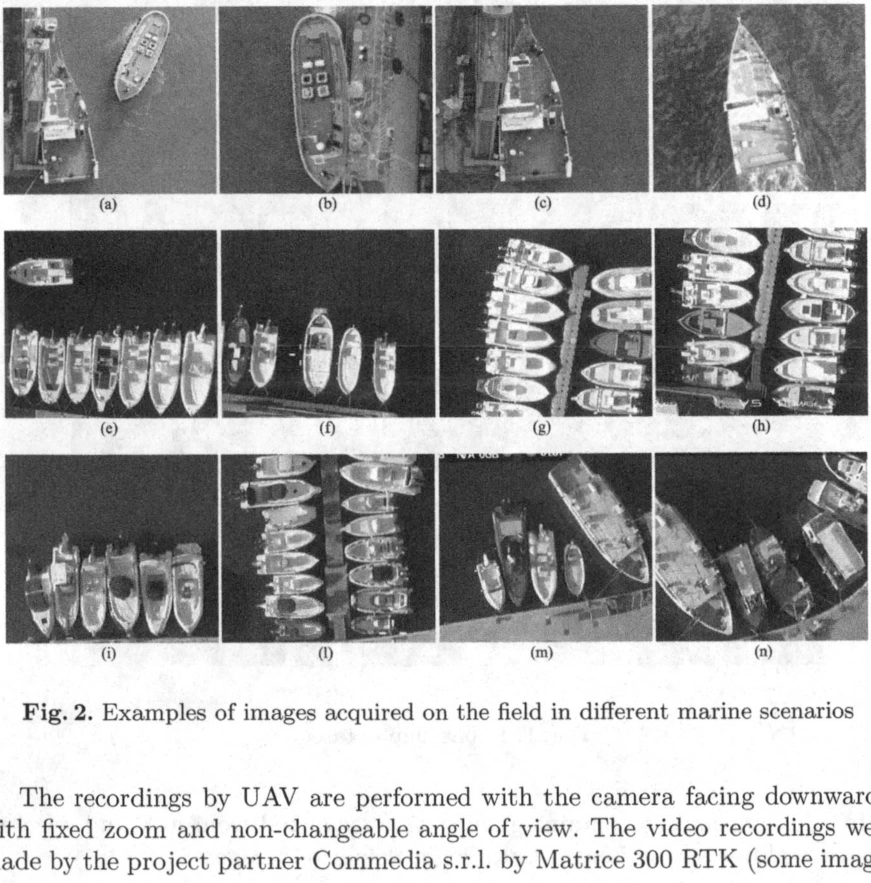

Fig. 2. Examples of images acquired on the field in different marine scenarios

The recordings by UAV are performed with the camera facing downwards, with fixed zoom and non-changeable angle of view. The video recordings were made by the project partner Commedia s.r.l. by Matrice 300 RTK (some images are shown in Fig. 2 (a–d)) and by Inspire 2 (some images are shown in Fig. 2(e–n)). As can be seen in Fig. 2, the size and type of ships presented are very different.

3.2 Data Preparation

The realized dataset consists of several images that include one or more ships in different conditions:

- ship in the open sea and isolated,
- ship in open sea close to others,
- ship in open sea with wake,
- ship in quay and isolated,
- ship in quay near others.

The field dataset is composed of the frames extracted from the videos recorded in the testing sites. It is used in the preliminary analysis of requirements and problems related to image processing and target identification. All images in the public dataset and the field dataset are cropped and resized to 608×608 pixels, a size compatible with the requirements of the neural network used for training. If necessary, the proportions of the images have been preserved by adding a background similar to that around the ship, to avoid distorting the proportions of the ships.

3.2.1 Labeling

For the ship detection task, the networks must be trained on an appropriate set of labeled images called the training dataset. After composing the starting dataset, we proceeded with the labeling of the objects belonging to the "ship" class. The goal of this activity is to draw a bounding box around the object of interest. PASCAL VOC [30] annotation guidelines are used during bounding box labeling. In this way, the dataset also contains the annotation files for the ships in txt format. The annotation files contain five parameters for each ship in image, with class (id) and references to bounding boxes (x_{min}, y_{min}, x_{max} and y_{max}).

3.2.2 Data Augmentation

It is necessary to make the ship detection system as robust as possible to cover the different scenarios. This is possible through a diversified representation of the ship object within the training dataset. As discussed, since the number of images containing aerial view ships is limited, this could cause overfitting where the trained network is unable to generalize and is constrained to only the scenarios present in the training set.

Data Augmentation allows increasing the size of the training set without acquiring new images. The idea is to duplicate images with some kind of variation so that the model can learn from more examples.

In this study, following data augmentation techniques were used:

- random flip (horizontally and vertical),
- random rotation (clockwise and counterclockwise),
- random brightness adjustment from −30% to +30%.

In this way, the final dataset includes over 1700 images obtained starting from a dataset of 500 images.

3.3 Data Splitting

The final dataset was split into a training and testing dataset. The training dataset consists of about 1700 images, which were taken from public datasets and fine-tuned. In the splitting operation, the high-resolution images of the public dataset were used in the training dataset, more information can be extracted in the training phase [31]. The test set is composed of 120 images, including images extracted by video acquired by the drone and a portion of images from Kaggle dataset.

4 Experimental Results and Analysis

4.1 Experimental Environment and Configuration

The experimental environment used is composed of the Darknet opensource framework. The hardware component used in this study is a Desktop PC with Windows 10 Pro operating system, CPU: Intel (R) Core (TM) i9-10900 CPU @ 2.80 GHz 2.81 GHz, RAM: 32 GB, GPU: NVIDIA GeForce RTX2060 SUPER. The authors perform fine-tuning network on their own custom dataset. The hyperparameters used for training the model were: filters = 18, max batches = 2000, steps = [1600, 1800], batch = 64, subdivision = 32, momentum = 0.949, decay = 0.0005, initial learning rate = 0.001.

4.2 Training Results

The YOLOv4 network was trained for a total of 2000 iterations. The loss function is shown in Fig. 3. The x axis represents the number of iterations and the y axis the loss value. The training phase ended after 2.50 h. The metrics obtained to evaluate the trained model are: Mean Average Precision: 99.5%, precision: 0.93, recall: 0.99, F1: 0.96, IoU: 81.12%.

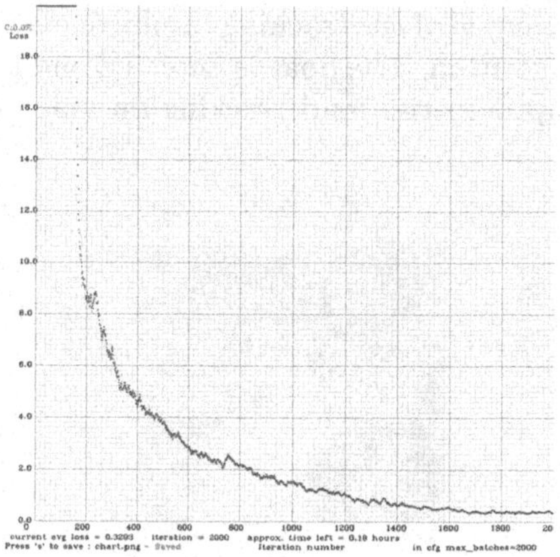

Fig. 3. YOLOv4 training loss function curve

4.3 Detection and Tracking Results

This section presents the results of the detection and tracking on data acquired by UAV. The chosen prediction threshold value is 0.5, in this way, the model does not draw ships with a predictive probability of less than 0.5. Figure 4 shows the ship detection results on test images using YOLOv4. As can be seen, YOLOv4 detected the different types of ship with high confidence scores. Ship detection on video captured in the field, with 1920×1080 px resolution, is performed with 35 FPS.

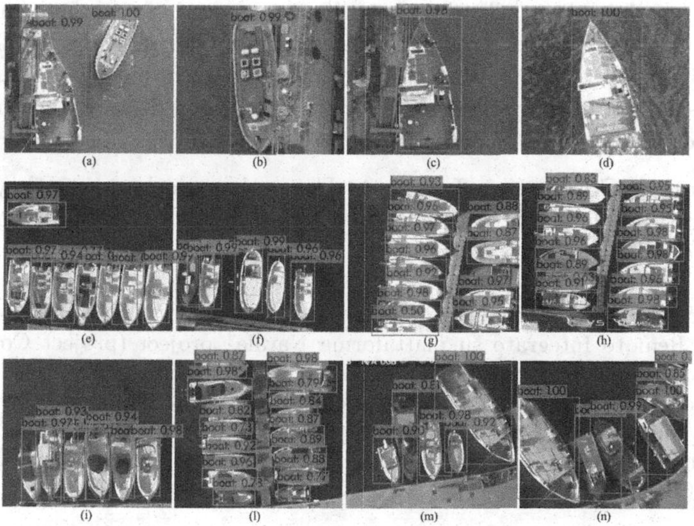

Fig. 4. Examples of YOLOv4 ship detection

Figure 5 shows an example of the tracking results on test video using YOLOv4 and the Deep SORT Algorithm. The boat is detected and tracked over time with the ID "boat-220". Ship detection and tracking on video is performed with 28 FPS.

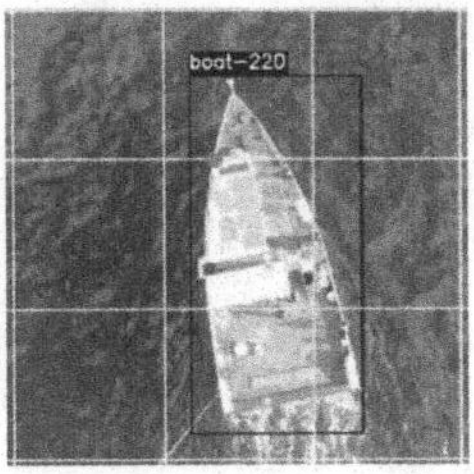

Fig. 5. Example of ship tracking

5 Conclusion

The detection and tracking of ships from aerial images is a hot topic and finds application in civil and military contexts. Ship detection in aerial imagery is a challenging task compared to general object detection, mainly due to the cost and lack of aerial data sets. This work focuses on the importance of having an appropriate dataset. It proposes a custom large data set based on aerial images. The experimental results obtained through the training of the YOLOv4 neural network on the custom dataset and the results of the detection and the tracking on data acquired by UAV in marine scenarios are presented. It is difficult to acquire marine data, but collecting a significant amount of real data for training deep learning models could ensure greater accuracy. One of the future tasks that authors might carry out is to perform further analysis on other real data in different scenarios to verify the adequacy of the training dataset and the correct behaviour of the network.

Declaration of Competing Interest. The authors declare that they have no known competing financial interests or personal relationships that could have appeared to influence the work reported in this paper.

Acknowledgements. This study is sponsored by the "MARIN - Monitoraggio Ambientale Remoto Integrato su piattaforma Navale" project (project Code: KAT-GSO3, "Programma operativo FESR 2014–2020 Obiettivo Convergenza" - Regolamento Regionale n. 17/2014 - Titolo II Capo 1 - "Aiuti ai programmi di investimento delle grandi imprese"), co-funded by Regione Puglia within the framework of "Contratti di Programma". Project beneficiaries are Fincantieri NexTech S.p.A., RINA Consulting S.p.A. and Co.M.Media s.r.l..

References

1. Thombre, S., et al.: Sensors and AI techniques for situational awareness in autonomous ships: a review. IEEE Trans. Intell. Transp. Syst. (2022). https://doi.org/10.1109/TITS.2020.3023957
2. Cruz, G., Bernardino, A.: Aerial detection in maritime scenarios using convolutional neural networks. In: Blanc-Talon, J., Distante, C., Philips, W., Popescu, D., Scheunders, P. (eds.) ACIVS 2016. LNCS, vol. 10016, pp. 373–384. Springer, Cham (2016). https://doi.org/10.1007/978-3-319-48680-2_33
3. Cruz, G., Bernardino, A.: Evaluating Aerial Vessel Detector in Multiple Maritime Surveillance Scenarios (2017)
4. Huixuan, F., Song, G., Wang, Y.: Improved YOLOv4 Marine Target Detection Combined with CBAM (2021). https://doi.org/10.3390/sym13040623
5. Gu, D., Xu, X., Jin, X.: Marine ship recognition algorithm based on faster-RCNN. Image Signal Process **7**, 136–141 (2018)
6. Qi L., et al.: Ship target detection algorithm based on improved faster R-CNN. Electronics (2019). https://doi.org/10.3390/electronics8090959
7. Zou Y., Zhao L., Qin S., Pan M., Li Z.: Ship target detection and identification based on SSD MobilenetV2. In: Proceedings of the 2020 IEEE 5th Information Technology and Mechatronics Engineering Conference (ITOEC), Chongqing, China, pp. 1676–1680, (2020). https://doi.org/10.1109/ITOEC49072.2020.9141734
8. Huang, H., Sun, D., Wang, R., Zhu, C., Liu, B.: Ship target detection based on improved YOLO network. Math. Probl. Eng. (2020). https://doi.org/10.1155/2020/6402149
9. Chen, X., et al.: Video-based detection infrastructure enhancement for automated ship recognition and behavior analysis (2020). https://doi.org/10.1155/2020/7194342
10. Huang, Z., Sui, B., Wen, J., Jiang, G.: An intelligent ship image/video detection and classification method with improved regressive deep convolutional neural network. Complexity (2020). https://doi.org/10.1155/2020/1520872
11. Yildirim, E., Kavzoglu, T.: Ship detection in optical remote sensing images using YOLOv4 and Tiny YOLOv4. In: Ben Ahmed, M., Boudhir, A.A., Kara, I.R., Jain, V., Mellouli, S. (eds.) Innovations in Smart Cities Applications Volume 5. SCA 2021. LNNS, vol. 393, pp. 913–924. Springer, Cham (2022). https://doi.org/10.1007/978-3-030-94191-8_74
12. Corbane C., Najman L., Pecoul E., Demagistri L., Petit M.: A complete processing chain for ship detection using optical satellite imagery. Int. J. Remote Sens. 5837–5854 (2010). https://doi.org/10.1080/01431161.2010.512310
13. Yang G., Li B., Ji S., Gao F., Xu Q.: Ship detection from optical satellite images based on sea surface analysis. IEEE Geosci. Remote Sens. Lett. 641–645 (2013). https://doi.org/10.1109/LGRS.2013.2273552
14. Tang, J., Deng, C., Huang, G., Zhao, B.: Compressed-domain ship detection on space-borne optical image using deep neural network and extreme learning machine. IEEE Trans. Geosci. Remote Sens. 1174–1185 (2014). https://doi.org/10.1109/TGRS.2014.2335751
15. Qi, S., Ma, J., Lin, J., Li, Y., Tian, J.: Unsupervised ship detection based on saliency and S-HOG descriptor from optical satellite images. IEEE Geosci. Remote Sens. Lett. 1451–1455 (2015). https://doi.org/10.1109/LGRS.2015.2408355
16. Zou, Z., Shi, Z.: Ship detection in spaceborne optical image with SVD networks. IEEE Trans. Geosci. Remote Sens. 5832–5845 (2016). https://doi.org/10.1109/TGRS.2016.2572736

17. Liu, Z., Hu, J., Weng, L., Yang, Y.: Rotated region based CNN for ship detection. In: IEEE International Conference on Image Processing (ICIP), pp. 900–904. IEEE (2017). https://doi.org/10.1109/ICIP.2017.8296411
18. Chang, Y.-L., Anagaw, A., Chang, L., Wang, Y.C., Hsiao, C.-Y., Lee, W-H.: Ship Detection Based on YOLOv2 for SAR Imagery (2019). https://doi.org/10.3390/rs11070786
19. Chen, Y., Zhang, C., Qiao, T., Xiong, J., Liu, B.: Ship Detection in Optical Sensing Images Based on Yolov5 (2021). https://doi.org/10.1117/12.2589395
20. Jie, Y., Leonidas, L.A., Mumtaz, F., Ali, M.: Ship Detection and Tracking in Inland Waterways Using Improved YOLOv3 and Deep SORT (2021). https://doi.org/10.3390/sym13020308
21. Zhang Sr., Y., Shu Sr., J., Hu Sr., L., Zhou Sr., Q., Du Sr., Z.: A ship target tracking algorithm based on deep learning and multiple features. In: Proceedings of the Twelfth International Conference on Machine Vision (ICMV 2019), Amsterdam, The Netherlands (2020). https://doi.org/10.1117/12.2559945
22. Lee, W.-J., et al.: Detection and tracking for the awareness of surroundings of a ship based on deep learning (2021). https://doi.org/10.1093/jcde/qwab053
23. Haghbayan, M.-H., et al.: An Efficient Multi-sensor Fusion Approach for Object Detection in Maritime Environments (2018). https://doi.org/10.1109/ITSC.2018.8569890
24. Redmon, J., Divvala, S., Girshick, R., Harhali, A.: You Only Look Once: Unified, Real-Time Object Detection (2016). https://arxiv.org/abs/1506.02640
25. Bochkovskiy, A., Wang, C.-Y., Liao, H.-Y.M.: YOLOv4: Optimal Speed and Accuracy of Object Detection (2020). https://arxiv.org/pdf/2004.10934.pdf
26. Bewley, A., Ge, Z., Ott, L., Ramos, F., Upcroft, B.: Simple Online and Realtime Tracking (2016). https://arxiv.org/abs/1602.00763
27. Wojke, N., Bewley, A., Paulus, D.: Simple Online and Realtime Tracking with a DEEP Association Metric (2017). https://arxiv.org/abs/1703.07402
28. Ribeiro, R., Cruz, G., Matos, J., Bernardino, A.: A Data Set for Airborne Maritime Surveillance Environments (2019). https://doi.org/10.1109/TCSVT.2017.2775524
29. Gallego, A.-J., Pertusa, A., Gil, P.: Automatic Ship Classification from Optical Aerial Images with Convolutional Neural Networks (2018). https://doi.org/10.3390/rs10040511
30. Everingham, M., Van Gool, L., Williams, C.K.I., Winn, J., Zisserman, A.: The PASCAL Visual Object Classes (VOC) challenge. Int. J. Comput. Vis. **88**, 303–338 (2010). https://doi.org/10.1007/s11263-009-0275-4
31. Ira, A., Parico, B., Ahamed, T.: Real time pear fruit detection and counting using YOLOv4 models and deep SORT (2021). https://doi.org/10.3390/s21144803

Prediction of Fish Location by Combining Fisheries Data and Sea Bottom Temperature Forecasting

Matthieu Ospici[1](✉), Klaas Sys[2], and Sophie Guegan-Marat[1]

[1] Atos, CVLab, 38130 Echirolles, France
{matthieu.ospici,sophie.gueganmarat}@atos.net

[2] Flanders Research Institute for Agriculture, Fisheries and Food, Animal Sciences Unit, Ankerstraat 1, 8400 Oostende, Belgium
klaas.sys@ilvo.vlaanderen.be

Abstract. This paper combines fisheries dependent data and environmental data to be used in a machine learning pipeline to predict the spatio-temporal abundance of two species (plaice and sole) commonly caught by the Belgian fishery in the North Sea. By combining fisheries related features with environmental data, sea bottom temperature derived from remote sensing, a higher accuracy can be achieved. In a forecast setting, the predictive accuracy is further improved by predicting, using a recurrent deep neural network, the sea bottom temperature up to four days in advance instead of relying on the last previous temperature measurement.

Keywords: Computer vision · Machine learning · Spatiotemporal modelling · Fisheries · Remote sensing data

1 Introduction

Spatio-temporal modeling of catch or landings per unit effort data has a long history in fisheries sciences. Such models are mainly used to standardize fisheries dependent (i.e. landing and effort data from commercial fisheries) or fisheries independent (catch data by haul from scientific research cruises) information in order to derive an index of abundance [10]. Such an index is typically used to inform population dynamic models used to assess the exploitation status of fish stocks.

Yet few applications exist in which spatio-temporal models, fitted to catch or landings and effort data, are used to forecast the spatio-temporal dynamics of fish species. Such applications could be very valuable in the context of dynamic ocean management, a new fisheries management paradigm defined 'management that changes rapidly in space and time in response to the shifting nature of the ocean and its users based on the integration of new biological, oceanographic, social and/or economic data in near real-time' [11]. A notable example of dynamic ocean management are Real Time Incentives (RTIs), a system that allows to

This work has received funding from the European Union's Horizon 2020 research and innovation program under grant agreement NO. 825355.

S. Sclaroff et al. (Eds.): ICIAP 2022, LNCS 13233, pp. 437–448, 2022.
https://doi.org/10.1007/978-3-031-06433-3_37

reduce by-catches of fish by providing weekly maps with tariffs set according the expected catches of the fishery for a given set of species [9]. Clearly, applications of dynamic ocean management typically build on the fusion of alternative data sources, i.e. remote sensing, and advanced analytical processing and modeling techniques.

In this paper, we combine common fisheries dependent data sources, being daily landings reported by fishers through electronic logbooks and data on vessel activity provided through the Vessel Monitoring System (VMS), and environmental data (sea bottom temperature). With these data, we use a computer vision and machine learning pipeline to predict the spatio-temporal abundance of two commercially important species of the Belgian fishery, sole (*Solea solea*) and plaice (*Pleuronectes platessa*).

2 Related Works

Fish Distribution Forecasting. Different applications exist that forecast the spatio-temporal distribution of fish in relation to environmental conditions. Many of these application build on suitable habitat models in which so called environmental envelopes are constructed from species occurrence data or expert knowledge. These envelopes, constrained to some predefined shape (i.e. trapezoidal [5], shape constrained GAMS [2]), show the probability of species occurrence in response to a set of environmental variables and can as such be used to predict suitable habitats. Typically, the environmental envelopes are rather static and are used to make long term prediction of changes in species distribution for a given climate scenario.

An alternative approach was developed for the EcoCast application in which machine learning models are used to predict the daily probability of occurrence of swordfish, sea lions, leatherback turtles and blue sharks off the coast of Oregon and California [3]. These individual species maps are combined to identify suitable areas for fishing. The approach applies boosted regression trees on historical observer (1991–2014) (presence/absence) and tracking (2001–2009) data derived from tagging experiments combined with multiple environmental data to model the probability of species occurrence. This model uses the most recent observations (often the previous day) of the environmental parameters used in the model to make predictions for the current day. As such, it is assumed that the most recent environmental conditions are similar to the current environmental conditions, a shortcoming we aim to address in this study.

Temperature Forecasting with Machine Learning. Many approaches have been specifically proposed to predict temperature from satellite data with recurrent deep learning methods [7,13,14,17]. Simultaneously, related approaches, designed to predict images from videos, have been proposed [15,16]. To our knowledge, none of these works have been applied to the prediction of water temperature at the sea bottom.

3 Proposed Framework

To predict the probability of fish presence at a given location at sea, we propose a framework that consists of two main building blocks. The first is a deep learning model, trained on satellite images, made to predict the sea bottom temperature. The second block is a machine learning (gradient boosting) model, which uses the predicted temperature as input plus a set of features from a dataset that comprises information on landings of particular fish species at particular locations.

3.1 Datasets Used

Satellite Data. The satellite data of sea bottom temperature used in this work has been collected using the Copernicus Marine Environment Monitoring Service (CMEMS). Sea bottom temperatures are not directly observed but generated by the NEMO (Nucleus for European Modelling of the Ocean) ocean model, using the 3DVar NEMOVAR system to assimilate observations.

The region of interest for our experiments is in a rectangular area with latitude ranging from 50.07 to 55.33 and longitude ranging from 0.22 to 8.56. The resolution of the latitude and longitude is 80×80. Figure 1 shows the location of this area on the earth. We collected data from the years 2006 to 2020, giving a total of 5295 successive days of bottom temperature.

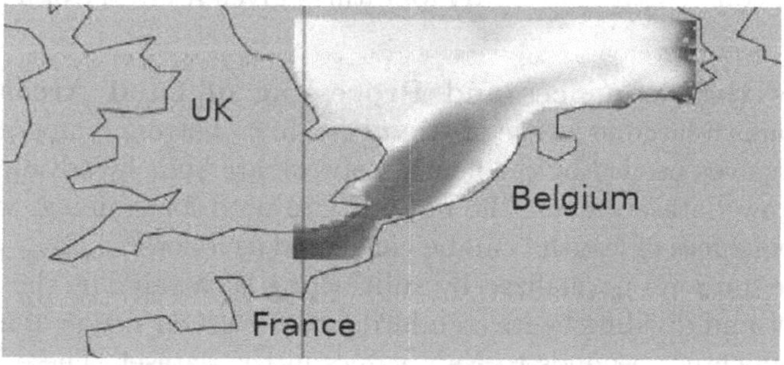

Fig. 1. The region of interest is an area located between north of France and south of UK (the area is in color in the picture). (Color figure online)

Fisheries Data. A dataset that contains species landings of the Belgian fishing fleet was compiled from electronic logbook and VMS data. The electronic logbook data comprise daily information on fish landings, a description of the fishing activity (gear and mesh size used, ICES statistical rectangle, fishing trip identifier, and day), and information on the fishing vessel (vessel identifier, vessel tonnage, engine power and length). The landing and effort data were combined according an agreed procedure as described in [4]. By merging both datasets, a final dataset with landing data at a higher spatio-temporal resolution is generated. Data was available from 2006 up to 2020. In total the dataset comprised 1 684 560 observations.

3.2 Predicting Bottom Temperature with Deep Learning

The first component of our framework is a deep learning model capable of forecasting the bottom temperature from a temperature history.

Problem Formulation. The bottom temperature for a given timestamp can be represented by a matrix BT of dimensions $M \times N$, where M denotes the resolution in longitude space and N the resolution in latitude space. Therefore, $BT[i][j]$ is the numerical value of the bottom temperature at coordinate i, j in degrees Celsius.

We can then construct a sequence of consecutive bottom temperatures for a given time period. Let BT_t be the temperature matrix at time t. The sequence is then the following:

$$BT_0, BT_1, \cdots, BT_{h-1}, BT_h, \cdots, BT_{h+p-1} \tag{1}$$

The length of the sequence is $l = p + h$ with p the number of matrices to predict and h the number of available histories. Our goal is therefore to predict p matrices of consecutive bottom temperatures given a history of h measures.

Creation of the Sequences and Processing of Land Areas. The raw dataset contains a large sequence of size $nb_R = 5295$ bottom temperature matrices. The sequences needed by the neural network are built by advancing step by step in the raw dataset. Thus, the dataset size used for training, which is the number of sequences of length l can be calculated as follows: $n_{sequence} = nb_R - l$.

Then, the data are normalized by subtracting each value by the mean value of the dataset and dividing by its standard deviation. Our region of interest contains land, which corresponds to NaN values in the dataset. These NaN values are replaced, after normalization, by a high negative value (-5 in our implementation).

Recurrent Neural Network. Our work is based on the framework PredRNN++ [16], we build our forecasting model on the building blocs introduced in [16] which are the recurrent units called Causal LSTM and the Gradient Highway Unit. We motivate our choice because this framework can handle short-term dependencies very efficiently and outperforms traditional approaches, often used in satellite data forecasting, such as LSTM or Convolutional LSTM [7,12,17].

3.3 Fish Prediction with Gradient Boosting

The fisheries dataset, which contains a substantial history of fishing operations in the North Sea, is used to build a machine learning model that aims to predict the probability of fish presence in the sea.

Since plaice and sole are bottom dwelling species that have a narrow thermal preference range, the temperature of the sea floor is assumed to influence the location of fish. Hence, we use the temperature prediction model presented in the previous section among the features used by our machine learning model. Therefore, the resolution of the fish prediction is the same as the resolution of the temperature map. As discussed in the previous section, the resolution of the temperature map is 80×80, so the resolution of the fish prediction is the same.

Gradient Boosting to Predict Potential Fishing Zone. Since the fisheries dataset contains many different types of features, we chose a decision tree-based machine learning algorithm. Specifically, we used the lightGBM [6] framework with GPU acceleration [18]. We build two models for two species of fish: one to predict the probability of presence of plaice, and one to predict the probability of presence of sole. The output of the models is then a probability of fish presence for each point of the 80×80 map.

Features Selection. From the original dataset, we build seven features to train the machine learning models: longitude (*i*), latitude (*ii*), bottom temperature (*iii*) and two features for the day (*iv*, *v*) and two for the year (*vi*, *vii*). Indeed, we transform day and month, which are cyclic in nature, into two new features using a sine and cosine transformation [1]. Furthermore, because the fish are predicted in an 80×80 resolution grid, the latitude and longitude correspond to the coordinates of the point in which the prediction is made and are therefore integers ranging from 0 to 79.

4 Experimental Settings

4.1 Train-Test-Evaluation Splits

The recurrent neural network is trained and evaluated on satellite data from sequences defined in Sect. 3.2. We follow the evaluation procedure described in [17], the training, validation, and test sets are chosen to follow each other in chronological order. In our study, the training set consists of the years from 2006 to 2018, and the evaluation set and test set are formed from the years 2019 to 2020. The plaice and sole gradient boosting models are trained on both satellite and fisheries dataset with a similar train-test-evaluation split. We take the years from 2019 to 2020 as test and evaluation sets.

4.2 Hyperparameters

The hyperparameters of the recurrent neural network for temperature forecasting are as follows. The input and output lengths are set to 4, which means that we predict 4 days with a history of 4 days. The optimization is performed with Adam [8] with a learning rate of 0.001. The loss function used in our experiments is the $L1 + L2$ function which gives the best performance compared to using $L1$ or $L2$ functions only. The batch size is set to 8. Following the neural network architecture introduced in [16], each layer is built with 4 causal LSTMs with 128, 64, 64, 64 channels and a 128 channel gradient highway unit. In all recurrent units, the convolution filter size is set to 5.

The two lightGBM models use the standard hyperparameters of the framework [6] with a number of leaves set to 23. The objective is binary classification and the loss used is the log-loss.

4.3 Metrics

We evaluate the temperature forecasting model using two metrics well suited to regression problems: MAE (mean absolute error) and RMSE (root-mean-square error). For the fish prediction models, we use the F1 score, which is well suited to binary classification problems.

$$\mathit{RMSE} = \sqrt{\frac{1}{mn}\sum_{i=1}^{m}\sum_{j=1}^{n}(x_{ij} - y_{ij})^2} \quad , \quad \mathit{MAE} = \frac{1}{mn}\sum_{i=1}^{m}\sum_{j=1}^{n}|x_{ij} - y_{ij}|$$

$$\mathit{F1} = \frac{2 * \mathit{Precision} * \mathit{Recall}}{\mathit{Precision} + \mathit{Recall}}$$

m and n correspond to the latitude and longitude resolutions ($m = n = 80$). Since we have land in our study area, we make sure that for both RMSE and MAE, the coordinates i, j correspond only to points that are in the sea.

5 Performance Evaluation

To measure the performance of our framework, we first evaluate the performance of its two component and then the performance of the entire framework. For the fish prediction models, we also verify that they have captured the link, which exists in the marine environment, between bottom temperature and fish abundance.

5.1 Temperature Forecasting Performance

We first ensure that the forecasting performance of our model is better than simply taking the last known day as a predictor of future days' temperature. Then, because we have chosen to set the temperature value of the areas with land to a high negative value, we validate this approach.

Last Know Day Estimator Compared to the Proposed Forecasting Model

Motivation. For a given sequence such as the one in Eq. (1), we call the last day estimator the temperature matrix BT_{h-1}. We can then calculate the error (RMSE and MAE) between this matrix and the matrices $BT_h \cdots BT_{h+p-1}$.

Similarly to [14], we perform this experiment because we observed that the variation of the temperature is extremely low between the different days. For this reason, it would be easy to develop a forecasting algorithm with a low error rate by taking only the last know day as prediction. It is then essential to ensure that our forecast model is able to outperform the last known day estimator.

Results. The errors obtained by simply using the last known day are shown in orange in Fig. 2a for MAE and Fig. 2b for RMSE. As expected, using the last known day as the temperature estimator for the following days results in a relatively small error (0.10 °C of MAE) for the first day. The error increases significantly for the following days (for predicting the fourth day, the error is 0.3 °C).

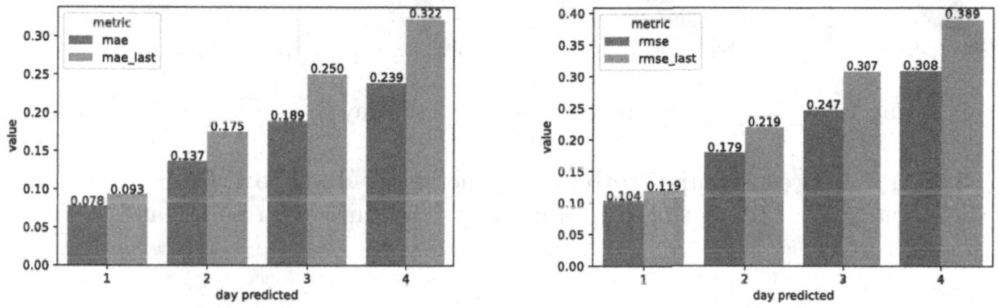

(a) MAE (b) RMSE

Fig. 2. For both figures, the error is calculated between the forecast and the ground truth (blue bars) and between the last day estimator and the ground truth (orange). The lowest values are the best. (Color figure online)

In both Figs. 2a and 2b, we plot the error for the day predicted by our model (in blue). We can observe here that our model outperforms the estimator of the last known day with both RMSE and MAE metrics for all days. Furthermore, the percentage of error between the estimator of the last known day and the value predicted by our model increases when we progress in the predictions. The error (in terms of RMSE) is 14.4% higher for the estimator of the last known day for day 1 and 26.3% for day 4.

Repartition of the Errors. During the preprocessing of the dataset, we replace the NaN values (which represent the land) by a high negative value.

To validate this approach we compute the RMSE between the predicted temperatures and the ground truth for days 1–4 and we show the RMSE error for each cell of the error matrix. Results are depicted in Figs. 3a to 3d.

For the predicted days, the errors are fairly well distributed in all areas. The errors around the interface between the sea and the land are comparable to those of the other zones, which validates our approach.

It should be noted, however, that these areas at the interface are among the areas that concentrate the most errors, especially when we advance in the prediction. This can be explained by the fact that these are shallow areas where the temperature variations are the most important from one day to another. We illustrate this on the Fig. 4. We calculate the location of the RMSE errors between the Last know day estimator and the ground truth of day 4. There is no prediction here, we only compare the ground-truth. The largest errors correspond to the largest temperature variations. This confirm that the areas where the prediction errors are the largest correspond to the areas where the temperature variations are the most important.

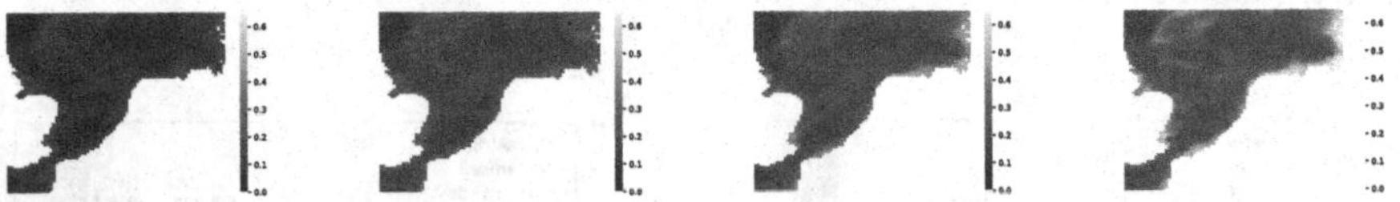

(a) day 1 (b) day 2 (c) day 3 (d) day 4

Fig. 3. RMSE between ground truth and prediction for day 1 to 4. RMSE is computed for each points (80 × 80) in the area of interest. The white area is the land.

5.2 Influence of Bottom Temperature for Fish Prediction

Link Between Temperature and Quality of Prediction. We train both fish prediction models with the fisheries dataset. To validate the relevance of using bottom temperature for the prediction of sole and plaice presence, we perform two experiments. In the first, we train and test a model with and without the "bottom temperature" feature. The results are depicted in Table 1 under the columns "w/ temp." and "w/o temp.". For this experiment, the bottom temperature is not predicted, we use the ground truth value from the Copernicus dataset.

According to the result presented Table 1, using bottom temperature for both plaice and sole models increases the quality of the prediction. This justify our approach of using bottom temperature for fish presence prediction.

Link Between Temperature and Abundance of Fish. To more precisely measure how the model has captured the link between temperature and presence of fish, the following experiment was conducted. We added an offset, from −2 °C to 7 °C, to the bottom temperature of each grid point on a given day and computed the mean of probability of fish presence in the study area using the sole and plaice model. Results are depicted in Fig. 5. As shown, when we decrease (resp. increase) the temperature, the model predicts a lower (resp. higher) occurrence probability of fish on average. The stronger positive correlation between species occurrence and seawater temperature for sole compared to plaice can be explained by the fact that sole is a Lusitanian species that prefers warmer water. Moreover, the study area comprises the northernmost habitat of the sole. In contrast, plaice is widely distributed over the entire North Sea which may explain why the response of species occurrence to temperature is less pronounced.

Table 1. F1 score for plaice (PLE) and sole (SOL) models

Fish type	w/ temp	w/o temp	pred. day 1	pred. day 2	pred. day 3	pred. day 4
SOL	82.8	81.4	82.7	82.7	82.6	82.4
PLE	82.0	80.6	81.8	81.8	81.8	81.8

Fig. 4. RMSE errors between the Last know day estimator and the ground truth of day 4.

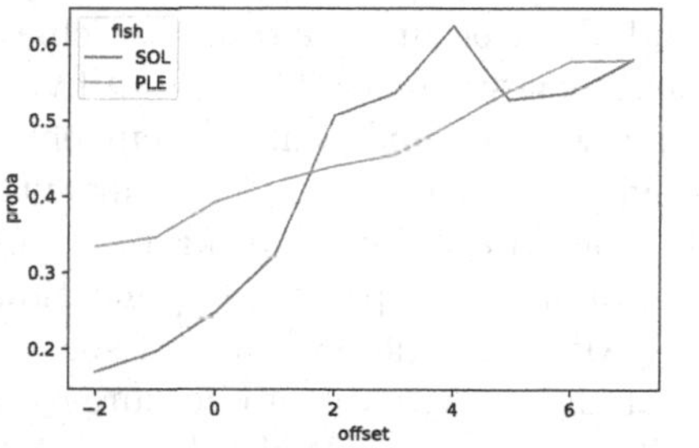

Fig. 5. Average probability of fish presence when an offset is added to bottom temperature.

5.3 Performance of the Complete Framework

The accuracy of the fish prediction is evaluated using the predicted bottom temperature as follows: for each day d of the fisheries data test set, we construct four predictions. For the prediction at day 1 (pred. day 1 in Table 1), we construct a sequence $BT_{d-4}, BT_{d-3}, \cdots, BT_{d-1}$ from the ground truth measurements. Thus, with our bottom temperature prediction model, we can predict day d by taking the first day of the predicted sequence. For the prediction of day 2, we build a sequence $BT_{d-5}, BT_{d-4}, \cdots, BT_{d-2}$ and take the second predicted day. We proceed in the same way for day 3 and day 4. We then measure the quality of the prediction in terms of F1 score when we use these different predictions as the temperature for day *d*. The results are presented in Table 1.

Our experiments show that, even the prediction of temperature contain small errors in term of MAE or RMSE, using this predicted value shows excellent results for 1,2,3 and 4 days ahead. With a 4-day prediction we have performance extremely close to using a ground truth value for both sole and plaice.

6 Discussion and Conclusion

Our study showed how a pipeline of advanced analytical tools and data fusion can be used to make a short term forecast of species occurrence in a highly variable environment such as the marine ecosystem.

We shown that a recurrent neural network building blocks designed to predict successive frames of a video can be efficient for sea bottom temperature forecasting. During the evaluation, we noticed that the variation in temperature is small and even using only the last known day as a predictor, we can have small errors. We therefore carefully ensured that our model outperformed this simple estimator. Moreover, our choice to replace the value of the coordinates containing land by a high negative value has been validated.

By enriching a fisheries dependent dataset with features on environmental conditions such as sea bottom temperature, a higher predictive accuracy of sole and plaice occurrence could be achieved with a gradient boosting algorithm. This accuracy was further increased by using the predicted sea bottom temperature that was inferred from a recurrent deep learning model. Since information on environmental conditions is currently available in near real time through satellite based earth monitoring programs such as Copernicus, this may provide an opportunity for practical applications of dynamic ocean management.

Although the accuracy of sole and plaice occurrence could be increased by adding an environmental feature to the fisheries dataset, the overall accuracy can still be improved. It should be noted however that the quality of the fisheries dependent data is limited. The landing data reported by fishers are estimated with a fault tolerance of 20%. Moreover, misreporting is known to occur wherewith species landings from one area are reported in another area. Finally, the landing data are reported on a daily basis and distributed over the GPS positions of a fishing vessel that were recorded during that day according [4]. To do so, an equal distribution of landings over all fishing position is assumed which is unlikely to be the case in reality. As more data will become available in the future, and alternative catch monitoring techniques, e.g. image analysis, will be implemented, the predictive accuracy of fish occurrence models relying on fisheries dependent information is likely to increase.

References

1. Chakraborty, D., Elzarka, H.: Advanced machine learning techniques for building performance simulation: a comparative analysis. J. Build. Perform. Simul. **12**, 1–15 (2018)
2. Citoresa, L., Ibaibarriaga, L., Lee, D.J., Brewer, M., Santos, M., Chust, G.: Modelling species presence-absence in the ecological niche theory framework using shape-constrained generalized additive models. Ecol. Model. **418**, 108926 (2020)
3. Hazen, E.L., et al.: A dynamic ocean management tool to reduce bycatch and support sustainable fisheries. Sci. Adv. **4**(5), 1–8 (2018)
4. Hintzen, N.T., et al.: VMStools: open-source software for the processing, analysis and visualization of fisheries logbook and VMS data. Fish. Res. **115–116**, 31–43 (2012)
5. Kaschner, K., Watson, R., Trites, A.W., Pauly, D.: Mapping world-wide distributions of marine mammal species using a relative environmental suitability (res) model. Marine Ecol. Progr. Ser. **316**, 285–310 (2006)
6. Ke, G., et al.: LightGBM: a highly efficient gradient boosting decision tree. In: Guyon, I., (eds.) Advances in Neural Information Processing Systems, vol. 30. Curran Associates, Inc. (2017)
7. Kim, M., Yang, H., Kim, J.: Sea surface temperature and high water temperature occurrence prediction using a long short-term memory model. Remote Sens. **12**(21), 3654 (2020)
8. Kingma, D.P., Ba, J.: Adam: a method for stochastic optimization. In: Bengio, Y., LeCun, Y. (eds.) 3rd International Conference on Learning Representations, ICLR 2015, San Diego, CA, USA, 7–9 May 2015, Conference Track Proceedings (2015)
9. Kraak, S.B., et al.: RTI ("real-time incentives") outperforms traditional management in a simulated mixed fishery and cases incorporating protection of vulnerable species and areas. Fish. Res. **172**, 209–224 (2015)
10. Maunder, M.N., Punt, A.E.: Standardizing catch and effort data: a review of recent approaches. Fish. Res. **70**(2–3), 141–159 (2004)
11. Maxwell, S.M., et al.: Dynamic ocean management: defining and conceptualizing real-time management of the ocean. Mar. Policy **58**, 42–50 (2015)
12. Moskolaï, W., Abdou, W., Dipanda, A., Kolyang: Application of deep learning architectures for satellite image time series prediction: a review. Remote Sens. **13**, 4822 (2021)
13. Qiao, B., Wu, Z., Tang, Z., Wu, G.: Sea surface temperature prediction approach based on 3D CNN and LSTM with attention mechanism. In: 2021 23rd International Conference on Advanced Communication Technology (ICACT), pp. 342–347, February 2021. (ISSN: 1738–9445)
14. Shi, X., et al.: Deep learning for precipitation nowcasting: a benchmark and a new model. In: Proceedings of the 31st International Conference on Neural Information Processing Systems, pp. 5622–5632. NIPS 2017, Curran Associates Inc., Red Hook, NY, USA (2017)
15. Su, J., Byeon, W., Kossaifi, J., Huang, F., Kautz, J., Anandkumar, A.: Convolutional tensor-train LSTM for spatio-temporal learning. In: Larochelle, H., Ranzato, M., Hadsell, R., Balcan, M.F., Lin, H. (eds.) Advances in Neural Information Processing Systems, vol. 33, pp. 13714–13726. Curran Associates, Inc. (2020)

16. Wang, Y., Gao, Z., Long, M., Wang, J., Yu, P.S.: PredRNN++: towards a resolution of the deep-in-time dilemma in spatiotemporal predictive learning. In: Dy, J., Krause, A. (eds.) Proceedings of the 35th International Conference on Machine Learning. Proceedings of Machine Learning Research, vol. 80, pp. 5123–5132. PMLR, 10–15 July 2018
17. Xiao, C., et al.: A spatiotemporal deep learning model for sea surface temperature field prediction using time-series satellite data. Environ. Modell. Softw. **120**, 104502 (2019)
18. Zhang, H., Si, S., Hsieh, C.J.: GPU-acceleration for large-scale tree boosting, p. 3 (2017)

Robust Human-Identifiable Markers for Absolute Relocalization of Underwater Robots in Marine Data Science Applications

Philip Herrmann[1,2], Sylvia Reissmann[1], Marcel Rothenbeck[1], Felix Woelk[2], and Kevin Köser[1](✉)

[1] GEOMAR Helmholtz Centre for Ocean Research Kiel, Kiel, Germany
kkoeser@geomar.de

[2] Kiel University of Applied Sciences, Kiel, Germany

Abstract. Since global navigation satellite systems (GNSS) for determining the absolute geolocation do not reach into the ocean, underwater robots typically obtain a GNSS position at the water surface and then use a combination of different sensors for estimating their pose while diving, including inertial navigation, acoustic doppler velocity logs, ultra short baseline localization systems and pressure sensors. When re-navigating to the same seafloor location after several days, months or years, e.g. for coastal monitoring, the absolute uncertainty of such systems can be in the range of meters for shallow water, and tens of meters for deeper waters in practice. To enable absolute relocalization in marine data science applications that require absolute seafloor positions in the range of centimeter precision, in this contribution we suggest to equip the monitoring area with visual markers that can be detected reliably even in case they are partially overgrown or partially buried by sediment, which can happen quickly in coastal waters. Inspired by patterns successful in camera calibration, we create robust markers that exhibit features at different scales, in order to allow detection, identification and pose estimation from different cameras and various altitudes as visibility (and therefore the maximum possible survey altitude) in coastal waters can vary significantly across seasons, tides and weather. The low frequency content of the marker resembles a human-readable digit, in order to allow easy identification by scientists. We present early results including promising initial tests in coastal waters.

Keywords: Absolute localization · Underwater robot · Robust marker

1 Introduction

Monitoring the extent of sea grass coverage, bacterial mats, deposition of sediment or drift of disposed munitions in coastal waters requires repeated measurements with sufficient precision. For areas that cannot be surveyed by satellites,

S. Sclaroff et al. (Eds.): ICIAP 2022, LNCS 13233, pp. 449–460, 2022.
https://doi.org/10.1007/978-3-031-06433-3_38

e.g. in deeper or turbid waters, optical observations using underwater cameras promise millimeter resolution and have the potential for detecting subtle changes. However, when repeatedly bringing a camera to the seafloor in such monitoring scenarios, not only the sensor's resolution is important, but also how good the sensor or the robotic platform can be localized. Detecting changes at centimeter resolution requires that also image data can be registered to older image data with centimeter accuracy. Although approaches such as differential GPS can theoretically provide absolute localization with such accuracy on land when using expensive devices and services, in practice off-the-shelf GNSS receivers without subscriptions to extra correction services have an accuracy in the range of few meters. Underwater robots employing such sensors typically geo-localize themselves at the surface just before the dive. When diving down, localization uncertainty increases since robots have to integrate accelerations and speed over ground in a dead reckoning approach. This leads to the problem that exactly revisiting a seafloor spot is very challenging, and often uncertainty of the absolute position at the seafloor can be in the range of several meters. In the deep ocean robots typically take hours to reach the seafloor increasing the uncertainty over time. In these scenarios position uncertainty can be an order of magnitude larger, even when techniques like acoustic localization with respect to surface vessels (e.g. ultra short baseline localization - USBL) is used, because these techniques suffer from refraction at water layers and multi-path propagation.

To facilitate absolute localization, or at least precise re-localization, we propose to deploy optical markers at the seafloor. In contrast to markers on land however, in coastal waters such markers have to be detectable under harsh conditions [7]. Particles floating in the water, as well as sediment or algae can partially occlude or bury the marker. In case automatic detection fails, also human identification of the marker should be possible as a fallback solution. Due to different visibility conditions and camera systems, markers should be detectable also at different scales. Consequently, in this contribution we present early results in exploring a base marker pattern successful in robot camera calibration, augmented by a human-readable code (see Fig. 1 for an example). We will review

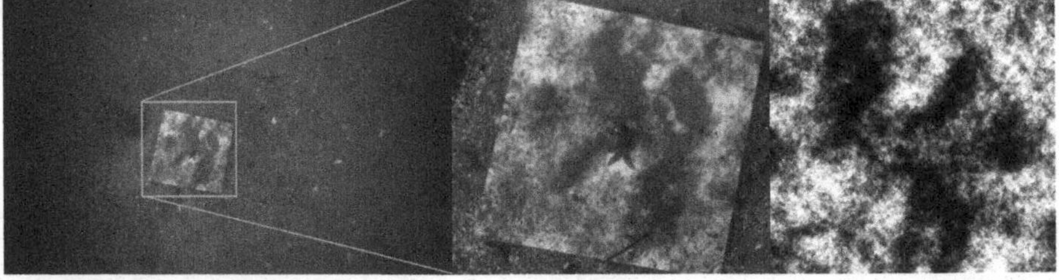

Fig. 1. The left image was captured by an underwater robot in coastal waters. It shows a marker (magnified, center) for pose estimation in a typical case where it is partially occluded (i.a. by a starfish) and overgrown. The right image shows the original marker texture coded with a human-readable four.

related work in the next section before we outline the marker design in Sect. 3 and present first experiments in Sect. 4.

2 Related Work

Several marker systems have been proposed for camera pose estimation such as ARToolkit [6], ARToolkitPlus [15], ARTag [3] or Aruco [5], to name just a few popular examples. Fiala [3] discusses design principles of such markers. Besides allowing pose estimation, the markers often encode also an ID that identifies which marker is actually detected, which facilitates distinguishing multiple markers. These IDs are typically encoded as a binary number using a black and white pattern. The encoding can contain redundancy to be robust in presence of some missing bits, however this is under the assumption that the entire marker is detected in general, and only a small fraction of the bits of the code are flipped or missing. The main applications of these approaches lie in augmented reality, where the scene is usually well illuminated and visibility is good. In commercial state-of-the art augmented reality (AR) application SDKs applications (e.g. ARCore, ARKit, Vuforia), tracking does no longer rely on predefined artificial black and white markers. Instead a generic image tracking algorithm for planar targets is provided, where the definition of the tracking target (i.e. the image) is handed over to the developer. The details of the algorithms are generally not disclosed, however most algorithms compute and save some abstract description of the image (i.e. features and descriptors for matching like SIFT [9] or ORB [12] or structures for tracking [1]) to a database beforehand and use this description in the actual tracking stage.

This assumes that the scene stays stable during application, which becomes more challenging in outdoor or even underwater applications in which long-term monitoring (days, months or years) is desired. Also for indoor simultaneous localization and mapping, markers can provide a fixed reference that also resolves the scale issue (see e.g. [10]). In this contribution we also suggest to extend the scene by adding artificial markers to enable long-term-stable pose estimation. Markers have been used for underwater pose estimation beforehand [2,13,16], however the studies focused on the optical effects of water and did not investigate occlusion due to sediment, algae or particles.

So, instead of using common square-shaped markers as discussed above, we require an approach that works under very challenging visibility conditions, in partial occlusion and at different scales. Towards this end, Li et al. [8] suggested to use a feature-rich texture as a calibration pattern and then to employ feature matching as well as robust estimation for calibration board detection. The authors reverse-engineer the feature detection process of the SIFT detector [9] that resembles a band-pass filter, employing a pyramidal image representation. They suggest to generate random image structures at different frequencies and combining these multiple resolutions into one image that afterwards contains features at multiple scales. We build on this approach by modifying the underlying pattern and using multiple patterns as markers rather than as a calibration board.

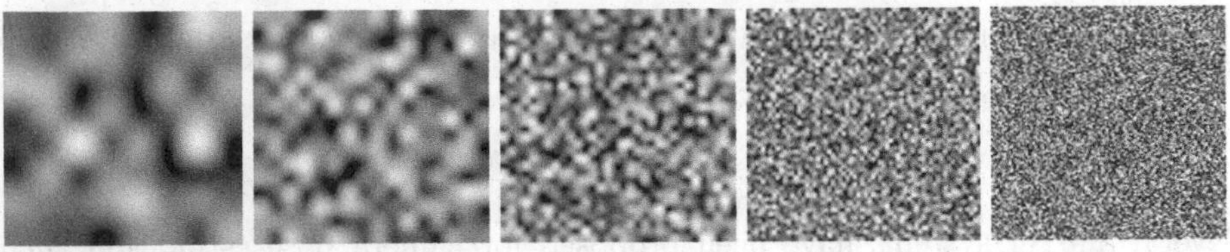

Fig. 2. Creating noise at different pyramid/resolution levels according to Li et al. [8].

Contributions: Since our application targets pose estimation in robotics, the camera is physically available, exactly known, and can usually be calibrated before the mission [11,14]. By using calibrated cameras, we can therefore transfer features detected in distorted fisheye images into rays of the local camera coordinate system and formulate the marker detection problem as a feature-based robust pose estimation problem, allowing for detection under heavy geometrical distortion. We can simply undistort features coordinates rather than undistorting entire images. Additionally, we modify the low frequency content of the marker in order to make it human-readable. The resulting markers are detectable under significant occlusion.

3 Robust Marker Design

Our marker design follows practical considerations for coastal underwater use, which are

1. The marker must be detectable from a wide range of distances and different cameras, including ultra wide-angle or medium-field of view cameras that can have a large variety of magnifications.
2. The marker must be detectable also in distorted fisheye lenses that let straight lines appear extremely curved toward the image boundary.
3. The marker should be detectable even when only partially visible in the image, e.g. because of occlusion by sediment or algae.
4. The marker should also support re-identification by humans, in order to manually screen mission data and to identify the image in which a certain marker is visible.

Requirement 1 is motivated by the fact that the water turbidity, and thus underwater visibility, can change significantly with season or weather, and so the survey altitude of the autonomous underwater vehicle (AUV) has to be adapted. Also, we would like to be able to continue the monitoring of a particular area also using another robot, or another camera or lens. Requirement 2 means that we cannot rely on detecting straight image lines. Additionally, fisheye images cannot easily be undistorted into a single pinhole image without loss of local resolution, and hence square marker detection in image space is more complex for fisheye images. Requirement 3, though self-explanatory, is the most important in marine applications, where the environment cannot be controlled and

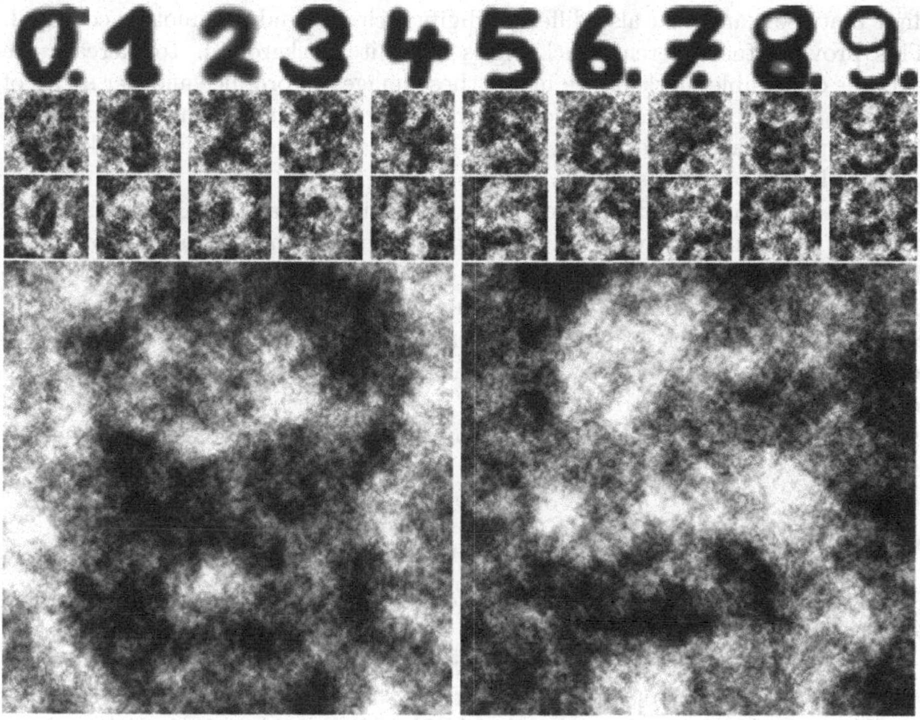

Fig. 3. Top row: Hand-drawn digits that are not rotationally self-similar and are not rotationally similar to any other digit. Rows 2+3: Marker design for ten different markers (0–9) black on white (top row) and white on black background. Bottom: The details view of marker 8 black on white (left) and marker 5 white on black (right).

can undergo extreme dynamics, due to currents, sedimentation or biological processes. Markers deployed here for a longer period of time will be very difficult to detect. The final requirement is a practical one. When looking at the calibration boards produced by the method of [8], we realized that they are really difficult to remember by a human, because they contain no semantic or regular structures. Due to the challenging underwater environment, it is possible that fully automated monitoring approaches might fail and might need human intervention or assistance to still make god use of expensively collected survey data. Also, for debugging it is convenient if a human can recognize the marker.

Creating the Random Marker Textures

We follow the approach of [8] to generate a cascade of images with different resolution and random textures (see Fig. 2) up to an image resolution of 2048×2048 pixels. In the resolution of 32×32 pixels the random dot image is replaced by a hand-drawn digit from 0 to 9 over uniform background. Care has to be taken that the pattern does not exhibit any rotational self-symmetries, e.g. for the digit 8 at 180°, and that also different digits such as 6 and 9 cannot be confused. This provides ten different black digits on white background. To increase the number of possible markers, we invert these images to produce another series of white digits on black background. The resulting markers and their hand-drawn 32×32 pixels layers are shown in Fig. 3. In case more different markers are needed, letters or other symbols could be used.

For each image a different random pattern is used at the higher resolution patterns and combined with the digit image in the same way as was proposed in [8] for the lowest resolution random pattern for the calibration board. The final pattern is printed at a size of 30 cm × 30 cm and coated onto an aluminum composite sheet, which is laminated afterwards. The sheet is then mounted onto a 30 cm × 30 cm paving stone, which is heavy enough to avoid motion such as drift of the marker underwater. The markers have been deployed in coastal waters in an area with little to moderate sea grass coverage but strong algae growth and significant currents.

Marker Detection and Pose Estimation

First, we detect features in the marker textures used for printing. Since the real world size of the markers is known, the detected features can be directly associated with 3D points on the marker surface, i.e. in the local marker coordinate system. These 3D points and their descriptors are stored in a dictionary.

When an image captured by the AUV is processed, the algorithm starts with an optional step of initial contrast improvement, since illumination can be very uneven across the image. Afterwards features are extracted from the image using the SIFT [9] feature detector. The features are matched against the offline created dictionary of features previously generated from the known marker pattern.

The matching algorithm provides the closest and second closest neighbor on the marker for each feature in the image and a distance between the descriptors for each match. As suggested by Lowe [9], only the features with a high distance ratio between closest and second closest neighbor are kept. The next step is to correct the distorted keypoint coordinates in the image with a camera specific algorithm to a normalized pinhole camera model with standard focal length of 1 (effectively to rays in the local camera coordinate system). At this point a set of 3D points on the marker plane, and their tentatively corresponding ideal image positions (viewing rays) in a normalized image are known (2D–3D correspondences). These correspondences are finally used to robustly estimate the pose using the RANSAC [4] method. Due to the construction of the markers, all 3D points are coplanar, allowing for simple homography estimation from a

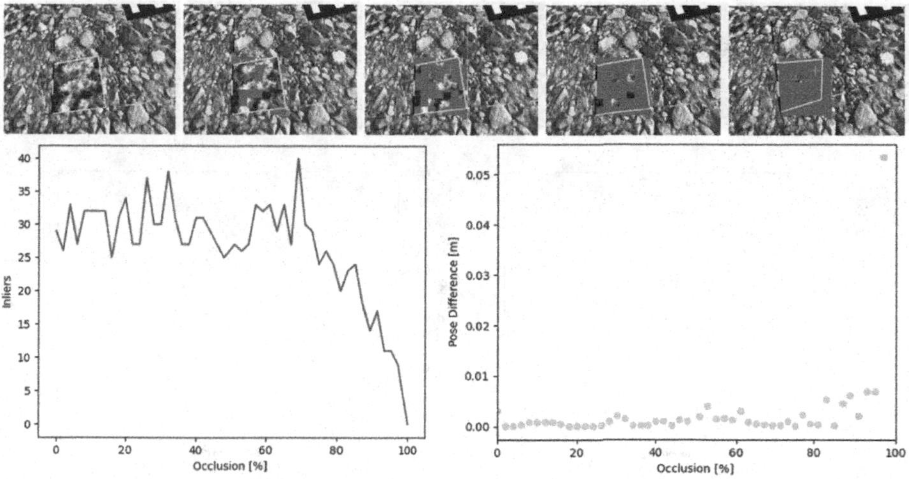

Fig. 4. Top: Artificially generated synthetic images of marker 7 to analyze the robustness of the detection approach to partial occlusion. The pose estimation breaks down when 48 of the 49 tiles are occluded. Bottom left: The number of inliers found by the RANSAC algorithm for pose estimation for different occlusions percentages. Bottom right: The absolute difference between the camera position from the un-occluded marker (top left image) and the partially occluded marker in meters is plotted against the occlusion percentage.

minimum set of 4 correspondences. The corresponding pose is extracted using a decomposition of the homography. Finally, the pose is optimized on all inliers using a Levenberg-Marquard algorithm. The marker is considered detected if a sufficient number of inliers have been found. This threshold controls accuracy versus specificity and can be set application dependent (either to detect anything that looks like a marker, or maybe for online scenarios, to only consider very safe detections).

4 Experiments

Occlusion. In order to evaluate the robustness with respect to occlusion, synthetic, clean, marker-images are artificially occluded in steps of increasing percentages. For this the marker area is subdivided into 7×7 tiles and an increasing portion of these tiles are randomly overwritten with uniform gray color (see top row of Fig. 4). To analyze the performance of the algorithm, each image is evaluated first by investigating the number of features used for pose estimation (i.e. inliers from RANSAC algorithm) and second by comparing the resulting pose for each image with the resulting pose of the corresponding un-occluded image. The same analysis has been conducted on a real-world image (Fig. 5).

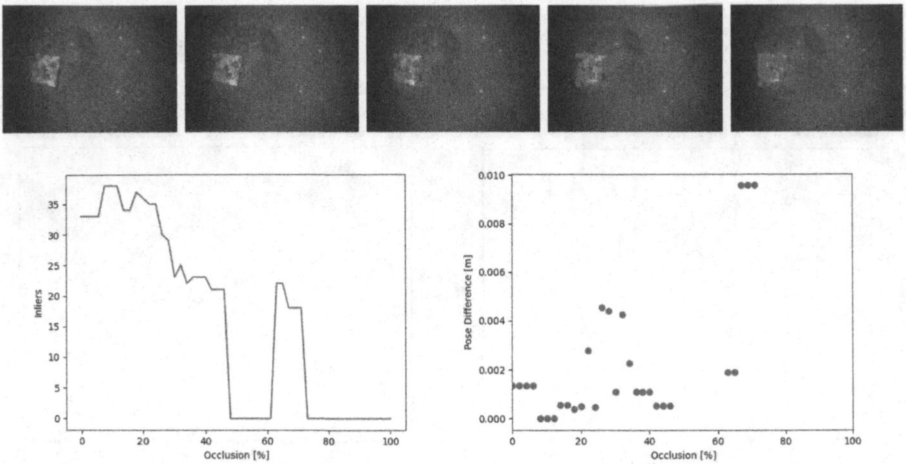

Fig. 5. (top) Underwater shot of the marker with the black four, which already has some natural occlusion. The pose can be detected until 20 of the 49 tiles are covered. (bottom left) The number of inliers found by the RANSAC algorithm for pose estimation for different occlusions percentages. (bottom right) The absolute difference between the camera position from the un-occluded marker (top left image) and the partially occluded marker in m is plotted against the occlusion percentage.

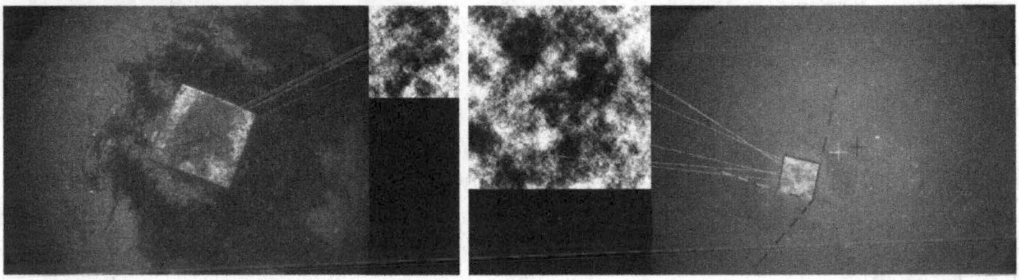

Fig. 6. Selected real world examples demonstrating the feasibility of the proposed approach. The poses have been successfully computed despite the partial occlusion or difficult visibility of the marker.

Additionally, we present the results on several real-world images that show different levels of occlusion (see Fig. 6). The markers have been deployed in shallow water in (location hidden for review). The images were taken by an AUV with a fish eye camera. To demonstrate the correctness of the extracted poses, the 3D borders of the marker have been projected together with a 3D coordinate system into the image. Each dash along the axes of the coordinate system corresponds to 10 cm in real world.

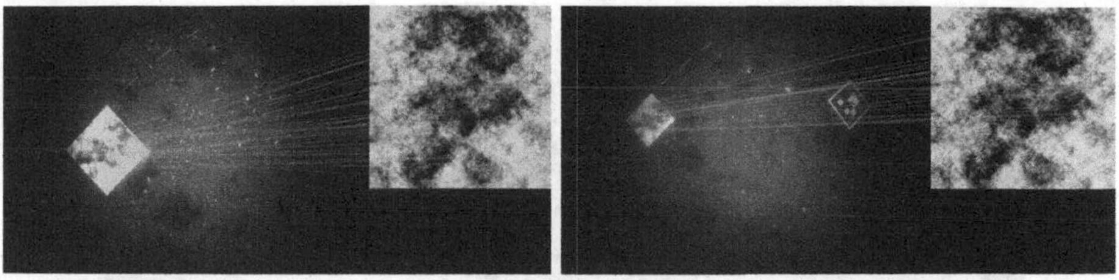

Fig. 7. A specific location with marker 7 has been revisited during different dives. Before each dive the AUV has resurfaced and obtained a new GPS-fix. The camera position has been projected to the seafloor (black cross). The difference between the two visually computed positions is 41 cm. The difference between the two positions from GPS fix and subsequent dead-reckoning is more than 3 m, probably due to differences in GPS initialization.

Geolocation Uncertainty. To demonstrate the advantages of the vision based position estimation given traditional localization, a certain location with a specific marker has been revisited in two subsequent dives (see Fig. 7). Before each dive, the AUV uses GPS at the sea surface to geo-localize itself. Afterwards, the robot dives down. The pose of the marker is estimated in the camera coordinate system. Although it can be seen that the marker is still located at the same seafloor spot, using the non-visual navigation more than 3 m difference is reported, which we attribute to the GPS uncertainty. The small black cross in the images (Fig. 7) represents the projection of the camera position to the plane defined by the marker. Assuming that the marker is located planar at the seafloor, the 3D plane defined by the marker is used as a rough approximation of the seafloor.

Success and Failure Cases. Since this is an early study, we only demonstrate some example cases of successful marker detection in Fig. 6 and of failures in Fig. 8. The first one shows that markers can in principle be detected under challenging conditions. However, parameters such as number of extracted features and RANSAC iterations have a huge impact on performance and so we believe that detection rates and accuracy will depend on the available computing resources: An online system might just detect the slightly occluded markers, whereas in postprocessing it is possible to also identify more heavily occluded markers. Still, as can be seen in Fig. 8, also the new markers suffer from common problems. For monitoring approaches where daily changes have to be found over a period of weeks, they are very promising. However, when deployed in coastal waters for months or years, they have to be cleaned in regular intervals (or, when monitoring often, the marker texture degradation could be learned).

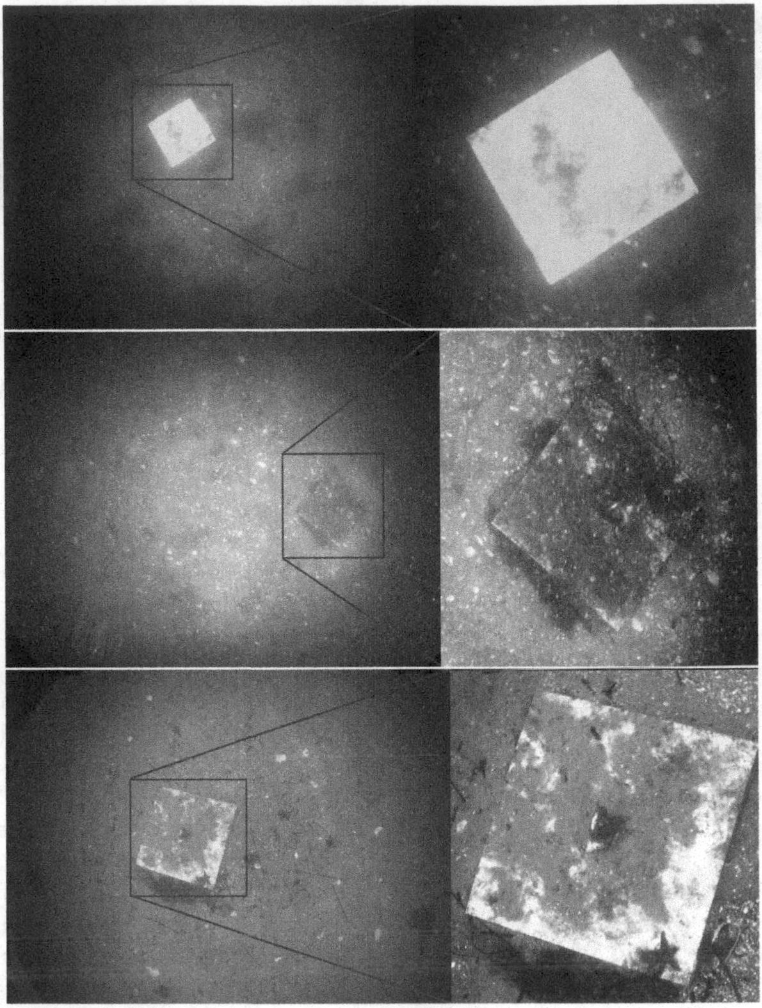

Fig. 8. Selected cases where detection is too challenging. Top row: Directly after deploying the markers they are extremely bright and reflective. Depending on flash settings, they appear much brighter than the surrounding and easily become overexposed. Center row: After 5 months between spring and autumn, the marker is heavily overgrown and cannot be detected anymore. Bottom row: Sediment deposited on the marker largely occludes it. The two bottom cases indicate that markers have to be coated or cleaned regularly in coastal waters.

5 Conclusion

In this contribution we argued that precise re-localization of underwater robots is required in many marine monitoring and data science applications that aim at tracking changes over time or detecting temporal patterns, introduced e.g.

by tidal cycles, currents or seasons. Since no absolute localization techniques such as GNSS are available underwater, we have proposed using optical markers that can be re-detected easily and serve as visual anchors. A key challenge is that markers will be partially occluded by vegetation or sediment or can be overgrown by algae, which makes the use of classical markers difficult. Instead, we have shown first promising results that our markers can be detected even in partial occlusion and under challenging visibility conditions. An early test showed that traditional navigation-based position of our robot differed by 3 m, which would be difficult to identify without absolute references at the seafloor. Future work should systematically analyze positioning quality using the markers, and also how their own absolute positions can be estimated using uncertain AUV navigation. Also, efficient ways are required to detect the markers online with the limited computing resources of the robot.

References

1. Bartczak, B., Köser, K., Woelk, F., Koch, R.: Extraction of 3d freeform surfaces as visual landmarks for real-time tracking. J. Real-Time Image Process. **2**, 81–101 (2007)
2. Ciaramella, A., Perrotta, F., Pappone, G., Aucelli, P., Peluso, F., Mattei, G.: Environment object detection for marine ARGO drone by deep learning. In: Del Bimbo, A., et al. (eds.) ICPR 2021. LNCS, vol. 12666, pp. 121–129. Springer, Cham (2021). https://doi.org/10.1007/978-3-030-68780-9_12
3. Fiala, M.: Designing highly reliable fiducial markers. IEEE Trans. Pattern Anal. Mach. Intell. **32**(7), 1317–1324 (2010). https://doi.org/10.1109/TPAMI.2009.146
4. Fischler, M.A., Bolles, R.C.: Random sample consensus: a paradigm for model fitting with applications to image analysis and automated cartography. Commun. ACM **24**(6), 381–395 (1981). https://doi.org/10.1145/358669.358692, http://doi.acm.org/10.1145/358669.358692
5. Garrido-Jurado, S., Muñoz-Salinas, R., Madrid-Cuevas, F., Marín-Jiménez, M.: Automatic generation and detection of highly reliable fiducial markers under occlusion. Pattern Recogn. **47**(6), 2280–2292 (2014). https://doi.org/10.1016/j.patcog.2014.01.005, https://www.sciencedirect.com/science/article/pii/S0031320314000235
6. Kato, H., Billinghurst, M.: Marker tracking and HMD calibration for a video-based augmented reality conferencing system. In: Proceedings 2nd IEEE and ACM International Workshop on Augmented Reality (IWAR 1999), pp. 85–94 (1999). https://doi.org/10.1109/IWAR.1999.803809
7. Köser, K., Frese, U.: Challenges in underwater visual navigation and SLAM. In: Kirchner, F., Straube, S., Kühn, D., Hoyer, N. (eds.) AI Technology for Underwater Robots. ISCASE, vol. 96, pp. 125–135. Springer, Cham (2020). https://doi.org/10.1007/978-3-030-30683-0_11
8. Li, B., Heng, L., Koser, K., Pollefeys, M.: A multiple-camera system calibration toolbox using a feature descriptor-based calibration pattern. In: 2013 IEEE/RSJ International Conference on Intelligent Robots and Systems, pp. 1301–1307 (2013). https://doi.org/10.1109/IROS.2013.6696517
9. Lowe, G.: SIFT-the scale invariant feature transform. Int. J. **2**(91–110), 2 (2004)

10. Muñoz-Salinas, R., Marín-Jimenez, M.J., Medina-Carnicer, R.: SPM-SLAM: simultaneous localization and mapping with squared planar markers. Pattern Recogn. **86**, 156–171 (2019). https://doi.org/10.1016/j.patcog.2018.09.003, https://www.sciencedirect.com/science/article/pii/S0031320318303224
11. Nocerino, E., et al.: Underwater calibration of dome port pressure housings. Int. Arch. Photogramm. Remote Sens. Spatial Inf. Sci. **40**, 127–134 (2016)
12. Rublee, E., Rabaud, V., Konolige, K., Bradski, G.: ORB: an efficient alternative to SIFT or SURF. In: 2011 International Conference on Computer Vision, pp. 2564–2571 (2011). https://doi.org/10.1109/ICCV.2011.6126544
13. dos Santos Cesar, D.B., Gaudig, C., Fritsche, M., dos Reis, M.A., Kirchner, F.: An evaluation of artificial fiducial markers in underwater environments. In: OCEANS 2015 - Genova, pp. 1–6 (2015). https://doi.org/10.1109/OCEANS-Genova.2015.7271491
14. She, M., Song, Y., Mohrmann, J., Köser, K.: Adjustment and calibration of dome port camera systems for underwater vision. In: Fink, G.A., Frintrop, S., Jiang, X. (eds.) DAGM GCPR 2019. LNCS, vol. 11824, pp. 79–92. Springer, Cham (2019). https://doi.org/10.1007/978-3-030-33676-9_6
15. Wagner, D., Schmalstieg, D.: ARToolkitPlus for pose tracking on mobile devices (2007)
16. Xu, Z., Haroutunian, M., Murphy, A.J., Neasham, J., Norman, R.: An underwater visual navigation method based on multiple ArUco markers. J. Mar. Sci. Eng. **9**(12), 1432 (2021). https://doi.org/10.3390/jmse9121432, https://www.mdpi.com/2077-1312/9/12/1432

Towards Cross Domain Transfer Learning for Underwater Correspondence Search

Patricia Schöntag[1(✉)], David Nakath[1], Stefan Röhrl[2], and Kevin Köser[1]

[1] Oceanic Machine Vision Group, GEOMAR Helmholtz Centre for Ocean Research Kiel, Kiel, Germany
paschoentag@geomar.de

[2] Chair for Data Processing, Technical University Munich, Munich, Germany

Abstract. Underwater images are challenging for correspondence search algorithms, which are traditionally designed based on images captured in air and under uniform illumination. In water however, medium interactions have a much higher impact on the light propagation. Absorption and scattering cause wavelength- and distance-dependent color distortion, blurring and contrast reductions. For deeper or turbid waters, artificial illumination is required that usually moves rigidly with the camera and thus increases the appearance differences of the same seafloor spot in different images. Correspondence search, e.g. using image features, is however a core task in underwater visual navigation employed in seafloor surveys and is also required for 3D reconstruction, image retrieval and object detection. For underwater images, it has to be robust against the challenging imaging conditions to avoid decreased accuracy or even failure of computer vision algorithms. However, explicitly taking underwater nuisances into account during the feature extraction and matching process is challenging. On the other hand, learned feature extraction models achieved high performance in many in-air problems in recent years. Hence we investigate, how such a learned robust feature model, D2Net, can be applied to the underwater environment and particularly look into the issue of cross domain transfer learning as a strategy to deal with the lack of annotated underwater training data.

Keywords: Underwater matching · Transfer learning · Physically based rendering · Image features

1 Introduction

Due to the lack of satellite navigation (GNSS) underwater, localization becomes a challenging problem. Pure inertial navigation systems (INS) tend to drift over time, and acoustical localization systems are limited by multi-path propagation, noise, refraction at water layers and often also suffer from suboptimal triangulation geometries (e.g. USBL). Jointly, this can lead to several meters uncertainty, in particular for deeper waters. Visual navigation can limit drift by recognizing

S. Sclaroff et al. (Eds.): ICIAP 2022, LNCS 13233, pp. 461–472, 2022.
https://doi.org/10.1007/978-3-031-06433-3_39

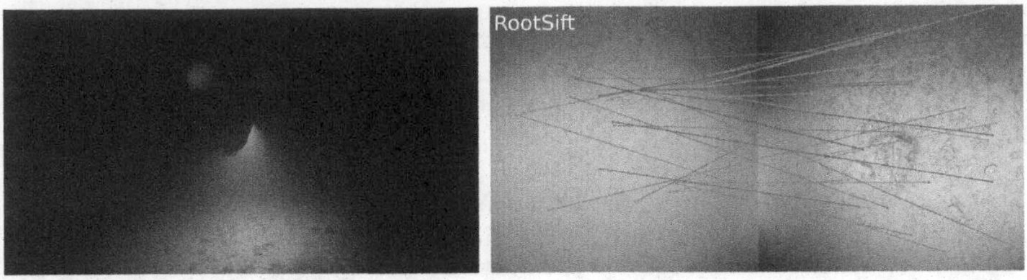

Fig. 1. Left: An AUV mapping the seafloor in coastal waters in about 100 m depth. The co-moving light cone, scattered and attenuated in the water, changes the appearance of seafloor spots for each image. Right: Feature matching in seafloor images is difficult, due to water effects (example using RootSIFT [1]).

places seen earlier, provides very good resolution and has the potential to detect changes at centimeter resolution. Consequently, underwater optical reconstructions and computer vision are becoming increasingly important, but still lag behind what is possible on land [9]. Feature matching is a fundamental step in computer vision tasks. But still, in the ocean or in coastal waters, it strongly suffers from the difficult observation conditions (see Fig. 1 for an example). On land and in air, several approaches have recently shown performance gains in feature matching by employing convolutional neural networks [2,5,15,22,25]. Learned models can be applied at different levels within the classical feature matching pipeline. Various approaches exist to either learn robust and repeatable keypoint detection [24] or distinctive descriptors [2,18], while others focus on combined end-to-end strategies [15,22,25]. However, to get a sufficient amount of underwater images to train a neural network is technically complex and costly. Therefore, we propose applying cross domain transfer learning as a strategy to build upon existing work using a smaller underwater data set. This work reports on our first findings on a fine-tuning approach for models trained on large in-air databases that already include significant appearance variations common on land, e.g. day-night illumination changes, motion blur or fog, to increase robustness for image distortions now due to water-light effects. A promising approach addressing robustness to challenging appearance changes was proposed with D2Net [5]. While most approaches that train the detection and description step use different training setups for two separate models, the D2Net strategy achieves improvements in matching features on challenging images by training a joint detector and descriptor within the same network. Detection and description can thus be optimized simultaneously with mutual influence. Hence, we will base our fine-tuning approach on the D2Net architecture.

Cross domain transfer learning with in-air images as source and underwater images as target domain was already applied for learning classifiers [7,23]. Although this strategy reduces the amount of required data significantly compared to learning a model from scratch, data acquisition still remains a major challenge, in particular for learning feature matching that additionally requires dense pixelwise ground-truth correspondences. Public databases like ImageNet [4] include some underwater related subsets, which are used e.g. by [7]. But their content is mostly limited to touristic photographies without any ground-truth information and thus not suitable to learn the appearance e.g. for scientific seafloor surveys including the variety of possible water effects. In [23] water effects are simulated by adding color distortions and haze onto a labeled in-air image set. Li *et al.* [10] applied a generative adversarial network that is trained to add water effects on in-air images that are similar to distortions of real underwater survey images. When in-air imagery with additional depth information is available, it is possible to augment it by adding water effects on top [16,17,21] using a rasterization-like approach with, at most, single scattering added on top.

The VAROS data set [26] uses ray-tracing and models an ideal AUV-trajectory and corresponding imagery in Blender in order to evaluate robotic pipeline inspection. Although it has been shown that Blender can also be used for correct underwater refraction effects [13], we found it difficult to mimick real radiometric optical ocean effects with blender, as it was designed to produce image that look plausible, but are not necessarily faithful simulations for a particular ocean spot.

Instead, we base our work on the scientific renderer Mitsuba II [14]. It allows control about virtually all physical aspects of light, medium and scene. Specifically, we employ its integrator for arbitrary output variables (AOV integrator) to obtain the required ground-truth information such as pixel world coordinates in combination with its volumetric integrator to obtain the actual image. Additionally, we use a wide variety of existing 3D models of natural landscapes with texture (sand, rocks, gravel etc.) that are also plausible underwater to study cross-domain transfer learning using D2Net and report our initial findings here.

Since the combination of the underwater data set with the D2Net architecture revealed training limitations arising from the loss function applied in D2Net, the next section will discuss these limitations and their implications for training. Once the prerequisites of our cross domain fine-tuning approach are determined, Sect. 3 focusses on the data set generation and transfer learning. The last part of the paper presents a preliminary evaluation and a summary of our findings.

2 D2Net Revisited for Underwater Training

The D2Net is based on a truncated VGG16 architecture [5,19]. The fully connected layers as well as the last subset of convolutional layers were cut off, so the output layer is a 3D tensor serving as multilayer detection score map and dense array of descriptor vectors at the same time [5]. The system employs triplet margin ranking loss [12] that tends to run in gradient or model collapses [3,6]. In the following paragraphs we will outline our findings from the perspective of underwater imagery-based training. More specifically, the loss function is an extension of the triplet margin ranking loss [12], which is designed to optimize the distinctiveness of descriptors. To account as well for a simultaneous optimization towards repeatable keypoint detections, Dusmanu *et al.* weighted the description term, i.e. the triplet loss, with the detection scores of the keypoints. However, since we observed that the detection scores have no significant impact on the problem of model collapses, we focus on the triplet part of the loss function. The following part introduces the triplet margin ranking loss and explains its connection with properties of the data set.

2.1 Triplet Margin Ranking Loss with Underwater Images

The triplet margin ranking loss aims to learn a descriptor that yields feature vectors which are similar for corresponding keypoints and distinguishable from feature vectors of non-corresponding keypoints, i.e. it optimizes distinctiveness. To this end, it *pulls* the two feature vectors of each correspondence c in the set of all true corresponding pixels C of an image pair close together in feature space, while *pushing* feature vectors of non-corresponding pixels away from one another. The geometrical distance between the feature vectors of corresponding pixels is termed positive distance $p(c)$, while the negative distance $n(c)$ refers to the distance between one of the ground-truth feature vectors and the most similar non-corresponding feature vector in the other image (hard negative). The difference between the positive and negative distance of a correspondence c gives a measure of distinctiveness. Hence, by minimizing the triplet loss $d(c)$, with

$$d(c) = \max\left[0, M + p(c)^2 - n(c)^2\right], \tag{1}$$

the descriptor model converges to optimal distinctiveness. The margin M keeps the loss positive and determines to which extent a negative distance can differ from the positive distance. In previous work [5,12] M always equals 1. Setting this number has some practical implications: If positive and negative distances have already a large difference because non-corresponding samples show only little similarity to true correspondences, Ha *et al.* [6] suggest a larger margin, because otherwise the gradients could collapse to zero [3,6]. However, in the underwater domain we observed that the initial feature vectors tend to be very similar rather than very different. We assume the few details on seafloor landscapes and the backscattering haze that additionally reduces pixel intensity variations are the reason. Therefore, we stick to a margin of 1. But nevertheless, with high similarity between feature vectors another boundary of the triplet loss needs to be considered.

Bui *et al.* [3,6], who apply triplet loss functions for learning image patch descriptors, report that gradients collapse to zero when the distances in feature space between ground-truth correspondences are close to the distances of false correspondences. Ha *et al.* [6] considers the tendency to model collapses, which is when all model outputs are mapped to the same value such that the model becomes useless. They report this problem can occur, when hard negative mining is applied, i.e. the similarity between true corresponding feature vectors is compared with the similarity to the hard negative feature vector. To identify implications of this limitation, the following paragraph gives an analysis of the behavior of the triplet loss $d(c)$ on our underwater data set.

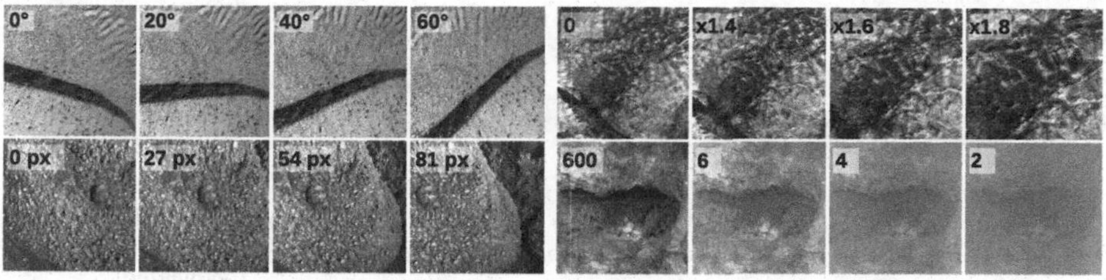

Fig. 2. Example transformations shown on four of the five different scenes in the test set. The test images have a resolution of 256 × 256 pixels. (Credits (from up left to bottom right): Epic_Tree_Store, PALEOV-UJA, sr-prod, Hawaii Aerial Visions LLC licensed under Creative Commons Attribution-NonCommercial (up right) or Creative Commons Attribution (others)).

2.2 Robustness Against Basic Image Transformations

The effect of model collapse and its relation to the similarity of feature vectors can be explained with a closer look on Eq. 1. The triplet loss is close to the margin M if the difference between positive and negative distances $p(c)^2 - n(c)^2$ is small. If the positive distance is even larger than the negative distance the triplet loss rises above M. Practically, that indicates for a correspondence c that there is a non-corresponding feature vector that is more similar than the actual corresponding vector. In other words, the model is not able to indentify corresponding pairs of pixels, because the pixels can not be distinguished by their feature vectors. In that case we observe that minimization of the triplet loss pushes $n(c)$ and $p(c)$ to the same value rather than pulling them apart. When all feature vectors are mapped to the same value, the model collapses and no distinctive features are learned. Consequently, we can read the risk of a model collapse from the inital values of the triplet losses. In our experience, convergence to a model that yields distinctive feature vectors can only be expected if the triplet losses at the starting point are predominantly below M.

Since we observed that the underwater images are challenging with respect to these limitations, we investigate the loss behavior regarding different basic image deformations in detail to avoid a data set with ill-posed starting conditions. On a small test set of five images, we evaluate the effect of translation, rotation and scaling without any water effects. A fourth experiment observes the influence of water effects without geometric transformation. Figure 2 shows samples of the test images. In Fig. 3 the triplet losses averaged over the test image pairs are presented. It can be observed that rotation angles about 40° and scaling factors larger than 1.6 yield a triplet loss higher than the margin $M = 1$. Regarding translations, the triplet loss also increases with stronger effects, which is likely due to border effects that influence descriptors of keypoints as they are pushed out from the image center. Nevertheless, the loss remains in an acceptable range to allow reasonable optimization. The influence of applied water effects on the triplet loss is measured by the mean free path, which is the average distance that a photon travels through a medium before it is absorbed or scattered and

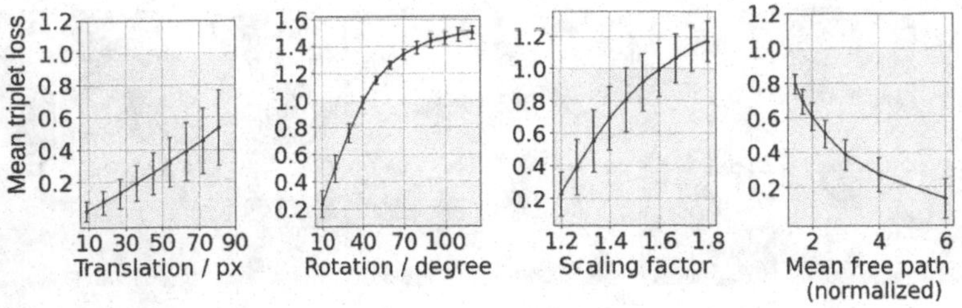

Fig. 3. Mean triplet loss with respect to translation, rotation, scaling and mean free path of light in modelled water volume.

is related to the underwater visibility. Since the water effect depends on the distance between camera and scene, the mean free path is normalized with the depth. As expected, Fig. 3 shows an increasing triplet loss for a shorter mean free path, supporting the intuition that feature vectors get less distinguishable with stronger impact of water effects. However, even up to very strong water effects the triplet loss remains below the threshold. In the following, we will discuss the implications on our transfer learning approach.

2.3 Consequences for Underwater Matching

We found that large rotations and scalings can cause model collapses. While many touristic in-air images are oriented in gravity direction or Exif data of photos provides camera orientation information, in underwater setups the viewing direction can be significantly different. In a downward looking camera, e.g. from a robot or a towsled observing the seafloor, the image can be arbitrarily "rolled". When the same seafloor spot is re-investigated the camera might approach from a different path and so the two images differ by a significant in-plane image rotation (up to 180°). In particular, if the goal is to detect loops, to geolocalize, or to perform image retrieval, images will not be automatically rotation-compensated. Our results show that a convolutional model such as D2Net can not be optimized on imagery including large rotations unless the model is specifically equipped with additional means to introduce rotation invariance. Consequently, such convolutional models require the knowledge of camera orientation to bring all samples to about the same orientation prior to training or model application. Similar considerations hold for scale changes, arising from different altitudes or camera zooming. However, in some scenarios this information is available from robotic navigation data. We consider convolutional models to be applicable to underwater tasks, when image orientation and scale is guaranteed to be limited to small amounts, or pre-normalized.

3 Cross-domain Transfer Learning

Common feature extraction models are learned from large scale data collections like ImageNet [4]. This way, they achieve high levels of generalisation from which also underwater feature extraction could benefit, especially regarding low level features. However, feature extraction on underwater images additionally requires robustness against blur, contrast reduction and color distortion. This can not be expected from classical land-based models, because those effects are usually not present in training data sets and not required for land-based feature extraction tasks. In order to achieve this robustness nonetheless, we propose fine-tuning pretrained models for the underwater image domain. As the D2Net model is already adapted to challenging imagery without water, we choose it as base model for our fine-tuning. A key challenge is that almost no underwater data sets (with ground-truth correspondences) exist that contain the aforementioned challenges. This is a chicken-and-egg-problem, since correspondence search is challenging on such images, and therefore corresponding images are hard to register. To break this vicious cycle, we render realistic underwater images with the physically based renderer Mitsuba II [14]. We parameterize the simulated water medium based on oceanographic measurements and choose high resolution 3D landscape models that are plausible in the ocean as a virtual sea floor, e.g. containing sand, rocks, pebbles, cliffs, and similar scenarios. The following paragraphs explain the implementation details of our approach.

3.1 Physical Simulation of Underwater Effects

Mitsuba II [14] is an open-source library that allows detailed parameterization of the camera, medium and light models. The base content of the simulated images is obtained from publicly available high-resolution landscape models. We have chosen 11 scenes of rocks, gravel and beach so that the images resemble plausible underwater landscapes. The following paragraphs describe the choice of rendering settings that add realistic underwater appearance to those landscape models. Figure 4 shows an overview over the resulting data set.

Camera Settings. A simple pinhole camera model defines all camera intrinsics. For the camera poses three to five different locations are selected per 3D scene, such that each shows semantically different parts of the scene to get a data set of high variety. For every base location a list of 60 camera poses is generated by randomly sampling the space of feasible transformations according to Sect. 2.

Medium Settings. The medium absorption and scattering is parameterized with three wide-band coefficients for the red, green and blue color channel. We rely on absorption and scattering coefficients that Solenko and Mobley [20] derived for the main ocean water types as described by Jerlov [8]. We use the 675 nm, 525 nm and 475 nm values as coefficients corresponding to the red, green and blue color channel. During the rendering procedure one of the Jerlov water types, ranging from clear ocean water to murky coastal waters is chosen randomly.

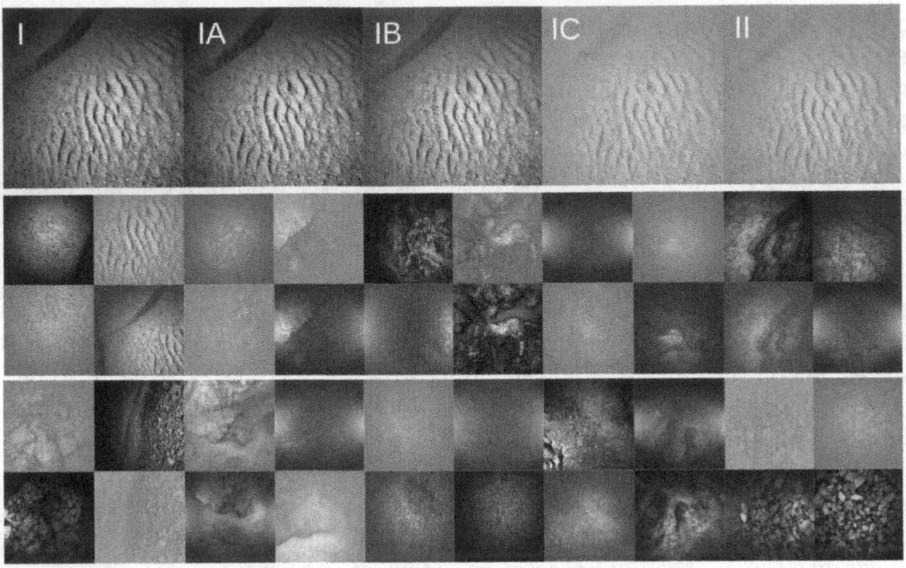

Fig. 4. Example images from the synthetic data set. Upper row: One example of each Jerlov water type applied. Lower rows: Each column shows two random images from two views of each of the 11 scenes. (Credits by (from up left to bottom right): Epic_Tree_Store, PALEOV-UJA, Kyle, callanbentley, 3dhdscan, GSXNet, Paul, Hawaii Aerial Visions LLC, lippo, sr-prod licensed under Creative Commons Attribution-NonCommercial (scene 2, 5 and 7) or Creative Commons Attribution (others)).

Light Settings. To include different typical artifical light patterns in the images, one of three light setups is chosen randomly. The light settings include one single light with narrow field of view and one single light with wide field of view. Both are placed one meter above the camera. The third light setting contains two lamps left and right of the camera, slightly pointing towards each other such that both light cones meet at the seafloor and strong backscattering effects appear in coastal waters at both sides of the image center.

3.2 Model Fine-tuning

The D2Net model is already a fine-tuned version of a truncated VGG16 architecture pretrained on ImageNet. To evaluate the benefit of the first D2Net fine-tuning stage, we apply our fine-tuning once on the plain truncated VGG16 model and once as additional fine-tuning stage on top of the D2Net fine-tuning. We refer to the first version as model A and call the underwater fine-tuning on top of the D2Net fine-tuning model B. Both, the D2Net fine-tuning and our approach optimize the last layer with a small gradually decreasing learning rate while weights of the deeper layers remain fixed.

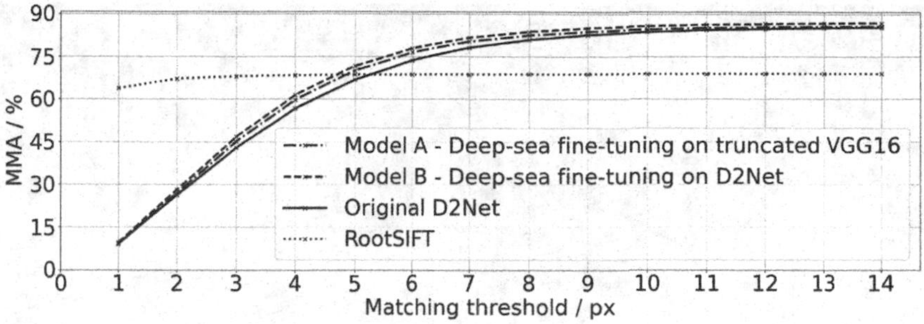

Fig. 5. Mean matching accuracy (MMA) with respect to reprojection error for the four evaluated feature extraction models.

Data Preprocessing. The preprocessing ensures, that training samples, matched by ground-truth information show actually the same content. To this end, the known D2Net preprocessing step checks for occluded correspondences. But in addition, we have to ensure that the water effects are checked for their impact on visibility such that structures that disappear in the distance or in the darkness are also excluded. This is checked by evaluating pixel variance around both pixels belonging to the candidate training sample. If the variance is too small, the training sample is excluded.

Training Details. The model is trained for 50 epochs on 735 images from 6 scenes including 21 different views. Per view 35 images are chosen randomly. In total, the training data set consists of 11 654 pairs resulting from all images. The training is validated by a separate validation set from 3 *different scenes* including a total of 8 locations. Overall, it contains 2 132 pairs.

4 Evaluation

The performance of feature extraction is evaluated based on the matching accuracy. To this end, the features extracted by the trained models are matched with a nearest neighbor approach, which is the same as in D2Net [5] such that results are comparable. The mean matching accuracy (MMA) is investigated like proposed in [11]. It is the fraction of valid correspondences compared to all received correspondences, while a correspondence is considered valid if its reprojection error is lower than a predefined threshold. We show in Fig. 5 that our fine-tuning improves matching performance on underwater images of three entirely different scenes. Both of our fine-tuned models achieve slightly better performance on the unseen images as compared to the D2Net. In particular, model B has superior performance, which confirms that the first fine-tuning stage towards more robust feature extraction on challenging in-air imagery performed by the authors of D2Net provided a beneficial initialization for our underwater model. Nevertheless, the improvements are small, so we see this study as a first step

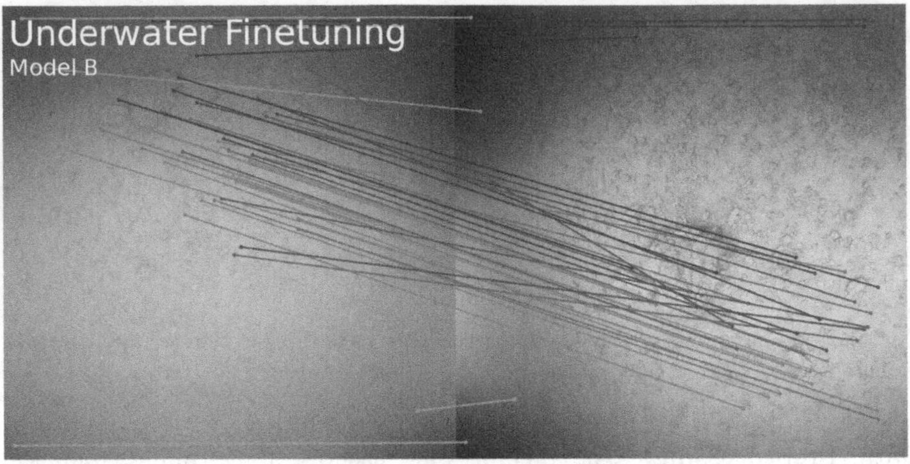

Fig. 6. Matching result with our underwater feature extraction model on images with intense backscattering effects, color distortions and artificial light pattern.

towards improving learned underwater matching. Figure 6 illustrates application on a real example image, which looks promising for learning underwater matching. A second main finding reveals a major drawback of all learned models. A suitable matching performance is only achieved if a high reprojection error can be tolerated. With a tolerated reprojection error of five pixels the best model, model B, shows 72% accuracy. For smaller errors, the learned models can not compede with the state-of-the-art non-learning method RootSIFT.

5 Conclusion

We investigated an in-air pretrained model in underwater matching tasks and how to improve performance by cross-domain transfer learning. To compensate the lack of training data, we synthesized underwater images using physically based simulations of Jerlov water types and different light source configurations in diverse virtual 3D landscape scenes. Because triplet loss functions imply to carefully choose training data in order to be able to optimize a model, we analysed our underwater training data regarding the loss behavior and found that poor rotation and scale invariance of in-air pretrained models result in model collapses. When restricting the data set to small rotations and scales (e.g. knowing altitude and heading in a robotic mission), our transfer learning approach showed improved matching performance on underwater images, in particular when the model was already pretrained on images with challenging appearance in air. Therefore, we think the cross domain transfer learning is a suitable approach towards learning underwater feature extraction models despite the lack of sufficient real training data. Still, this study is only a first step and issues such as limited image resolution and reprojection errors remain.

References

1. Arandjelović, R., Zisserman, A.: Three things everyone should know to improve object retrieval. In: 2012 IEEE Conference on Computer Vision and Pattern Recognition, pp. 2911–2918. IEEE (2012)
2. Balntas, V., Riba, E., Ponsa, D., Mikolajczyk, K.: Learning local feature descriptors with triplets and shallow convolutional neural networks. In: British Machine Vision Conference, vol. 1, p. 3 (2016)
3. Bui, T., Ribeiro, L., Ponti, M., Collomosse, J.: Compact descriptors for sketch-based image retrieval using a triplet loss convolutional neural network. Comput. Vision Image Underst. **164**, 27–37 (2017)
4. Deng, J., Dong, W., Socher, R., Li, L.J., Li, K., Fei-Fei, L.: Imagenet: a large-scale hierarchical image database. In: 2009 IEEE Conference on Computer Vision and Pattern Recognition, pp. 248–255. IEEE (2009)
5. Dusmanu, M., et al.: D2-net: a trainable CNN for joint description and detection of local features. In: Proceedings of the IEEE/CVF Conference on Computer Vision and Pattern Recognition, pp. 8092–8101 (2019)
6. Ha, M.L., Blanz, V.: Deep ranking with adaptive margin triplet loss. arXiv preprint arXiv:2107.06187 (2021)
7. Irfan, M., Jiangbin, Z., Iqbal, M., Arif, M.H.: A novel lifelong learning model based on cross domain knowledge extraction and transfer to classify underwater images. Inf. Sci. **552**, 80–101 (2021)
8. Jerlov, N.: Irradiance optical classification. In: Optical Oceanography, pp. 118–120. Elsevier (1968)
9. Köser, K., Frese, U.: Challenges in underwater visual navigation and SLAM. In: Kirchner, F., Straube, S., Kühn, D., Hoyer, N. (eds.) AI Technology for Underwater Robots. ISCASE, vol. 96, pp. 125–135. Springer, Cham (2020). https://doi.org/10.1007/978-3-030-30683-0_11
10. Li, J., Skinner, K.A., Eustice, R.M., Johnson-Roberson, M.: WaterGAN: unsupervised generative network to enable real-time color correction of monocular underwater images. IEEE Robot. Autom. Lett. **3**(1), 387–394 (2017)
11. Mikolajczyk, K., et al.: A comparison of affine region detectors. Int. J. Comput. Vision **65**(1), 43–72 (2005)
12. Mishchuk, A., Mishkin, D., Radenović, F., Matas, J.: Working hard to know your neighbor's margins: Local descriptor learning loss. In: Advances in Neural Information Processing Systems, pp. 4827–4838 (2017)
13. Nakath, D., She, M., Song, Y., Köser, K.: An optical digital twin for underwater photogrammetry. PFG J. Photogramm. Remote Sens. Geoinf. Sci. **90**, 69–81 (2022)
14. Nimier-David, M., Vicini, D., Zeltner, T., Jakob, W.: Mitsuba 2: a retargetable forward and inverse renderer. Trans. Graph. **38**(6), 1–7 (2019)
15. Ono, Y., Trulls, E., Fua, P., Yi, K.M.: LF-Net: learning local features from images. In: Proceedings of the 32nd International Conference on Neural Information Processing Systems, pp. 6237–6247 (2018)
16. Sedlazeck, A., Koch, R.: Simulating deep sea underwater images using physical models for light attenuation, scattering, and refraction. In: Eisert, P., Hornegger, J., Polthier, K. (eds.) Vision, Modeling, and Visualization. The Eurographics Association (2011)
17. Shepelev, D.A., Bozhkova, V.P., Ershov, E.I., Nikolaev, D.P.: Simulation of underwater color images using banded spectral model. In: ECMS Conference on modelling and simulation, pp. 11–18 (2020)

18. Simo-Serra, E., Trulls, E., Ferraz, L., Kokkinos, I., Fua, P., Moreno-Noguer, F.: Discriminative learning of deep convolutional feature point descriptors. In: Proceedings of the IEEE International Conference on Computer Vision, pp. 118–126 (2015)
19. Simonyan, K., Zisserman, A.: Very deep convolutional networks for large-scale image recognition (2015)
20. Solonenko, M.G., Mobley, C.D.: Inherent optical properties of Jerlov water types. Appl. Opt. **54**(17), 5392–5401 (2015)
21. Song, Y., Nakath, D., She, M., Elibol, F., Köser, K.: Deep sea robotic imaging simulator. In: Del Bimbo, A., et al. (eds.) ICPR 2021. LNCS, vol. 12662, pp. 375–389. Springer, Cham (2021). https://doi.org/10.1007/978-3-030-68790-8_29
22. Yi, K.M., Trulls, E., Lepetit, V., Fua, P.: LIFT: learned invariant feature transform. In: Leibe, B., Matas, J., Sebe, N., Welling, M. (eds.) Computer Vision – ECCV 2016. LNCS, vol. 9910, pp. 467–483. Springer, Cham (2016). https://doi.org/10.1007/978-3-319-46466-4_28
23. Yu, X., Xing, X., Zheng, H., Fu, X., Huang, Y., Ding, X.: Man-made object recognition from underwater optical images using deep learning and transfer learning. In: 2018 IEEE International Conference on Acoustics, Speech and Signal Processing (ICASSP), pp. 1852–1856. IEEE (2018)
24. Zhang, L., Rusinkiewicz, S.: Learning to detect features in texture images. In: Proceedings of the IEEE Conference on Computer Vision and Pattern Recognition, pp. 6325–6333 (2018)
25. Zheng, L., Yang, Y., Tian, Q.: SIFT meets CNN: a decade survey of instance retrieval. IEEE Trans. Pattern Anal. Mach. Intell. **40**(5), 1224–1244 (2017)
26. Zwilgmeyer, P.G.O., Yip, M., Teigen, A.L., Mester, R., Stahl, A.: The VAROS synthetic underwater data set: towards realistic multi-sensor underwater data with ground truth. In: Proceedings of the IEEE/CVF International Conference on Computer Vision, pp. 3722–3730 (2021)

On the Evaluation of Two Methods Applied to the Morphometry of Linear Dunes

Tatiana Taís Schein[1], Leonardo R. Emmendorfer[2(✉)], Fabiano Nobre Mendes[1], Bárbara D. A. Rodriguez[3], Luis Pedro Almeida[4], and Vinícius Menezes de Oliveira[2]

[1] Programa de Pós-graduação em Modelagem Computacional, Universidade Federal do Rio Grande, Rio Grande City, Brazil

[2] Centro de Ciências Computacionais, Universidade Federal do Rio Grande, 96203-900 Rio Grande City, Brazil
leonardo.emmendorfer@gmail.com, vinicius@ieee.org

[3] Instituto de Matemática, Estatística e Física, Universidade Federal do Rio Grande, 96203-900 Rio Grande City, Brazil
barbararodriguez@furg.br

[4] CoLAB+ATLANTIC, Edifício LACS, Estrada da Malveira da Serra 920, 2750-834 Cascais, Portugal
pedro.almeida@colabatlantic.com

Abstract. The evolution of dunes is a relevant topic in many areas, including the study of the marine environment. In particular, the widespread linear dunes are characterized by roughly parallel ridges, which extend for long distances. The identification of the main morphometric characteristics of linear dunes is a relevant topic, with wide applicability. This work evaluates two approaches for linear dune morphometry. The methods are built over the Radon transform, the Discrete Fourier Transform, and autocorrelation. The evaluation is performed over a set of publicly available dune images. Results show that the performance of the methods varies over the morphometric characteristics considered.

Keywords: Dune morphology · Morphometry · Discrete Fourier Transform · Radon transform · Autocorrelation

1 Introduction

Patterns arising from the accumulation of wind-blown sand in dunes are among the most dynamic and intriguing phenomena in nature [1]. Tracking morphological changes in dunes is a relevant topic in environmental sciences [4]. Sand migration, for instance, might present serious threats to populations around the

This study was financed in part by the Coordenação de Aperfeiçoamento de Pessoal de Nível Superior - Brazil (CAPES) - Finance Code 001.

S. Sclaroff et al. (Eds.): ICIAP 2022, LNCS 13233, pp. 473–479, 2022.
https://doi.org/10.1007/978-3-031-06433-3_40

world [1]. Migrating dunes can potentially invade forests, bury roads and buildings, among other harmful effects.

Dunes arise as a result of erosive soil and sediment transport performed by the wind [2]. The formation dynamics of patterns in dunes is essentially complex because of the discreteness of the system on a phenomenological scale, at the level of sand grains [9]. Transported sediments settle when obstacles are reached in the path. These obstacles accelerate the formation of the dune and define its morphology [7].

Conditions such as dry or wet sand, flat surface, and grain size, as well as the presence or absence of vegetation, are relevant for the locomotion of sediments. Sand grains with a diameter between 0.062 and 2.0 mm are not cohesive, whereas grains smaller than 0.02 mm of silt and/or clay are cohesive and can resist wind migration [1]. Dunes are generally composed of sediments between 0.125 and 0.250 mm in diameter, as smaller sediments are carried by the force of winds over longer distances [1].

Parallel (or longitudinal) dunes are composed of parallel ridges which elongate in the direction of the prevailing winds [2]. Ridges might extend for hundreds of meters. The term "linear dune", is both morphological and descriptive, which can be used even when formative processes are not known [3,4]. Linear dunes are both the most widespread and controversial forms of dune [11].

Several tools are adopted for understanding the evolution of dunes and other landforms [3]. Morphological changes in dune systems over time can be induced by natural causes (sea-level rise, storms, the decline of submerged sandbanks, among others) and/or by human activity [4]. The presence of vegetation prevents the migration of sand and thus promotes stability to the dunes system.

This paper investigates the automatic estimation of two morphological features of linear dunes: the orientation, which is the angle α between the parallel ridges and a reference vector $\mathbf{i}$ in a standard basis $(\mathbf{i}, \mathbf{j})$ [6], and the distance between parallel ridges, or dune period p. Simple methods such as the Fourier transform, Radon transform, and autocorrelation will be adopted, since those approaches play a relevant role in geomorphometry, usually when applied to cross-sectional views (transects).

This paper aims to evaluate the adoption of the Radon transform for determining the orientation and period of linear dunes from images. Two alternative approaches will be considered, by combining the Radon transform with both the Fourier Transform and autocorrelation. In both cases, transects are avoided by performing projections of the dune image, using the Radon transform for this aim. Public images of linear are used in the evaluations.

This work is organized as follows: Sect. 2 describes the methodology, including the techniques adopted in the work. Section 3 describes the results that were achieved. Finally, Sect. 4 presents the conclusions.

2 Materials and Methods

A total of 20 publicly available linear dune images were selected for the evaluation of the methods. Figure 1 illustrates the images adopted. The annotation of the period was performed by computing the average distance between subsequent ridges in each dune image, whereas the annotation of the orientation is obtained by computing the average orientation of each ridge in a linear dune image. Only ridges that stretch over at least half the dimension of the whole image (vertically, horizontally, or diagonally at any angle) are considered for the annotation process, as illustrated in Fig. 2, which shows image #16.

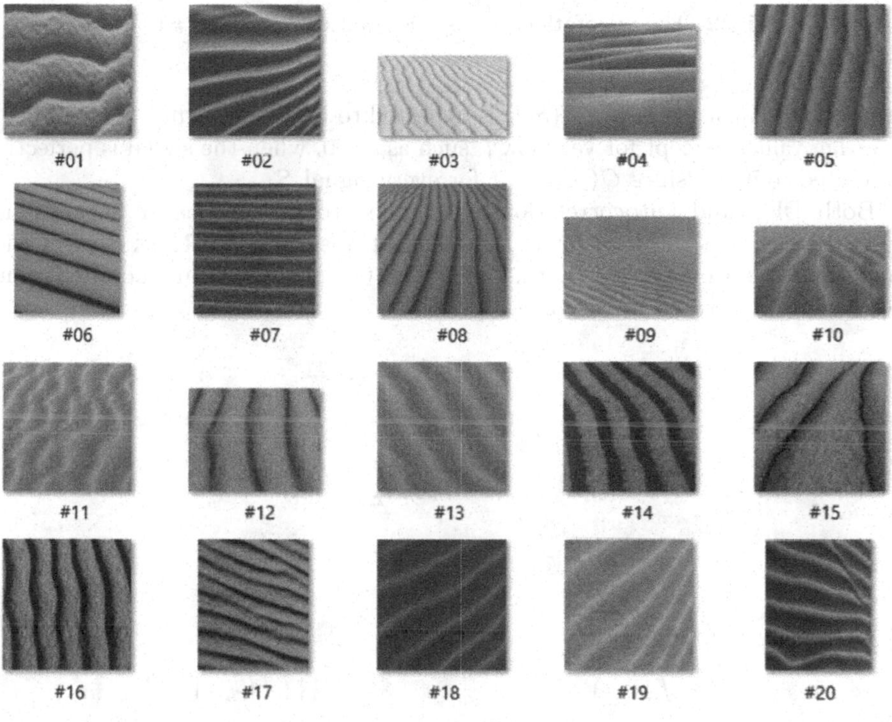

Fig. 1. Linear dune images adopted in the experiments.

The detection of morphometric parameters from the grayscale version of a linear dune image $I(\cdot,\cdot)$ is performed by the combination of the Radon transform [10] and one out of two methods applied for periodicity detection: the single-dimensional Discrete Fourier Transform (DFT) [5], and autocorrelation [8].

Autocorrelation [8] is also able to extract relevant information about the periodicity of a signal S. It is computed as $C(S,l)$, which is the correlation between a signal $S = (s_1, s_2, \cdots, s_n)$ and its lagged version $S = (s_{1-l}, s_{2-l}, \cdots, s_{n-l})$, given a nonnegative lag parameter l. For lag values similar to the period of the

Fig. 2. The annotation process, as adopted for image #16.

signal s, the autocorrelation $C(S, l)$ is expected to be higher, when compared to other lag values, except for very low l, such as $l = 0$, when the signal is perfectly correlated to itself, since $C(S, 0) = 1$ for every signal S.

Both DFT and autocorrelation are applied to projections of linear dune images, which are obtained using the Radon transform [10] $\mathbf{R}(I(\cdot,\cdot), \phi)$ of the image $I(\cdot,\cdot)$, parameterized by the angle ϕ of the projection line adopted. The approach is illustrated in Fig. 3.

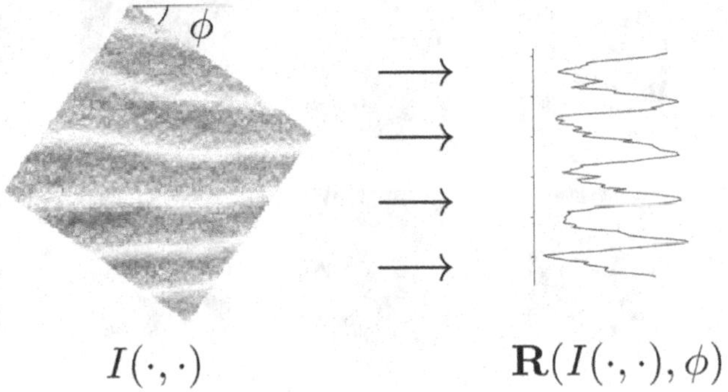

Fig. 3. The single-dimensional projection $\mathbf{R}(I(\cdot,\cdot), \phi)$ (right) of a grayscale dune image $I(\cdot,\cdot)$ (left) as obtained using the Radon transform, given an angle ϕ.

The first method to be described here arises from the combination of the Radon transform and DFT. The orientation $\hat{\alpha}_{DFT}$ is obtained by maximizing, concerning the projection angle ϕ, the amplitude resulting from the DFT of the Radon transform of the image $I(\cdot,\cdot)$, considering all n frequencies obtained from the DFT:

$$\hat{\alpha}_{DFT} = \underset{\phi}{\operatorname{argmax}} \left(\max \left(\mathbf{A}(\mathbf{R}(I(\cdot,\cdot),\phi))_1, \mathbf{A}(\mathbf{R}(I(\cdot,\cdot),\phi))_2, \cdots, \mathbf{A}(\mathbf{R}(I(\cdot,\cdot),\phi))_n \right) \right) \quad (1)$$

The numerical solution to (1) is obtained by an exhaustive search over all integer values $\phi \in [0, 360]$ (in degrees).

The dune period corresponding to the orientation given by $\hat{\alpha}$ can be computed from the frequency, obtained using the DFT, corresponding to the maximal amplitude as adopted in (1). Initially, the array position $\hat{k}$ in $\mathbf{A}(\mathbf{R}(\mathbf{I}(\cdot,\cdot),\hat{\alpha}))$ of the maximal amplitude for $\phi = \hat{\alpha}$ can be expressed obtained as:

$$\hat{k} = \underset{i}{\operatorname{argmax}}\, \mathbf{A}(\mathbf{R}(I(\cdot,\cdot),\hat{\alpha}))_i \quad (2)$$

The period $\hat{p}_{DFT}$ can be obtained by inverting the frequency corresponding to the maximal amplitude, which is in $\mathbf{F}(\mathbf{R}(I(\cdot,\cdot),\hat{\alpha}))$ at the same array position $\hat{k}$ computed in (2):

$$\hat{p}_{DFT} = \frac{1}{\mathbf{F}(\mathbf{R}(I(\cdot,\cdot),\hat{\alpha}))_{\hat{k}}} \quad (3)$$

Similarly, the next method adopted consists in applying autocorrelation to the Radon transform of a linear dune image. The optimal orientation is found by maximizing, with respect to the angle ϕ, the autocorrelation of the Radon transform of the image $I(\cdot,\cdot)$, considering all possible values for the lag l:

$$\hat{\alpha}_{autocorrelation} = \underset{\phi}{\operatorname{argmax}} \left(\max_{l} \left(\mathbf{C}(\mathbf{R}(I(\cdot,\cdot),\phi), l), l \in \{l_0, l_0+1, \cdots, n\} \right) \right) \quad (4)$$

where l_0 is the minimal lag to be considered. This parameter is relevant since lower values for the lag must not be considered, since the autocorrelation approaches to 1 as the lag decreases[1].

The period of the dune image corresponds to the lag that maximizes the autocorrelation in (4), as:

$$\hat{p}_{autocorrelation} = \underset{l}{\operatorname{argmax}} \left(\mathbf{C}(\mathbf{R}(I(\cdot,\cdot),\hat{\alpha}_{autocorrelation}), l), l \in \{l_0, l_0+1, \cdots, n\} \right) \quad (5)$$

The methods proposed here are denoted as Radon+DFT (Eqs. (1)–(3)) and Rado+Autocorrelation (Eqs. (4)–(5)). Both methods are applied to the 20 linear dune images, and the quality of the results is evaluated by comparing the computed morphometric parameters to the annotated values using the Root Mean Squared Error (RMSE):

$$RMSE = \sqrt{\frac{1}{m}\sum_{i=1}^{m}\left\{t_i - v_i\right\}^2}$$

where m is the number of images, t_i is an annotated ground-truth reference, and v_i is a computed, estimated value.

[1] The value $l_0 = 5$ pixels was adopted in the experiments.

Table 1. Period (in pixels) and orientation (in degrees) resulting from both morphometry methods used, for the 20 dune images adopted. The best result, in terms of absolute error when compared to the respective annotation, for each case, is shown in bold.

Image #	Period (px)			Orientation (°)		
	Annotation	Radon+ DFT	Radon+ Autocorrelation	Annotation	Radon+ DFT	Radon+ Autocorrelation
01	155.7	**146**	141	7.0	0	**8**
02	136.7	**140**	130	−14.3	−4	**−12**
03	113.5	106	**116**	−27.9	−45	**−32**
04	17.2	35	**17**	−3.8	**−4**	−7
05	111.9	**109**	106	10.1	19	**15**
06	48.5	56	**54**	16.6	**15**	20
07	26.3	29	**25**	0.9	**1**	2
08	111.5	108	**111**	11.6	21	**15**
09	18.8	10	**19**	12.6	**12**	28
10	137.0	170	**116**	16.9	51	**17**
11	135.2	**117**	106	18.2	**14**	11
12	195.4	**174**	118	−6.0	−8	**−5**
13	242.5	**215**	205	−33.7	**−33**	−29
14	142.5	**143**	126	−21.6	−20	**−23**
15	350.0	**237**	109	19.6	40	**38**
16	136.2	**140**	153	−6.2	**−8**	−4
17	123.7	**120**	146	18.8	**18**	16
18	134.0	**127**	116	−29.8	−35	−35
19	136.0	**125**	124	−29.5	**−31**	−54
20	125.5	133	**121**	10.2	**10**	−1
RMSE		28.5	58.6		10.6	8.7

3 Results and Discussion

Table 1 shows the results from the experiments. The Radon+DFT method was superior at computing the dune period for 12 out of 20 images, while the Radon+Autocorrelation is better for the remaining 8 images. When considering the estimation of the orientation, each method is superior for a subset of 10 out of 20 images.

The Radon+DFT method also achieved a better RMSE when estimating the period of the dunes, since the RMSE is 28.5 for this method, while Radon+ Autocorrelation achieved an RMSE of 58.6. However, at the estimation of orientation, the Radon+Autocorrelation is superior when compared to Radon+DFT, when considering the RMSE: Radon+ Autocorrelation achieved an RMSE of 8.7, while Radon+DFT reached a slightly worse RMSE equal to 10.6.

4 Conclusion

This work revealed that both DFT and autocorrelation are useful in determining orientation and period from linear dune images, after the adoption of the Radon transform. The superiority of each method depends on the characteristic under consideration. The Radon+DFT method was superior for determining the period of the linear dunes, while the alternative Radon+Autocorrelation method was better at finding the correct orientation of the dunes.

As further work, both methods should be adopted for determining dynamical aspects from dunes, by computing the speed and direction of spread from sequences of dune images over time. Additionally, the methods described here could be useful in similar applications, such as the morphometry of ocean waves.

References

1. Bhadra, B., Rehpade, S.B., Meena, H., Rao, S.S.: Analysis of parabolic dune morphometry and its migration in Thar desert area, India, using high-resolution satellite data and temporal DEM. J. Indian Soc. Remote Sens. **47**(12), 2097–2111 (2019)
2. Börgel, R.: Las dunas litorales en Chile: Teoría y Aplicación (1963)
3. Bourke, M., Balme, M., Beyer, R., Williams, K., Zimbelman, J.: A comparison of methods used to estimate the height of sand dunes on mars. Geomorphology **81**(3–4), 440–452 (2006)
4. Cabrera-Vega, L., et al.: Morphological changes in dunes as an indicator of anthropogenic interferences in arid dune fields. J. Coast. Res. **65**(10065), 1271–1276 (2013)
5. Fisher, R., Perkins, S., Walker, A., Wolfart, E.: Hypermedia Image Processing Reference. John Wiley & Sons Ltd., England, pp. 118–130 (1996)
6. Gao, X., Narteau, C., Rozier, O., Du Pont, S.C.: Phase diagrams of dune shape and orientation depending on sand availability. Sci. Rep. **5**(1), 1–12 (2015)
7. Guimaraes, P.V., Pereira, P.S., Calliari, L.J., Ellis, J.T.: Behavior and identification of ephemeral sand dunes at the backshore zone using video images. Anais da Acad. Brasileira de Ciências **88**, 1357–1369 (2016)
8. Hyndman, R.J., Athanasopoulos, G.: Forecasting: principles and practice. OTexts (2018)
9. Nishimori, H., Ouchi, N.: Formation of ripple patterns and dunes by wind-blown sand. Phys. Rev. Lett. **71**(1), 197 (1993)
10. Radon, J.: On the determination of functions from their integral values along certain manifolds. IEEE Trans. Med. Imaging **5**(4), 170–176 (1986). https://doi.org/10.1109/TMI.1986.4307775
11. Tsoar, H.: Linear dunes-forms and formation. Progr. Phys. Geogr. **13**(4), 507–528 (1989)

Underwater Image Enhancement Using Pre-trained Transformer

Abderrahmene Boudiaf[1(✉)], Yuhang Guo[1], Adarsh Ghimire[1], Naoufel Werghi[1], Giulia De Masi[1,2], Sajid Javed[1], and Jorge Dias[1]

[1] Khalifa University, Abu Dhabi, UAE
100058322@ku.ac.ae

[2] ARRC, Technology Innovation Institute, Abu Dhabi, UAE

Abstract. The goal of this work is to apply a denoising image transformer to remove the distortion from underwater images and compare it with other similar approaches. Automatic restoration of underwater images plays an important role since it allows to increase the quality of the images, without the need for more expensive equipment. This is a critical example of the important role of the machine learning algorithms to support marine exploration and monitoring, reducing the need for human intervention like the manual processing of the images, thus saving time, effort, and cost. This paper is the first application of the image transformer-based approach called "Pre-Trained Image Processing Transformer" to underwater images. This approach is tested on the UFO-120 dataset, containing 1500 images with the corresponding clean images.

Keywords: Vision transformer · Underwater imaging · Image enhancement

1 Introduction

The ocean covers more than the 70% of our planet, being a large source of energy, food and materials. Yet, it is still in large part unknown, given the technological challenges that have to be faced during any underwater mission: poor visibility, high hydrostatic pressure, poor communications (essentially still based on acoustic systems), absence of GPS with consequent challenging localizaton and navigation. Despite these challenges, many kinds of underwater missions are necessary to preserve the environment, explore deep sea and support human activities. Scientific missions are oriented to exploration and monitoring of underwater geology and ecosystems; other missions aim to find and explore submerged archeological sites (like sheep wrecks along commercial sea routes from ancient times, that can improve the knowledge of past civilations). Technological missions are oriented to exploitation of underwater resources (oil, natural gas, minerals). Nowadays, offshore wind farms are becoming more and more common, providing a source of clean energy. Fish farms installations are also increasing, to provide additional

S. Sclaroff et al. (Eds.): ICIAP 2022, LNCS 13233, pp. 480–488, 2022.
https://doi.org/10.1007/978-3-031-06433-3_41

source of food; this phenomenon is also expected to increase, given the predicted worldwide shortage of food in the coming decades. Finally, rescue missions are very relevant, given the intense marine traffic for transportation of goods and people that occasionally may lead to accidents, often causing natural disasters and risk of loss of life of men overboard. All these activities become even more challenging in harsh environments, like very deep ocean or polar areas.

For all the above reasons, underwater data, images and videos collection and processing become more and more important to support marine operations at sea. Underwater imaging is particularly challenging because the images suffer from light absorption, light backscattering due to suspended particles, color attenuation that drastically affect the quality of the image, especially at greater depths (where the natural light is poor), or in dirty water. Colours and contours are both lessen in the underwater environment producing a blurry image with indistinct background, where the blue color is dominating as it can be easily explained based on the optical spectrum of the water. These are specific drawbacks affecting the performance of operative tasks like image-based underwater navigation algorithms, object detection or fish tracking. On the other hand, due to the limited and low-bandwidth communications channels that are nowadays available, the image processing has to be performed on board of underwater robots, enhancing the need for fast and efficient algorithms.

Image processing has seen rapid improvement since the use of Convolutional Neural Networks (CNNs) for many tasks such as super-resolution, image denoising [14], dehazing [1], deraining [6] and other image enhancement tasks. Also in the marine context, many CNN-based networks have been developed to enhance the quality of underwater images [4,5,7,9,10,12]. Meanwhile, image transformer networks have proven to be a strong competitor to the ordinary CNNs in many tasks as shown in a recent work [2]. The pre-trained image transformer [2] network has been trained to perform four main tasks: denoising, dehazing, and two levels of up-scaling. The results of the Image Processing Transformer (IPT) network showed that the trained IPT in most cases has outperformed the other specialized CNNs in the mentioned tasks. Transformer networks [11,15] usually deal with sequential data. The input sequence consists of a number of indexed units. The sequence is fed to the network which applies attention on the input. The attention operation is done by calculating the product between each two units in the sequence. This results in an exponentially growing number of computations as the length of the input sequence increases. For instance, increasing the resolution of the image from $100 * 100$ to $200 * 200$ increases the number of pixels by a factor of 4 $[100^2 = 10,000, 200^2 = 40,000]$, while the number of computation required by the attention network will increase by a factor of 16 $[(100^2)^2 = 100 \times 10^6, (200^2)^2 = 1600 \times 10^6]$. Due to the required large number of operations when dealing with images, local attention is used where the image is divided into clusters and the attention is calculated within the cluster [3]. The process of relating each pixel with other pixels showed high performance in many tasks that vary from natural language processing to image processing.

In this work, we utilize the pre-trained image transformer IPT [2] for creating a model that performs image enhancement for underwater images based on the dataset UFO-120 [9]. The results show good performance (see an example in Fig. 1) indicating the potential of using a vision transformer for image enhancement. The paper is organized as follows. Section 2 discusses the related work for underwater image enhancement. In Sect. 3 we discuss the methodology of our work. Section 4 describes the experimental setup. Section 5 shows the results of this work and finally, Sect. 6 concludes the paper.

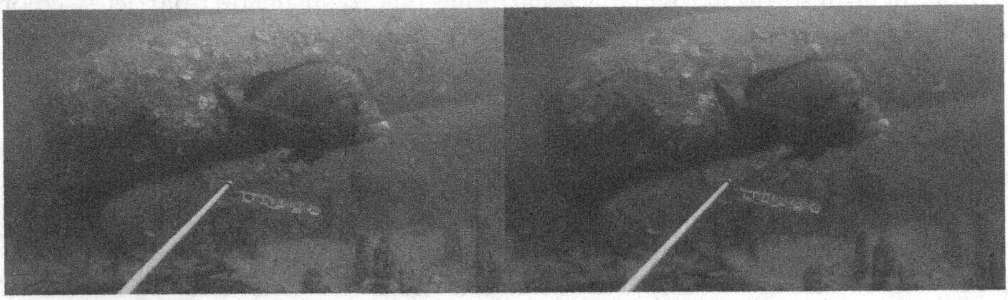

Fig. 1. Example of the input and output of the trained IPT model.

2 Related Work

There have been many proposed works that performed image dehazing/enhancement using different approaches ranging from simple math-based technqiues to more complex systems using CNNs. One of the works that build up upon a basic yet effective concept is [5] which utilizes the dark channel prior approach to estimate and remove the haze from an image. This method is based on the assumption that in an image with haze, some pixels (usually background pixels) will have a low-intensity value in at least one of the color channels that represent the haze in the image. By finding those pixels we can estimate the haze in the image and thus remove it considering it as an additive noise to the image. One drawback of this technique is that it depends on the estimation of the atmospheric light which might be invalid in some underwater lighting conditions. The work [12] also proposes a math-based approach but instead of relying on the dark channel prior, it utilizes the image blurriness map to estimate background light, transmission map, and depth estimation. Using those three estimations, an enhanced image can be produced from the underwater image. This method showed a noticeable improvement over the dark channel prior method. Since CNNs are continuously proving its robustness in many domains, some works have utilized it to perform underwater image enhancement. The work [10] presents a "Water-Net" network that processes images at a deeper level. The network applies White balance, Histogram equalization, and Gamma Correction on the inputted image individually. The output of each processes is then fed into a CNN network with 8 layers. The same output of the three processes will also be fed to three Feature Transformation Units (FTU) which have the purpose of decreasing the color casts and artifacts created by the these processes. The outputs of the CNN for each process will be then combined with the corresponding output of the FTU to form the final enhanced image. This network showed the ability to adapt to different types of image distortion introduced by different underwater conditions. In a different approach, [4] proposed a Generative Adversarial Networks (GANs) which utilize the concept of game theory to create a network for enhancing underwater images. The network consists of two parts, the generator, and the discriminator. The generator will always work to create "fake" aiming to fool the discriminator while the discriminator will always try to differentiate between the "real" and "fake" data. In the case of underwater image enhancement, the generator will try to generate a copy of the haze image that looks similar to the clean image and the discriminator will try to distinguish between the clean image and the enhanced haze image thus it drives the generator to generate a further enhanced images that look as close to the clean images, and so on. The network was tested on a diver tracking algorithm where it showed an improvement in the tracking performance compared to when it is used with images without any enhancement. In a similar manner, [7] proposed a GAN approach. The generator is based on the U-Net [13] and it consists of an encoder-decoder network with skip-connections between the mirrored layers followed by Leaky-ReLU non-linearity and Batch Normalization. As for the discriminator, it was designed based on Markovian Patch-GAN architecture [8] which assumes the independence of pixels thus improving the effectiveness of capturing high-frequency features in addition to requiring fewer parameters to compute. The proposed network showed higher performance for algorithms of object detection, saliency prediction, and human pose estimation. Last but not least, [9] presents a CNN-based approach called Deep SESR that performs the best among the discussed methods on the UFO-120 dataset. The work proposes an end-to-end architecture that performs underwater image cleaning and up-sampling. The network mainly consists of Feature Extraction Network (FENet) block followed by an Auxiliary Attention Network (AAN) for saliency and convolutional layers for up-sampling. The FENet block aims to learn locally dense features while keeping shallow global architecture to guarantee fast feature extraction. The FENet consists of convolutional layers in addition to Residual Dense Blocks (RDBs) which in turn, consists of three sets of convolutional layers while the input and the output between each layer are concatenated. Such a design of the RDBs improves the ability to learn hierarchical features. The robust design of the Deep SESR network has achieved the highest PSNR and SSIM values on the UFO-120 dataset.

3 Methodology

The pre-trained image transformer [2] showed good results in denoising, deraining, and super-resolution that outperformed some state-of-art algorithms which was the reason for choosing this network. The IPT network has four main parts:

heads, transformer encoder, transformer decoder, and tails. The first part contains four heads, each head consists of three convolutional layers and handles one task (denoising, deraining, *2 up-scale and *4 up-scale). Each head will generate a feature map and flatten it into patches before passing it to the transformer encoder. The transformer encoder has the same structure as in the work [15] which is based on a multi-head self-attention module with a feed-forward network. It follows that the transformer decoder that has the same structure as the encoder with the difference being the use of task-specific embedding as additional input. The output of the transformer decoder will be reshaped into the original input dimensions before passing to the next step. Last but not least, four tails for each of the four tasks similar to the heads. Patch normalization was not used. The general network can be seen in Fig. 2.

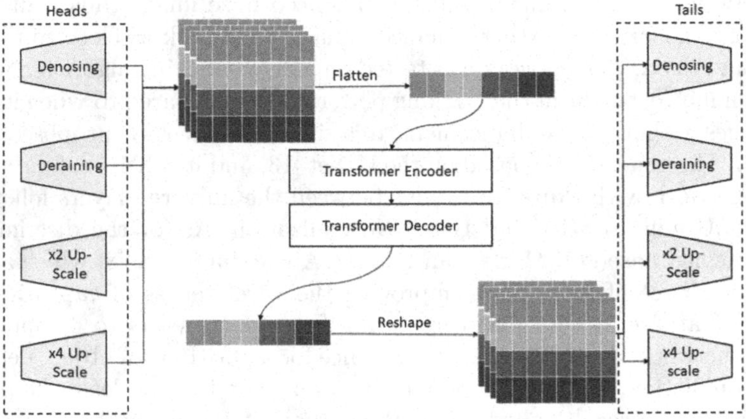

Fig. 2. The overall process of the proposed work [2].

The proposed network was trained on images from the ImageNet dataset. For each task, the images from the dataset were modified to train the network for each of the target tasks. For instance, to train the network for denoising, an image will be taken from the dataset regardless of its label. A Gaussian noise will be applied to the image, then the image with noise will be fed to the network as an input while the original image without the noise will be the ground truth. The same concept is applied to deraining where a "rain noise" is applied to the image. As for up-scaling, the inputted image is a down-scaled copy of the original image. The work presented a number of pre-trained models for two levels of noise, deraining, and super-resolution. The hardware used for creating the trained models consist of 32 NVIDIA Tesla V100 cards.

Since the ground truth image is available during the training, a supervised fashion loss function was used as follow:

$$L_{supervised} = \sum_{i=1}^{N_t} L_1(IPT(I^i_{corrupted}), I_{clean}) \tag{1}$$

where I_{clean} is the ground truth image and $I^i_{corrupted}$ is the corrupted while i is one of the four tasks which indicate that the training of the different tasks occurs simultaneously. L_1 is the L_1 norm. As for the optimizer, the Adam optimizer was used.

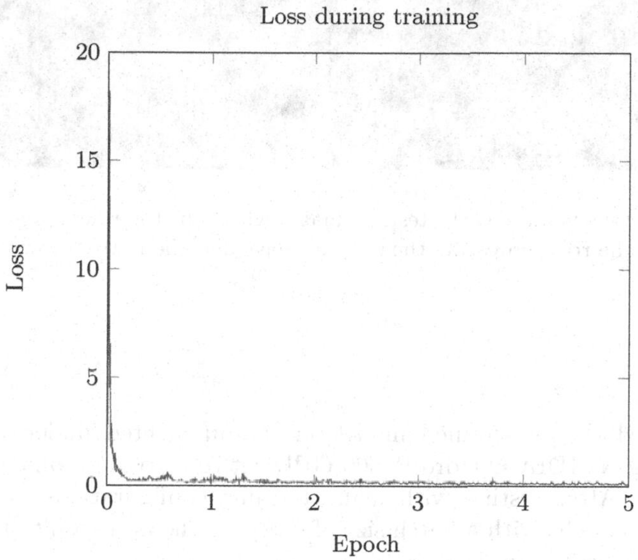

Fig. 3. The loss values during the training process of 5 epochs.

4 Experiment

The pre-trained transformer came with pre-trained models for the four previously mention tasks (denosing, deraining, 2× up-scaling and 4× up-scaling). Since we have limited processing power, we will fine-tune one of the pre-trained models. The model that gave the best initial results was the model for denosing with Gaussian noise of ($\Sigma = 30$) which make sense since the distortion in underwater images is basically a noise in the image, we fine tune the pre-trained model with images from underwater dataset. The dataset that will be used is UFO-10 [9]. This dataset was created for three main tasks: image enhancement, super-resolution and object segmentation. The dataset consists of 1500 pairs of clear and non-clear underwater images as well as a mask for the objects in the image. The clear images come with higher resolution in comparison to the non-clear images for the up-scaling task. Since we do not attempt to train for up-scaling, we applied a smoothing filter with a size of 3 × 3 so both clear and non-clear images have the same level of details. In addition to that, data augmentation was applied in the form of rotation of 5 different angles (0, 45, 135, 225 and 315) to make the total number of images pairs (haze & ground truth) 7500. 6000 images pairs were taken for training while the rest 1500 images pairs were reserved for testing.

Fig. 4. Some of the results on the testing images where the top row represents the input image, the middle row represents the output image and the bottom row represents the ground truth.

5 Results

We fine tuned the pre-trained model on the augmented underwater dataset resulted using NVIDIA Quadro P5000 GPU with 16 Gb of graphic memory and 80 Gb RAM. After testing with many training configurations we found that training for 6 epochs with a batch size of one gave the best result. The resulting loss graph can be seen in Fig. 3.

We report some qualitative results in Fig. 4. The top row represents the input images, the middle row represents the output images and the bottom row represents the ground truth. We noticed that the trained model works best with images that contain green tint where it removes most of the color distortion. In addition to that, we have noticed that the output images have slightly less details compared to the input images. This slight loss of details is due to the smoothing effect of the base model that was used for training. The pre-trained transformer

Table 1. The Peak Signal to Noise Ratio (PSNR) and Structural Similarity Index Measurement (SSIM) values for different methods on the UFO-120 [9] dataset. The highest values is marked with bold.

Method	PSNR	SSIM
DCP [5]	18.20	0.71
IBLA [12]	17.50	0.65
WaterNet [10]	22.46	0.79
U-GAN [4]	23.45	0.80
FUnIE-GAN [7]	25.15	0.82
Deep SESR [9]	**27.15**	0.84
Our approach	23.14	**0.90**

network came with up-scaling model which could be used to restore some of the details in the image. However, that will almost double the processing time which reduce the ability of implementing it in a real-time system. On average, processing one image (320 * 240) took around 12 to 13 s which is not practical for real-time processing. In a future work, we will try different approached to reduce the processing time to for real-time applications.

For a quantitative comparison with other approaches for underwater imaging enhancing, we tested our algorithm on UFO-120 dataset. We selected two measurements: Peak Signal to Noise Ratio (PSNR) and Structural Similarity Index Measurement (SSIM). Table 1 shows the PSNR and SSIM values for the most relevant approaches found in the literature compared to our work.

It is evident that our approach outperforms the others in terms of SSIM, and came in fourth place in terms of PSNR. We believe that the difference if performance between SSIM and PSNR is due the fact that the trained model perform well in restoring the structure and color of the image. However, the smoothing effect reduced the details on the image hence the relatively low PSNR value.

6 Conclusions

Underwater operations, such as exploration, monitoring and recovery, performed by autonomous or semiautonomos robots strongly rely on computer vision. However, the underwater environment is specifically challenging, due to poor visibility induced by relevant light absorption and scattering caused by suspended particles that make the images blurry, bluish and with wide opaque background. For all these reasons, image enhancement of underwater images play a critical role for any mission of marine robotics. Many techniques have been proposed in the literature, mostly based on convolutional neural networks. In this work, we propose a different solution for image enhancement of underwater images, based on the image transformer network (IPT) [2]. The underwater imaging dataset UFO-120 [9] has been selected for this study. A process of data augmentation allowed to increase the original size of 1500 images to 7500 images. The augmented dataset has been used for training the transformer.

We evaluated our results using two measurements (peak-signal-to-noise ratio (PSNR) and structure similarity index measurement (SSIM). This allowed us to compare the results with the same indexes reported in similar works using different approaches. We find that our approach outperforms the previous methodologies proposed in the literature in terms of SSIM and it is fourth in terms of PSNR.

In the future work we will test our approach on different datasets from the underwater environments and we will also evaluate the effects on specific operative tasks, like object recognition and object tracking, when applying these technologies in an operative scenario, based on the use of semi-autonomous/autonomous robots.

Acknowledgement. This work acknowledges the support provided by the Khalifa University of Science and Technology under awards No. RC1-2018-KUCARS, and CIRA-2019-047. The second author, Sajid Javed, of this publication is supported by the FSU-2022-003 Project under Award No. 000628-00001.

References

1. Cai, B., Xu, X., Qing, C., Tao, D.: DehazeNet: an end-to-end system for single image haze removal. IEEE Trans. Image Process. **25**(11), 5187–5198 (2016)
2. Chen, H., et al.: Pre-trained image processing transformer (2021)
3. Dosovitskiy, A., et al.: An image is worth 16×16 words: Transformers for image recognition at scale (2021)
4. Fabbri, C., Islam, M.J., Sattar, J.: Enhancing underwater imagery using generative adversarial networks. In: 2018 IEEE International Conference on Robotics and Automation (ICRA), pp. 7159–7165 (2018). https://doi.org/10.1109/ICRA.2018.8460552
5. He, K., Sun, J., Tang, X.: Single image haze removal using dark channel prior. IEEE Trans. Pattern Anal. Mach. Intell. **33**(12), 2341–2353 (2011). https://doi.org/10.1109/TPAMI.2010.168
6. Hu, X., Fu, C.W., Zhu, L., Heng, P.A.: Depth-attentional features for single-image rain removal. In: 2019 IEEE/CVF Conference on Computer Vision and Pattern Recognition (CVPR), pp. 8014–8023 (2019). https://doi.org/10.1109/CVPR.2019.00821
7. Islam, M.J., Xia, Y., Sattar, J.: Fast underwater image enhancement for improved visual perception. IEEE Robot. Autom. Lett. **5**, 3227–3234 (2020)
8. Isola, P., Zhu, J.Y., Zhou, T., Efros, A.A.: Image-to-image translation with conditional adversarial networks (2018)
9. Jahidul Islam, M., Luo, P., Sattar, J.: Simultaneous enhancement and super-resolution of underwater imagery for improved visual perception. Robotics: Science and Systems XVI (2020). https://doi.org/10.15607/rss.2020.xvi.018
10. Li, C., et al.: An underwater image enhancement benchmark dataset and beyond. IEEE Trans. Image Process. **29**, 4376–4389 (2020). https://doi.org/10.1109/TIP.2019.2955241
11. Parmar, N., et al.: Image transformer (2018)
12. Peng, Y.T., Cosman, P.C.: Underwater image restoration based on image blurriness and light absorption. IEEE Trans. Image Process. **26**(4), 1579–1594 (2017). https://doi.org/10.1109/TIP.2017.2663846
13. Ronneberger, O., Fischer, P., Brox, T.: U-net: Convolutional networks for biomedical image segmentation (2015)
14. Tian, C., Xu, Y., Li, Z., Zuo, W., Fei, L., Liu, H.: Attention-guided CNN for image denoising. Neural Netw. **124**, 117–129 (2020). https://doi.org/10.1016/j.neunet.2019.12.024, https://www.sciencedirect.com/science/article/pii/S0893608019304241
15. Vaswani, A., et al.: Attention is all you need. In: Guyon, I., et al. (eds.) Advances in Neural Information Processing Systems, vol. 30. Curran Associates, Inc. (2017). https://proceedings.neurips.cc/paper/2017/file/3f5ee243547dee91fbd053c1c4a845aa-Paper.pdf

Author Index

Printed in the United States
by Baker & Taylor Publisher Services

Printed in the United States
by Baker & Taylor Publisher Services